Plasma Scattering of Electromagnetic Radiation

Plasma Scattering of Electromagnetic Radiation:

Theory and Measurement Techniques

Dustin H. Froula
Plasma & Ultrafast Physics Group Leader
Laboratory for Laser Energetics
University of Rochester
250 East River Road
Rochester, NY 14623

Siegfried H. Glenzer
Plasma Physics Group Leader
Lawrence Livermore National Laboratory
P.O. Box 808, L-399
Livermore, Ca 94551

Neville C. Luhmann, Jr.
Department of Applied Science
3001 Engineering III
University of California
One Shields Ave. Davis, Ca 95616

John Sheffield
Institute for Secure and Sustainable Environment
University of Tennessee
315 Conference Center Building
Henley Street, Knoxville, TN 37996-4138

ELSEVIER

AMSTERDAM • BOSTON • HEIDELBERG • LONDON
NEW YORK • OXFORD • PARIS • SAN DIEGO
SAN FRANCISCO • SINGAPORE • SYDNEY • TOKYO
Academic Press is an imprint of Elsevier

Academic Press is an imprint of Elsevier
30 Corporate Drive, Suite 400, Burlington, MA 01803, USA
525 B Street, Suite 1900, San Diego, CA 92101-4495, USA
84 Theobald's Road, London WC1X 8RR, UK
Radarweg 29, PO Box 211, 1000 AE Amsterdam, The Netherlands

First edition 2011

Library of Congress Cataloging-in-Publication Data
Plasma scattering of electromagnetic radiation : experiment, theory and computation/John Sheffield ... [et al.]. – 1st ed.
p. cm.
Includes bibliographical references and index.
ISBN 978-0-12-374877-5 (alk. paper)
1. Plasma diagnostics. 2. Electromagnetic waves–Scattering. I. Sheffield, John, 1936-
QC718.5.D5P546 2011
530.4'4–dc22

2010030308

British Library Cataloguing in Publication Data
A catalogue record for this book is available from the British Library

ISBN: 978-0-12-374877-5

For information on all Academic Press publications
visit our website at *books.elsevier.com*

Contents

Preface

In this update of the 1975 book, we have added extensive material on industrial and warm dense matter plasmas, energetic particle measurements and instrumentation, collective Thomson scattering, and techniques. We have also expanded the discussion on relativistic theory and added an appendix on computation.

The form of the spectrum of electromagnetic radiation scattered by a charged particle depends on the mass, charge, position, and velocity of the particle. Similarly, the spectrum of radiation scattered by a plasma, an assembly of free charges, depends on the properties of the plasma. By suitable choice of a source of radiation, we can in principle probe any plasmas and measure quantities such as electron and ion densities and temperatures, ionic charge, magnetic fields, wave amplitudes, and plasma instabilities. It is of particular importance that these measurements can usually be made without significant perturbation of the plasma. In order that we may isolate the effect of the scattering, we require an essentially monochromatic source. Because the scattering cross section is small, the source must be powerful. The growing importance, over the years, of scattering as a diagnostic technique has been associated to a great extent with the development of lasers and microwave sources that satisfy these requirements.

In this monograph, we have attempted to cover both theory and experiment in a consistent and detailed fashion. We have made an effort toward providing quantitative results in which units are clearly defined, and there are no 4π's and c's lurking in the background. We review the underlying mathematical techniques.

The book is biased on the experimental side in the direction of laboratory high-temperature plasmas because this is our field of research. To illustrate the application of the technique, we describe a number of significant laboratory experiments that have been made over the last few decades. Scattering is a basic tool of the plasma physicist, and to help those who are not spectroscopists, we have included a section on the most commonly used dispersion instrumentation.

This text is written as an introduction text and reference book for graduate students and researchers. We hope that the readers will find answers to their questions at the level to which they are prepared to go.

ACKNOWLEDGMENTS

John Sheffield started writing the original version of this book during a most enjoyable time at the University of Texas in Austin. He completed the book while working at the Culham Laboratory in England. He owes a great debt to the many colleagues from these two institutions who helped him. As always, he remains truly appreciative of the support of his wife Dace.

Dustin Froula compiled this second edition while on sabbatical at University of California at Los Angeles and would like to thank his colleagues and friends at Lawrence Livermore for the novel ideas and collaborations that have lead to his understanding of Thomson scattering and plasma physics in general. In his life, there have been many influential teachers and mentors but none greater than his mother and father who have truly provided a solid foundation for both life and science. The love and support of his wife, Lynette, and kids, Scott, Adelae, and Victoria, have been invaluable.

Neville Luhmann would like to dedicate this to his parents who instilled in him a love of science and technology.

Siegfried Glenzer would like to thank his family, his wife Anja and his children Helena and Arend, for support during the time when he worked on this book. He is also grateful to his colleagues at the Lawrence Livermore National Laboratory, Los Alamos National Laboratory, Laboratory for Laser Energetics, Rutherford Appleton Laboratory, Ruhr-Universität Bochum and Deutsches Elektronen Synchrotron where he performed most of his Thomson scattering experiments described here. He also likes to thank the Alexander-von-Humboldt foundation and colleagues at the University of Rostock for enabling his engagement with free electron laser experiments.

The authors are very appreciative of the advice and support of Roland Behn, Henrik Bindslev, Alan Costley, Laurent Divol, Tilo Doeppner, Tony Donné, Carsten Fortmann, Alan Garscadden, Gianluca Gregori, Takaki Hatae, Jerry Hughes, Dave Johnson, Chan Joshi, Robert Kaita, Otto Landen, Lynette Lombardo, Atsushi Mase, David Montgomery, Tobin Munsat, Katsunori Muraoka, Allan Offenberger, John Palastro, Hyeon Park, Brad Pollock, Sean Prunty, Ronald Redmer, James Steven Ross, Christophe Rousseaux, Kenji Tanaka, Th. Tschentscher, Kiichiro Uchino, Michael Walsh, Henri Weisen, Paul Woskov, and Chang-Xuan Yu.

We are deeply indebted to our Elsevier colleagues who provided continuing support in the preparation of this book, Gavin Becker, Mohana Natarajan, and Patricia Osborn.

Chapter 1

Introduction

Chapter Outline Head

DOI: 10.1016/B978-0-12-374877-5.00001-4

1.1 INTRODUCTION

It is well known that electromagnetic radiation is emitted by an accelerating charge. An important example of this phenomenon occurs when acceleration is induced by an electromagnetic wave. This interaction, when the incident radiation is of sufficiently low frequency ω that $\hbar\omega$ is much less than $m_e c^2$, the rest energy of the charge, is generally referred to as Thomson scattering.[1] It is the extension of the theory to include the simultaneous scattering from a large number of free positive and negative charges, that is, a plasma, and the experimental application of scattering that are the topics discussed in this book.

For a single charge, the angular distribution of intensity, the frequency, and the phase of the scattered radiation depend on the orbit of that charge relative to the observer. Equally, for a large group of charges, the scattered spectrum is related to the orbits of all those charges, or rather in practice, to some average taken over the probable behavior of the group. Anticipating the results derived below, we find that from the spectrum of radiation scattered from a plasma, we may in principle determine the electron and ion temperature, the ionization state, densities the ionization state, the direction and magnitude of a magnetic field in the plasma, and in general, information about all the fluctuations (waves, instabilities) within the plasma. In reality, we are of course limited by the radiation sources available to us; the cross section for scattering is so small that measurements on laboratory plasmas were limited until the advent of high-power lasers. The first measurements were by the scattering of radio waves from the ionosphere in the late 1950s. The history of the subject is discussed briefly in Appendix E.

CHAPTER 1

The purpose of this introductory chapter is to remind the reader of some basic properties of a plasma and of the interaction of radiation with a plasma. The conditions under which radiation will penetrate a plasma are established. The

---

[1] Mie scattering is that obtained from a conducting sphere in the long wavelength limit. The dependence of the cross section on ω^4 is known as *Rayleigh's law*. When scattering from atoms and molecules, the scattered component at the incident frequency is due to Rayleigh (Mie) scattering. In addition, there can be satellite lines. The *Raman* satellites (Stokes and anti-Stokes lines) derive from the linear interaction of the incident wave with electronic transitions in the atom or molecules. The *Brillouin* lines come from a linear interaction with thermal sound waves in the medium. These two effects are discussed in Chapter 12.

Thomson scattering is commonly referred to as scattering off free electrons and is strictly speaking the limiting case of Compton scattering applicable at sufficiently low frequencies that the photon energy is much less than the rest energy of the charge. For high energy photons, the scattering spectrum shows the Compton effect where the momentum transfer to the electrons results in the Compton downshifted feature.

discussion is then restricted to situations where the radiation is primarily transmitted. In this situation, we can reasonably deal with the interaction of each charge in the plasma taken independently. The response of a single charge to radiation is then evaluated. The scattered power is found to be inversely proportional to the mass of a charge, and thus, we can see immediately that the scattering is essentially only from the electrons.

Finally, the problem of adding up the scattered waves from the large number of electrons in the scattering volume is discussed in general terms. The scattered spectrum is found to have two parts. The first is the spectrum that would be obtained if there were no charge interactions, the "noncollective spectrum." The second is a result of these interactions (collective effects).

CHAPTER 2

In this chapter, a general relationship is derived between the scattered power spectrum and the fluctuations in plasma density. The spectral density function $S(\mathbf{k}, \omega)$ is introduced.

CHAPTER 3

The general scattered spectrum for an unmagnetized quasi-equilibrium plasma is derived. The effect of collisions on the result is determined.

CHAPTER 4

A derivation is given of the noncollective spectrum for a plasma, including a steady magnetic field. The application of the results is discussed.

CHAPTER 5

The results of the general scattered spectrum (Chapter 3) in the collective regime are analyzed, and their experimental applications are discussed. These are illustrated by reference to some of the significant experimental work in the field.

CHAPTER 6

The constraints and problems that arise in the application of scattering as a diagnostic technique are discussed.

CHAPTER 7

The characteristic performance of various dispersion elements, image dissectors, and detectors is reviewed.

CHAPTER 8

Some interesting applications of scattering are discussed to illustrate the use of Thomson scattering.

CHAPTER 9

Industrial plasmas, scattering from energetic ions, and fusion plasmas are discussed.

CHAPTER 10

A derivation is given of the general scattering spectrum for a magnetized plasma, and the application of the results is discussed.

CHAPTER 11

The use of hard x-rays (high energy photons) is discussed for probing warm-dense matter and dense plasmas. Recent applications in the Compton and plasmon scattering regime are presented.

CHAPTER 12

Work on the scattering from unstable plasmas is reviewed, e.g., the scattering from enhanced fluctuations driven by plasma wave turbulence or laser-plasma interaction.

APPENDIX A

A brief review is given of relevant mathematical techniques.

APPENDIX B

The kinetic theory of plasmas is reviewed.

APPENDIX C

A derivation of the general dispersion relation for a hot, magnetized uniform plasma is presented.

APPENDIX D

A brief discussion of computational techniques used to calculate the Thomson scattering spectrum is given. A simple computational method for solving for the scattering spectrum is presented.

APPENDIX E

A historical review of work on the scattering of radiation from plasmas is given.

APPENDIX F

This appendix contains a list of physical constants and significant formulas. The latter includes a list of the various scattered spectra obtained under different approximations.

1.2 PLASMAS

A plasma is an assembly of free electrons and positive ions, which is essentially neutral. Thus, while locally there may be a charge imbalance, in the assembly as a whole, there are to a very good approximation equal number of electrons and positive ions. The term "plasma" was used by Langmuir (1928) as a description of the ionized state found in an arc discharge. For gases at temperatures[2] >1 eV (11,600 K), there are many particles in the high-energy tail of the distribution function with sufficient energy to ionize, and consequently large numbers of free charges. The ideal plasma state may be characterized by the following ordering of characteristic scale lengths:

$$r_c \ll n^{-1/3} \ll \lambda_{De} \ll \lambda_c, L_p, \tag{1.2.1}$$

where $r_c = q^2/\kappa T$ is the distance at which the potential and kinetic energies are equal when two like charges q approach each other; $n^{-1/3}$ is the average interparticle separation; n is the number density of charges; λ_{De}, the "Debye length," is the characteristic distance over which the potential of a charge is shielded by neighboring charges; λ_c is the mean collision length, $\lambda_c = 1/4\pi r_c^2 n$ for simple 90° Rutherford scattering; and L_p is a representative dimension of the plasma.

The key features are the freedom of the charges to move, the interaction via the long-range Coulomb force, and as a consequence the collective interaction of the charges within a Debye sphere of a given charge. In order that this collective interaction will play a role, we have required that $N_D = \frac{4}{3}\pi n_e \lambda_{De}^3 \gg 1$, i.e., $\lambda_{De} \gg n_e^{-1/3}$.

Within the framework of this definition, there are a wide range of plasma situations apart from that found in the high-temperature gaseous state. The free electrons and "holes" in a solid and the free ions in a liquid (e.g., salt solution) also constitute a plasma. While it is the application of scattering to gaseous

---

[2] In this region, it is customary to use the electron volt as the unit of temperature. That is, we equate the mean translational kinetic energy with the equivalent energy which a unit charge e would receive from a potential drop V, $eV = \kappa T$. Thus, 1 electron volt (eV) is equivalent to 11,600 K, $e = 1.6 \times 10^{19}$ C; $\kappa = 1.38 \times 10^{-23}$ J/K.

plasmas that is emphasized in this book, most of the results will of course have relevance to work in these other plasmas.

The plasma state has a wide variety of applications; flames and fluorescent lights are plasmas, and the plasma state is involved in gas lasers, magnetic hydrodynamics (MHD) generators, inertial confinement fusion (ICF), and plasma propulsion systems. Further from the earth are the plasmas of the ionosphere, magnetosphere, solar wind, sun, and stars. Plasmas play an important role in the release of fusion energy. The nuclei of the lighter elements (H, ^{2}H, ^{3}H,...) can combine following a collision to produce a heavier element. In this fusion, some of the binding energy is released, as for example in the reactions

$$
\begin{aligned}
{}^2\mathrm{H}+{}^2\mathrm{H} \; & \begin{array}{l} \nearrow {}^3\mathrm{He}+{}^1n+3.27\,\mathrm{MeV} \\ \searrow {}^3\mathrm{H}+{}^1p+4.03\,\mathrm{MeV} \end{array} \\
{}^2\mathrm{H}+{}^3\mathrm{H} & \rightarrow {}^4\mathrm{He}+{}^1n+17.58\,\mathrm{MeV} \\
{}^2\mathrm{H}+{}^3\mathrm{He} & \rightarrow {}^4\mathrm{He}+{}^1\mathrm{H}+18.34\,\mathrm{MeV}
\end{aligned}
\tag{1.2.2}
$$

Unless the nuclei are moving with sufficient velocity to overcome their Coulomb repulsion, they will simply scatter off each other. (This particle scattering is not to be confused with the scattering of radiation from charges.) For the useful release of fusion power, it is necessary to raise a quantity of some light element to a high enough temperature that fusion can compete with particle scattering. That is, we need a high-temperature plasma of, for example, deuterium or deuterium and tritium. There are currently two main stream approaches to producing fusion plasmas in the laboratory. In magnetic fusion energy (MFE), magnetic fields are used to contain hot plasma from material walls for a long enough time that the nuclei can fuse. Work in this field indicates that a fusion reactor will probably have a plasma of deuterium and tritium of density 10^{14}–10^{15} charges/cm^3 at a temperature of 10–20 keV, and that this plasma will be contained by a magnetic field of 2–10 T. Inertial confinement fusion (ICF) is an approach to fusion that relies on the inertia of the fuel mass to provide confinement. Energy from a driver is delivered rapidly to a capsule that heats up and expands. The outward expansion drives an inward force that compresses the fusion material to electron densities above 10^{25} charges/cm^3 and temperatures of 10–20 keV.

At the other end of the density scale, the ionospheric plasma has densities of $\sim 10^5$ charges/cm^3 and electron temperatures < 1 eV. In between are a wide-range of plasma devices and plasmas that occur in nature. The range of plasmas of interest is indicated in Fig. 1.1.

One problem is common to the investigation of all these plasmas. How do we diagnose the plasma without disturbing it? Conventional diagnostic devices,

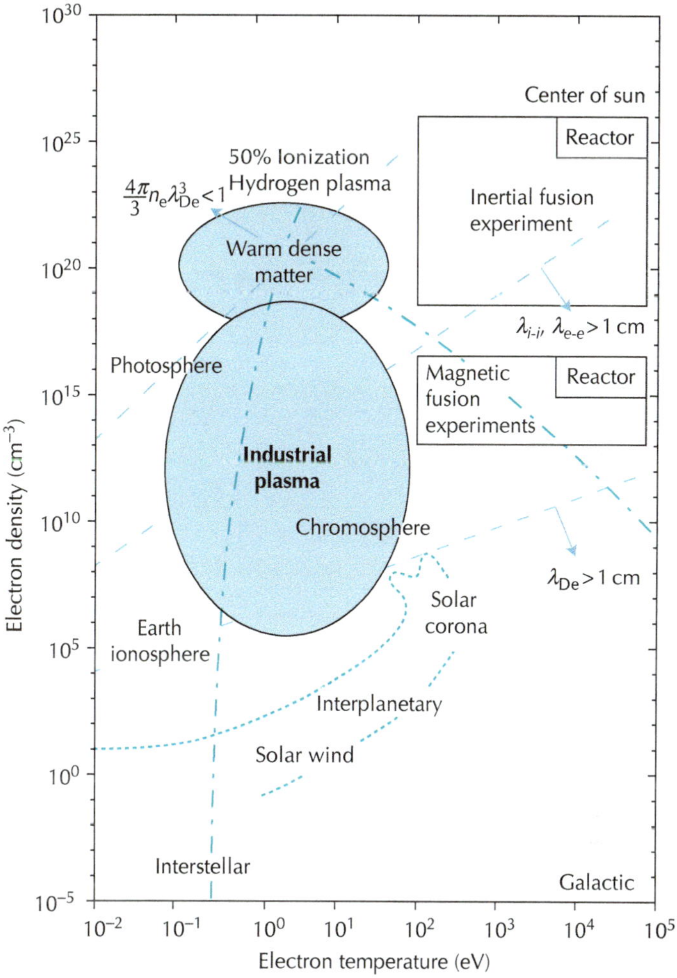

FIGURE 1.1 The density and temperature ranges of typical plasmas in nature and in the laboratory.

such as probes for measuring electrostatic and magnetic fields, not only contaminate the plasma but are often too large for the investigation of the microscopic structure of the plasma. In addition, they are not usable in dense plasmas encountered in warm dense matter and ICF research.

It is natural to consider the use of a probing beam of electromagnetic radiation. Ideally, the intensity will be kept sufficiently low so as not to perturb the plasma. Further, in principle, we can choose radiation of a wavelength to fit the characteristic lengths of the plasma under investigation.

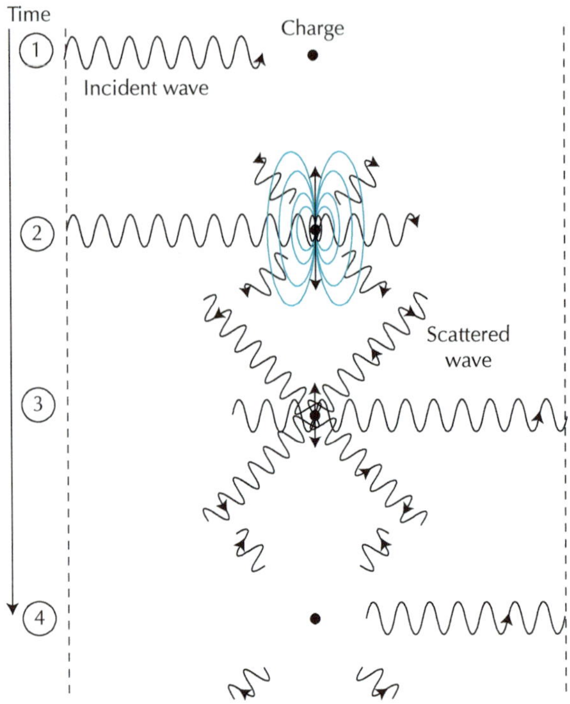

FIGURE 1.2 The scattering of radiation by a free charge.

The oscillating electric field of the incident beam accelerates each of the charged particles in the plasma, and these charges subsequently reradiate (Fig. 1.2). The interaction is primarily concerned with the electrons because the acceleration of the ions is less by virtue of their greater mass. The spectrum of the scattered radiation depends on the electron density, electron and ion temperatures, the magnetic field, and the plasma oscillations. Before discussing this valuable diagnostic technique, we will discuss the system of units that will be used and then review some basic features of the plasma state and of the interaction of an electromagnetic wave with it.

1.3 SYSTEM OF UNITS

The Gaussian system of units will be used for all the theoretical calculations. This choice was made because most of the significant work in this field has been done in this system. However, for the benefit of the reader, all the important results will be given in both Gaussian and rationalized MKS system (see Jackson, 1998, "Conversion of Equations and Amounts between SI Units and Gaussian Units." Appendix 4, pp. 783–784). The approach to the definition of unit charge is used to illustrate the difference between the two systems.

1.3.1 The Gaussian System

Here, the unit charge q is defined by Coulomb's law

$$\mathbf{F}=(q_1 q_2/r_{12}^2)\hat{r}_{12}, \tag{1.3.1}$$

where $\hat{r}_{12}$ is the unit vector in the direction from charge one to charge two.

A unit charge placed a distance $r_{12}=1$ cm from a like charge in vacuum will repel it with a force of 1 dyne. The charge on an electron is 4.8×10^{-10} statcoulomb.

The electric field $\mathbf{E}$ of a charge q is defined as the force per unit charge:

$$\mathbf{E}=(q/r^2)\hat{r}. \tag{1.3.2}$$

This system is unrationalized, i.e., no factor 4π has been introduced into Eqn (1.3.2) to bring out the spherical symmetry of the expression.

1.3.2 The SI System

Here, the unit charge is defined from the unit current, the ampere. The ampere is the current which when flowing in each of two infinitely long parallel wire of negligible cross-sectional area separated by 1 m in vacuum causes a transverse force per unit length of 2×10^{-7} N/m to act between the wires.

With this definition, Coulomb's law becomes

$$\mathbf{F}=\left(\frac{1}{4\pi\varepsilon_0}\frac{q_1q_2}{r_{12}^2}\right)\hat{r}_{12}, \tag{1.3.3}$$

where the constant is

$$\varepsilon_0=\frac{1}{4\pi c^2}\times10^7\ \frac{\mathrm{C}^2}{\mathrm{N\,m}^2}$$

in rationalized units, $\mathbf{F}$ is in Newtons, q_1 and q_2 is in coulombs, and r_{12} is in meters. Maxwell's equations and the Lorentz force are given below for both systems.

1.3.3 Maxwell's Equations and Lorentz Force on a Unit Charge q

Gaussian

$$\mathbf{D}=\mathbf{E}+4\pi\mathbf{P},\qquad \nabla\cdot\mathbf{D}=4\pi\rho,\qquad \nabla\times\mathbf{H}=\frac{4\pi\mathbf{J}}{c}+\frac{1}{c}\frac{\partial\mathbf{D}}{\partial t},$$
$$\mathbf{F}=q\left(\mathbf{E}+\frac{\mathbf{v}}{c}\times\mathbf{B}\right),$$
$$\mathbf{H}=\mathbf{B}-4\pi\mathbf{M},\qquad \nabla\times\mathbf{E}=-\frac{1}{c}\frac{\partial\mathbf{B}}{\partial t},\qquad \nabla\cdot\mathbf{B}=0. \tag{1.3.4}$$

SI System

$$
\begin{aligned}
&\mathbf{D}=\varepsilon_0\mathbf{E}+\mathbf{P}, \quad \nabla\cdot\mathbf{D}=\rho, \quad \nabla\times\mathbf{H}=\mathbf{J}+\frac{\partial\mathbf{D}}{\partial t},\\
&\mathbf{F}=q(\mathbf{E}+\mathbf{v}\times\mathbf{B}),\\
&\mathbf{H}=\frac{1}{\mu_0}\mathbf{B}-\mathbf{M}, \quad \nabla\times\mathbf{E}=-\frac{\partial\mathbf{B}}{\partial t}, \quad \nabla\cdot\mathbf{B}=0,\\
&\varepsilon_0=\frac{1}{4\pi c^2}\times 10^7\ \frac{\mathrm{F}}{\mathrm{m}}, \quad \mu_0=4\pi\times 10^{-7}\ \frac{\mathrm{H}}{\mathrm{m}},\\
&\mathbf{P}=\text{polarization}, \quad \mathbf{M}=\text{magnetization}.
\end{aligned}
\tag{1.3.5}
$$

(For all the discussions in this book, $\mathbf{M}=0$.)

1.4 CHARACTERISTIC LENGTHS AND TIMES IN A PLASMA[3]

The main difference between the plasma state and other states involving macroscopically neutral but microscopically charged systems rests in the freedom of motion of the individual charges. As a consequence of this freedom, charges are able to adjust their positions to screen mutually the electric fields of each other which can result in collective effects that can be probed by techniques described in this book. The potential ϕ around a given charge q in a plasma has the form, for $r > q^2/\kappa T$,

$$\varphi(r)=(q/r)\cdot\exp\left(-r/\sqrt{2}\lambda_{\mathrm{De}}\right). \tag{1.4.1}$$

The potential falls away in the characteristic distance, the "Debye length" λ_{Dq},

$$
\begin{aligned}
\lambda_{\mathrm{Dq}} &= \left(\kappa T_q/4\pi e^2 n_q\right)^{1/2}, \quad \text{Gaussian},\\
&= \left(\varepsilon_0\kappa T_q/e^2 n_q\right)^{1/2}, \quad \text{SI},\\
&= 743\left[T_q(\mathrm{eV})/n_q\left(\mathrm{cm}^{-3}\right)\right]^{1/2} \quad \mathrm{cm}.
\end{aligned}
\tag{1.4.2}
$$

If a charge in a plasma is displaced, the resultant net electrostatic force is a restoring force. The charge oscillates under the action of this force with the characteristic "plasma frequency":

$$
\begin{aligned}
\omega_{\mathrm{p}q} &= \left(4\pi n_q q^2/m_q\right)^{1/2}\ \mathrm{rad/s}, \qquad \text{Gaussian},\\
&= \left(\frac{n_q e^2}{m_q\varepsilon_0}\right)^{1/2}\ \mathrm{rad/s}, \qquad \text{SI}.
\end{aligned}
\tag{1.4.3}
$$

The electron plasma frequency

$$\omega_{\mathrm{pe}} = 5.64\times 10^4\left(n_{\mathrm{e}}\left(\mathrm{cm}^{-3}\right)\right)^{1/2}\ \mathrm{rad/s}.$$

---

[3] See Appendix B.

We see that $\omega_{pq}\lambda_{Dq} = \left(\kappa T_q/m_q\right)^{1/2} = \bar{v}_{th}$, the characteristic thermal speed of the particular charge of mass m_q.

Finally, we must add the characteristic lengths and times associated with the gyration of the charges when the plasma contains a magnetic field. The cyclotron (gyration) frequency is given by

$$\begin{aligned}\Omega_q &= qB/m_q c \quad \text{rad/s}, \quad \text{Gaussian}, \\ &= qB/m_q \quad \text{rad/s}, \quad \text{SI}.\end{aligned} \tag{1.4.4}$$

The electron cyclotron frequency $\Omega_e = 1.76 \times 10^7 B(\text{G})\,\text{rad/s}$. The cyclotron radius is

$$\rho_q = v_\perp/\Omega_q. \tag{1.4.5a}$$

The electron cyclotron radius is

$$\rho_e = \frac{0.57 \times 10^{-7} v_\perp\,(\text{cm/s})}{\text{B(G)}} \quad \text{cm}, \tag{1.4.5b}$$

where $v_\perp$ is the velocity of the charge perpendicular to the magnetic field.

1.5 SCATTERING OF ELECTROMAGNETIC RADIATION BY A PLASMA

The complete calculation of the scattering of radiation by a plasma would be extremely involved, and it is fortunate that for most cases of interest, we are able to decouple the various parts of the problem. This occurs partly because with a wide range of sources available to us, we can avoid the parameter regions where, for example, absorption, reflection, and multiple scattering are important.

Consider a volume V of plasma containing N electrons, charge $-e$, and N/Z ions of charge Ze. The behavior of this assembly of charges may be described by the Klimontovich equation for each species q (see Section B.7):

$$\frac{\partial F_q}{\partial t} + \mathbf{v} \cdot \frac{\partial F_q}{\partial \mathbf{r}} + \mathbf{a} \cdot \frac{\partial F_q}{\partial \mathbf{v}} = 0, \tag{1.5.1}$$

where

$$F_q(\mathbf{r}, \mathbf{v}, t) = \sum_{j=1}^{N_q} \delta\left(\mathbf{r} - \mathbf{r}_j(t)\right) \delta\left(\mathbf{v} - \mathbf{v}_j(t)\right)$$

is the microscopic distribution function, and $F_q d\mathbf{r} d\mathbf{v}$ represents the number of charges in the range $\mathbf{r} \to \mathbf{r} + d\mathbf{r}$, $\mathbf{v} \to \mathbf{v} + d\mathbf{v}$ at the time t.

This is a complicated equation, because the acceleration $\mathbf{a}$ involves all the interparticle forces, as well as the effect of external forces. For terrestrial plasmas, we may reasonably restrict the forces to electric and magnetic; therefore, we may use Maxwell's equations and the Lorentz equation (1.3.4) to determine $\mathbf{a}$. The charges in the volume, external charges, and the electromagnetic wave are all sources of **E** and **B**. The following results that are touched on in the next sections help us to simplify the calculation.

1. As a result of the Debye shielding, the influence of each charge is limited so that we may neglect the simultaneous interaction of three or more charges when conditions (1.2.1) are met. This is discussed in more detail in Appendix B.
2. When the incident radiation has a high frequency, $\omega_i \gg \omega_{pe}, \Omega_e$, the electromagnetic wave is primarily transmitted, and the attenuation owing to scattering and absorption is small. If the scattering volume is optically thin (no significant losses in crossing it), then we may treat the interaction with each charge in the scattering volume separately. In addition, we limit the incident power so that it will not change the plasma conditions. The total scattered electric field is then obtained as the sum of the individually scattered fields.

1.6 RADIATION BY A MOVING CHARGE

This topic is elegantly discussed in numerous books (Jackson, 1998; Landau & Lifshitz, 1962; Brau, 2004), and the reader is referred to them for a detailed treatment.

We may combine Maxwell's equations 1.3.5 to obtain

$$\nabla \times (\nabla \times \mathbf{E}) + \frac{1}{c^2}\frac{\partial^2 \mathbf{E}}{\partial t^2} = -\frac{4\pi}{c}\frac{\partial \mathbf{J}}{\partial t}. \tag{1.6.1}$$

Now, we need to determine the electric field **E** when **J** is the current density from a single charge q, i.e.,

$$\mathbf{J} = q\mathbf{v}(t'). \tag{1.6.2}$$

It must be remembered that the electric field at a distance R from the charge (see Fig. 1.3) at a time t is related to the behavior of the charge at the previous time t', the retarded time, where

$$t' = t - (R'/c). \tag{1.6.3}$$

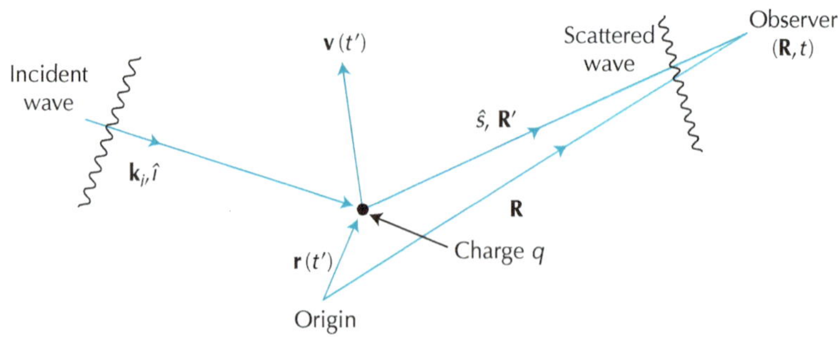

FIGURE 1.3 The scattering coordinate system.

The solution of Eqn (1.6.1) (see, e.g., Jackson, 1998, p. 664) is

$$\mathbf{E}(\mathbf{R}',t) = q\left[\frac{(\hat{s}-\boldsymbol{\beta})(1-\beta^2)}{(1-\hat{s}\cdot\boldsymbol{\beta})^3 R'^2}\right]_{\text{ret}} + \frac{q}{c}\left[\frac{\hat{s}\times\{(\hat{s}-\boldsymbol{\beta})\times\dot{\boldsymbol{\beta}}\}}{(1-\hat{s}\cdot\boldsymbol{\beta})^3 R'}\right]_{\text{ret}}, \tag{1.6.4}$$
$$\mathbf{B}(\mathbf{R}',t) = n(\hat{s}\times\mathbf{E}),$$

where $\boldsymbol{\beta} = \mathbf{v}/c$, $\hat{s}$ is the unit vector pointing from the charge to the observer, and the term $n = ck/\omega$ is the refractive index. Generally, it is assumed that $n = 1$; however, this is not always the case, for example, for high densities, see Chapter 11. The fields are to be evaluated at the retarded time, i.e.,

$$\dot{\boldsymbol{\beta}} = \frac{1}{c}\frac{d\mathbf{v}}{dt'}.$$

In scattering calculations, the condition $R' \gg L$ is normally made, and in experiment, the condition is normally met. That is, the observing point P is at a large distance from the charge compared to the characteristic length L over which we observe the motion of the charge.

The first term is therefore dropped in all calculations that follow, and in the denominator of Eqn (1.6.4), we make the approximation $R' \simeq R$. It is, however, important to note that this approximation may not be made in Eqn (1.6.3). The retarded time may be approximately written as

$$t' \simeq t - (|R - \hat{s}\cdot\mathbf{r}|/c), \tag{1.6.5}$$

where $\hat{s}$ is now constant in time.

The quantity that is measured is the scattered power, and the scattered power per unit solid angle is given by

$$\frac{dP_s}{d\Omega} = R^2\mathbf{S}\cdot\hat{s}, \tag{1.6.6}$$

where the Poynting vector $\mathbf{S}$ is defined as

$$\mathbf{S} \equiv \frac{c}{4\pi}\mathbf{E}\times\mathbf{B}.$$

Substituting from Eqn (1.6.6), we obtain

$$\frac{dP_s}{d\Omega} = \frac{R^2 c}{4\pi}E_s^2. \tag{1.6.7}$$

For low-velocity charges, where $|v/c| \ll 1$, the electric field Eqn (1.6.4) is

$$\mathbf{E}_s(R,t) = \frac{q}{cR}\left[\hat{s}\times(\hat{s}\times\dot{\boldsymbol{\beta}})\right]_{\text{ret}}. \tag{1.6.8}$$

The scattered power per unit solid angle is

$$\frac{dP_s}{d\Omega} = \frac{q^2}{4\pi c}\left[\hat{s}\times(\hat{s}\times\dot{\boldsymbol{\beta}})\right]^2_{\text{ret}}. \tag{1.6.9}$$

The variation of $dP_s/d\Omega$ with direction has the characteristic donut shape illustrated in Fig. 1.4a.

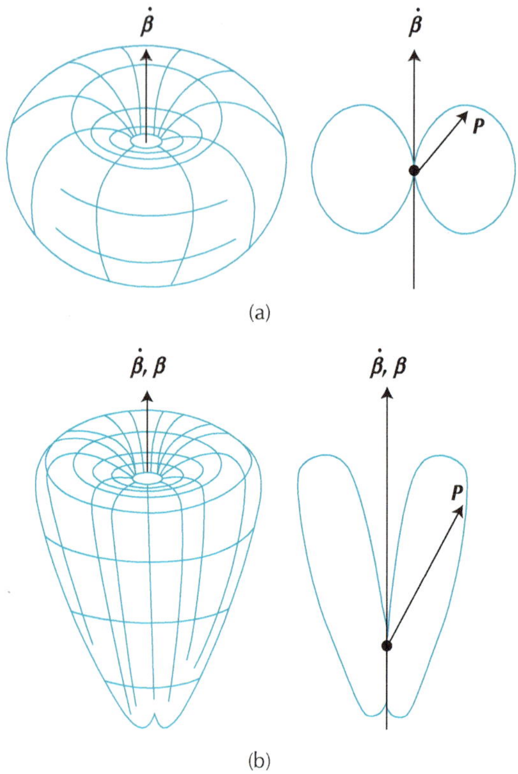

FIGURE 1.4 The angular variation of the power radiated by an accelerated charge. (a) Stationary charge; (b) charge moving with $\beta || \dot{\beta}$.

At higher velocities, the full expression for the second term of Eqn (1.6.4) is used, and the most noticeable change is in the increase of power scattered into the direction of **v** (see Fig. 1.4b). The effect is understood when we realize that the scattered power must still have the symmetrical donut shape in the electron frame of reference. These high-velocity corrections have been observed in scattering from plasmas and are discussed further in Chapters 3, 4, and 5.

1.7 ACCELERATION OF A CHARGE BY AN ELECTROMAGNETIC WAVE

Radiation is scattered by a charge, because the charge is accelerated by the oscillating fields of the radiation and an accelerating charge itself radiates (see Fig. 1.2). Here, we derive the formula for the electric field scattered by a single free electron, but at the same time, we investigate briefly some of the conditions that must be met for the charge to appear to be free.

Consider the case of a plane monochromatic wave incident upon a charge q, whose position and velocity are $\mathbf{r}(t')$ and $\mathbf{v}(t')$. For the sake of consistency, we write these as functions of the retarded time. The incident wave is

$$\mathbf{E}_\mathrm{i}(\mathbf{r},t') = \mathbf{E}_\mathrm{i0}\cos(\mathbf{k}_\mathrm{i}\cdot\mathbf{r}-\omega_\mathrm{i}t'),\quad \mathbf{B}_\mathrm{i}(\mathbf{r},t') = n(\hat{\imath}\times\mathbf{E}_\mathrm{i}); \tag{1.7.1}$$

$k_\mathrm{i} = 2\pi/\lambda_\mathrm{i}$, λ_i, and ω_i are the incident wave number, wavelength, and frequency, respectively. Note that $\mathbf{E}_\mathrm{i0}$ is not time dependent.

1.7.1 Low-Velocity Charge ($v/c \ll 1$) with No Other Forces Acting

For a charge with low velocity, we may neglect the effect of the field $\mathbf{B}_\mathrm{i}$, and when no other forces act, the equation of motion is

$$m_\mathrm{e}\left(\frac{d\mathbf{v}}{dt'}\right) - q\,\mathbf{E}_\mathrm{i0}\cos\left[\mathbf{k}_\mathrm{i}\cdot\mathbf{r}(t')-\omega_\mathrm{i}t'\right]. \tag{1.7.2}$$

In determining the orbit of the charge, we neglect the influence of the electromagnetic wave, and the unperturbed orbit is simply

$$\mathbf{r}(t') = \mathbf{r}(0)+\mathbf{v}t'. \tag{1.7.3}$$

We substitute from Eqn (1.7.3) in Eqn (1.6.5) and obtain

$$t' = \left(t-\frac{R}{c}+\frac{\hat{s}\cdot\mathbf{r}(0)}{c}\right)\Big/\left(1-\hat{s}\cdot\boldsymbol{\beta}\right). \tag{1.7.4}$$

With the aid of Eqns (1.7.3) and (1.7.4), we see that

$$\mathbf{k}_\mathrm{i}\cdot\mathbf{r}(t')-\omega_\mathrm{i}t' = k_\mathrm{i}\frac{(1-\hat{\imath}\cdot\boldsymbol{\beta})}{(1-\hat{s}\cdot\boldsymbol{\beta})}R-\omega_\mathrm{i}\frac{(1-\hat{\imath}\cdot\boldsymbol{\beta})}{(1-\hat{s}\cdot\boldsymbol{\beta})}t-k_\mathrm{i}\frac{(1-\hat{\imath}\cdot\boldsymbol{\beta})}{(1-\hat{s}\cdot\boldsymbol{\beta})}\hat{s}\cdot r_0+\mathbf{k}_\mathrm{i}\cdot\mathbf{r}(0). \tag{1.7.5}$$

Substitution of Eqns (1.7.5) and (1.7.2) in Eqn (1.6.8) now yields the scattered electric field

$$\mathbf{E}_\mathrm{s}(R,t) = \left(\frac{q^2}{c^2 m_\mathrm{e}R}\right)\left[\hat{s}\times(\hat{s}\times\mathbf{E}_\mathrm{i0})\right]\cos\left[k_\mathrm{s}R-\omega_s t-(\mathbf{k}_\mathrm{s}-\mathbf{k}_\mathrm{i})\cdot\mathbf{r}(0)\right], \tag{1.7.6}$$

that is, the charge radiates the Doppler-shifted electromagnetic wave whose frequency and wave vector are

$$\omega_\mathrm{s} - \omega_\mathrm{i}(1-\hat{\imath}\cdot\boldsymbol{\beta})/(1-\hat{s}\cdot\boldsymbol{\beta}) \quad \text{and} \quad \mathbf{k}_\mathrm{s} = \omega_\mathrm{s}\hat{s}. \tag{1.7.7}$$

It is convenient to work in terms of the shift in frequency ω and the shift in wave number $\mathbf{k}$. From Eqn (1.7.7) (see Fig. 1.5),

$$\omega = \omega_\mathrm{s}-\omega_\mathrm{i} = (\mathbf{k}_\mathrm{s}-\mathbf{k}_\mathrm{i})\cdot\mathbf{v} = \mathbf{k}\cdot\mathbf{v}, \tag{1.7.8}$$

$$\mathbf{k} = \mathbf{k}_\mathrm{s}-\mathbf{k}_\mathrm{i}. \tag{1.7.9}$$

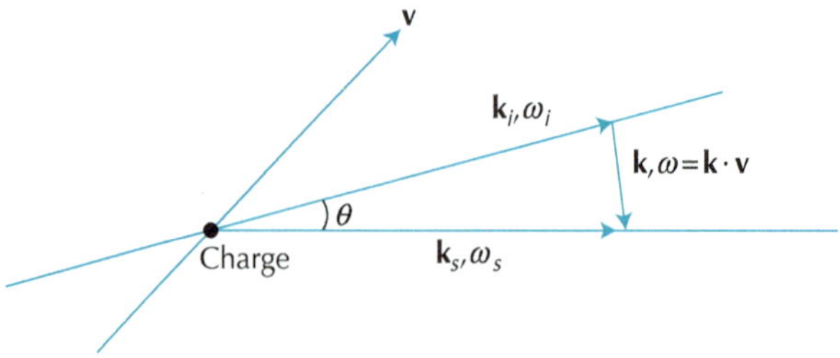

FIGURE 1.5 Wave vector diagram for the scattering of radiation.

Equations (1.7.8) and (1.7.9) are statements of the conservation of energy and momentum; they are valid in the long wavelength regions where Compton scattering may be neglected.[4]

The scattering angle is denoted by θ, and it can be seen from Fig. 1.5 that

$$|k| = \left(k_s^2 + k_\mathrm{i}^2 - 2k_s k_\mathrm{i} \cos\theta\right)^{1/2}. \tag{1.7.10}$$

For a charge for which $\mathrm{v}/c \ll 1$, this reduces to

$$|k| \cong 2\,|k_\mathrm{i}|\sin(\theta/2). \tag{1.7.11}$$

The scattered power in the unit solid angle $d\Omega$ from (1.7.6) and (1.6.7) is

$$P_\mathrm{s}(\mathbf{R})d\Omega = \frac{q^4 E_{\mathrm{i}0}^2 d\Omega}{4\pi m_\mathrm{e}^2 c^3}\left[\hat{s} \times \left(\hat{s} \times \hat{E}_{\mathrm{i}0}\right)\right]^2 \cos^2\left[k_\mathrm{s} R - \omega_\mathrm{s} t - \mathbf{k} \cdot \mathbf{r}(0)\right]. \tag{1.7.12}$$

It is clear from the mass dependence that we may neglect the scattering from the positive ions in comparison to that from the electrons.

For an electron, the time-average scattered power in the solid angle $d\Omega$ is

$$\overline{P_\mathrm{s}(\mathbf{R})}d\Omega = \left(\frac{cE_{\mathrm{i}0}^2 r_0^2 d\Omega}{8\pi}\right)\left[\hat{s} \times \left(\hat{s} \times \hat{E}_{\mathrm{i}0}\right)\right]^2. \tag{1.7.13}$$

---

[4] Strictly, we should allow for the particle nature of the photon. If we assign the incident photon the energy $\hbar\omega_i$, and momentum $\hbar\omega_i/c$ then the conservation of energy equation (1.7.8) becomes

$$\omega_\mathrm{s} \cong \omega_\mathrm{i} + \mathbf{k} \cdot \mathbf{v} - \frac{2\hbar\omega_i^2}{m_e c^2}\sin^2\left(\frac{\theta}{2}\right)\left(1 - \frac{\hat{i} \cdot \mathbf{v}}{c}\right); \quad |\omega_\mathrm{s} - \omega_\mathrm{i}| \ll \omega_\mathrm{i}.$$

It can be seen that the loss of energy owing to the electron recoil is very small provided $2\hbar\omega_i/m_\mathrm{e}c^2 \ll 1$, or $\lambda_i \gg 0.1$ Å and $\mathrm{v}/c \not\simeq 1$ so that the inequality is well satisfied. The correction may be written in terms of a wavelength shift; this is very small:

$$\Delta\lambda = \left(\frac{4\pi\hbar}{m_e c}\right)\sin^2(\theta/2) = 0.048\,\sin^2(\theta/2)\ \text{Å}.$$

These effects are discussed further in Chapter 11.

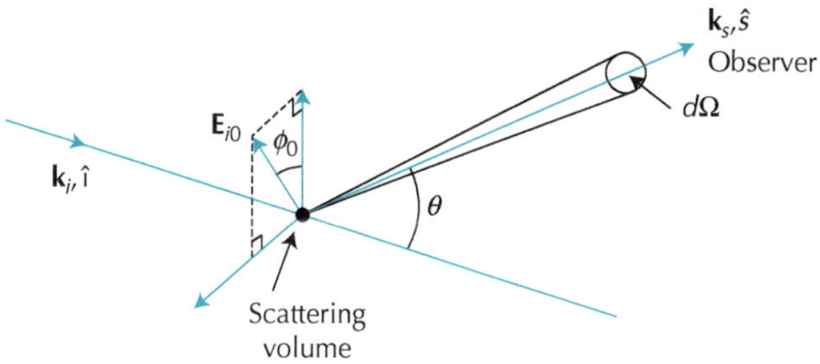

FIGURE 1.6 Scattering geometry, showing the relative orientation of the incident polarization (E vector) and the observer.

The over-bar denotes a time average. The classical electron radius is $r_o = e^2/m_e c^2 - 2.82 \times 10^{-13}$ cm. If the radiation is polarized (see Fig. 1.6),

$$\left[\hat{s} \times \left(\hat{s} \times \hat{E}_{\mathrm{i}0}\right)\right]^2 = 1 - \sin^2\theta \cdot \cos^2\varphi_0. \qquad (1.7.14)$$

If it is not polarized, we may average over ϕ_o and

$$\left[\hat{s} \times \left(\hat{s} \times \hat{E}_{\mathrm{i}0}\right)\right]^2 = 1 - \frac{1}{2}\left(\sin^2\theta\right). \qquad (1.7.15)$$

The total scattering cross section, the ratio of the total scattered power to the incident power, is called the *Thomson cross section* and is given by

$$\sigma_T = (8\pi/3)\, r_0^2.$$

For high frequencies, the Compton effect is important, and scattering must be treated quantum mechanically. The asymptotic forms of the total cross section are given by Klein–Nishina formulas,

$$\sigma_{\mathrm{KN}} = r_0^2 \begin{cases} \dfrac{8\pi}{3}\left(1 - \dfrac{2h\nu_{\mathrm{i}}}{m_{\mathrm{e}}c^2} + \cdots\right), & h\nu_{\mathrm{i}} \ll m_{\mathrm{e}}c^2 \\[2ex] \pi \dfrac{m_{\mathrm{e}}c^2}{h\nu_{\mathrm{i}}}\left[\ln\left(\dfrac{2h\nu_{\mathrm{i}}}{m_{\mathrm{e}}c^2}\right) - \dfrac{1}{2}\right], & h\nu_{\mathrm{i}} \gg m_{\mathrm{e}}c^2. \end{cases}$$

Quantum effects will not be considered in most of this book because $h\nu_i \ll m_e c^2$, and for high-velocity charges, the Doppler shift can be large and $v/c \to 1$. However, the effects can be important in warm-dense matter plasmas, as discussed in Chapter 11.

1.7.2 High-Velocity Charge with No Other Forces Acting

The magnetic field $\mathbf{B}_i$ of the incident wave may not be neglected, and the equation of motion has the form

$$\frac{dm_e\mathbf{v}}{dt'} = \frac{d}{dt'}\frac{m_e\mathbf{v}}{\left[1-\left(\mathrm{v}^2/c^2\right)\right]^{1/2}} = q\left(\mathbf{E}_\mathrm{i} + \frac{\mathbf{v}}{c}\times\mathbf{B}_\mathrm{i}\right) \tag{1.7.16}$$

or

$$\gamma m_e\dot{\mathbf{v}} + \gamma^3 m_e\mathbf{v}\left(\frac{\mathbf{v}\cdot\dot{\mathbf{v}}}{c^2}\right) = q\left(\mathbf{E}_\mathrm{i} + \frac{\mathbf{v}}{c}\times\mathbf{B}_\mathrm{i}\right),$$

where $\gamma = \left[1-\left(\mathrm{v}^2/c^2\right)\right]^{-1/2}$. The scalar product with $\mathbf{v}$ is taken and after suitable rearrangement

$$\frac{d\mathbf{v}}{dt'} = \frac{q}{m_e}\left(1-\frac{\mathrm{v}^2}{c^2}\right)^{1/2}\{\mathbf{E}_{\mathrm{i}0} + \boldsymbol{\beta}\times\mathbf{B}_{\mathrm{i}0} - \boldsymbol{\beta}(\boldsymbol{\beta}\cdot\mathbf{E}_{\mathrm{i}0})\}\cdot\cos\left[\mathbf{k}_\mathrm{i}\cdot\mathbf{r}(t') - \omega_\mathrm{i}t'\right]. \tag{1.7.17}$$

Compare this with Eqn (1.7.2). The additional terms contribute scattered radiation polarized in a plane different from the plane of polarization of the incident wave. These high-velocity effects are discussed in more detail in Chapters 3, 4, and 5.

In summary, the electric field of a wave scattered by an electron is given by

$$\mathbf{E}_s(\mathbf{R},t) = -\frac{e}{cR}\left[\frac{\hat{s}\times\left[(\hat{s}-\boldsymbol{\beta})\times\dot{\beta}\right]}{(1-\hat{s}\cdot\boldsymbol{\beta})^3}\right], \tag{1.7.18}$$

which can be expanded as

$$\begin{aligned}\mathbf{E}_\mathrm{s}(R,t) = {}& \frac{e^2}{m_e c^2 R}\frac{(1-\beta^2)^{1/2}}{\left(1-\hat{s}\cdot\boldsymbol{\beta}\right)^3}\{\hat{s}\times\left[(\hat{s}-\boldsymbol{\beta})\right.\\ & \times\left(\mathbf{E}_{\mathrm{i}0} + \boldsymbol{\beta}\times\left\{n\left(\hat{i}\times\mathbf{E}_{\mathrm{i}0}\right)\right\} - \boldsymbol{\beta}(\boldsymbol{\beta}\cdot\mathbf{E}_{\mathrm{i}0})\right)\left.\right]\} \times\cos\left(k_\mathrm{i}\cdot r(t') - \omega_\mathrm{i}t'\right).\end{aligned} \tag{1.7.19}$$

Note: Normally $\mathbf{r}(t')$ is taken as the solution of Eqn (1.7.16) *where the interaction with the incident wave is neglected. Also, the refractive index n is not always unity.*

In the case that is normally considered, the incident wave is polarized with its electric field $\mathbf{E}_{io}$ perpendicular to the scattering plane; consequently $\mathbf{E}_{io}\cdot\hat{s} = 0$ and by definition $\mathbf{E}_{io}\cdot\hat{i} \equiv 0$. Then, if a polarizer is used to select $\mathbf{E}_s$ parallel to $\hat{e}$, Eqn 1.7.19 reduces to

$$\hat{e}\cdot\mathbf{E}_s(\mathbf{R},t) = \frac{r_o E_{io}}{R}\frac{\left(1-\beta^2\right)^{1/2}}{(1-\beta_s)^3}\left|\left[1-n\left(\hat{s}\cdot\hat{i}\right)\right]\beta_E^2 - (1-n\beta_i)(1-\beta_s)\right|_{\mathrm{ret}}, \tag{1.7.20}$$

where $\beta_E = \mathbf{E}_{io}\cdot\boldsymbol{\beta}$, $\beta_i = \hat{i}\cdot\boldsymbol{\beta}$, $\beta_s = \hat{s}\cdot\boldsymbol{\beta}$, and $\left(\hat{s}\cdot\hat{i}\right) = \cos\theta$.

1.7.3 Low-Velocity Charge ($v/c \ll 1$) with Additional Forces Acting

It is common practice to neglect all other forces when computing the scattering of radiation from a plasma. The example below is given as a warning that the acceleration is not always given by Eqn (1.7.16) and the effects of all forces must be carefully weighed. In general, the equation of motion of a charge is

$$\frac{d(m_e\mathbf{v})}{dt'} = q\left[\mathbf{E}_i + \mathbf{E}_o + \frac{\mathbf{v}}{c} \times (\mathbf{B}_i + \mathbf{B}_o)\right] + \mathbf{F},$$

where $\mathbf{E}_o$ and $\mathbf{B}_o$ are additional applied fields (they could be externally applied or result from the interaction of the charges in the plasma) and $\mathbf{F}$ represents any other forces acting. Consider the case where B_{oz} is a steady externally applied field, and $\mathbf{F}$ is the result of collisions:

$$\mathbf{F}_c = -m_e \nu \mathbf{v},$$

where ν is the collision frequency for momentum transfer. Let $v/c \ll 1$. The solution (see Tanenbaum, 1967, p. 70) is

$$\dot{\mathbf{v}} = R_e \begin{vmatrix} W_\perp & W_H & 0 \\ -W_H & W_\perp & 0 \\ 0 & 0 & W_z \end{vmatrix} \cdot \mathbf{E}_i \cdot \exp\left[i\left(\mathbf{k}_i \cdot \mathbf{r} - \omega_i t'\right)\right], \tag{1.7.21}$$

where

$$W_\perp = \frac{q}{m}\frac{\omega^2 + i\nu\omega_i}{\left(\omega_i^2 + 2i\omega_i\nu - \nu^2 - \Omega^2\right)}, \quad W_H = \frac{q}{m}\frac{i\omega_i\Omega}{\left(\omega_i^2 + 2i\omega_i\nu - \Omega^2\right)}, \tag{1.7.22}$$

$$W_z = \frac{\omega_i}{(\omega_i + i\nu)}.$$

It can be seen that the magnetic field B_{oz} may be neglected in the mobility tensor provided $\Omega = qB_{oz}/m_e c \ll \omega_i$. Collisions may be neglected if $\omega_i \gg \nu$. However, the effect of the magnetic field and of collisions may not be ignored in the exponent, for they modify the orbit of each charge. We see, for example, that the scattered radiation received by the observer will be coherent only for the time between successive collisions.

In the calculations that follow, these effects will be retained where appropriate in the determination of the orbits of the charges but will not be considered in the calculation of the mobility tensor $|W|$.

1.7.4 High-Velocity Charge with Applied Magnetic Fields

In the situation in which there is an applied magnetic field B_o, an additional term appears in Eqn (1.7.17).

$$\frac{q}{m_e}\left(1 - \frac{v^2}{c^2}\right)^{1/2} (\boldsymbol{\beta} \times \mathbf{B}_o). \tag{1.7.23}$$

The only part of the term connected to the incident wave is that part of beta that is driven by the wave. The restrictions applied to scattering calculations in this book, Section 1.8.1, require this incremental velocity to be negligible. Therefore, this term is ignored in calculating the scattering power. This term leads to cyclotron emission, and these contributions to the background radiation will be important for sources operating near the cyclotron frequency, $f_{ce} = 2.8 \times 10^{10}\text{B[T]}$ Hz. Magnetic fusion energy plasmas typically operate at fields in the range of 1 to 10 T and $f_{ce} = 2.8 \times 10^{10}$ to $f_{ce} = 2.8 \times 10^{11}$, i.e., $\lambda \sim 10^{-2}$ to 10^{-3} m. Note that the applied magnetic field does affect the charged particle orbits and hence the fluctuating density see Chapters 4 and 10.

1.8 GENERAL RESTRICTIONS APPLIED TO CALCULATIONS IN THIS BOOK

1.8.1 Macroscopic Interaction of Radiation with a Plasma

It is advisable first to determine the conditions under which the electromagnetic wave will penetrate the plasma. For $\omega_i < \omega_{pe}$ and $\lambda_i, \lambda_{De} \ll L$, the typical scale of the plasma, only the surface layer of electrons interacts with the incident radiation, and the beam is reflected. This is because the charges can respond on a timescale $\sim 1/\omega_{\text{pe}}$ and are able to cancel the incident fields. Setting $\omega_{\text{i}_{\text{crit}}} = \omega_{\text{pe}}$, and rearranging, we find

$$\lambda_{\text{i}_{\text{crit}}} = \frac{3.33 \times 10^6}{\left[n_{\text{e}}\left(\text{cm}^{-3}\right)\right]^{1/2}} \text{cm}. \tag{1.8.1}$$

The transmission and reflection regions are indicated in Fig. 1.7. This result is calculated for the simple case where the ion motion is neglected and the electrons are treated as a conducting fluid of density n_e. Consider an incident plane

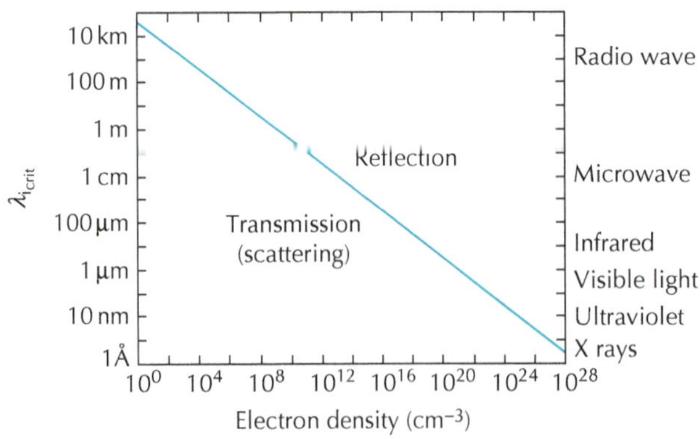

FIGURE 1.7 The dependence of the critical wavelength for transmission on the electron density using Eqn (1.8.1).

electromagnetic wave propagating into a plasma of density n_e:

$$\mathbf{E}_\mathrm{i} = \mathbf{E}_{i0} \exp[i(\mathbf{k}_\mathrm{i} \cdot \mathbf{r} - \omega_\mathrm{i} t)]. \tag{1.8.2}$$

We have Maxwell's equations (1.3.4)

$$\nabla \times \mathbf{E} = -\frac{1}{c}\frac{\partial \mathbf{H}}{\partial t}, \quad \nabla \times \mathbf{H} = \frac{4\pi}{c}\mathbf{J} + \frac{1}{c}\frac{\partial \mathbf{E}}{\partial t},$$

and if we ignore the motion of the more massive ions,

$$\mathbf{J} = -n_e e \mathbf{u}, \tag{1.8.3}$$

where the drift velocity $\mathbf{u}$ of the electron fluid under the influence of the incident wave for $\mathbf{u}/c \ll 1$ is simply given by

$$m_\mathrm{e}\frac{d\mathbf{u}}{dt} = -e\mathbf{E}_\mathrm{i}.$$

Combining these equations, we have

$$\nabla \times (\nabla \times \mathbf{E}_\mathrm{i}) + \frac{1}{c^2}\frac{\partial^2 \mathbf{E}_\mathrm{i}}{\partial t^2} + \frac{4\pi n_e e^2}{c^2 m_\mathrm{e}}\mathbf{E}_\mathrm{i} = 0. \tag{1.8.4}$$

With the substitution of $\mathbf{E}_\mathrm{i}$ from Eqn (1.8.2), this becomes

$$k_\mathrm{i}^2 c^2 - \omega_\mathrm{i}^2 + \omega_\mathrm{pe}^2 = 0$$

or

$$k_\mathrm{i} = \left(\omega_\mathrm{i}^2 - \omega_\mathrm{pe}^2\right)^{1/2}/c. \tag{1.8.5}$$

For $\omega_\mathrm{i}^2 > \omega_\mathrm{pe}^2$, the wave number k_i is real, that is, the wave is transmitted; for $\omega_\mathrm{i}^2 < \omega_\mathrm{pe}^2$, the wave number k_i is imaginary. (These are often referred to as *underdense* and *overdense* states, respectively.) Substitution of this imaginary k_i into the exponent of Eqn (1.8.2) shows a wave which is damped as it propagates into the plasma. The wave is reflected provided the plasma is many wavelengths thick.

The treatment may readily be extended to include the positive ions, the effect of collision, and the effect of a magnetic field in the plasma (see, e.g., Tanenbaum, 1967, Chapter 2) and Section 4.6.

The scattering calculations presented in this book are restricted to situations where certain conditions are met. Some of these have been discussed previously; to help the reader, they are listed below.

1.8.2 General Restrictions

1. We consider a scattering volume V that contains on the average N electrons and N/Z ions of charge Ze.
2. The relativistic case, which brings in terms of order $(\mathrm{v}/c)^2$ and higher, is discussed in Chapter 3, 4, and 5. The correction terms of first order in v/c are discussed in Chapter 2.

3. Quantum effects are also omitted except in Chapter 11; this means that the results are strictly valid only for incident wavelengths, $\lambda_i \gg 0.1$ Å, and for nonrelativistic plasmas.
4. We neglect scattering from the positive ions because they are more massive than electrons and scatter relatively little.
5. We study the scattered radiation at a distance R from the scattering volume such that $R \gg V^{1/3}, \lambda_i$.
6. We require $\omega_i > \omega_{pe}$, and the scale of the plasma small enough that the incident wave is transmitted with negligible attenuation. Each electron in the scattering volume then sees the incident wave at the same intensity. The plasma dimensions are also to be small enough that no multiple scattering occurs.
7. Refraction of the incident and scattered radiation must be taken into account in some cases, notably for laboratory plasmas in the mm range and near plasma cutoffs. Discussions may be found in Tatari *et al.* (1999), Section 5, Bindslev (1992), Chapter 5, Woskov *et al.* (1993), Mazzucato *et al.* (2003), and Smith *et al.* (2008), Section E. The incident electric field is kept small enough that it will not perturb the plasma. This situation certainly requires that

$$\mathrm{v} = \frac{eE_i}{m_e \omega_i} \ll \left(\frac{2\kappa T_e}{m_e} \right)^{1/2} = a.$$

Now, the electric field is related to the incident power by

$$\frac{P_i}{A} = \frac{cE_i^2}{8\pi} \quad \text{Gaussian units,}$$

where A is the cross-sectional area of the incident beam. Therefore, we require that

$$\frac{P_i}{A} \ll \left(\frac{m_e^2 a^2}{8\pi e^2} \right) \omega_i^2$$

or

$$\frac{P_i}{A} \ll 1.3 \times 10^{-9} \left(\frac{a}{c} \right)^2 \omega_i^2 \ \mathrm{W\,cm^{-2}}.$$

Under these conditions, the total scattered electric field $\mathbf{E}_s^T$ is simply the vector sum of the scattered fields of the individual electrons within the scattering volume. The time-averaged scattered power per unit solid angle

$$\frac{dP_s}{d\Omega} = \frac{cR^2}{4\pi} \left\langle \sum_{j=1}^{N} \mathbf{E}_{js} \sum_{l=1}^{N} \mathbf{E}_{ls} \right\rangle. \tag{1.8.6}$$

1.8.3 Noncollective and Collective Spectra

Equation (1.8.6) may be rewritten as

$$\frac{dP_s}{d\Omega} = \frac{cR^2}{8\pi} NE_s^2 + \frac{cR^2}{4\pi} N(N-1)\overline{(\mathbf{E}_j \cdot \mathbf{E}_l)}_{j \neq l}, \tag{1.8.7}$$

where we have separated out the terms for which, respectively, $j = l$ and $j \neq l$.

First Term ($j = l$)

The average field for each electron is the same, $R \gg V^{1/3}$, and we denote it by E_s. This term represents the scattering we would obtain from N randomly distributed electrons, i.e., noncollective. It is convenient to introduce at this point the parameter α,

$$\alpha \equiv 1/k\lambda_{De}.$$

For $\lambda \ll \lambda_{De}$, i.e., $k\lambda_{De} \ll 1$, the wave "sees" the charges on a scale length in which they appear free. If their positions are random, as might be expected for a system in thermodynamic equilibrium, the first term dominates, and noncollective scattering requires

$$\alpha = \frac{1.08 \times 10^{-4} \lambda_i \text{ cm}}{\sin\theta/2} \left[\frac{n_e(\text{cm}^{-3})}{T_e(\text{eV})} \right]^{1/2} \ll 1. \tag{1.8.8}$$

The important feature of this noncollective scattering regime is that spectral shape reflects the electron velocity distribution. This is reasonably obvious from an inspection of Eqn (1.7.8), which shows that the frequency shift ω is proportional to the electron velocity component v_k which lies along the vector $\mathbf{k}$. As a first approximation, we may take $\mathbf{k}$ as constant, and then, the intensity in a given frequency interval is proportional to the number of electrons with velocities in the corresponding velocity range around v_k. In addition, we see from Eqn (1.8.7) that if the scattering volume is known, we can in principle, after suitable calibration, arrive at one method to determine the density $n_e = N/V$.

The noncollective spectrum, which we obtain when this term dominates, is discussed separately and in detail in Chapter 4 and in addition is discussed as a part of the complete spectrum in Chapters 5 and 10 .

Second Term ($j \neq l$)

This takes the form of Eqn (1.8.7) because the electrons are indistinguishable from one another, and there are N ways in which we can pick electron j leaving $(N-1)$ ways to pick electron l, subject to $l \neq j$.

If the electrons are randomly distributed, the second term is zero, because for a given E_j, the E_l, with a given phase, is as likely to be negative as positive. The term makes a contribution only when the electron positions are correlated.

We expect, and in fact obtain, a considerable contribution from the second term when the scattering wavelength λ is comparable to or greater than Debye length, i.e., $\alpha \geq 1$. In this situation, the incident wave interacts with the shielded charges, that is, we have "cooperative" or collective scattering.

In the collective scattering regime ($\alpha > 1$), each electron sees the collective nature of the surrounding electrons providing a correlation in the phases, and the second term in Eqn (1.8.7) provides a nonzero result. The details of the scattered power are discussed in detail in Chapter 3 where the scattered power is shown to scale with N.

Collective scattering can occur from coherent plasma waves when a particular thermal fluctuation is enhanced above the broad background fluctuation spectrum. These periodic fluctuations "waves" can be driven by various unstable situations (e.g., plasma turbulence, laser-plasma instabilities), which are discussed further in Chapter 12. Scattering from a particular wave leads to a considerable enhancement of the level of scattered power in the direction that satisfies conservation of momentum and the power scales as N^2.

The dependence of the form of the complete spectrum on α is determined in Chapter 5 for an unmagnetized plasma and in Chapter 10 for a magnetized plasma. The scattering from unstable plasmas is discussed in Chapter 12.

PROBLEMS

1.1. In the majority of scattering experiments, $\lambda_{De} \gg n_e^{-1/3}$. Calculate the condition under which $\lambda_{De} < n_e^{-1/3}$, and discuss in which plasmas it might occur.

1.2. Compton scattering can be important when x-rays are used (see Chapter 11). The frequency shift becomes $\omega = -\hbar k^2/2m_e \pm \mathbf{k}\cdot\mathrm{v}$. Calculate the effect for the example case in which $|\mathbf{k}| = 4\times10^{10}\,\mathrm{m}^{-1}$, the x-ray energy is $E_0 = 4.75$ keV (helium-α radiation), $T_e = 50$ eV, and $|\mathbf{v}|$ is taken as the mean electron thermal speed.

1.3. Calculate λ_{De}, ω_{pe}, Ω_e, and N_D (let the electron perpendicular velocity be equal to the mean electron thermal speed) for
the ionosphere ($n_e = 10^5\,\mathrm{cm}^{-3}$, $T_e = 0.1\,\mathrm{eV}$, and $B = 4\times10^{-5}$ T);
an industrial plasma ($n_e = 10^{12}\,\mathrm{cm}^{-3}$, $T_e = 2$ eV, and $B = 0.1$ T),
a magnetic fusion plasma ($n_e = 10^{14}\,\mathrm{cm}^{-3}$, $T_e = 10$ keV, and $B = 6$ T);
and a high-energy density plasma [$n_e = 5\times10^{23}\,\mathrm{cm}^{-3}$, $T_e = 12$ eV, $B = 0$ (but note that very high magnetic fields may be generated in HED plasmas, e.g., 100 T)].

1.4. Show that the cutoff for wave propagation with $\mathbf{E}$ perpendicular to $\mathbf{B}_0$ (X-mode polarization) occurs at a density $1 + \frac{\omega_{ce}}{\omega}$ times higher than the cutoff for propagation with $\mathbf{E}$ parallel to $\mathbf{B}_0$ (O-mode polarization).

1.5. Show that

$$\mathbf{E}(\mathbf{x}, t) = e\int\left[\frac{\mathbf{s}}{R^2}\cdot\delta\left(t' + \frac{R}{c-t}\right) + \frac{1}{cR}\cdot(\boldsymbol{\beta} - \mathbf{s})\delta'\left(t' + \frac{R}{c} - t\right)\right]dt'$$

may be rewritten in the form of Eqn (1.6.4) for the case of scattering of electromagnetic radiation.

1.6. Derive Eqn (1.7.17).

1.7. Show how the use of E_{i0} perpendicular to the scattering plane and the use of a polarizer to select E_s parallel to $\hat{e}_i$ allows Eqn (1.7.18) to be simplified to obtain Eqn (1.7.19).

1.8. In the fluid approximation, derive the dispersion relation for the ordinary mode ($\mathbf{E}||\mathbf{B}_0, k\perp\mathbf{B}_0$) in a cold plasma ($\kappa T_e=0$) where the effects of collisions are accounted for through a constant collision frequency ν. Assuming real frequency ω, show that the index of refraction $n=c/\mathrm{v}_{ph}$ and the wave number are now complex. Find expressions for the attenuation length $\delta=[\mathrm{Im}(k)]^{-1}$ in the three limits (a) low frequency $\omega\ll\nu\ll\omega_{pe}$ (b) $\nu<\omega<\omega_{pe}$, and (c) high frequency $\omega\gg\omega_{pe}$ and $\nu^2\ll\omega^2-\omega_{pe}^2$ and $\nu^2\ll\omega^2\frac{\left(\omega^2-\omega_{pe}^2\right)^2}{\omega_{pe}^2}$.

1.9. What are the wavelengths for which $\omega_i\sim\Omega_e$? Consider magnetic fields in the range 1 to 10 T. Note that multiples of the cyclotron frequency ($s\Omega_e$) can be important (See Chapter 10 and Section 9.2). Why is 60 GHz a source frequency for scattering on ITER a better choice than 170 GHz?

1.10. Derive the dispersion relation for high-frequency transverse waves (electrostatic waves) in an unmagnetized plasma with electron temperature $T_e\neq 0$ in the fluid approximation.

1.11. Derive again the dispersion relation for electromagnetic waves propagating along the magnetic field $\mathbf{B}_0\,(=B_0\hat{z})$ in a cold-fluid plasma. However, this time include the ion motion. Show that there is a separate resonance now at $\omega=\Omega_c$. Sketch $1/n_e^2$ versus ω for the R- and L- waves. Now, assuming $\omega\lesssim\Omega_c\ll\omega_c$, find the slow-wave dispersion relation. What is the wave polarization for $\omega\ll\Omega_c$ and $\omega\lesssim\Omega_c$?

SOLUTIONS TO ODD PROBLEMS

1.1. $\lambda_{De}=7.4\times10^3[T_e(\mathrm{eV})/n_e(\mathrm{m}^{-3})]^{1/2}\,\mathrm{m}$, and $\lambda_{De}<n_e^{-1/3}$ when $n_e>1.64\times10^{23}[T_e(\mathrm{eV})]^3\,\mathrm{m}^{-3}$ (see the discussion of warm dense matter plasmas in Chapter 11).

1.3.

Plasma	λ_{De}(cm)	ρ_e(cm)	ω_e(rad s^{-1})	Ω_e(rad s^{-1})	N_D
Ionosphere	0.74	2.7	1.8×10^7	7.0×10^6	1.7×10^5
Industrial	1.1×10^{-3}	4.8×10^{-3}	5.6×10^{10}	1.8×10^{10}	4.8×10^3
Magnetic Fusion	7.4×10^{-3}	5.7×10^{-3}	5.6×10^{11}	1.1×10^{12}	1.7×10^8
HED	3.6×10^{-9}	1.2×10^{-5}	4.0×10^{16}	1.8×10^{13}	0.10

1.5. Following J.D. Jackson's (1998) approach in the chapter Radiation by Moving Charges in his book *Classical Electrodynamics*, the first step is to change the variable of integration to $f(t') = t' + R(t')/c$.

Now, $df/dt' = 1 + (1/c)dR/dt' = 1 - \mathbf{s}\cdot\boldsymbol{\beta} = \kappa$, and $dt' = df/\kappa$. Consequently, the equation becomes

$$\mathbf{E}(\mathbf{x},t) = e\int\left[\frac{\mathbf{s}}{\kappa R^2}\cdot\delta(f-t) + \frac{1}{ckR}\cdot(\boldsymbol{\beta}-\mathbf{s})\delta'(f-t)\right]df.$$

Integrating the derivative of the delta function by parts, recalling that $udv/df = d(uv)/df - vdu/df$, and replacing $d(\boldsymbol{\beta}-\mathbf{s})/df$ by $d(\boldsymbol{\beta}-\mathbf{s})/\kappa\, dt'$, we obtain

$$\mathbf{E}(\mathbf{x},t) = e\left[\mathbf{s}/\kappa R^2 + 1/c\kappa\cdot d\left\{(\mathbf{s}-\boldsymbol{\beta})/\kappa R\right\}/dt'\right]_{\text{ret}}.$$

The rate of change of the unit vector $\mathbf{s}$ with time is due the motion of the charge perpendicular to $\mathbf{R}$ and may be written as $1/c\cdot d\mathbf{s}/dt' = \mathbf{s}\times(\mathbf{s}\times\boldsymbol{\beta})/R = \{(\mathbf{s}\cdot\boldsymbol{\beta})\mathbf{s} - (\mathbf{s}\cdot\mathbf{s})\boldsymbol{\beta}\}/R$,

$$1/c\kappa\cdot d\{\mathbf{s}/\kappa R\}/dt' = 1/\kappa^2R^2\cdot(\mathbf{s}\cdot\boldsymbol{\beta})\mathbf{s} - \boldsymbol{\beta}/(\kappa R)^2 + (\mathbf{s}/c\kappa)d(1/\kappa R)/dt',$$

$$\begin{aligned}1/c\kappa\cdot d\{(\mathbf{s}-\boldsymbol{\beta})/\kappa R\}/dt' &= 1/\kappa^2R^2\cdot(\mathbf{s}\cdot\boldsymbol{\beta})\mathbf{s} - \boldsymbol{\beta}/\kappa^2R^2\\ &\quad + (\mathbf{s}/c\kappa)d(1/\kappa R)/dt' - 1/c\kappa\{d(\boldsymbol{\beta}/\kappa R)dt',\end{aligned}$$

and $\mathbf{s}/\kappa R^2 + 1/\kappa^2R^2\cdot(\mathbf{s}\cdot\boldsymbol{\beta})\mathbf{s} = \mathbf{s}/\kappa^2R^2$ leading to

$$\begin{aligned}\mathbf{E}(\mathbf{x},t) = e\Big[&\mathbf{s}/\kappa^2R^2 - \boldsymbol{\beta}/\kappa^2R^2\\ &+ (\mathbf{s}/c\kappa)d(1/\kappa R)/dt' - 1/c\kappa\{d(\boldsymbol{\beta}/\kappa R)/dt'\}\Big]_{\text{ret}}.\end{aligned}$$

Now, $1/c\cdot d(\kappa R)/dt' = R/c\cdot d(1-\mathbf{s}\cdot\boldsymbol{\beta})/dt' + (1-\mathbf{s}\cdot\boldsymbol{\beta})(1/c)dR/dt' = \boldsymbol{\beta}^2 - \mathbf{s}\cdot\boldsymbol{\beta} - (R1c)\mathbf{s}\cdot d\boldsymbol{\beta}/dt'$, and $1/c\cdot d(1/\kappa R)/dt' = -(1/\kappa^2R^2)(1/c)d(\kappa R)/dt'$ Substituting in the equation for $\mathbf{E}$ leads to

$$\begin{aligned}\mathbf{E}(\mathbf{x},t) = e[&(\mathbf{s}-\boldsymbol{\beta})/\kappa^2R^2 - \{(\mathbf{s}-\boldsymbol{\beta})/\kappa^3R^2\}\{\boldsymbol{\beta}^2 - \mathbf{s}\cdot\boldsymbol{\beta} - (R/c)\mathbf{s}\cdot d\boldsymbol{\beta}/dt'\}\\ &-(1/c\kappa^2R)\cdot d(\boldsymbol{\beta}/dt']_{\text{ret}}\end{aligned}$$

Rearranging and recalling that $\kappa = 1 - \mathbf{s}\cdot\boldsymbol{\beta}$, we obtain Eqn (1.6.4)

1.7. Use $\mathbf{A}\times\mathbf{B}\times\mathbf{C} = (\mathbf{A}\cdot\mathbf{C})\mathbf{B} - (\mathbf{A}\cdot\mathbf{B})\mathbf{C}$.

$$\begin{aligned}\mathbf{E}_s &\propto \hat{s}\times(\hat{s}-\boldsymbol{\beta})\times\{\mathbf{E}_{i0} + \boldsymbol{\beta}\times\hat{\imath}\times\mathbf{E}_{i0} - \boldsymbol{\beta}(\boldsymbol{\beta}\cdot\mathbf{E}_{i0})\}\\ &= [\hat{e}_i\{-\hat{s}\cdot\hat{s} + \hat{s}\cdot\boldsymbol{\beta} + (\hat{s}\cdot\hat{s})(\boldsymbol{\beta}\cdot\hat{\imath}) - (\hat{s}\cdot\boldsymbol{\beta})(\boldsymbol{\beta}\cdot\hat{\imath})\}\\ &\quad + \hat{s}\{\hat{s}\cdot\hat{e}_i + (\hat{s}\cdot\hat{\imath})((\boldsymbol{\beta}\cdot\hat{e}_i) - (\hat{s}\cdot\hat{e}_i)(\boldsymbol{\beta}\cdot\hat{\imath}) - (\hat{s}\cdot\boldsymbol{\beta})(\boldsymbol{\beta}\cdot\hat{e}_i)\}\\ &\quad + \hat{\imath}\{-(\hat{s}\cdot\hat{s})(\boldsymbol{\beta}\cdot\hat{e}_i) + (\hat{s}\cdot\boldsymbol{\beta})(\boldsymbol{\beta}\cdot\hat{e}_i)\} + \boldsymbol{\beta}\{-(\hat{S}\cdot\hat{\imath})(\boldsymbol{\beta}\cdot\hat{e}_i) - (\hat{s}\cdot\hat{e}_i)\\ &\quad + (\hat{s}\cdot\hat{e}_i)(\boldsymbol{\beta}\cdot\hat{\imath}) + (\hat{s}\cdot\hat{s})(\boldsymbol{\beta}\cdot\hat{e}_i)\}]\mathbf{E}_{i0}.\end{aligned}$$

With the orientation chosen for $\mathbf{E}_{i0}, \hat{e}_i \cdot \hat{s} = 0$, the equation reduces to

$$\begin{aligned}\mathbf{E}_s &\propto \hat{s} \times (\hat{s} - \boldsymbol{\beta}) \times \{\mathbf{E}_{i0} + \boldsymbol{\beta} \times \hat{\imath} \times E_{i0} - \boldsymbol{\beta}(\boldsymbol{\beta} \cdot \mathbf{E}_{i0})\} \\ &= [\hat{e}_i\{-\hat{s}\cdot\hat{s} + \hat{s}\cdot\boldsymbol{\beta} + (\hat{s}\cdot\hat{s})(\boldsymbol{\beta}\cdot\hat{\imath}) - (\hat{s}\cdot\boldsymbol{\beta})(\boldsymbol{\beta}\cdot\hat{\imath})\} \\ &\quad + \hat{s}\{(\hat{s}\cdot\hat{\imath})((\boldsymbol{\beta}\cdot\hat{e}_i) - (\hat{s}\cdot\boldsymbol{\beta})(\boldsymbol{\beta}\cdot\hat{e}_i)\} + \hat{\imath}\{-(\hat{s}\cdot\hat{s})(\boldsymbol{\beta}\cdot\hat{e}_i) \\ &\quad + (\hat{s}\cdot\boldsymbol{\beta})(\boldsymbol{\beta}\cdot\hat{e}_i)\} + \boldsymbol{\beta}\{-(\hat{s}\cdot\hat{\imath})((\boldsymbol{\beta}\cdot\hat{e}_i) + (\hat{s}\cdot\hat{s})(\boldsymbol{\beta}\cdot\hat{e}_i)\}]\mathbf{E}_{i0}.\end{aligned}$$

When a polarizer is used to select E_s parallel to $\hat{e}_i, \hat{e}_i \cdot \hat{\imath} = 0$, consequently

$$\begin{aligned}\mathbf{E}_s \cdot \hat{e}_i &\propto \hat{s} \times (\hat{s} - \boldsymbol{\beta}) \times \{\mathbf{E}_{i0} + \boldsymbol{\beta} \times \hat{\imath} x E_{i0} - \boldsymbol{\beta}(\boldsymbol{\beta} \cdot \mathbf{E}_{i0})\} \cdot \hat{e}_i \\ &= [\{-\hat{s}\cdot\hat{s} + \hat{s}\cdot\boldsymbol{\beta} + (\hat{s}\cdot\hat{s})(\boldsymbol{\beta}_i) - (\hat{s}\cdot\boldsymbol{\beta})(\boldsymbol{\beta}\cdot\hat{\imath})\} + \boldsymbol{\beta}\cdot\hat{e}_i\{-(\hat{s}\cdot\hat{\imath})(\boldsymbol{\beta}\cdot\hat{e}_i) \\ &\quad + (\hat{s}\cdot\hat{s})(\boldsymbol{\beta}\cdot\hat{e}_i)\}]E_{i0}.\end{aligned}$$

Substituting $\boldsymbol{\beta}_E = \boldsymbol{\beta} \cdot \hat{e}_i$, $\boldsymbol{\beta}_i = \boldsymbol{\beta} \cdot \hat{\imath}$, and $\boldsymbol{\beta}_s = \boldsymbol{\beta} \cdot \hat{s}$, with $\hat{s} \cdot \hat{\imath} = \cos\theta$, yields

$$\begin{aligned}\mathbf{E}_s \cdot \hat{e}_i = E_s &\propto \hat{s} \times (\hat{s} - \boldsymbol{\beta}) \times \{\mathbf{E}_{i0} + \boldsymbol{\beta} \times \hat{\imath} \times \mathbf{E}_{i0} - \boldsymbol{\beta}(\boldsymbol{\beta} \cdot \mathbf{E}_{i0})\} \cdot \hat{e}_i \\ &= [(1 - \cos\theta)\boldsymbol{\beta}_E^2\}) - (1 - \beta_i)(1 - \beta_s)]E_{i0}\end{aligned}$$

1.9. $s\Omega_e = 1.76. \times 10^{11} sB(T)$ rad $\cdot$ s^{-1}. To satisfy $\omega_i - s\Omega_e, \lambda_i = 2\pi c/\omega_i = 0.011/sB(T)$ m and $\nu_i = 2.73 \times 10^{10} s(T)$ Hz.

At 1 T, $\lambda = 0.011/s$ (m) and $\nu_i = 1 \times 2.73 \times 10^{10}$ Hz.
At 10 T, $\lambda = 0.0011/s$ (m) and $\nu_i = 1 \times 2.73 \times 10^{11}$ Hz.
At the higher fields, the low cyclotron harmonics will be important, as for example in ITER, where it is proposed to avoid the cyclotron resonances by using a 60 GHz gyrotron rather than a 170 GHz gyrotron source for scattering from a plasma with a field around 5 T, where $s\Omega_e = s \times 1.37 \times 10^{11}$ Hz.

1.11. We consider the case of $\mathbf{k}||\mathbf{B}_0$ where ion dynamics are included. We assume a cold plasma with $\omega \lesssim \Omega_c \ll \omega_c$.

The ion and electron equations of motion are given by:

$$\mathrm{v}_{ex} = \frac{e}{im_e\omega}\left(E_x - \frac{i\omega_c}{\omega}E_y\right)\left(1 - \frac{\omega_c^2}{\omega^2}\right)^{-1}$$

$$\mathrm{v}_{ey} = \frac{e}{im_e\omega}\left(E_y + \frac{i\omega_c}{\omega}E_x\right)\left(1 - \frac{\omega_c^2}{\omega^2}\right)^{-1}$$

$$\mathrm{v}_{ix} = \frac{ie}{m_e\omega}\left(E_x + \frac{i\Omega_c}{\omega}E_y\right)\left(1 - \frac{\Omega_c^2}{\omega^2}\right)^{-1}$$

$$\mathrm{v}_{iy} = \frac{ie}{m_e\omega}\left(E_y + \frac{i\Omega_c}{\omega}E_x\right)\left(1 - \frac{\omega_c^2}{\omega^2}\right)^{-1}$$

The linearized wave equation gives us:

$$
\begin{aligned}
E_x(\nabla\times\mathbf{E}) = -\mathbf{k}\times(\mathbf{k}\times\mathbf{E}) = k^2\mathbf{E}-\mathbf{k}(\mathbf{k}\cdot\mathbf{E}) &= \frac{4\pi}{c^2}i\omega\mathbf{j}+\frac{\omega^2}{c^2}\mathbf{E} \\
&= \frac{4\pi}{c^2}n_e e i\omega(\mathbf{v}_i-\mathbf{v}_e)+\frac{\omega^2}{c^2}\mathbf{E}
\end{aligned}
$$

and for our electromagnetic waves $\mathbf{k}\cdot\mathbf{E}=0$ we have

$$
\begin{aligned}
(k^2c^2-\omega^2)E_x = 4\pi n_e e i\omega\Bigg\{&\frac{ie}{m_i\omega}\left(E_x+\frac{i\Omega_c}{\omega}E_y\right)\left(1-\frac{\Omega_c^2}{\omega^2}\right)^{-1} \\
&-\frac{e}{im_e\omega}\left(E_x-\frac{i\omega_c}{\omega}E_y\right)\left(1-\frac{\omega_c^2}{\omega^2}\right)^{-1}\Bigg\} \\
(k^2c^2-\omega^2)E_y = 4\pi n_e e i\omega\Bigg\{&\frac{ie}{m_i\omega}\left(E_y-\frac{i\Omega_c}{\omega}E_x\right)\left(1-\frac{\Omega_c^2}{\omega^2}\right)^{-1} \\
&-\frac{e}{im_e\omega}\left(E_y+\frac{i\omega_c}{\omega}E_x\right)\left(1-\frac{\omega_c^2}{\omega^2}\right)^{-1}\Bigg\}
\end{aligned}
$$

Collecting terms we have:

$$
E_x\left[k^2c^2-\omega^2+\frac{\Omega_{pe}^2}{\left(1-\frac{\Omega_c^2}{\omega^2}\right)}+\frac{\omega_{pe}^2}{\left(1-\frac{\omega_c^2}{\omega^2}\right)}\right]+E_y\left[\frac{i\Omega_c}{\omega}\frac{\Omega_p^2}{\left(1-\frac{\Omega_c^2}{\omega^2}\right)}-\frac{i\omega_c}{\omega}\frac{\omega_{pe}^2}{\left(1-\frac{\omega_c^2}{\omega^2}\right)}\right]=0
$$

$$
E_x\left[-\frac{i\Omega_c}{\omega}\frac{\Omega_p^2}{\left(1-\frac{\Omega_c^2}{\omega^2}\right)}+\frac{i\omega_c}{\omega}\frac{\omega_{pe}^2}{\left(1-\frac{\omega_c^2}{\omega^2}\right)}\right]+E_y\left[k^2c^2-\omega^2+\frac{\Omega_p^2}{\left(1-\frac{\Omega_c^2}{\omega^2}\right)}+\frac{\omega_{pe}^2}{\left(1-\frac{\omega_c^2}{\omega^2}\right)}\right]=0
$$

Requiring the determinant of the coefficients to vanish gives:

$$
\left[\omega^2-k^2c^2-\frac{\Omega_p^2}{1-\frac{\Omega_c^2}{\omega^2}}-\frac{\omega_{pe}^2}{1-\frac{\omega_c^2}{\omega^2}}\right]^2=\left[\frac{\omega_c}{\omega}\frac{\omega_{pe}^2}{1-\frac{\omega_c^2}{\omega^2}}-\frac{\Omega_c}{\omega}\frac{\Omega_p^2}{1-\frac{\Omega_c^2}{\omega^2}}\right]^2
$$

which gives

$$
\omega^2-k^2c^2-\frac{\Omega_p^2}{1-\frac{\Omega_c^2}{\omega^2}}-\frac{\omega_{pe}^2}{1-\frac{\omega_c^2}{\omega^2}}=\pm\left(\frac{\omega_c}{\omega}\frac{\omega_{pe}^2}{1-\frac{\omega_c^2}{\omega^2}}-\frac{\Omega_c}{\omega}\frac{\Omega_p^2}{1-\frac{\Omega_c^2}{\omega^2}}\right)
$$

The upper sign gives:

$$\omega^2 - c^2k^2 - \frac{\Omega_p^2}{1+\frac{\Omega_c}{\omega}} - \frac{\omega_{pe}^2}{1-\frac{\omega_c}{\omega}} = 0 \quad \text{R-wave}$$

The lower sign:

$$\omega^2 - c^2k^2 - \frac{\Omega_p^2}{1-\frac{\Omega_c}{\omega}} - \frac{\omega_{pe}^2}{1-\frac{\omega_c}{\omega}} = 0 \quad \text{L-wave}$$

and for $\omega \lesssim \Omega_c \ll \omega_c$ these give

$$\frac{k^2c^2}{\omega^2} \approx 1 - \frac{\Omega_p^2}{\omega(\omega+\Omega_c)} + \frac{\omega_{pe}^2}{\omega\omega_c} \quad \text{R-wave}$$
$$\frac{k^2c^2}{\omega^2} \approx 1 - \frac{\Omega_p^2}{\omega(\omega-\Omega_c)} - \frac{\omega_{pe}^2}{\omega\omega_c} \quad \text{L-wave}$$

For $\omega \leq \Omega_c$ the L-wave is the slow wave ($\omega/k \to 0$ at $\omega = \Omega$ resonance). Now from our original equations we have:

$$\frac{E_x}{E_y} = -\left[\frac{\frac{i\Omega_c}{\omega}\frac{\Omega_p^2}{\left(1-\frac{\Omega_c^2}{\omega^2}\right)} - \frac{i\omega_c}{\omega}\frac{\omega_{pe}^2}{\left(1-\frac{\omega_c^2}{\omega^2}\right)}}{k^2c^2 - \omega^2 + \frac{\Omega_p^2}{1-\frac{\Omega_c^2}{\omega^2}} + \frac{\omega_{pe}^2}{1-\frac{\omega_c^2}{\omega^2}}}\right]$$

Now for the L-wave we have:

$$\frac{E_x}{E_y} = -\left[\frac{\frac{i\Omega_c}{\omega}\frac{\Omega_p^2}{\left(1-\frac{\Omega_c^2}{\omega^2}\right)} - \frac{i\omega_c}{\omega}\frac{\omega_{pe}^2}{\left(1-\frac{\omega_c^2}{\omega^2}\right)}}{-\frac{\Omega_p^2}{1-\frac{\Omega_c}{\omega}} - \frac{\omega_{pe}^2}{1+\frac{\omega_c}{\omega}} + \frac{\Omega_p^2}{1-\frac{\Omega_c^2}{\omega^2}} + \frac{\omega_{pe}^2}{1-\frac{\omega_c^2}{\omega^2}}}\right] = i \quad \text{unless}$$

Now for $\omega \ll \Omega_c$ our equations reduce to:

$$\frac{k^2c^2}{\omega^2} \approx 1 + \frac{\Omega_p^2}{\omega\Omega_c} - \frac{\omega_{pe}^2}{\omega\omega_c}$$

and the denominator becomes

$$\frac{\omega\Omega_p^2}{\Omega_c} - \frac{\omega\omega_{pe}^2}{\omega_c} - \frac{\omega^2\Omega_p^2}{\Omega_c} - \frac{\omega^2\omega_{pe}^2}{\omega_c^2} = \frac{\omega\Omega_p^2}{\Omega_c}\left(1 - \cancel{\frac{\omega}{\Omega_c}}\right) - \frac{\omega\omega_{pe}^2}{\omega_c}\left(1 + \cancel{\frac{\omega}{\omega_c}}\right)$$
$$= \omega\left(\frac{\Omega_p^2}{\Omega_c} - \frac{\omega_{pe}^2}{\omega_c}\right) = 0$$

This therefore justifies a standard assumption in the Alfven wave derivation that $E_y = 0$ (i.e., linear polarization) and the wave is linearly polarized for $\omega \ll \Omega_c$. We also have:

$$\frac{k^2c^2}{\omega^2} \simeq 1 + \frac{\Omega_p^2}{\Omega_c^2} + \frac{\omega_{pe}^2}{\omega_c^2} \simeq 1 + \frac{4\pi n_e m_c^2}{B_0^2} = 1 + \frac{c^2}{v_A^2}.$$

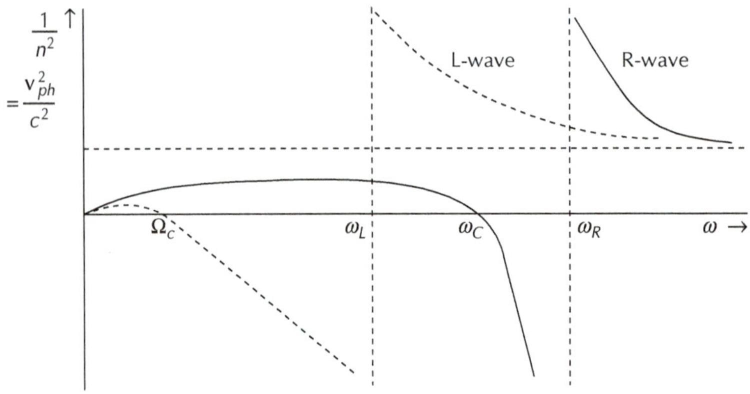

Chapter 2

Scattered Power Spectrum

Chapter Outline Head

2.1 SPECTRAL DENSITY FUNCTION $S(\mathbf{k}, \omega)$

2.1.1 Ensemble Average

The scattered electric field due to a single electron is given by Eqn (1.7.6) or more generally by Eqn (1.7.19) and is a function only of the trajectory of the electron within the scattering volume V. The electron position determines the relative phase of the scattered wave at the observer. [This is the factor $\mathbf{k} \cdot \mathbf{r}(0)$ of Eqn (1.7.12).] The electron velocity fixes the scattered frequency. If we know where the electrons are and how they are moving, we can obtain the scattered spectrum. Often in discussions of scattering theory, the following question is asked: "Well, what is the scattered spectrum if we have all the electrons arranged in a cubic lattice?" Well, if you do have this situation, your problem is to explain,

DOI: 10.1016/B978-0-12-374877-5.00002-6

certainly for a plasma, how it occurred. The scattered power is simply given by Eqns (1.7.19) and (1.8.7). In our case, the questioner had the problem backward. Our problem is to determine the most probable positions for the electrons. For example, let $V = 1\ \text{cm}^3$ and the electron density $10^{14}\ \text{cm}^3$, then $N = 10^{14}$. Given the basic conditions under which the plasma is held, we find that our system can exist happily with many different arrangements of the electrons and ions. Experimentally we of course deal with just one system. Theoretically we must consider a large number of systems that represent the possible states of the one system. In computing the value of a parameter of the system, we take an average over this ensemble of systems. This ensemble average, which will be denoted by angular brackets, is defined as (see, Huang, 1963, p. 142)

$$\langle X \rangle \equiv \frac{\text{all}q \int dq X(q) P(q)}{\text{all}q \int dq P(q)};$$

$P(q)$ is the probability of finding the system in state q. In order to compare theory with experiment, we must assume that the system is stationary on the timescale of the measurement. That is, the ensemble average is independent of time. The ensemble average and time average are the same.[1] In our case, we must average over all the possible electron positions **r** and velocities **v**.

2.1.2 Fluctuating Electron Density $n_e(\mathbf{k}, \omega)$

If the electrons were completely uniformly distributed and their charge smeared out, there would be no net scattering out of the beam, because for each scattered field component in one direction we could always find an equal but opposite component to cancel it. Here, we are ignoring multiple scattering because the plasma is optically thin, and we are ignoring edge effects.

Now, on the average, the electron density is uniform, but on a fine scale, the local level is fluctuating; the net scattering is from the fluctuations. As a first step, we will relate the scattered power spectrum to the expectation value of the Fourier transform of the electron density $\langle |n_e(\mathbf{k}, \omega)|^2 \rangle$. We will then obtain $n_e(\mathbf{k}, \omega)$ by solving the kinetic equations for our system of charges.

The differential scattering vector $\mathbf{k} = \mathbf{k}_s - \mathbf{k}_i$ and the frequency shift of the scattered radiation $(\mathbf{k}_s, \omega_s)$ from the incident radiation $(\mathbf{k}_i, \omega_i)$, $\omega = \omega_s - \omega_i$, were introduced in Section 1.7. The former is a statement of the conservation of momentum in the interaction of the electromagnetic wave with the charges. The latter deals with the conservation of energy.

[1] The relevant timescale of the measurement in this context is usually the integration time τ_I of the measuring system, provided $\tau_I \leq T$, the total time of the measurement. We require the smaller of T and τ_I to be significantly longer than τ_c, the correlation time of waves in the plasma. Thus, the ensemble average need not be completely independent of time, but it may change on a timescale longer than τ_c.

The significance of the quantities **k** and ω may be explained in the following way. If we take a cut through our scattering volume in the direction **k** at some instant in time, we will see that the local density varies with position about some mean level. Clearly we may Fourier analyze this density profile and determine the amplitude of the component with wave number **k**, i.e., $n_e(\mathbf{k})$. This specific density component gives us the scattered wave in the direction θ, where $k^2 = k_s^2 + k_i^2 - 2k_s k_i \cos\theta$.

In similar fashion, if we observe the density at a given point in the plasma, we will see it fluctuating in time. We therefore take a Fourier or Laplace time transform and obtain $n(\mathbf{k}, \omega)$.

2.1.3 Definition of $S(\mathbf{k}, \omega)$

Experimentally we measure the time-average value of the fluctuating density:

$$\overline{|n_e(\mathbf{k}, \omega)|^2} = (1/T) \int_{-T/2}^{+T/2} |n_e(\mathbf{k}, \omega)|^2 dt.$$

We consider a stationary system and equate this with the theoretical ensemble average $\langle |n_e(\mathbf{k}, \omega)|^2 \rangle$. The spectrum is normally written in terms of the spectral density function; this may be defined as

$$S(\mathbf{k}, \omega) \equiv \lim_{V\to\infty, T\to\infty} \frac{1}{VT} \left\langle \frac{n_e(\mathbf{k}, \omega), n_e^*(\mathbf{k}, \omega)}{n_{eo}} \right\rangle. \quad (2.1.1)$$

This is discussed in Appendix A; n_{eo} is the mean electron density, and $N = n_{eo}V$ is the number of electrons within the scattering volume V.

In Section 2.3.3, we show that the scattered power, in the frequency range $\omega_s \to \omega_s + d\omega_s$ and in the solid angle $d\Omega$ centered on **R**, is

$$P(\mathbf{R}, \omega_s) d\Omega d\omega_s = \frac{P_i r_o^2}{A} d\Omega \frac{d\omega_s}{2\pi} \left(1 + \frac{2\omega}{\omega_i}\right) |\hat{s} \times (\hat{s} \times \hat{E}_{io})|^2 N S(\mathbf{k}, \omega).$$

For convenience in the first step, we will relate the scattered power to the spectral density function Eqn (2.1.1). We will then exchange this function for an equivalent one which is defined in terms of Fourier space but now Laplace time transforms. This allows us to follow the approach of Salpeter (1960) in dealing with our final problem, namely the derivation of $n_e(\mathbf{k}, \omega)$ from the plasma equations. The derivation of the plasma equations is discussed at some length in Appendix B; a brief resume is given in Section 2.2.

2.2 KINETIC EQUATIONS FOR A PLASMA

2.2.1 Klimontovich Equations

In our scattering system, we consider a beam of monochromatic radiation to be incident on a plasma. We collect radiation scattered at an angle θ from a

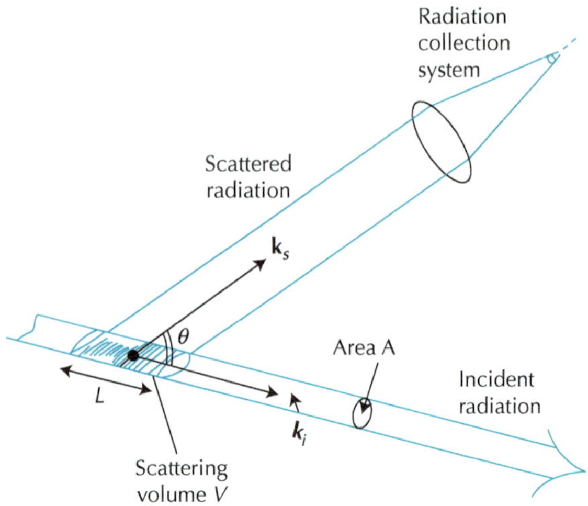

FIGURE 2.1 The relation of the scattering volume to the radiation collection system.

length L of this beam. The collection system defines the scattering volume V (see Fig. 2.1). In the volume V, there are N electrons and N/Z ions of charge Ze. Overall, the volume is electrically neutral, but locally the net charge fluctuates about the mean level. The scattering comes from these fluctuations.

We describe the system of particles q within the scattering volume by the distribution function

$$F_q(\mathbf{r},\mathbf{v},t)=\sum_{s=1}^{N}\delta(\mathbf{r}-\mathbf{r}_s(t))\delta(\mathbf{v}-\mathbf{v}_s(t)). \qquad (2.2.1)$$

This function gives the number of particles of species q per unit volume in phase space, with velocity $\mathbf{v}$ at position $\mathbf{r}$ at time t, and it satisfies the Klimontovich equation (Section B.7)

$$\frac{\partial F_q}{\partial t}+\mathbf{v}\cdot\frac{\partial F_q}{\partial \mathbf{r}}+\mathbf{a}\cdot\frac{\partial F_q}{\partial \mathbf{v}}=0.$$

For our case, we may write the acceleration as

$$\mathbf{a}=\frac{q}{m}\left(\mathbf{E}+\frac{\mathbf{v}}{c}\times\mathbf{B}\right).$$

In addition, we have Maxwell's equations

$$\nabla\cdot\mathbf{E}=\sum_q 4\pi q\int d\mathbf{v}F_q, \qquad \nabla\cdot\mathbf{B}=0$$

$$\nabla\times\mathbf{E}=-\frac{1}{c}\frac{\partial\mathbf{B}}{\partial t}, \qquad \nabla\times\mathbf{B}=\frac{1}{c}\frac{\partial\mathbf{E}}{\partial t}+\sum_q 4\pi q\int d\mathbf{v}\mathbf{v}F_q,$$

where $q=-e$, $m=m_e$ for electrons; $q=Ze$, $m=m_i$, $N\rightarrow N/Z$ for ions. In this form, we are still discussing the motion and interaction of N electrons and N/Z ions, and no solution is possible. Inevitably we must make assumptions or rather determine conditions under which the equations may be simplified to a usable form. This is discussed in Appendix B. Stated briefly, we expand the distribution function, using as a first approximation the ensemble-average value, i.e.,

$$F_q=\langle F_q\rangle+F_{1q}=F_{oq}+F_{1q}.$$

For a stationary, homogeneous system, the ensemble-averaged function is related to the average one-particle distribution function by

$$F_{oe}=n_{e0}f_{oe}(\mathbf{v}) \quad \text{and} \quad F_{oi}=\frac{n_{e0}}{Z}f_{oi}(\mathbf{v}),$$

where n_{e0} and n_{e0}/Z are the mean electron and ion densities, respectively. For thermodynamic equilibrium, we have Maxwellian velocity distributions and

$$f_o(\mathbf{v})=\left(\frac{m}{2\pi\kappa T}\right)^{-3/2}\exp[-(m\mathrm{v}^2/2\kappa T)].$$

We find that for a nonrelativistic plasma ($\mathbf{B}_1=0$), with no steady electric field ($\mathbf{E}_o=0$), the equations may be written as

$$\frac{\partial F_{0q}}{\partial t}+\mathbf{v}\cdot\frac{\partial F_{0q}}{\partial \mathbf{r}}+\frac{q}{m}(\mathbf{v}\times\mathbf{B}_0)\cdot\frac{\partial F_{0q}}{\partial \mathbf{v}}$$
$$=\left(\frac{\delta F_{0q}}{\delta t}\right)_c=-\frac{q}{m}\left\langle\mathbf{E}_1\cdot\frac{\partial F_{1q}}{\partial \mathbf{v}}\right\rangle, \qquad (2.2.2)$$

$$\frac{\partial F_{1q}}{\partial t}+\mathbf{v}\cdot\frac{\partial F_{1q}}{\partial \mathbf{r}}+\frac{q}{m}(\mathbf{v}\times\mathbf{B}_0)\cdot\frac{\partial F_{1q}}{\partial \mathbf{v}}+\frac{q}{m}\mathbf{E}_1\cdot\frac{\partial F_{0q}}{\partial \mathbf{v}}$$
$$=\left(\frac{\delta F_{1q}}{\delta t}\right)_c=-\frac{q}{m}\mathbf{E}_1\cdot\frac{\partial F_{1q}}{\partial \mathbf{v}}+\frac{q}{m}\left\langle\mathbf{E}_1\cdot\frac{\partial F_{1q}}{\partial \mathbf{v}}\right\rangle, \qquad (2.2.3)$$

$$\nabla\cdot\mathbf{E}_1=\sum_q 4\pi q\int d\mathbf{v}F_{1q}. \qquad (2.2.4)$$

The terms on the far right of (2.2.2) and (2.2.3) are the correct form of the plasma "collision terms" in this situation. More generally, we must add a charge-neutral collision term.

2.2.2 Collisions in a Plasma Close to Equilibrium

The phrase "close to equilibrium" is used because we will consider situations with $T_e\neq T_i$ and with a relative drift of electrons and ions. Strictly in this situation, we should include the collision term $(\delta F_{oq}/\delta t)_c$ in Eqn (2.2.2), as this acts to equalize temperatures and remove drifts. However, when the timescale of the measurement is less than the relaxation time for the distribution functions, we may neglect this term. We will restrict ourselves to situations where this procedure is valid, and set $(\delta F_{oq}/\delta t)_c = 0$. This allows us to treat F_{oq} as constant in time while we solve Eqns (2.2.3) and (2.2.4) for F_{1q}.

In the event that the measurement time is greater than the time for F_{oq} to change, as for example is often the case in scattering from shock waves, then we may interpret our results by using it in a suitable time-average value of F_{oq}.

Collisionless Case

In this context, "collisionless" now means that we also set $(\delta F_{1q}/\delta t)_c = 0$ so that there is no collisional damping of the charge fluctuations.

Collisional Case

We will investigate the effect of collisions using the BGK (Bhatnagar-Gross-Krook) collision term

$$\left(\frac{\delta F_{1q}}{\delta t}\right)_c = -\nu[F_{1q} - N_{1q} f_{oq}(\mathbf{v})], \tag{2.2.5}$$

where ν is the collision frequency and $N_{1q}(\mathbf{r}, t) = \int F_{1q}(\mathbf{r}, \mathbf{v}, t) d\mathbf{v}$. This simple form allows us to make rough estimates of collisional effects without too much unwieldy mathematics. While the term conserves the number of particles, it has the weakness that it cannot simultaneously represent both the transfer of momentum and energy with accuracy. This term is discussed in Section 3.7. More exotic forms of the term are also discussed in Appendix B.

Coulomb Collisions[2]

For the interaction of two charges, types α and β,

$$\nu_{\alpha\beta} = \frac{(m_\alpha + m_\beta)}{3\pi^{3/2}(m_\alpha^2 m_\beta)} \frac{q_\alpha^2 q_\beta^2}{\varepsilon_0^2} n_\beta \left(\frac{2\kappa T_\alpha}{m_\alpha} + \frac{2\kappa T_\beta}{m_\beta}\right)^{-3/2} \ln\Lambda \quad \text{(MKS)}$$

$$\Lambda \cong \frac{12\pi\varepsilon_0 \kappa T}{q_\alpha q_\beta}\left(\frac{\varepsilon_0 \kappa T}{e^2 n_e}\right)^{1/2} \quad \text{(MKS)}, \tag{2.2.6}$$

---

[2](See Tanenbaum, 1967, Chapter 5.5). The electron–ion collision frequency usually has the basic form

$$\nu_{ei} = A\omega_{pe} \ln(B n_e \lambda_{\mathrm{De}}^3)/n_e \lambda_d^3,$$

but in the literature, one can find various values for the parameters A and B. This occurs because of the different approximations used for the various intended ranges of validity. Tanenbaum gives $A = [(2\pi)^{1/2} 6\pi]^{-1}$, $B = 12\pi$. For the high-frequency corrections, see the work of Dawson and Oberman (1962), and for quantum effects, see Dubois and Gilinsky (1964), Bekefi (1966), and Sections 3.5 and 3.6.

where $T = T_\alpha \cong T_\beta$. For electrons and singly charged ions $(T_\alpha \cong T_\beta, m_\alpha \ll m_\beta)$,

$$\begin{aligned} \nu_{\text{ei}} &= 2.92 \times 10^{-6} n_{\text{i}}[\text{cm}^{-3}] \cdot (T_{\text{e}}[\text{eV}])^{-3/2} \ln \Lambda \, \text{s}^{-1} \\ \Lambda &= 1.53 \times 10^{10} (T_{\text{e}}[\text{eV}])^{3/2} / (n_{\text{e}}[\text{cm}^{-3}])^{1/2}. \end{aligned} \tag{2.2.7}$$

Charge-Neutral Collisions

$$\nu_{q\beta} = \frac{8\pi^{1/2}}{3} \frac{m_\beta}{(m_q + m_\beta)} n_\beta \sigma^2 \left(\frac{2\kappa T_q}{m_q} + \frac{2\kappa T_\beta}{m_\beta} \right)^{1/2}, \tag{2.2.8}$$

where σ is the sum of the effective radii of the interacting particles. For electron-neutral collisions $(m_\alpha \ll m_\beta)$,

$$\nu_{en} = 2.8 \times 10^8 (r_n[\text{cm}])^2 n_n[\text{cm}^{-3}](T_e[\text{eV}])^{1/2} \, \text{s}^{-1}, \tag{2.2.9}$$

where r_n is the effective radius of the neutral and n_n is the neutral density (typically $r_n \simeq 10^{-8}$ cm).

2.3 POWER SCATTERED $[P_s(\mathbf{R}, \omega_s)]$

2.3.1 General Solution $P_s(\mathbf{R}, \omega_s)$

We consider a volume V that contains N electrons and N/Z ions of charge Ze. The plane monochromatic incident wave is written as

$$\mathbf{E}_i = \mathbf{E}_{io} \cos(\mathbf{k}_i \cdot \mathbf{r} - \omega_i t).$$

The restrictions are $L = V^{1/3} \gg \lambda_i, \lambda$; $|R| \gg L$; $\omega_i \gg \omega_{pe}, \nu, \Omega_e, \omega$; $\lambda_i \gg 0.1$ Å; $eE_{io}/m_e\omega_i \ll \text{v}_{th}$; $E_s \ll E_i$.

For a given electron, the scattered field at the observer $(\mathbf{r}, t)$ is determined by the motion of the electron at the retarded time t' where (see Fig. 2.2)

$$t' = t - \frac{|\mathbf{R} - \mathbf{r}(t')|}{c} \simeq t - \frac{R}{c} + \frac{\hat{s} \cdot \mathbf{r}}{c}. \tag{2.3.1}$$

This approximation is valid for $|R| \gg L$.

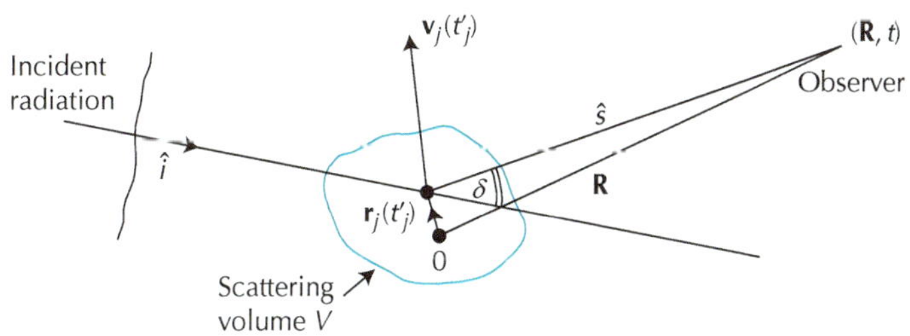

FIGURE 2.2 The scattering coordinate system.

The electron orbit is, for $B=0$,

$$\mathbf{r}_j(t')=\mathbf{r}_j(0)+\mathbf{v}_j t'; \tag{2.3.2}$$

for $B\neq 0$,

$$\mathbf{r}_j(t')=\mathbf{r}_j(0)+\rho_{ej}\hat{x}\cos\phi+\rho_{ej}\hat{y}\sin\phi+\mathrm{v}_{\|j}t', \tag{2.3.3}$$

where $\mathrm{v}_{\|j}=\mathbf{v}_j\cdot\hat{B}_{oz}$, $\rho_{ej}=\mathrm{v}_{\perp j}\Omega_e$ (see Fig. 4.12), and

$$\phi t'=\Omega_e t'+\phi(0).$$

For simplicity, we will write $\hat{s}\cdot\mathbf{v}_j/c=\beta_{sj}$, $\hat{i}\cdot\mathbf{v}_j/c=\beta_{ij}$, and substituting from Eqns (2.3.2) and (2.3.3) into (2.3.1), we see that

$$B=0,\quad t=(1-\beta_{sj})t'+\frac{R}{c}-\frac{\hat{s}\cdot\mathbf{r}_j(0)}{c}$$

$$B\neq 0,\quad t=(1-\beta_{s\|j})t'-\hat{s}\cdot\hat{x}\frac{\beta_{\perp j}}{\Omega_e}\cos\phi-\hat{s}\cdot\hat{y}\frac{\beta_{\perp j}}{\Omega_e}\sin\phi+\frac{R}{c}-\frac{\hat{s}\cdot\mathbf{r}_j(0)}{c}$$

[for all cases, differentiate Eqn (2.3.1), $dt=(1-\beta_{sj})dt'$].

In this section, we are interested in the nonrelativistic relations for the scattered power; therefore, we will eventually put $\beta\rightarrow 0$. Nevertheless, in establishing the result, it is useful to look first at the relativistic equations and then make this approximation. The acceleration of the electron by the incident wave is given by Eqn (1.7.17) (remember $\mathbf{B}_i=\hat{i}\times\mathbf{E}_i$ at conditions with the index of refraction is assumed to be one)

$$\dot{\beta}_j=-\frac{e}{cm_e}(1-\beta_j^2)^{1/2}[\mathbf{E}_i+\boldsymbol{\beta}_j\times\mathbf{B}_i-\boldsymbol{\beta}_j(\boldsymbol{\beta}_j\cdot\mathbf{E}_i)].$$

The scattered electric field in the wave zone ($R\gg\lambda_i$)(see Section 1.6) is

$$\mathbf{E}_{sj}(\mathbf{R},t)=-\frac{e}{cR}\left[\frac{\hat{s}\times[(\hat{s}-\boldsymbol{\beta}_j)\times\dot{\boldsymbol{\beta}}_j]}{(1-\hat{s}\cdot\boldsymbol{\beta}_j)^3}\right]_{ret}. \tag{2.3.4}$$

The approximation $R'\simeq R\gg L$ has been used in the denominator. In this relativistic form, we see that the amplitude and polarization of the scattered field are functions of the electron velocity. We denote by $F_e(\mathbf{r},\mathbf{v},t')$ the quantity

$$\sum_{j=1}^{N}\delta(\mathbf{r}-\mathbf{r}_j(t'))\delta(\mathbf{v}-\mathbf{v}_j(t'))\delta\left(t'-t+\frac{R}{c}-\frac{\hat{s}}{c}\cdot\mathbf{r}_j\right)$$

The total scattered field is, thus,

$$E_s^T(\mathbf{R},t)=-\frac{e}{cR}\int_V d\mathbf{r}\int d\mathbf{v}F_e(\mathbf{r},\mathbf{v},t')\left[\frac{\hat{s}\times[(\hat{s}-\boldsymbol{\beta})\times\dot{\boldsymbol{\beta}}]}{(1-\hat{s}\cdot\boldsymbol{\beta}_s)^3}\right]. \tag{2.3.5}$$

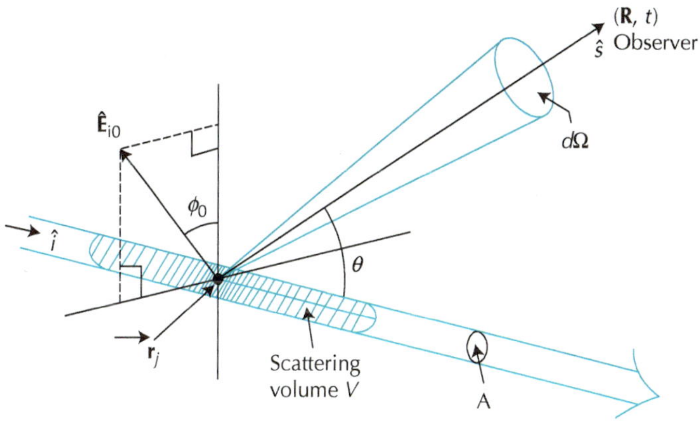

FIGURE 2.3 The scattering geometry.

The time-averaged scattered intensity at **R**, which includes all scattered frequencies, is

$$\overline{I_s(\mathbf{R})} = \frac{c}{4\pi}\frac{1}{T}\int_{-T/2}^{+T/2} dt|E_s^T|^2 = \frac{c}{4\pi}\lim_{T\to\infty}\frac{1}{T}\int_{-\infty}^{+\infty} dt|E_s^T|^2.$$

The limit $T \to \infty$ may be taken if T is much greater than the coherence time of fluctuations in the plasma (see Born & Wolf, 1965, p. 495).

The time-averaged scattered power in the solid angle $d\Omega$, centered on **R** (see Fig. 2.3), is

$$P_s(\mathbf{R})d\Omega = \frac{cR^2}{4\pi}d\Omega \lim_{T\to\infty}\frac{1}{T}\int_{-\infty}^{+\infty} dt|E_s^T|^2. \tag{2.3.6}$$

Experimentally we will use a detector that accepts radiation at some scattered frequency ω_s in some interval $-d\omega_s/2 \to +d\omega_s/2$. We therefore need an expression for the variation of the scattered power with frequency. We define the Fourier transform of the scattered field by

$$E_s^T(\omega_s) = \int_{-\infty}^{\infty} dt E_s^T(t)e^{-i\omega_s t}.$$

With the aid of Parseval's theorem (see Appendix A), we obtain

$$P_s(\mathbf{R}, \omega_s) d\Omega d\omega_s = \frac{cR^2 d\Omega}{4\pi} \lim_{T\to\infty} \frac{1}{\pi T} \int_{\omega_s - d\omega_s/2}^{\omega_s + d\omega_s/2} d\omega_s \left| \int_{-\infty}^{\infty} dt E_s^T(t) e^{-i\omega_s t} \right|^2, \tag{2.3.7}$$

where ω_s now takes on only positive values.

In the integral over t, E_s^T is given in Eqn (2.3.5) in terms of the retarded time t'. It is in fact convenient to work in terms of t' as given by Eqn (2.3.1). The important point to remember is that $dt = (1-\beta_s)dt'$; thus, in the high-temperature case,

$$\int_{-\infty}^{+\infty} dt E_{\mathrm{s}}^{\mathrm{T}}(t) e^{i\omega_s t} = -\frac{e}{cR} \int_{-\infty}^{+\infty} dt' \int d\mathbf{r} \int d\mathbf{v} F_{\mathrm{e}}(\mathbf{r}, \mathbf{v}, t') \times \exp\left[-i\omega_{\mathrm{s}}\left(t' - \frac{\hat{s}\cdot\mathbf{r}}{c} + \frac{R}{c}\right)\right] \left[\frac{\hat{s} \times \{(\hat{s}-\boldsymbol{\beta}) \times \dot{\boldsymbol{\beta}}\}}{(1-\beta_{\mathrm{s}})^2}\right]. \tag{2.3.8}$$

2.3.2 Low-Temperature Plasmas $(\mathbf{v}/c) \ll 1$

In this situation, we do not need to differentiate between t and t' except for a constant factor. We retain the electron motion when we determine the Doppler shift in frequency, but otherwise, we treat the electron as stationary.

$$\left[\frac{\hat{s} \times \{\hat{s}-\boldsymbol{\beta}) \times \dot{\boldsymbol{\beta}}}{(1-\beta_{\mathrm{s}})^2}\right]_{ret} \Rightarrow \hat{s} \times \left(\hat{s} \times \frac{\dot{\mathbf{v}}}{c}\right)$$

and

$$\dot{\mathbf{v}} = -(e/m_{\mathrm{e}}) \cdot \mathbf{E}_{\mathrm{i}}(t), \qquad \int d\mathbf{v} F_{\mathrm{e}} = n_{\mathrm{e}}(\mathbf{r}, t). \tag{2.3.9}$$

Substituting in Eqns (2.3.8) and (2.3.7), we obtain

$$P_s(\mathbf{R}, \omega_s) d\Omega d\omega_s = \frac{cR^2 d\Omega}{4\pi} \left[\hat{s} \times (\hat{s} \times \hat{E}_{io})\right] \lim_{T\to\infty} \frac{d\omega_s}{\pi T} \left| \int_{-\infty}^{\infty} dt \frac{e^2}{m_{\mathrm{e}} c^2} E_{io} \times \int_V d\mathbf{r} n_{\mathrm{e}}(\mathbf{r}, t) \exp\left[-i\omega_s\left(t' - \frac{\hat{s}\cdot\mathbf{r}}{c} + \frac{R}{c}\right)\right] \times \cos(\mathbf{k}_i \cdot \mathbf{r} - \omega_i t) \right|^2. \tag{2.3.10}$$

We substitute for electron density in terms of the Fourier space and time transform:

$$n_e(\mathbf{r}, t) = \int \frac{d\mathbf{k}}{(2\pi)^3} \int \frac{d\omega}{2\pi} n_e(\mathbf{k}, \omega) e^{-i(\mathbf{k}\cdot\mathbf{r} - \omega t)}. \qquad (2.3.11)$$

Now, $\cos a = \frac{1}{2}(e^{-ia} + e^{+ia})$, and in the exponent of the first term in Eqn (2.3.10), we have

$$-i\left[\mathbf{k}\cdot\mathbf{r} - \omega t + \omega_s t - \frac{\omega_s}{c}\hat{s}\cdot\mathbf{r} + \frac{\omega_s}{c}R + \mathbf{k}_i\cdot\mathbf{r} - \omega_i t\right]$$
$$= i\left[(\omega - (\omega_s - \omega_i))t - \left(\mathbf{k} - \left(\frac{\omega_s}{c}\hat{s} - \mathbf{k}_i\right)\right)\cdot\mathbf{r} - \frac{\omega_s}{c}R\right]. \qquad (2.3.12)$$

In the second term,

$$i\left[(\omega - (\omega_s + \omega_i))t - \left(\mathbf{k} - \left(\frac{\omega_s}{c}\hat{s} + \mathbf{k}_i\right)\right)\cdot\mathbf{r} - \frac{\omega_s}{c}R\right].$$

First Term

Integration over time t produces the delta function $2\pi\delta[\omega - (\omega_s - \omega_i)]$; therefore, when we subsequently integrate over ω, we simply have to substitute $\omega = \omega_s - \omega_i$. Now, the scattered radiation is electromagnetic; therefore, $\omega_s\hat{s}/c = \mathbf{k}_s$, and the integration over $\mathbf{r}$ produces the delta function $(2\pi)^3\delta[\mathbf{k} - (\mathbf{k}_s - \mathbf{k}_i)]$. Thus, finally, when we integrate over $\mathbf{k}$, we set $\mathbf{k} = \mathbf{k}_s - \mathbf{k}_i$. The phase factor $\exp[-i(\omega_s/c)R]$ may be dropped.

Second Term

We find by a similar procedure that $\omega = \omega_s + \omega_i, \mathbf{k} = \mathbf{k}_s + \mathbf{k}_i$. We now take an ensemble average, and for the average scattering power in the frequency range $\omega_s \to \omega_s + d\omega_s$ and within the solid angle $d\Omega$ centered on $\mathbf{R}$ (see Fig. 2.3), we have

$$P_s(\mathbf{R}, \omega_s)d\Omega d\omega_s = \frac{P_i r_o^2}{A2\pi} d\Omega d\omega_s \left|\hat{s} \times (\hat{s} \times \hat{E}_{io})\right|^2 NS(\mathbf{k}, \omega), \qquad (2.3.13)$$

where the spectral density function is defined as

$$S(\mathbf{k}, \omega) \equiv \lim_{T\to\infty, V\to\infty} \frac{1}{TV} \left\langle \frac{|n_e(\mathbf{k}, \omega)|^2}{n_{eo}} \right\rangle. \qquad (2.3.14)$$

The mean electron density is $n_{eo} = N/V\ \mathrm{cm}^{-3}$, $\mathbf{k} = \mathbf{k}_s - \mathbf{k}_i$, $\omega = \omega_s - \omega_i$, $k_s = \omega_s/c$, and ω_s takes both positive and negative values.[3]

---

[3]The two terms involving the fluctuating density have been combined; these were both obtained with ω_s positive:

$$\left\langle|n_e(\mathbf{k}_s - \mathbf{k}_i, \omega_s - \omega_i)|^2\right\rangle + \left\langle|n_e(\mathbf{k}_s + \mathbf{k}_i, \omega_s + \omega_i)|^2\right\rangle = \left\langle|n_e(\mathbf{k}_s - \mathbf{k}_i, \omega_s - \omega_i)|^2\right\rangle,$$

where now ω_s takes both positive and negative values. Note $|n_e(\mathbf{k}, \omega)|^2 = |n_e(-\mathbf{k}, -\omega)|^2$.

2.3.3 High-Temperature Plasma $(v/c)^2 \ll 1$

For a high-temperature plasma, the terms of order v/c must be included. Here, we include terms of order (v/c), and in Section 3.5, a fully relativistic treatment is presented. This first-order treatment is a fairly good approximation for electron temperatures, $T_e \lesssim 5$ keV. The ions do not present a problem because of their greater mass. To first order in v/c, we may rewrite Eqn (2.3.8), when the index of refraction is assumed to be unity ($n_i = 1$), as

$$\int_{-\infty}^{+\infty} dt E_s^T(t) e^{i\omega_s t} = -\frac{e}{cR}\int_{-\infty}^{+\infty} dt' \int d\mathbf{r} \int_{-\infty}^{+\infty} d\mathbf{v} F_e(\mathbf{r},\mathbf{v},t') \exp\left[-i\omega_s\left(t' - \frac{\hat{s}\cdot\mathbf{r}}{c}\right)\right]$$
$$\times [(1-\beta_i+\beta_s)\hat{E}_{i0} - \{(1-\beta_i+2\beta_s)\cos\eta + \beta_E\cos\theta\}\hat{s}$$
$$+ \beta_E \hat{i} + \cos\eta\boldsymbol{\beta}]\cos(\mathbf{k}_i\cdot\mathbf{r} - \omega_i t'), \quad (2.3.15)$$

$$\text{where} \quad \beta_E = \hat{E}_{i0}\cdot(\mathbf{v}/c) \quad \text{and} \quad \cos\eta = \hat{s}\cdot\hat{E}_{i0}. \quad (2.3.16)$$

To first order in β, it is a good approximation to take the velocity limits as $\pm\infty$, instead of $\pm c$. The constant phase factor $\exp[-i(\omega_s/c)/R]$ has been dropped.

For the case considered in Section 1.7.2, in which $\mathbf{E_{i0}}\cdot\hat{s} = 0$ and $\mathbf{E_s}||\mathbf{E_{i0}}$, we find that the second term in the square brackets in Eqn (2.3.15) is actually $\frac{(1-\beta_i)}{(1-\beta_s)}\mathbf{E_{i0}} \equiv \frac{\omega_s}{\omega_i}\mathbf{E_{i0}}$. Since $\frac{\omega_s}{\omega_i}$ is not a function of $\mathbf{v}$, it may be taken outside the integral. When the electric field is squared to get power, all the cross terms are zero, and the remaining term, $\beta_E^2\cos^2\theta$, is second order in beta. To first order in beta, the remaining term is $\left(1+\frac{2\omega}{\omega_i}\right)$. In this case, Eqn (2.3.7) becomes

$$P_s(\mathbf{R},\omega_s)d\Omega d\omega_s = \frac{P_i r_o^2}{A2\pi} d\Omega d\omega_s \left(1+2\frac{\omega}{\omega_i}\right)\left|\hat{s}\times(\hat{s}\times\hat{E}_{io})\right|^2 NS(\mathbf{k},\omega). \quad (2.3.17)$$

The additional term, $(1+2\omega/\omega_i)$, first derived by Pogutse (1964) is a result of two effects. The first effect is due to the relativistic aberration, also referred to as the relativistic "headlight" effect, where light is preferentially directed in the emitters direction of propagation [see Fig. 1.4(b)]. The second effect is a result of the electron motion in the direction of the incident light vector interacting with the magnetic field of the probing beam. The resulting $\mathbf{v}\times\mathbf{B_i}$ force is parallel to the force of the incident electric field. When the electron is moving toward the detector, the increased force enhances the velocity and as a result the scattered power. When the electron is moving away from the detector, the force is in the opposite direction, and the scattered power is reduced (Ross *et al.*, 2010).

2.3.4 $S(\mathbf{k},\omega)$ Fourier–Laplace Transforms and Collisions

We will follow the approach of Salpeter (1960) and use Fourier space but Laplace time transforms. The Fourier–Laplace components of the electron density are given by

$$n_e(\mathbf{k},\omega - i\gamma) = \int_{-\infty}^{+\infty} d\mathbf{r} e^{+i\mathbf{k}\cdot\mathbf{r}} \int_{-\infty}^{+\infty} dt e^{-i(\omega - i\gamma)t} n_e(\mathbf{r},t), \tag{2.3.18}$$

where γ is a small positive quantity. The spectral density function Eqn (2.3.14) now takes the form (see Appendix A)

$$S(\mathbf{k},\omega) = \lim_{\gamma \to 0} \frac{2\gamma}{V} \left\langle \frac{|n_e(\mathbf{k},\omega - i\gamma)|^2}{n_{eo}} \right\rangle. \tag{2.3.19}$$

BGK Collision Term

When we use a BGK collision term Eqn (2.2.5) and take a time transform of Eqn (2.2.3), the effect of the component νF_{1q} is to change the ω of the collisionless case to $(\omega + i\nu)$. There is now a damping term $e^{-\nu t}$ in the time dependence of the fluctuations. The form of the collision term conserves the fluctuations but does not allow us to follow each individual particle, and straight application of Eqn (2.3.19) leads to the not surprising result that

$$S(\mathbf{k},\omega) \to 0 \quad \text{when} \quad 1/\gamma \simeq T \gg 1/\nu.$$

The simplest approach is to average the fluctuations over a collision period $1/\nu$. The form for $S(\mathbf{k},\omega)$ is then

$$S(\mathbf{k},\omega) = \frac{2\nu_q}{V} \left\langle \frac{|n_e(\mathbf{k},\omega)|^2}{n_{eo}} \right\rangle, \tag{2.3.20}$$

where ν_q is the electron damping frequency in the ensemble average over the initial electron conditions and represents the ion damping frequency for the average for the ions.

PROBLEMS

2.1. Calculate the collision frequency ν_{ei} and compare to ω_{pe} for the ionosphere ($n_e = 10^5\,\text{cm}^{-3}$, $T_e = 0.1\,\text{eV}$, and $4 \times 10^{-5}\,\text{T}$); an industrial hydrogen plasma ($n_e = 10^{12}\,\text{cm}^{-3}$ and $T_e = 2\,\text{eV}$); a fusion plasma using 50:50 deuterium-tritium (D-T) ($n_e = 10^{14}\text{cm}^{-3}$ and $T_e = 10\,\text{keV}$); and a high-energy density plasma using hydrogen ($n_e = 10^{23}\,\text{cm}^{-3}$, $T_e = 12\,\text{eV}$, and $Z = 1$). Comment on when collisions might be important.

What about effects around ω_{pi}?

Would neutrals play a role in low temperature plasmas ($n_n \sim n_e$), such as industrial and/or HED plasmas?

SOLUTIONS TO ODD PROBLEM

2.1.

Plasma	$n_e(\mathrm{cm}^{-3})$	$T_e\,(\mathrm{eV})$	$\nu_{ei}\,(s^{-1})$	$\omega_{pe}\,(\mathrm{rad}\,s^{-1})$
Ionosphere	10^5	0.1	1.3×10^2	1.8×10^7
Industrial	10^{12}	2	1.1×10^7	5.7×10^{10}
Fusion	10^{14}	10^4	6.2×10^3	5.7×10^{14}
HED	10^{23}	12	2.0×10^{17}	4.9×10^{15}

Since $\omega_{pi} \ll \omega_{pe}$ charge particle collisions would play a role in this region in industrial and HED plasmas. For the example case, $\nu_{en} \sim \nu_{ei}$ and electron–neutral collisions might play a role in HED plasmas.

Chapter 3

Scattering Spectrum from a Plasma Theory

Chapter Outline Head

3.1 INTRODUCTION

The scattering calculations presented in this book are made for the case where the incident wave is primarily transmitted. In this situation, each charge in

DOI: 10.1016/B978-0-12-374877-5.00003-8

the plasma "sees" the same incident field. In addition, a restriction is made to situations where the plasma as a whole is not disturbed by the wave.[1] In the interaction of the wave with the plasma, each charge may then be treated separately. The scattered field seen by the observer is a function of the orbits of the charges. The plasma effects enter in the calculation of these orbits.

In Chapter 2, it was shown that for a high-temperature plasma ($(\mathrm{v}/c)^2 \ll 1$), the average scattered power in the frequency range $\omega_s \to \omega_s + d\omega_s$ and within the solid angle $d\Omega$ centered on $\mathbf{R}$ [see Fig. 2.3 and Eqns (2.3.17) and (2.3.19)] is

$$P_s(\mathbf{R}, \omega_s) d\Omega d\omega_s = \frac{P_i r_o^2 d\Omega}{A 2\pi} d\omega_s \left(1 + \frac{2\omega}{\omega_i}\right) |\hat{s} \times (\hat{s} \times \hat{E}_{io})|^2 N S(\mathbf{k}, \omega), \qquad (3.1.1)$$

where the spectral density function is given by

$$S(\mathbf{k}, \omega) = \lim_{\gamma \to 0, V \to \infty} \frac{2\gamma}{V} \left\langle \frac{|n_e(\mathbf{k}, \omega - i\gamma)|^2}{n_{eo}} \right\rangle.$$

The Fourier space–Laplace time transform of the electron density is

$$n_e(\mathbf{k}, \omega - i\gamma) = \int_{-\infty}^{+\infty} d\mathbf{r} e^{i(\mathbf{k}\cdot\mathbf{r})} \int_0^{\infty} dt e^{-i(\omega - i\gamma)t} n_e(\mathbf{r}, t),$$

where

$$(\mathbf{k} = \mathbf{k}_s - \mathbf{k}_i, \quad \omega = \omega_s - \omega_i).$$

In this chapter, the form of the spectral density function is derived from a collisionless, unmagnetized plasma, in thermodynamic equilibrium using the approach of Salpeter (1960). The effect of collisions on the result is obtained by repeating the calculation with a BGK collision term included in the kinetic equations. At the end of the chapter, some brief comments are made on the derivation of the scattering spectrum from the fluctuation dissipation theorem. The effect of a magnetic field is discussed in Chapter 10.

This calculation is extended to involve scattering from a volume V of plasma that contains N electrons and N/Z ions of charge Ze. The total scattered field is simply the vector sum of the individual scattered fields of the N electrons (the ions do not contribute significantly because of their greater mass). The problem remaining is to determine the orbits of these electrons. It is not possible to follow the detailed motion of all the charges, therefore, we use a statistical approach and describe the behavior of each species by a probability distribution function. The distribution function used here has the form

$$F_q(\mathbf{r}, \mathbf{v}, t) = \sum_{j=1}^{N_q} \delta[(\mathbf{r} - \mathbf{r}_j(t))] \delta[(\mathbf{v} - \mathbf{v}_j(t))]. \qquad (3.1.2)$$

---

[1]The condition for this situation to hold are discussed in Section 1.8.1; we certainly require $\omega_i > \omega_{pe}, \Omega_e, \nu$, and $eE_i/m_e\omega_i \ll (2\kappa T_e/m_e)^{1/2}$.

This gives the number of particles of species q per unit volume in phase space with velocity $\mathbf{v}$ at position $\mathbf{r}$ and at time t.

The charge number density is obtained by summing over all velocities

$$n_q(\mathbf{r}, t) = \sum_{v} d\mathrm{v} F_q(\mathbf{r}, \mathrm{v}, t) = \sum_{j=1}^{N_q} \delta[\mathbf{r} - \mathbf{r}_j(t)]. \tag{3.1.3}$$

The density has a mean level $n_{qo} = N_q/V$, but locally, the density is fluctuating about this level. The power spectrum of radiation scattered from a plasma depends on these microscopic density fluctuations.

3.2 DERIVATION OF $n_e(\mathbf{k}, \omega)$ FOR $B=0, v=0$

3.2.1 Basic Equations

The probability distribution function $F_q(\mathbf{r}, \mathbf{v}, t)$ [Eqn (3.1.2)] satisfies the Klimontiovich equation

$$\frac{\partial F_q}{\partial t} + \mathbf{v} \cdot \frac{\partial F_q}{\partial \mathbf{r}} + \mathbf{a} \cdot \frac{\partial F_q}{\partial \mathbf{v}} = 0, \tag{3.2.1}$$

where for the purpose of the calculation below, we may set

$$\mathbf{a} = (q/m)\mathbf{E} \quad \text{and} \quad \nabla \cdot \mathbf{E} = \textstyle\sum_q 4\pi q \int d\mathbf{v} F_q. \tag{3.2.2}$$

As they stand, these equations cannot be solved because they contain all the detailed multiparticle interactions. Fortunately, it can be shown that the behavior of the plasma is dominated by the two particle interactions under normal circumstances, i.e., $n_{eo}\lambda_{\mathrm{De}}^3 \gg 1$. We therefore set $F_q = F_{oq} + F_{1q}$, where F_{oq} represents the average state of the system and F_{1q} represents the microscopic fluctuations about the mean level.

$$F_{1q}(\mathbf{r}, \mathbf{v}, t) = \sum_{j=1}^{N_q} \delta[\mathbf{r} - \mathbf{r}_j(t)]\delta[\mathbf{v} - \mathbf{v}_j(t)] - F_{oq}. \tag{3.2.3}$$

and

$$n_{1q}(\mathbf{r}, t) = n_q - n_{oq}. \tag{3.2.4}$$

With this expanded distribution function, we form a hierarchy of equations in which each successive equation brings in a higher-order correlation (see Appendix B.7). The hierarchy is terminated at the two-particle interaction level, and for a stationary, homogeneous plasma at low-temperatures, with no magnetic field, the system is described by the equations (see Section 2.2)

$$\frac{\partial F_{0q}}{\partial t} + \mathbf{v} \cdot \frac{\partial F_{0q}}{\partial \mathbf{r}} = 0 \tag{3.2.5}$$

$$\frac{\partial F_{1q}}{\partial t} + \mathbf{v} \cdot \frac{\partial F_{1q}}{\partial \mathbf{r}} + \frac{q}{m}\mathbf{E}_1 \cdot \frac{\partial F_{0q}}{\partial \mathbf{v}} = \left(\frac{\partial F_{1q}}{\partial t}\right)_c \tag{3.2.6}$$

$$\nabla \cdot \mathbf{E}_1 = \sum_q 4\pi q \int d\mathbf{v} F_{1q}, \quad \mathbf{E}_0 = 0, \tag{3.2.7}$$

where for the electrons

$$q = -e, \quad m = m_e, \quad N_q = N, \quad \text{and} \quad F_{0e} = n_{e0} f_{0e}(\mathbf{v}) \tag{3.2.8}$$

and for the ions

$$q = Ze, \quad m = m_i, \quad N_q = N/Z, \quad \text{and} \quad F_{0i} = (n_{e0}/Z) f_{0i}(\mathbf{v}), \tag{3.2.9}$$

f_{0e}, f_{0i} are the average one-particle distribution functions.

Equation (3.2.5) tells us that to a first approximation, the charges move in straight lines. There are however small perturbations about this motion owing to the interaction of the charges with each other, via the Coulomb field. These correlation effects enter through the microscopic field $\mathbf{E}_1$, which is given by Poisson equation (3.2.7). The more drastic deflections from the sum of many small deflections enter through the collision term $(\partial F_{1q}/\partial t)_c$. The effect of collisions is discussed in Section 3.7, and for the moment, we set $(\partial F_{1q}/\partial t)_c = 0$; this term acts to equalize the temperature of the species and to "Maxwellize" the distribution functions. The restriction implies, therefore, that we are either working on a timescale less than the time for the distributions to change or considering some time-averaged distributions.

3.3 THE FLUCTUATING DENSITY $n_e(\mathbf{k}, \omega)$

(a) The Fourier–Laplace transform of the distribution function is given by

$$F_{1q}(\mathbf{k}, \mathbf{v}, \omega) = \int_0^\infty dt e^{-(i\omega+\gamma)t} \int_{-\infty}^{+\infty} d\mathbf{r} F_{1q}(\mathbf{r}, \mathbf{v}, t) e^{i\mathbf{k}\cdot\mathbf{r}}.$$

This quantity is the contribution to the kth, ω^{th} component of the fluctuating density from charges whose velocities lie in the limited range $\mathbf{v} \to \mathbf{v} + d\mathbf{v}$.

The kth, ω^{th} component of the density is

$$n_{1q}(\mathbf{k}, \omega - i\gamma) = \int_{o}^{\infty} dt e^{-(i\omega+\gamma)t} \int_{-\infty}^{+\infty} d\mathbf{r} n_{1q}(\mathbf{r}, t) e^{i\mathbf{k}\cdot\mathbf{r}}$$

$$= \sum_{\mathrm{v}} d\mathbf{v} F_{1q}(\mathbf{k}, \mathbf{v}, \omega). \tag{3.3.1}$$

(b) From Poisson's equation (3.2.7), the potential $\phi(\mathbf{r}, t)$ is given by

$$\nabla^2 \phi(\mathbf{r}, t) = -4\pi(Zen_{1i} - en_{1e}) \equiv -4\pi\rho_1(\mathbf{r}, t) = \frac{4\pi}{(2\pi)^3} \int_{-\infty}^{+\infty} d\mathbf{k} e^{-\mathbf{k}\cdot\mathbf{r}} \rho_1(\mathbf{k}, t).$$

This equation may be integrated to give

$$\phi(\mathbf{r}, t) = \frac{4\pi}{(2\pi)^3} \int_{-\infty}^{+\infty} \frac{d\mathbf{k}}{k^2} e^{-\mathbf{k}\cdot\mathbf{r}} \rho_1(\mathbf{k}, t). \tag{3.3.2}$$

This result is easily demonstrated by differentiating Eqn (3.3.2) with respect to $\mathbf{r}$. Finally,

$$\mathbf{E}_1(\mathbf{r}, t) = -\nabla\phi = \frac{4\pi i}{(2\pi)^3} \int \frac{d\mathbf{k}}{k^2} \mathbf{k} \rho_1(\mathbf{k}, t) e^{-i\mathbf{k}\cdot\mathbf{r}}. \tag{3.3.3}$$

(c) This electric field Eqn (3.3.3) is now substituted in Eqn (3.2.6) with $(\partial F_{1q}/\partial T)_c = 0$, and we take the Fourier–Laplace transform of this equation to obtain[2]

$$-F_{1q}(\mathbf{k}, \mathbf{v}, 0) + [i\omega + \gamma - i\mathbf{k}\cdot\mathbf{v}] F_{1q}(\mathbf{k}, \mathbf{v}, \omega)$$

$$= -\frac{q}{m} \int d\mathbf{r} \int \frac{d\mathbf{k}'}{(k')^2} \frac{4\pi i}{(2\pi)^3} \rho_1(\mathbf{k}', \omega) e^{-i(\mathbf{k}'-\mathbf{k})\cdot\mathbf{r}} \mathbf{k}' \cdot \frac{\partial F_{oq}}{\partial \mathbf{v}}. \tag{3.3.5}$$

---

[2] Use has been made of the condition that the system is stationary; therefore, F_{oq} is not a function of time. Further, note that

$$\int_{-\infty}^{+\infty} d\mathbf{r} e^{i\mathbf{k}\cdot\mathbf{r}} \frac{\partial F_{1q}}{\partial \mathbf{r}} = F_{1q} e^{i\mathbf{k}\cdot\mathbf{r}} \Big|_{-\infty}^{+\infty} - i\mathbf{k} \int_{-\infty}^{+\infty} d\mathbf{r} e^{i\mathbf{k}\cdot\mathbf{r}} F_{1q} = -i\mathbf{k} F_{1q}(\mathbf{k}, \mathbf{v}, t) \tag{3.3.4}$$

and

$$\int_0^{\infty} dt e^{-(i\omega+\gamma)t} \frac{\partial F_{1q}}{\partial t} = F_{1q}(\mathbf{k}, \mathbf{v}, t) e^{-(i\omega+\gamma)t} \Big|_{-\infty}^{+\infty} + \int_0^{\infty} dt F_{1q} F_{1q}(\mathbf{k}, \mathbf{v}, t) e^{-(i\omega+\gamma)t} (i\omega + \gamma)$$

$$= F_{1q}(\mathbf{k}, \mathbf{v}, 0) + (i\omega + \gamma) F_{1q}(\mathbf{k}, \mathbf{v}, \omega).$$

For a homogeneous plasma, only the $\mathbf{k}' = \mathbf{k}$ part survives the integration, and Eqn (3.3.5) may be rearranged to give

$$F_{1q}(\mathbf{k},\mathbf{v},\omega) = \frac{-F_{1q}(\mathbf{k},\mathbf{v},0) - (4\pi q/mk^2)\rho_1(\mathbf{k},\omega)\mathbf{k}\cdot\partial F_{oq}/\partial\mathbf{v}}{\omega - \mathbf{k}\cdot\mathbf{v} - i\gamma}, \quad (3.3.6)$$

where

$$\rho_1(\mathbf{k},\omega) = Zen_{1i}(\mathbf{k},\omega) - en_{1e}(\mathbf{k},\omega).$$

(d) We substitute from Eqns (3.2.8) and (3.2.9) into Eqn (3.3.6) and sum over all velocities to obtain the transform of the fluctuating electron and ion densities, respectively,[3]

$$n_{1e}(\mathbf{k},\omega - i\gamma) = -i\sum_{j=1}^{N}\frac{e^{i\mathbf{k}\cdot\mathbf{r}_j(0)}}{\omega - \mathbf{k}\cdot\mathbf{v}_j(0) - i\gamma} + \frac{\chi_e(\mathbf{k},\omega)}{e}\rho_1(\mathbf{k},\omega) \quad (3.3.7)$$

$$n_{1i}(\mathbf{k},\omega - i\gamma) = -i\sum_{l=1}^{N/Z}\frac{e^{i\mathbf{k}\cdot\mathbf{r}_l(0)}}{\omega - \mathbf{k}\cdot\mathbf{v}_l(0) - i\gamma} - \frac{\chi_i(\mathbf{k},\omega)}{Ze}\rho_1(\mathbf{k},\omega), \quad (3.3.8)$$

where in the last terms, the sums have been replaced by integrals over velocity, and we have defined the electron and ion susceptibilities,

$$\chi_e(\mathbf{k},\omega) = \int_{-\infty}^{+\infty} d\mathbf{v}\frac{4\pi e^2 n_{eo}}{m_e k^2}\frac{\mathbf{k}\cdot\partial f_{eo}/\partial\mathbf{v}}{\omega - \mathbf{k}\cdot\mathbf{v} - i\gamma} \quad (3.3.9)$$

$$\chi_i(\mathbf{k},\omega) = \int_{-\infty}^{+\infty} d\mathbf{v}\frac{4\pi Z^2 e^2 n_{io}}{m_i k^2}\frac{\mathbf{k}\cdot\partial f_{io}/\partial\mathbf{v}}{\omega - \mathbf{k}\cdot\mathbf{v} - i\gamma}. \quad (3.3.10)$$

These functions are components of the longitudinal dielectric function

$$\epsilon(\mathbf{k},\omega) = 1 + \chi_e(\mathbf{k},\omega) + \chi_i(\mathbf{k},\omega).$$

(e) We now multiply Eqn (3.3.7) by $-e$ and Eqn (3.3.8) by Ze and add them. This gives, after rearranging, an expression for $\rho_1(\mathbf{k},\omega)$, which we

---

[3]

$$\sum_{\mathbf{v}}\frac{d\mathbf{v}F_{1e}(\mathbf{k},\mathbf{v},0)}{(\omega - \mathbf{k}\cdot\mathbf{v} - i\gamma)} = \sum_{\mathbf{v}} d\mathbf{v}\sum_{j=1}^{N}\frac{\delta[\mathbf{v} - \mathbf{v}_j(0)]}{\omega - \mathbf{k}\cdot\mathbf{v} - i\gamma}\int d\mathbf{r}\delta[\mathbf{r} - \mathbf{r}_j(0)]$$
$$= \sum_{j=1}^{N}\frac{e^{i\mathbf{k}\cdot\mathbf{r}_j(0)}}{\omega - \mathbf{k}\cdot\mathbf{v}_j(0) - i\gamma}.$$

substitute back into Eqn (3.3.7) to obtain

$$n_{1e}(\mathbf{k}, \omega - i\gamma) = -i\left[\sum_{j=1}^{N} \frac{e^{i\mathbf{k}\cdot\mathbf{r}_j(0)}}{\omega - \mathbf{k}\cdot\mathbf{v}_j(0) - i\gamma} - \frac{\chi_e}{\epsilon}\sum_{j=1}^{N} \frac{e^{i\mathbf{k}\cdot\mathbf{r}_j(0)}}{\omega - \mathbf{k}\cdot\mathbf{v}_j(0) - i\gamma} + \frac{Z\chi_e}{\epsilon}\sum_{l=1}^{N/Z} \frac{e^{i\mathbf{k}\cdot\mathbf{r}_l(0)}}{\omega - \mathbf{k}\cdot\mathbf{v}_l(0) - i\gamma}\right]. \quad (3.3.11)$$

3.3.1 First Term

This is the contribution from the electrons as they follow unperturbed orbits, i.e., the free-electron contribution. Now, as we shall see below, χ_e/ϵ is proportional to $\alpha^2 = (1/k\lambda_{De})^2$ for the equilibrium case, and for $\alpha^2 \ll 1$, this term dominates and leads to noncollective scattering.

3.3.2 Second Term

This is the contribution resulting from the correlation of the electrons with each other. Each test electron j causes a perturbation of the background electron density. The perturbation of the motion of each background electron is of order $1/n_e\lambda_{De}^3$; however, there are $\sim n_e\lambda_{De}^3$ electrons involved in shielding each test electron; therefore, this dressing term is important. If one looks in detail at the spectral density function derived in the following section, it will become evident that this term leads to the high-frequency collective features observed when $\alpha > 1$; this is investigated further in Chapter 5. We must always remem ber that the plasma equations which we are using are restricted to situations where $n_e\lambda_{De}^3 \gg 1$. For $n_e\lambda_{De}^3$ close to unity, we might expect three particle correlations to play a role in which case the expression for $n_e(\mathbf{k}, \omega)$ could take a more complex form.

3.3.3 Third Term

This is the contribution from the electron dressing on each ion, where the ion is treated as a test charge. This term is responsible for the "ion feature" in the scattering spectrum. When observing the shielding of the ion by the attraction of the electrons and other ions, we will see in Chapter 5 that this term leads to the ion thermal spread when the ions cannot respond collectively and a low-frequency collective regime when the ions play a major role in the shielding. Again for $n_e\lambda_{De}^3 \simeq 1$, we might expect modifications to the result above; however, measured profiles of the ion feature for $n_e\lambda_{De}^3 \simeq 3$ obtained by Röhr (1968) and Kato (1972) agree very well with the spectral profile derived from the above form.

3.4 THE SPECTRAL DENSITY FUNCTION $S(\mathbf{k},\omega)$ FOR A COLLISIONLESS PLASMA

3.4.1 Arbitrary Function

$$
\begin{aligned}
S(\mathbf{k},\omega) &= \lim_{\gamma\to 0, V\to\infty} \frac{2\gamma}{V}\left\langle \frac{|n_e(\mathbf{k},\omega-i\gamma)|^2}{n_{e0}}\right\rangle \\
&= \lim_{\gamma\to 0, V\to\infty} \frac{2\gamma}{Vn_{e0}}\left\{\left|1-\frac{\chi_e}{\varepsilon}\right|^2\left\langle \sum_{j=1}^{N}\frac{e^{i\mathbf{k}\cdot\mathbf{r}_j}}{\omega-\mathbf{k}\cdot\mathbf{v}_j-i\gamma}\sum_{m=1}^{N}\frac{e^{-i\mathbf{k}\cdot\mathbf{r}_m}}{\omega-\mathbf{k}\cdot\mathbf{v}_m+i\gamma}\right\rangle\right. \\
&\quad + Z\left|\left(1-\frac{\chi_e}{\varepsilon}\right)\left(\frac{\chi_e}{\varepsilon}\right)^*\right|\left\langle \sum_{j=1}^{N}\frac{e^{i\mathbf{k}\cdot\mathbf{r}_j}}{\omega-\mathbf{k}\cdot\mathbf{v}_j-i\gamma}\sum_{l=1}^{N/Z}\frac{e^{-i\mathbf{k}\cdot\mathbf{r}_l}}{\omega-\mathbf{k}\cdot\mathbf{v}_l+i\gamma}\right\rangle \\
&\quad + Z\left|\left(1-\frac{\chi_e}{\varepsilon}\right)^*\left(\frac{\chi_e}{\varepsilon}\right)\right|\left\langle \sum_{m=1}^{N/Z}\frac{e^{-i\mathbf{k}\cdot\mathbf{r}_m}}{\omega-\mathbf{k}\cdot\mathbf{v}_m+i\gamma}\sum_{s=1}^{N/Z}\frac{e^{i\mathbf{k}\cdot\mathbf{r}_s}}{\omega-\mathbf{k}\cdot\mathbf{v}_s-i\gamma}\right\rangle \\
&\quad \left. + \left|Z\frac{\chi_e}{\varepsilon}\right|^2\left\langle \sum_{l=1}^{N/Z}\frac{e^{i\mathbf{k}\cdot\mathbf{r}_l}}{\omega-\mathbf{k}\cdot\mathbf{v}_l-i\gamma}\sum_{s=1}^{N/Z}\frac{e^{-i\mathbf{k}\cdot\mathbf{r}_s}}{\omega-\mathbf{k}\cdot\mathbf{v}_s+i\gamma}\right\rangle\right\}. \qquad (3.4.1)
\end{aligned}
$$

In taking the ensemble averages, we now use as the weighting factor the most probable initial state of the velocity [i.e., $f_{eo}(\mathbf{v},0)$ and $f_{io}(\mathbf{v},0)$] and of position. Consider the first term; it may be separated into two parts for which $j=m$ and $j\neq m$, respectively.

$$
\begin{aligned}
&\lim_{\gamma\to 0}\gamma\left\langle \sum_{j=1}^{N}\frac{e^{i\mathbf{k}\cdot\mathbf{r}_j}}{\omega-\mathbf{k}\cdot\mathbf{v}_j-i\gamma}\sum_{m=1}^{N}\frac{e^{i\mathbf{k}\cdot\mathbf{r}_m}}{\omega-\mathbf{k}\cdot\mathbf{v}_m-i\gamma}\right\rangle \\
&= \lim_{\gamma\to 0}\gamma N\int_{-\infty}^{+\infty}\frac{d\mathbf{v}f_{eo}(\mathbf{v})}{(\omega-\mathbf{k}\cdot\mathbf{v})^2+\gamma^2} \qquad \text{[see Eqn (A.1.15)]} \\
&\quad + \lim_{\gamma\to 0}\gamma N(N-1)\int\!\!\int_{j\neq m}\frac{d\mathbf{v}_j d\mathbf{v}_m f_{eo}(\mathbf{v}_j)f_{eo}(\mathbf{v}_m)\langle e^{i\mathbf{k}\cdot(\mathbf{r}_j-\mathbf{r}_m)}\rangle_{j\neq m}}{(\omega-\mathbf{k}\cdot\mathbf{k}_j-i\gamma)(\omega-\mathbf{k}\cdot\mathbf{v}_m+i\gamma)} \\
&= \frac{\pi N}{k}f_{eo}\left(\frac{\omega}{k}\right) \\
&\quad + \lim_{\gamma\to 0}\gamma N(N-1)\int\!\!\int_{j\neq m}\frac{d\mathbf{v}_j d\mathbf{v}_m f_{eo}(\mathbf{v}_j)f_{eo}(\mathbf{v}_m)\langle e^{i\mathbf{k}\cdot(\mathbf{r}_j-\mathbf{r}_m)}\rangle_{j\neq m}}{(\omega-\mathbf{k}\cdot\mathbf{k}_j-i\gamma)(\omega-\mathbf{k}\cdot\mathbf{v}_m+i\gamma)}, \qquad (3.4.2)
\end{aligned}
$$

where $f_{eo}(\omega/k)$ is now the one-dimensional velocity distribution in the direction $\mathbf{k}$. The second part of this term requires some discussion.

(a) If the charges are initially uncorrelated, then

$$\langle \exp(i\mathbf{k}\cdot[\mathbf{r}_j(0)-\mathbf{r}_m(0)])\rangle_{j\neq m}=0.$$

Equally the cross terms in Eqn (3.4.1) and the ion dressing cross term with $l\neq s$ are also zero.

(b) The effect of correlations and a nonequilibrium velocity distribution in the initial state are commented on briefly in Section 3.6 where it is shown that the cross terms may be dropped provided the plasma is initially close to equilibrium, and then

$$S(\mathbf{k},\omega)=\frac{2\pi}{k}\left|1-\frac{\chi_e}{\epsilon}\right|^2 f_{eo}\left(\frac{\omega}{k}\right)+\frac{2\pi Z}{k}\left|\frac{\chi_e}{\epsilon}\right|^2 f_{io}\left(\frac{\omega}{k}\right). \qquad (3.4.3)$$

See Appendix D for computational techniques used to solve Eqn 3.4.3.

3.5 POWER SCATTERED FROM A HOT PLASMA

The fully relativistic scattered power for noncollective scattering was derived first by Pechacek & Trivelpiece (1967). There has been some debate about their introduction of the so-called finite transit time effect. Kukushkin (1981) concludes that the introduction of a factor $(1-\beta_s)$ for the transit time effect simply corrected for an error in the use of the square of a delta function. The discussion of this factor in the original book, Sheffield (1975), was also in error. The finite transit time factor is quite small as discussed in Section 4.4.

In a nonrelativistic development of the scattering spectrum, the relative motion between the electron and the incident field only affects the phase of the incident radiation experienced by the electron. The acceleration of the electron is assumed to be independent of velocity (other than the phase), which neglects any modification of the electron trajectory from the Lorentz force and relativity.

In noncollective scattering experiments, these modifications have little consequence on the scattered spectrum unless the electron temperature becomes a sizable fraction of the electron rest energy. In collective Thomson scattering, however, the phase velocity of the electron plasma wave can approach c at moderate or even low electron temperatures. As the Thomson scattering process involves electrons traveling at the phase velocity of the electron plasma wave, this leads to a situation where relativistic effects can be observe (Ross *et al.*, 2010) in a plasma traditionally considered nonrelativistic ($T_e \ll m_e c^2$ and $\mathrm{v}_{osc}/c \ll 1$); see Section 5.4.1.

3.5.1 Fully Relativistic Power Spectrum

Following the general expression for scattered power developed in Section 2.3, a fully relativistic form factor was developed by Palastro *et al.* (2010) and is valid to all orders in the normalized electron velocity, $\boldsymbol{\beta}=\mathbf{v}/c$, and for all scattering angles perpendicular to the incident polarization. The fully relativistic power spectrum is given by combining Eqns (2.3.7) and (2.3.8) and can be written in analogy to the nonrelativistic case derived in Section 2.3.2 as

$$P_s(\mathbf{k},\omega)=\frac{NP_i r_e^2}{2\pi A}NS(\mathbf{k},\omega), \tag{3.5.1}$$

where the relativistic fluctuation spectrum is given by

$$S(\mathbf{k},\omega)=\lim_{\gamma\to 0, V\to\infty}\frac{2\gamma}{V}\left\langle\frac{1}{n_{e0}}\left|\int d\mathbf{v}\left[\frac{\hat{s}\times\{(\hat{s}-\boldsymbol{\beta})\times\dot{\boldsymbol{\beta}}\}}{(1-\beta_s)^2}\right]F_e(\mathbf{k},\mathbf{v},\omega+i\gamma)\right|^2\right\rangle. \tag{3.5.2}$$

In the noncollective regime where one expects uncorrelated electrons, the scattered power can be calculated by simply squaring the electric field before doing the integral over velocity, which simplifies the expressions, see Chapter 4, but in the general case developed below, it is necessary to evaluate the complete term before the integration. We, therefore, break the velocity integral into four parts,

$$\int d\mathbf{v}\left[\frac{\hat{s}\times\{(\hat{s}-\boldsymbol{\beta})\times\dot{\boldsymbol{\beta}}\}}{(1-\beta_s)^2}\right]F_e(\mathbf{k},\mathbf{v},\omega+i\gamma)=(\mathbf{H_e}+\mathbf{H_s}+\mathbf{H_i}+\mathbf{H_p}), \tag{3.5.3}$$

where we define

$$\mathbf{H_e}\equiv\hat{e}\int d\mathbf{v}\frac{(1-\beta^2)^{1/2}}{(1-\beta_s)}(1-n_i\beta_i)F_e(\mathbf{k},\mathbf{v},\omega), \tag{3.5.4}$$

$$\mathbf{H_s}\equiv-\hat{s}\int d\mathbf{v}\beta_E\frac{(1-\beta^2)^{1/2}}{(1-\beta_s)^2}(n_i\cos\theta-\beta_s)F_e(\mathbf{k},\mathbf{v},\omega), \tag{3.5.5}$$

$$\mathbf{H_i}\equiv\hat{i}\int d\mathbf{v}n_i\beta_E\frac{(1-\beta^2)^{1/2}}{(1-\beta_s)}F_e(\mathbf{k},\mathbf{v},\omega), \tag{3.5.6}$$

$$\mathbf{H_p}\equiv-(1-n_i\cos\theta)\int d\mathbf{v}\beta_E\boldsymbol{\beta}\frac{(1-\beta^2)^{1/2}}{(1-\beta_s)^2}F_e(\mathbf{k},\mathbf{v},\omega), \tag{3.5.7}$$

where $n_i=\frac{ck_i}{\omega_i}$ is the index of refraction for the incident radiation and $\mathbf{H_e}$, $\mathbf{H_s}$, $\mathbf{H_i}$ are the components of the scattered electric field in the $\hat{e}$, $\hat{s}$, and $\hat{i}$ directions, respectively, for the electrons evolving under the classical Lorentz force of the incident wave, while $\mathbf{H_p}$ is the additional field component resulting from the relativistic motion of the electrons, which leads to a polarization rotation of the scattered light. Assuming a symmetric equilibrium distribution function, these terms can be written as

$$\mathbf{H_e} = -\hat{e}i\sum_{j=1}^{N}\frac{\left(1-\beta^2\right)^{1/2}(1-n_i\beta_i)}{(1-\beta_s)}\frac{e^{-i\mathbf{k}\cdot\mathbf{r}_j(0)}}{\omega-\mathbf{k}\cdot\mathbf{v}_j-i\gamma} - \hat{e}\frac{X_e}{e}\rho_1(\mathbf{k},\omega), \tag{3.5.8}$$

$$\mathbf{H_s} = \hat{s}i\sum_{j=1}^{N}\frac{\left(1-\beta^2\right)^{1/2}}{(1-\beta_s)^2}\frac{\beta_E(n_i\cos\theta-\beta_s)e^{-i\mathbf{k}\cdot\mathbf{r}_j(0)}}{\omega-\mathbf{k}\cdot\mathbf{v}_j-i\gamma}, \tag{3.5.9}$$

$$\mathbf{H_i} = \hat{i}i\sum_{j=1}^{N}\frac{\left(1-\beta^2\right)^{1/2}}{(1-\beta_s)}\frac{n_i\beta_E e^{-i\mathbf{k}\cdot\mathbf{r}_j(0)}}{\omega-\mathbf{k}\cdot\mathbf{v}_j-i\gamma}, \tag{3.5.10}$$

$$\mathbf{H_p} = i(1-n_i\cos\theta)\sum_{j=1}^{N}\frac{\left(1-\beta^2\right)^{1/2}}{(1-\beta_s)^2}\frac{\beta_E\boldsymbol{\beta}e^{-i\mathbf{k}\cdot\mathbf{r}_j(0)}}{\omega-\mathbf{k}\cdot\mathbf{v}_j-i\gamma} + \frac{\mathbf{X_p}}{e}\rho_1(\mathbf{k},\omega), \tag{3.5.11}$$

where the spectral fluctuation density is

$$\rho_1(\mathbf{k},\omega) = -\frac{ie}{\epsilon}\left[\sum_{j=1}^{N}\frac{e^{-\mathbf{k}\cdot\mathbf{r}_j(0)}}{\omega-\mathbf{k}\cdot\mathbf{v}_j-i\gamma} - Z\sum_{l=1}^{N/Z}\frac{e^{-\mathbf{k}\cdot\mathbf{r}_l(0)}}{\omega-\mathbf{k}\cdot\mathbf{v}_l-i\gamma}\right] \tag{3.5.12}$$

and where the sum over j is for ions. The functions X_e and $\mathbf{X_p}$ can be considered relativistic analogs of the electron susceptibility:

$$X_e \equiv \frac{4\pi e^2}{m_e k^2}\int d\mathbf{p}\frac{\left(1-\beta^2\right)^{1/2}(1-n_i\beta_i)}{(1-\beta_s)}\frac{\mathbf{k}\cdot\partial f_{0q}/\partial\mathbf{p}}{\omega-\mathbf{k}\cdot\mathbf{v}-i\gamma}, \tag{3.5.13}$$

$$\mathbf{X_p} \equiv (1-n_i\cos\theta)\frac{4\pi e^2}{m_e k^2}\int d\mathbf{p}\beta_E\boldsymbol{\beta}\frac{\left(1-\beta^2\right)^{1/2}}{(1-\beta_s)^2}\frac{\mathbf{k}\cdot\partial f_{0q}/\partial\mathbf{p}}{\omega-\mathbf{k}\cdot\mathbf{v}-i\gamma}. \tag{3.5.14}$$

For small beta, X_e reduces to the classical susceptibility, Eqn (3.3.9), $X_e \to \chi_e$, while $\mathbf{X_p} \to 0$.

The scattered power is then determined by substituting these results into Eqn (3.5.2) and combining with Eqn (3.5.1). Using our conditions that $\hat{e}\cdot\mathbf{k}=0$, $\hat{e}\cdot\hat{i}=0$, and $\hat{e}\cdot\hat{s}=0$, there are eight terms that need to be considered. The sum of these components allows for grouping and simplifies the overall expression into a "few" integrals allowing the fully relativistic power scattered in an unmagnetized plasma to be written as

$$P_s(\mathbf{k},\omega) = \frac{NP_i r_0^2}{2\pi A}\left\{\left|\frac{X_e-\mathbf{X_p}}{\epsilon}\right|^2 I_{1e} + \left|\frac{X_e-\mathbf{X_p}}{\epsilon}\right|^2 I_{1i} - 2\mathrm{Re}\left[\frac{X_e-\mathbf{X_p}}{\epsilon}\right]I_2 + I_3\right\}, \tag{3.5.15}$$

where A is the cross-sectional area of the beam and

$$I_{1q} \equiv \pi \int d\boldsymbol{\beta} f_{q0}(\boldsymbol{\beta}) \delta\left(\beta_k - \hat{\beta}_k\right), \tag{3.5.16}$$

$$I_2 \equiv \pi \int d\boldsymbol{\beta} \frac{\left(1-\beta^2\right)^{1/2}}{(1-\beta_s)} \left[1 - n_i \beta_i - \left(\frac{1-n_i\cos\theta}{1-\beta_s}\right)\beta_E^2\right] f_{e0}(\boldsymbol{\beta}) \delta\left(\beta_k - \hat{\beta}_k\right), \tag{3.5.17}$$

$$I_3 \equiv -\pi n_i \int d\boldsymbol{\beta} \frac{\left(1-\beta^2\right)}{(1-\beta_s)^2} \left[(1-n_i\beta_i)^2\left(1-\beta_E^2\right) - \left(\frac{(1-n_i\cos\theta)^2\left(1-\boldsymbol{\beta}^2\right)\beta_E^2}{(1-\beta_s)^2}\right)\beta_E^2\right] f_{e0}(\boldsymbol{\beta}) \delta\left(\beta_k - \hat{\beta}_k\right). \tag{3.5.18}$$

Here, $\hat{\beta}_k = \omega/ck\,\hat{k}$ and $\hat{k}$ is the unit vector along $\mathbf{k}$. For $\beta^2 \ll 1$, $\mathbf{X_p} \sim 0$. Taylor expanding and retaining terms of $O(\boldsymbol{\beta})$ and setting $n_i = 1$, Eqn (3.5.15) reduces to the first-order beta result Eqn (2.3.17).

3.5.2 Second Order in Beta

For experiments that are only mildly relativistic, it is useful to have a simple expression for the next order corrections to the form factor. Here, we maintain terms to order β^2, but we use a Maxwellian, $f_e(\beta) = \pi^{-3/2}(c/a)^3 \exp[-(c\beta/a)^2]$, for evaluation. The error introduced when using the Maxwellian is generally small at laboratory temperatures $< 10\,\text{keV}$. In addition, use of the Maxwellian greatly simplifies the final expression. By using the Maxwellian, however, we cannot treat second-order thermal corrections, which arise in the deviation between the Maxwellian and Maxwell-Juttner distributions. We again set $n_i = 1$. Using the relation $\beta_i - \beta_s = -(\omega/\omega_i)(1-\beta_s)$, see Eqn (1.7.7), the various components of Eqn (3.5.15) can be expressed as follows:

$$I_2 = \frac{\pi}{c} f_{e0}(\boldsymbol{\beta}_k)\left(1 + \frac{\omega}{\omega_i} - \frac{1}{2}\boldsymbol{\beta}_k^2\right), \tag{3.5.19}$$

$$I_3 = \frac{\pi}{c} f_{e0}(\boldsymbol{\beta}_k)\left[\left(1 + \frac{\omega}{\omega_i}\right)^2 - \boldsymbol{\beta}_k^2\right], \tag{3.5.20}$$

$$X_e \simeq \left[1 + \frac{\omega}{\omega_i} - \frac{1}{2}\boldsymbol{\beta}_k^2 + (\hat{s}\cdot\hat{k})\left(\frac{\mathrm{v}_p}{c}\right)\left(\frac{\omega}{\omega_i}\right)\right]\chi_e - \frac{1}{2}\left(\frac{\omega_p}{\omega}\right)^2 \boldsymbol{\beta}_k^2, \tag{3.5.21}$$

and we have used the fact that

$$\frac{4\pi e^2}{m_e k^2}\int d\mathbf{v}\,\boldsymbol{\beta}_k^2 \frac{\mathbf{k}\cdot\nabla_{\mathrm{v}} f_{e0}}{\omega - \mathbf{k}\cdot\mathbf{v}} = \left(\boldsymbol{\beta}_k^2 - \frac{3}{2}\frac{a^2}{c^2}\right)\chi_e + \left(\frac{\omega_p}{\omega}\right)^2 \boldsymbol{\beta}_k^2. \tag{3.5.22}$$

We then find that the Thomson scattering power to order β^2 can be expressed as

$P_s(\omega, \mathbf{k}) = P_{es}(\omega, \mathbf{k}) + P_{is}(\omega, \mathbf{k})$ where

$$P_{es}(\omega, \mathbf{k}) = \frac{NP_i r_0^2}{2ckA} f_{e0}(\boldsymbol{\beta}_k) \left\{ \left[\left(1 + \frac{\omega}{\omega_i}\right)^2 - \boldsymbol{\beta}_k^2 + (\hat{k} \cdot \hat{s}) \left(\frac{\omega}{\omega_i}\right) \boldsymbol{\beta}_k \right] \left|1 - \frac{\chi_e}{\epsilon}\right|^2 \right.$$
$$\left. + (\hat{k} \cdot \hat{s}) \left(\frac{\omega}{\omega_i}\right) \boldsymbol{\beta}_k \left|\frac{\chi_i}{\epsilon}\right|^2 + \left(\frac{\omega}{\omega_i}\right)^2 \boldsymbol{\beta}_k^2 \mathrm{Re}\left[\frac{1+\chi_i}{|\epsilon|^2}\right] - \frac{1}{2}\boldsymbol{\beta}_k^2 \right\},$$
$$P_{is}(\omega, \mathbf{k}) = \frac{NP_i r_0^2}{2ckA} f_{i0}(\boldsymbol{\beta}_k) \left\{ \left[\left(1 + \frac{\omega}{\omega_i}\right)^2 - \boldsymbol{\beta}_k^2 + 2(\hat{k} \cdot \hat{s}) \left(\frac{\omega}{\omega_i}\right) \boldsymbol{\beta}_k \right] \left|\frac{\chi_e}{\epsilon}\right|^2 \right.$$
$$\left. - \left(\frac{\omega}{\omega_i}\right)^2 \boldsymbol{\beta}_k^2 \mathrm{Re}\left[\frac{\chi_e}{|\epsilon|^2}\right] \right\}. \tag{3.5.23}$$

3.6 COMMENTS ON THE EFFECTS OF VARIOUS INITIAL CONDITIONS

There is an apparent inconsistency in the above calculations, namely that while we have maintained that correlations play a larger part in forming the scattered spectrum, we have set the initial conditions as correlationless. In fact as we shall see below, we obtain the same average answer provided the averaging (integration) time $T \gg \tau_{pe} \simeq 1/\omega_{pe}$. In addition, we look briefly at the effects of having nonequilibrium initial conditions.

3.6.1 Charges Initially Correlated

For a plasma in thermodynamic equilibrium, the conditional probability that an electron has the position $\mathbf{r}$ is given by

$$P(\mathbf{r}) = e^{e\phi(\mathbf{r})/\kappa T_e}. \tag{3.6.1}$$

From Eqns (3.3.2) and (3.1.3), we have

$$\phi(\mathbf{r}) = \frac{4\pi}{(2\pi)^3} \int \frac{d\mathbf{k}}{k^2} \left\{ Ze \sum_{l=1}^{Z/N} e^{i\mathbf{k}\cdot\mathbf{r}_l} - e \sum_{j=1}^{N} e^{i\mathbf{k}\cdot\mathbf{r}_j} \right\} e^{-i\mathbf{k}\cdot\mathbf{r}}.$$

This potential is now substituted in Eqn (3.6.1), and the exponential is expanded[4]

$$P(\mathbf{r}) \simeq 1 + \frac{4\pi}{(2\pi)^3} \frac{e}{\kappa T_e} \int \frac{d\mathbf{k}}{k^2} \left\{ Ze \sum_{l=1}^{Z/N} e^{i\mathbf{k}\cdot\mathbf{r}_l} - e \sum_{j=1}^{N} e^{i\mathbf{k}\cdot\mathbf{r}_j} \right\}.$$

[4]We interpret $\mathbf{r}$ as $\mathbf{r}_m$, $m \neq j$; therefore, typically the smallest separation $|r_m - r_j|$ will be the mean interparticle spacing, i.e., $1/n_e^{1/3}$, and $\phi_{\max} \simeq e n_e^{1/3}$. Therefore, $(e\phi/\kappa T_e)_{\max} \simeq (1/n_e \lambda_{De}^3)^{2/3}$, and this is a small quantity for the conditions which we are considering.

Now let $\mathbf{r}=\mathbf{r}_m$, and it is clear that there is some correlation in the positions of the mth and j^{th} electron. Now substitute into the ensemble average of the second part of Eqn (3.4.2), i.e., use this probability as the weighting function, and

$$\langle e^{i\mathbf{k}\cdot(\mathbf{r}_m\cdot\mathbf{r}_j)}\rangle_{j\neq m}=\frac{\underset{j\neq m}{\int d\mathbf{r}_j\int d\mathbf{r}_m}\left\{1-\frac{4\pi}{(2\pi)^3}\frac{e^2}{\kappa T_e}\int\frac{d\mathbf{k}'}{k'^2}\sum_{j=1}^{N}e^{i\mathbf{k}'\cdot(\mathbf{r}_m-\mathbf{r}_j)}\right\}\cdot e^{i\mathbf{k}\cdot(\mathbf{r}_m-\mathbf{r}_j)}}{\int d\mathbf{r}_j\int d\mathbf{r}_m\{1\}}.$$

In the denominator, we may drop the second term in $P(\mathbf{r})$ as it averages to zero. We integrate over $\mathbf{r}_m$, and this brings down the delta function $(2\pi)^3\delta(\mathbf{k}'+\mathbf{k})$ so that when we finally integrate over $\mathbf{r}_m$ and then $\mathbf{r}_j$, we obtain

$$\langle e^{-i\mathbf{k}\cdot(\mathbf{r}_m-\mathbf{r}_j)}\rangle_{j\neq m}=\frac{-4\pi e^2}{k^2\kappa T_e V}.$$

The second term in Eqn (3.4.2) then becomes

$$\lim_{\gamma\to 0}\left[-\frac{\gamma N^2}{V}\frac{4\pi e^2}{k^2\kappa T_e}\int\underset{j\neq m}{\int}\frac{d\mathbf{v}_j d\mathbf{v}_m f_{eo}(\mathbf{v}_j)f_{eo}(\mathbf{v}_m)}{(\omega-\mathbf{k}\cdot\mathbf{v}_j-i\gamma)(\omega-\mathbf{k}\cdot\mathbf{v}_m-i\gamma)}\right].$$

For thermodynamic equilibrium, we may use a Maxwellian velocity distribution in the integrals

$$f_{eo}(\mathbf{v})=\left(\frac{m_e}{2\pi\kappa T_e}\right)^{3/2}e^{\frac{-m_e v^2}{2\kappa T_e}},$$

and from Eqn (A.1.14), we have

$$\frac{1}{(\pi a^2)^{3/2}}\int\frac{d\mathbf{v}e^{(-v^2/a^2)}}{\omega-\mathbf{k}\cdot\mathbf{r}\pm i\gamma}$$

$$=\frac{\exp(-(\omega/ka)^2)}{ka}\int_0^{\omega/ka}\exp(p^2)dp\mp i\frac{\pi^{1/2}}{ka}\exp(-(\omega/ka)^2) \tag{3.6.2}$$

$$=\frac{1}{ka}(Y_{\pm}), \tag{3.6.3}$$

where $a^2=2\kappa T_e/m_e$.

In total, the term in Eqn (3.4.2) becomes

$$\frac{\pi N}{k}f_{eo}\left(\frac{-\omega}{k}\right)-\lim_{\gamma\to 0}\gamma N\frac{1}{(k\lambda_{De})^2}\left(\frac{1}{ka}\right)(Y_-)(Y_+). \tag{3.6.4}$$

Thus, in the limit $\gamma\to 0$, the second part clearly disappears even if we start with the charges correlated as they are in equilibrium.

More generally, if we make the interpretation that $\gamma\simeq 1/T$ and take $k\simeq 1/\lambda_{De}$, then the ratio of the second part of Eqn (3.6.4) to the first part is of order τ_{pe}/T, and in the limit $T\to\infty(\gamma\to 0)$, the second part may be dropped.

3.6.2 Nonequilibrium Initial Conditions

We must distinguish between mild departures from equilibrium, for example, when $T_e \neq T_i$ or when the electrons are drifting through the ions at a speed much less than the ion acoustic speed $(z\kappa T_e/m_i)^{1/2}$, and extreme departures as when the drift velocity exceeds the sound speed or the distribution functions are grossly non-Maxwellian.

In the former case, the functions will relax gradually by simple collisions toward equilibrium, and we can allow for this in our calculation by taking suitable average conditions appropriate to the relation of the averaging time T to the relaxation time τ_R.

In the latter case, the system may be unstable, and sections of the background noise spectrum can be enhanced. In this situation, we must retain the nonlinear term $-(q/m)\mathbf{E}_1 \cdot \partial F_{1q}/\partial \mathbf{v}$ in Eqn (2.2.3) for at least some section of the spectrum. We must allow for the interaction of the waves back on the average distribution function F_{oq} through the term $-(q/m)\langle \mathbf{E}_1 \partial F_{1q}/\partial \mathbf{v}\rangle$ in Eqn (2.2.2), because the system will no longer be stationary. These topics are discussed further in Chapter 12.

3.7 $S(\mathbf{k},\omega)$ FOR A COLLISIONAL PLASMA $B=0$

3.7.1 Arbitrary Distribution Functions

A discussion of various approximations to the collision terms is given in Appendix B.4. In this section, the simplest collision term, which is obtained from the BGK particle conserving model, is used to investigate the effect of collisions on the density fluctuations, and through them on the scattered spectrum. This calculation was first made by Dougherty (1963) and Taylor & Comisar (1963). The BGK collision term is primarily designed to represent electron-neutral and ion-neutral collisional effects, but it may also be sensibly applied in some cases to represent electron–ion collisional effects, e.g., in the damping of electron plasma waves.

(a) The kinetic equations with the BGK collision term have the form (see Dougherty & Farley, 1963b)

$$\frac{\partial F_{oq}}{\partial t} + \mathbf{v} \cdot \frac{\partial F_{oq}}{\partial \mathbf{r}} = 0 \tag{3.7.1}$$

$$\frac{\partial F_{1q}}{\partial t} + \mathbf{v} \cdot \frac{\partial F_{1q}}{\partial \mathbf{r}} + \frac{q}{m}\mathbf{E}_1 \cdot \frac{\partial F_{oq}}{\partial \mathbf{v}} = -\nu_q[F_{1q} - n_{1q} f_{oq}(\mathbf{v})], \tag{3.7.2}$$

where

$$n_{1q}(\mathbf{r},t) = \int d\mathbf{v} F_{1q}(\mathbf{r},\mathbf{v},t). \tag{3.7.3}$$

We have set $(\partial F_{oq}/\partial t)_c = 0$, which implies that the average state of the system does not change significantly while we investigate the effect of $(\partial F_{oq}/\partial t)_c$. The final term in Eqn (3.7.2) acts to conserve particles, i.e., the rest of the equation shows that particles move in approximately straight lines, with small perturbation owing to collective Coulomb interactions until after a time $\sim 1/\nu_e$ they have suffered a significant deflection; nevertheless on the average, there are always the same number of particles following a given orbit, and the last term allows for this by compensating for the damping (loss) term $-\nu F_{1q}$.

The appropriate form for the spectral density function (see Section 2.3.4) is

$$S(\mathbf{k},\omega)=\frac{2\nu_q}{V}\frac{\langle|n_{1e}(\mathbf{k},\omega)|^2\rangle}{n_{eo}}, \tag{3.7.4}$$

where ν_q is the effective collision frequency appropriate to the species over which we are taking an ensemble average. A small problem arises when we determine $n_{1e}(\mathbf{k},\omega)$ by taking a Fourier–Laplace transform of Eqn (3.7.4), namely the particle-conserving term copes adequately with the long timescale evolution or average state of F_{1q}, but it is not satisfactory for short times $t \ll 1/\nu_q$. This shows when we compare the spectral density function Eqn (3.7.4) with the function derived from the fluctuation-dissipation theorem (Section 3.8). It is necessary to incorporate in the Fourier–Laplace transform of Eqn (3.7.2) the initial value of the particle-conserving term, i.e., $n_{1q}(\mathbf{k},0)F_{oq}(\mathbf{v})$, and

$$\begin{aligned}&-F_{1q}(\mathbf{k},\mathbf{v},0)+[i\omega+\nu_q-i(\mathbf{k}\cdot\mathbf{v})]F_{1q}(\mathbf{k},\mathbf{v},\omega)\\&\quad=-\frac{i4\pi q}{mk^2}\rho_1(\mathbf{k},\omega)\mathbf{k}\cdot\frac{\partial F_{oq}}{\partial\mathbf{v}}+\nu_q n_{1q}(\mathbf{k},\omega)F_{oq}-n_{1q}(\mathbf{k},0)F_{oq}. \end{aligned}\tag{3.7.5}$$

(b) We now substitute from Eqns (3.2.8) and (3.2.9) into Eqn (3.2.5) and sum over all velocities, using Eqn (3.7.3) and rearranging to obtain

$$\begin{aligned}&n_{1e}(\mathbf{k},\omega)\\&=-\frac{i}{(1+D_e)}\sum_{j=1}^{N}e^{i\mathbf{k}\cdot\mathbf{r}_j(0)}\left[\frac{1}{\omega-\mathbf{k}\cdot\mathbf{v}_j(0)-i\nu_e}-\frac{D_e}{i\nu_e}\right]+\frac{C_e(\mathbf{k},\omega)}{e}\rho_1(\mathbf{k},\omega),\end{aligned}\tag{3.7.6}$$

$$\begin{aligned}&n_{1i}(\mathbf{k},\omega)\\&=-\frac{i}{(1+D_i)}\sum_{l=1}^{N/Z}e^{i\mathbf{k}\cdot\mathbf{r}_l(0)}\left[\frac{1}{\omega-\mathbf{k}\cdot\mathbf{v}_l(0)-i\nu_i}-\frac{D_i}{i\nu_i}\right]+\frac{C_i(\mathbf{k},\omega)}{Ze}\rho_1(\mathbf{k},\omega),\end{aligned}\tag{3.7.7}$$

where

$$D_q=i\nu_q\int_{-\infty}^{+\infty}\frac{d\mathbf{v}f_{oq}}{\omega-\mathbf{k}\cdot\mathbf{v}-i\nu_q} \tag{3.7.8}$$

and

$$C_e(\mathbf{k}, \omega) = \frac{1}{(1+D_e)} \int_{-\infty}^{+\infty} d\mathbf{v} \frac{4\pi e^2 n_{oe}}{m_e k^2} \frac{\mathbf{k} \cdot \partial f_{oe}/\partial \mathbf{v}}{\omega - \mathbf{k} \cdot \mathbf{v} - i\nu_e}$$

$$C_i(\mathbf{k}, \omega) = \frac{1}{(1+D_i)} \int_{-\infty}^{+\infty} d\mathbf{v} \frac{4\pi Z^2 e^2 n_{oi}}{m_i k^2} \frac{\mathbf{k} \cdot \partial f_{oi}/\partial \mathbf{v}}{\omega - \mathbf{k} \cdot \mathbf{v} - i\nu_i}. \tag{3.7.9}$$

(c) n_{1i} is now eliminated between Eqns (3.7.6) and (3.7.7), and an expression for $n_{1e}(\mathbf{k}, \omega)$ is obtained

$$\begin{aligned} n_{1e}(\mathbf{k}, \omega) &= -i \left[\left(1 - \frac{C_e}{\epsilon(\nu_e, \nu_i)}\right) \times \left(\frac{1}{1+D_e} \sum_{j=1}^{N} e^{i\mathbf{k}\cdot\mathbf{r}(0)} \left\{ \frac{1}{\omega - \mathbf{k} \cdot \mathbf{v}_j(0) - i\nu_e} - \frac{D_i}{i\nu_e} \right\} \right) \right. \\ &\left. + \frac{ZC_e}{\epsilon(\nu_e, \nu_i)} \left(\frac{1}{1+D_i} \sum_{l=1}^{N} e^{i\mathbf{k}\cdot\mathbf{r}_l(0)} \left\{ \frac{1}{\omega - \mathbf{k} \cdot \mathbf{v}_l(0) - i\nu_i} - \frac{D_i}{i\nu_i} \right\} \right) \right]. \end{aligned} \tag{3.7.10}$$

The dielectric function is

$$\epsilon(\mathbf{k}, \omega, \nu_e, \nu_i) = 1 + C_e + C_i.$$

Compare Eqn (3.7.10) with the collisionless case Eqn (3.3.11).

(d) The spectral density function is obtained by a method similar to that used in Section 3.7.1.

$$S(\mathbf{k}, \omega) = 2 \left| \frac{(1+C_i)}{\epsilon} \right|^2 B_e + 2Z \left| \frac{C_e}{\epsilon} \right|^2 B_i, \tag{3.7.11}$$

where

$$B_q = \frac{\nu_q}{|1+D_q|^2} \left[\int_{-\infty}^{+\infty} \frac{d\mathbf{v} f_{oq}(\mathbf{v})}{(\omega - \mathbf{k} \cdot \mathbf{v})^2 + \nu_q^2} - \frac{|D_q|^2}{\nu_q^2} \right]. \tag{3.7.12}$$

3.7.2 Solution with Maxwellian Velocity Distributions

Integrals of the type Eqns (3.7.8), (3.7.9), and (3.7.12) are discussed in Appendix A. For future reference, their solutions for Maxwellian velocity distributions are noted here. A comparison of these collisional spectra with the corresponding "collisionless" spectra is made in Chapter 5.

$$f_{e0}(\mathbf{v}) = \exp(-\mathrm{v}^2/a^2)/(\pi a^2)^{3/2}, \quad f_{i0}(\mathbf{v}) = \exp(-\mathrm{v}^2/b^2)/(\pi a^2)^{3/2},$$
$$a = (2\kappa T_e/m_e)^{1/2}, \quad b = (2\kappa T_i/m_i)^{1/2}.$$

The functions D_q become

$$D_e = \frac{i\nu_e}{ka}\left[2e^{-y_e^2}\int_0^{y_e} e^{p^2}\,dp + i\pi^{1/2}e^{-y_e^2}\right],$$

$$D_i = \frac{i\nu_i}{kb}\left[2e^{-y_i^2}\int_0^{y_i} e^{p^2}\,dp + i\pi^{1/2}e^{-y_i^2}\right], \tag{3.7.13}$$

where

$$y_e = (\omega - i\nu_e)/ka, \qquad y_i = (\omega - i\nu_i)/kb, \tag{3.7.14}$$

$$B_e = \frac{1}{\pi^{1/2}ka|1+D_e|^2}\,\mathrm{Im}\left\{\int_{-\infty}^{+\infty} d\left(\frac{\mathrm{v}}{a}\right)\frac{\exp\left[-(\mathrm{v}/a)^2\right]}{(y_e - \mathrm{v}/a)}\right\} - \frac{|D_e|^2}{\nu_e|1+D_e|^2}$$

$$= \frac{1}{ka|1+D_e|^2}\,\mathrm{Im}\left\{2\exp\left(-y_e^2\right)\int_0^{y_e}\exp(p^2)dp + i\pi^{1/2}\exp\left(-y_e^2\right)\right\}$$

$$- \frac{|D_e|^2}{\nu_e|1+D_e|^2} \tag{3.7.15}$$

$$B_i = \frac{1}{\pi^{1/2}kb|1+D_i|^2}\,\mathrm{Im}\left\{\int_{-\infty}^{+\infty} d\left(\frac{\mathrm{v}}{b}\right)\frac{\exp\left[-(\mathrm{v}/b)^2\right]}{(y_i - \mathrm{v}/b)}\right\} - \frac{|D_i|^2}{\nu_i|1+D_i|^2}.$$

Apart from the factor $(1+D_q)^{-1}$, the functions C_q are the same as the functions χ_q [Eqns (5.2.5) and (5.2.6)] with $\omega - i\nu_q$.

$$C_e = \frac{\alpha^2}{(1+D_e)}\left[1 - 2y_e e^{-y_e^2}\int_0^{y_e} e^{p^2}\,dp - i\pi^{1/2}y_e e^{-y_e^2}\right]$$

$$C_i = \frac{ZT_e}{T_i}\frac{\alpha^2}{(1+D_i)}\left[1 - 2y_i e^{-y_i^2}\int_0^{y_i} e^{p^2}\,dp - i\pi^{1/2}y_i e^{-y_i^2}\right]. \tag{3.7.16}$$

Solution for $\nu_e, \nu_i \to 0$

We may expand the functions $D_q(y), B_q(y)$, and $C_q(y)$ about their values when $\nu_q = 0$, i.e., the values $D_q(x), B_q(x)$, and $C_q(x) = X_q(x)$, where $x_e = \omega/ka$ and $x_i = \omega/kb$.

To first order in ν_e/ka and ν_i/kb, we find

$$D_e \cong -\frac{i\nu_e}{\omega}\left[\frac{G_e(x_e)}{\alpha^2}-1\right], \quad D_i \cong -\frac{i\nu_i}{\omega}\left[\frac{T_i}{ZT_e\alpha^2}G_i(x_i)-1\right] \tag{3.7.17}$$

$$B_e \cong \frac{1}{ka}\left\{\pi^{1/2}\exp\left(-x_e^2\right)-\frac{2\nu_e}{ka}\left(1-2x_e\exp\left(-x_e^2\right)\int_0^{x_e}\exp(p^2)dp\right)\right\} - \frac{\nu_e}{\omega^2}\left|\frac{G_e(x_e)}{\alpha^2}-1\right|^2$$

$$B_i \cong \frac{1}{kb}\left\{\pi^{1/2}\exp\left(-x_i^2\right)-\frac{2\nu_i}{kb}\left(1-2x_i\exp\left(-x_i^2\right)\int_0^{x_i}\exp(p^2)dp\right)\right\} - \frac{\nu_i}{\omega^2}\left|\frac{T_iG_i(x_i)}{ZT_e\alpha^2}-1\right|^2 \tag{3.7.18}$$

and

$$C_e \cong \chi_e(x_e)\left(1+\frac{2i\nu_e}{ka}\cdot x_e\right)+\frac{i\nu_e}{\omega}\alpha^2\left(1-\frac{\chi_e(x_e)}{\alpha^2}\right)^2$$

$$C_i \cong \chi_i(x_i)\left(1+\frac{2i\nu_i}{kb}\cdot x_i\right)+\frac{i\nu_i}{\omega}\frac{ZT_e}{T_i}\alpha^2\left(1-\frac{T_i\chi_i(x_i)}{ZT_e\alpha^2}\right)^2. \tag{3.7.19}$$

Note that when $\nu_e, \nu_i = 0$, we recover the collisionless result.

$$D_e = 0, \quad B_e - \frac{\pi^{1/2}}{ka}\exp\left(-x_e^2\right), \quad B_i = \frac{\pi^{1/2}}{kb}\exp\left(-x_i^2\right).$$

$$C_e = \chi_e, \quad C_i = \chi_i.$$

3.8 $S(\mathbf{k},\omega)$ FROM THE FLUCTUATION-DISSIPATION THEOREM

It was shown by Nyquist (1928), for the case of an ideal transmission line, that in thermodynamic equilibrium, the voltage fluctuations on the line (both the mean square level and the frequency spectrum of this "noise") are related to the impedance of the line. It has since been demonstrated (Callen & Welton, 1951; Landau & Lifshitz, 1958) that this simple relationship between the fluctuation in a force acting on a system and the dissipative properties of the system applies generally to linear dissipative systems in equilibrium.

The application of this generalized Nyquist theorem to plasmas[5] is discussed by Dougherty & Farley (1960), Farley *et al.* (1961), and Sitenko (1967). It

[5]It might appear that the theorem would not be applicable to a "collisionless" plasma; however, even in the "collisionless" case, there is dissipation due to Landau damping.

is possible to apply the theorem to an isotropic plasma with unequal electron and ion temperatures. In this quasi-equilibrium state, although the electron and ion density fluctuations are interdependent, the plasma may be treated as if it contained two separate components, respectively, dependent on the electron temperatures and the ion temperature. Sitenko (1967, Eqns (1.2.7), (5.4.2), and (10.20)) show that in this situation,

$$S(\mathbf{k},\omega) = -\frac{2}{\alpha^2\omega}\left|\frac{1+C_i}{\epsilon}\right|^2 \mathrm{Im}(C_e) - \frac{2T_i}{T_e\alpha^2\omega}\left|\frac{C_e}{\epsilon}\right|^2 \mathrm{Im}(C_i), \tag{3.8.1}$$

where the longitudinal dielectric function is $\epsilon = 1 + C_e + C_i$ and C_e and C_i are related to the conductivity (i.e., dissipation) by

$$C_e = \sigma_e/4\pi i\omega, \quad C_i = \sigma_i/4\pi i\omega.$$

Consider the case of a collisional plasma with Maxwellian velocity distributions. We see by comparing Eqns (3.8.1) and (3.7.2) that to demonstrate the equivalence of the two results, we must prove that $-\mathrm{Im}(C_e)/\alpha^2\omega = B_e$.

From Eqns (3.7.16), (3.7.13), and (3.7.14),

$$-\frac{\mathrm{Im}(C_e)}{\alpha^2\omega} = -\frac{1}{\omega}\mathrm{Im}\left\{\alpha^2 - \frac{x_e\left[2\exp(-y_e^2\int_0^{y_e}\exp(p^2) + i\pi^{1/2}\exp(-y_e^2)\right]}{(1+D_e)}\right\}$$

$$= \frac{1}{ka|1+D_e|^2}\mathrm{Im}\left\{2\exp(-y_e^2)\int_0^{y_e}\exp(p^2)dp + i\pi^{1/2}\exp(-y_e^2)\right\}$$

$$-\frac{|D_e|^2}{\nu_e|1+D_e|^2} = B_e \quad [\text{Eqn}(6.7.41)]. \tag{3.8.2}$$

The agreement of the two results indicates that it is necessary to add the initial value of the particle-conserving term to the Fourier–Laplace transformed kinetic equation if we are to use $S(\mathbf{k},\omega)$ as defined in Eqn (3.7.4).

PROBLEMS

3.1. Comment on why in the case of $n_e\lambda_{\mathrm{De}}^3 \sim 1$ measured scattered profiles might not reflect three particle correlation effects.

3.2. As a first step in understanding the implications of the Vlasov theory, we wish to utilize some simplified equilibria. We therefore wish to derive the time asymptotic high-frequency dispersion relation for the following field-free, collisionless plasmas assuming an electrostatic perturbation.

(a) $f_0(\mathrm{v}) = \delta(\mathrm{v})$

(b) $f_0(\mathrm{v}) = \delta(\mathrm{v} - \mathrm{v}_0)$

(c) $f_0(\mathrm{v}) = \begin{cases}(2a)^{-1} & |\mathrm{v}| < a \\ 0 & |\mathrm{v}| > a\end{cases}$

(d) $f_0(\mathrm{v}) = \frac{a}{\pi}\frac{1}{\mathrm{v}^2 + a^2}$

3.3. Show approximately why the lower hybrid resonance, $\omega_{LH} = [\omega_{pi}^2/(1 + \omega_{pe}^2/\Omega_e^2)]^{1/2}$, might impact the measurement of fusion produced 3.5 MeV alpha particles, as they slow down, by comparing v_{ph} versus v_α. Use the case of a 60-GHz input beam with scattering at 60° from a deuterium-tritium plasma of density $1 \times 10^{20}\,\mathrm{m}^{-3}$ in a field of 5 T.

3.4. We wish to continue our exploration of Vlasov dispersion relations by first assuming simple Cauchy or Lorentzian distributions that are amenable to simple solution. We therefore consider an infinite, field free plasma with

$$\hat{f}_{0e} = \frac{\hat{a}_e}{\pi} \frac{1}{v^2 + a_e^2} \quad \text{and} \quad \hat{f}_{0i} = \frac{\hat{a}_i}{\pi} \frac{1}{v^2 + a_i^2}.$$

Derive the dispersion relation for ion-acoustic waves and show that the damping is only weak for $T_e \gg T_i$.

3.5. We are finally ready to tackle a more physical problem. Consider an infinite, field free plasma with Maxwellian ion and electron distributions with corresponding temperatures T_i and T_e. Derive the expressions for the real and imaginary parts of the ion-acoustic wave frequency assuming

$$b \ll \frac{\omega}{k} \ll a, \quad \text{and} \quad T_e \gg T_i.$$

3.6. In deriving the dispersion relation for high-frequency electromagnetic waves in an unmagnetized, fluid plasma, one often neglects the $\mathbf{v}_e \times \mathbf{B}$ term, where $\mathbf{v}_e$ is the perturbed electron velocity and $\mathbf{B}$ is the wave magnetic field. Justify this neglect as well as indicate conditions where the assumption is not valid.

3.7. Show that when the refractive index (n_i) is not unity

$$\begin{aligned}\mathbf{E}_s(R,t) = \frac{-e^2 E_{i0}(1-\beta^2)^{\frac{1}{2}}}{\{(m_e c^2 R)(1-\hat{s}\cdot\boldsymbol{\beta})^3\}} &\cdot [(1-n_i\beta_i)(1-\beta_s) \\ &- \{(1-n_i\beta_i)\cos\eta + (n_i\cos\theta - \beta_s)\beta_E\}\hat{s} + n_i\beta_E(1-\beta_s)\hat{\imath} \\ &+ \{(1-n_i\beta_i)\cos\eta - (1-n_i\cos\theta)\beta_E\}\boldsymbol{\beta}]_{\text{ret}} x \cos(\mathbf{k}_i\cdot\mathbf{r}(t') - \omega_i t').\end{aligned}$$

3.8. Consider a cold, magnetized, neutralized fluid plasma ($\mathbf{B}_0 = \hat{z}B_0$) with two ion species (densities and charges n_1, n_2, Z_1e, and Z_2e, respectively). In such a plasma, there is a so-called ion–ion hybrid resonance involving only the ions that can occur for $\mathbf{k}\perp\mathbf{B}_0$. For the sake of definiteness, take $\mathbf{k} = \hat{x}k$. Using the fact that at resonance the perturbation is purely electrostatic derive the resonance condition by requiring that the ion space charges cancel each other as the ions oscillate. For singly charged ions, show that the frequency is given by

$$\omega^2 = \Omega_j \cdot \Omega_k \frac{x_j\Omega_k + x_k\Omega_j}{x_j\Omega_j + x_k\Omega_k},$$

where x_j and x_k are the fractional densities of the two ion species and Ω_j and Ω_k are their cyclotron frequencies. Note that alternatively you can derive the resonance condition by simply requiring $\Omega_p^2\Omega_c^2 \gg 1$ and taking $\mathbf{k}\perp\mathbf{B}_0$.

SOLUTIONS TO ODD PROBLEMS

3.1. In order to perform a scattering measurement, it will be necessary for the scattering volume to contain, for example, at least 10^{10} electrons. In this situation, the dimensions ofthe scattering volume will be much greater than λ_{De}, and the fact that each electron is contributing to the shielding of more than one ion might not be detected.

3.3. $\omega_{pi} = 8.33 \times 10^9$ rad s^{-1}, $\omega_{pe} = 5.65 \times 10^{11}$ rad s^{-1}, $\Omega_e = 8.8 \times 10^{11}$ rad s^{-1}, and $\omega_{LH} = 7.0 \times 10^9$ rad s^{-1}. For 60 GHz radiation, $k_i = 1.26 \times 10^3$m^{-1}, and $|k| \approx 2|k_i| \sin\theta/2 = 1.26 \times 10^3$m^{-1}, and then in the region of the lower hybrid resonance, $v_{ph} = 5.6 \times 10^6$ m s^{-1} The alpha velocity at birth is $v_\alpha = [2eE_\alpha/m_\alpha]^{1/2} = 1.29 \times 10^7$m s^{-1}.

3.5. Our dispersion can be readily expressed in terms of the plasma dispersion function or Z-function (see Eqns (5.2.5) and (5.2.6)):

$$\epsilon(k,\omega) = 1 - \frac{1}{2}\frac{k^2\lambda_{De}^2}{k^2}Z'\left(\frac{\omega}{ka}\right) - \frac{1}{2}\frac{k^2\lambda_{Di}^2}{k^2}Z'\left(\frac{\omega}{ka}\right) = 0$$

To proceed, we make use of

$$b \ll \frac{\omega}{k} \ll a, \quad T_e \gg T_i \quad \text{(Weak Damping)}$$

We can then write

$$\epsilon(k,\omega) \approx \epsilon(k,\omega_r) + i\omega_i \frac{\partial\epsilon(k,\omega_r)}{\partial\omega_r}$$

However,

$$\epsilon(k,\omega_r) = \epsilon_r(k,\omega_r) + i\epsilon_i(k,\omega_r)$$
$$\Rightarrow \quad \epsilon_r(k,\omega_r) = 0$$

and

$$\omega_i = -\frac{\epsilon_i(k,\omega_r)}{\frac{\partial\epsilon_r(k,\omega_r)}{\partial\omega_r}}.$$

Looking at the real part, we have

$$\mathcal{E}_r = 1 - \frac{1}{2}\frac{k^2\lambda_{De}^2}{k^2}(-2) - \frac{1}{2}\frac{k^2\lambda_{Di}^2}{k^2}\left(\frac{1}{\xi_i^2}\right) = 0$$

$$\Rightarrow \quad 1 + \frac{k^2\lambda_{De}^2}{k^2} - \frac{k^2\lambda_{Di}^2}{2\omega_r^2/b^2} = 0$$

or

$$\omega_r^2 = \frac{k^2b^2k^2\lambda_{Di}^2k^2\lambda_{De}^2}{(1+k^2\lambda_{De}^2)} = \frac{k^2b^2\frac{\lambda_{De}^2}{\lambda_{Di}^2}}{(1+k^2\lambda_{De}^2)} = \frac{k^2b^2\frac{T_e}{T_i}}{(1+k^2\lambda_{De}^2)}$$

$$\Rightarrow \quad \omega_r^2 = \frac{k^2c_s^2}{1+k^2\lambda_{De}^2} \text{ where } c_s = \sqrt{\kappa T_e/m_i} \text{ and } Z = 1 \text{ for simplicity.}$$

Now,

$$Z' = -2(1+\xi Z)$$

$$\Rightarrow \operatorname{Im} Z'_e = -2\xi_e \pi^{1/2} e^{-\xi_e^2}$$

and

$$\operatorname{Im} Z'_i = -2\xi_i \pi^{1/2} e^{-\xi_i^2}$$

$$\Rightarrow \quad \epsilon_i = -\frac{1}{2}\frac{k^2\lambda_{\mathrm{De}}^2}{k^2}(-2\xi_e)\pi^{1/2}e^{-\xi_e^2} - \frac{1}{2}\frac{k^2\lambda_{Di}^2}{k^2}(-2\xi_i)\pi^{1/2}e^{-\xi_i^2}.$$

But $\xi_e \ll 1$

$$\Rightarrow \quad \epsilon_i \approx \frac{\pi^{1/2}}{k^2}\left[k^2\lambda_{\mathrm{De}}^2\xi_e + k^2\lambda_{Di}^2\xi_i e^{-\xi_i^2}\right]$$

$$= \frac{\pi^{1/2}k^2\lambda_{\mathrm{De}}^2}{k^2}\left[\underset{\frac{\omega}{ka}=\frac{\omega\sqrt{m_e}}{k\sqrt{2\kappa a}}}{\underset{\uparrow}{\xi_e}} + \underset{\frac{\omega}{kb}=\frac{\omega\sqrt{m_i}}{k\sqrt{2\kappa b}}}{\underset{\uparrow}{\xi_i}}\frac{T_e}{T_i}e^{-\xi_i^2}\right]$$

$$= \frac{\pi^{1/2}k^2\lambda_{\mathrm{De}}^2}{k^2}\frac{\omega_r}{k}\frac{1}{\sqrt{2}}\frac{1}{c_s}\left[\sqrt{m_\mathrm{e}/m_\mathrm{i}} + \left(\frac{T_e}{T_i}\right)^{3/2} e^{-\xi_i^2}\right].$$

In the above, we used $k^2\lambda_{\mathrm{De}}^2 = \dfrac{4\pi n_\mathrm{e} e^2}{\kappa T_e}$ and $k^2\lambda_{Di}^2 = \dfrac{4\pi n_\mathrm{e} e^2}{\kappa T_i}$.

For the ions, we have

$$\xi_i^2 = \frac{\omega_r^2}{k^2 \mathrm{v}_{Tl}^2} = \frac{k^2 c_s^2}{k^2(1+k^2\lambda_{\mathrm{De}}^2)^2} = \frac{T_e/T_i}{2(1+k^2\lambda_{\mathrm{De}}^2)}.$$

Also

$$\epsilon_r = 1 + \frac{k^2\lambda_{\mathrm{De}}^2}{k^2} - \frac{k^2\lambda_{Di}^2 a^2}{2\omega_r^2}$$

$$\Rightarrow \quad \frac{\partial\epsilon_r}{\partial\omega_r} = +\frac{k^2\lambda_{Di}^2 b^2}{\omega_r^3}$$

$$\Rightarrow \quad \omega_i = \frac{-\pi^{1/2}k^2\lambda_{\mathrm{De}}^2\omega_r}{\sqrt{2}k^3 c_s \frac{k^2\lambda_{Di}^2 \mathrm{v}_{Ti}^2}{\omega_r^3}} \quad [\ldots\ldots]$$

$$= \frac{-\pi^{1/2}k^2\lambda_{\mathrm{De}}^2\omega_r^4}{\sqrt{2}k^3 c_s k^2\lambda_{Di}^2 b^2} \quad [\ldots\ldots].$$

We recognize that $\dfrac{k^2\lambda_{\mathrm{De}}^2}{k^2\lambda_{Di}^2} = \dfrac{T_i}{T_e}$

$$\Rightarrow \quad \omega_i = \frac{-\pi^{1/2}\omega_r^4}{\sqrt{2}k^3 c_s (T_e/T_i)\mathrm{v}_{Ti}^2} \quad [\ldots..]$$

$$= \frac{-\pi^{1/2}\omega_r^4}{\sqrt{2}\sqrt{4}k^3 c_s^3} \quad [\ldots..]$$

$$\Rightarrow \quad \omega_t = \frac{-\omega_r\sqrt{\pi/8}}{(1+k^2\lambda_{\mathrm{De}}^2)^{3/2}}\left[\left(\frac{T_e}{T_i}\right)^{3/2}\exp\left(\frac{-T_e/T_i}{2(1+k^2\lambda_{\mathrm{De}}^2)}\right)+\sqrt{\frac{m_{\mathrm{e}}}{m_{\mathrm{i}}}}\right].$$

The above is the familiar expression for weakly damped ion-acoustic waves.

3.7. $d/dt'\{(m_{\mathrm{e}}\mathbf{v}/(1-\mathrm{v}^2/c^2)\} = q(\mathbf{E}_{i0} + (\mathbf{v}/c)\times\mathbf{B}_{i0})$
$= m_{\mathrm{e}}d\mathbf{v}/dt'(1-\mathrm{v}^2/c^2)^{1/2} - (-2\mathbf{v}\cdot d\mathbf{v}/dt'/c^2)m_{\mathrm{e}}/2(1-\mathrm{v}^2/c^2)^{-1/2}\mathbf{v}\}/(1-\mathrm{v}^2/c^2)$

Take scalar product with $\mathbf{v}$ and
$m_{\mathrm{e}}(d\mathbf{v}/dt')\cdot\mathbf{v}(1-\mathrm{v}^2/c^2+\mathrm{v}^2/c^2)/(1-\mathrm{v}^2/c^2)^{3/2} = q(\mathbf{E}_{i0}\cdot\mathbf{v})$ because $(\mathbf{v}/c)\times\mathbf{B}_{i0}\times\mathbf{v} = 0$.

Then, $(d\mathbf{v}/dt')\cdot\mathbf{v} = (q/m_{\mathrm{e}})(\mathbf{E}_{i0}\cdot\mathbf{v})(1-\mathrm{v}^2/c^2)^{3/2}$ and substituting back leads to

$d\mathrm{v}/dt' = q/m_{\mathrm{e}}(1-\mathrm{v}^2/c^2)^{1/2}\{\mathbf{E}_{i0} + \boldsymbol{\beta}\times\mathbf{B}_{i0} - \boldsymbol{\beta}(\boldsymbol{\beta}\times\mathbf{E}_{i0})\}\cos(\mathbf{k}_i\cdot\mathbf{r}(t') - \omega_i t')$.

But and $\mathbf{B}_{i0} = n_i(\hat{1}\times\mathbf{E}_{i0})$

And $\mathbf{E}_s(R,t) = e^2(1-\beta^2)^{1/2}/\{(m_{\mathrm{e}}c^2R)(1-\hat{s}\cdot\boldsymbol{\beta})^3\}\cdot[\hat{s}\times(\hat{s}-\boldsymbol{\beta})\times\{\mathbf{E}_{i0} + n_i\boldsymbol{\beta}\times\hat{1}\times\mathbf{E}_{i0} - \boldsymbol{\beta}(\boldsymbol{\beta}\cdot\mathbf{E}_{i0})]_{\mathrm{ret}}\times\cos(\mathbf{k}_i\cdot\mathbf{r}(t') - \omega_i t')$.

And using $\mathbf{A}\times\mathbf{B}\times\mathbf{C} = (\mathbf{A}\cdot\mathbf{C})\mathbf{B} - (\mathbf{A}\cdot\mathbf{B})\mathbf{C}$.

$$\begin{aligned}
&[\hat{s}\times(\hat{s}-\boldsymbol{\beta})\times\{\mathbf{E}_{i0} + n_i\boldsymbol{\beta}\times\hat{1}\times\mathbf{E}_{i0} - \boldsymbol{\beta}(\boldsymbol{\beta}\cdot\mathbf{E}_{i0})] \\
&\quad = E_{i0}[\{-\hat{s}\cdot\hat{s} + \hat{s}\cdot\boldsymbol{\beta} + (\hat{s}\cdot\hat{s})(n_i\boldsymbol{\beta}\cdot\hat{1}) + (\hat{s}\cdot\boldsymbol{\beta})(n_i\boldsymbol{\beta}\cdot\hat{1})\}\hat{e}_i \\
&\qquad + \{+\hat{s}\cdot\hat{e}_i + (\hat{s}\cdot\hat{1})(n_i\boldsymbol{\beta}\cdot\hat{e}_i) - \hat{s}\cdot\hat{e}_i(n_i\boldsymbol{\beta}\cdot\hat{1}) - (\hat{s}\cdot\boldsymbol{\beta})(\boldsymbol{\beta}\cdot\hat{e}_i)\}\hat{s} \\
&\qquad + \{-(\hat{s}\cdot\hat{s})(n_i\boldsymbol{\beta}\cdot\hat{e}_i) + (\hat{s}\cdot\boldsymbol{\beta})(n_i\boldsymbol{\beta}\cdot\hat{e}_i)\}\hat{1} \\
&\qquad + \{-(\hat{s}\cdot\hat{1})(n_i\boldsymbol{\beta}\cdot\hat{e}_i) - \hat{s}\cdot\hat{e}_i\boldsymbol{\beta} + \hat{s}\cdot\hat{e}_i(n_i\boldsymbol{\beta}\cdot\hat{1}) + (\hat{s}\cdot\hat{s})(\boldsymbol{\beta}\cdot\hat{e}_i)\}\boldsymbol{\beta}]
\end{aligned}$$

Substituting $\hat{s}\cdot\hat{1} = \cos\theta$, $\hat{s}\cdot\hat{e}_i = \cos\eta$, $\hat{e}_i = \hat{E}_{i0}$, $\boldsymbol{\beta}\cdot\hat{1} = \beta_i$, $\hat{s}\cdot\boldsymbol{\beta} = \beta_s$, and $\boldsymbol{\beta}\cdot\hat{e}_i = \beta_E$ and rearranging yields

$$\begin{aligned}
\mathbf{E}s(R,t) = &-e^2E_{i0}(1-\beta^2)^{1/2}/\{(m_{\mathrm{e}}c^2R)(1-\hat{s}\cdot\boldsymbol{\beta})^3\}\cdot[(1-n_i\boldsymbol{\beta}_i)(1-\beta_s)\hat{e}_i \\
&-\{(1-n_i\beta_i)\cos\eta + (n_i\cos\theta - \beta_s)\beta_E\}\hat{s} + n_i\beta_E(1-\beta_s)_{\hat{1}} \\
&+\{(1-n_i\beta_i)\cos\eta - (1-n_i\cos\theta)\beta_E\}\boldsymbol{\beta}]_{\mathrm{ret}}\times\cos(\mathbf{k}_i\cdot\mathbf{r}(t') - \omega_i t').
\end{aligned}$$

Chapter 4

Noncollective Scattering

Chapter Outline Head

DOI: 10.1016/B978-0-12-374877-5.00004-X

4.1 INTRODUCTION

The scattering of radiation from a group of charges that are randomly distributed within the scattering volume is called "noncollective scattering." The conditions under which the charges may be treated as having a random distribution are determined by a full calculation of scattering from a plasma determined in Chapter 3. It is shown that noncollective scattering occurs for a plasma in thermodynamic equilibrium when

$$\alpha = \frac{1}{k\lambda_{\mathrm{De}}} \ll 1. \tag{4.1.1}$$

That is, when the scattering wavelength (λ) is much less than the Debye length λ_{De} Eqn (1.4.2).

The noncollective power spectrum can be derived by dropping the first-order fluctuations in the general expression for the scattered power developed in Section 2.3; $F_e \to f(\mathbf{v})\delta(\mathbf{v}\cdot\mathbf{v}-\omega)$ where for high-temperature plasmas in thermal equilibrium, the Maxwell-Juttner velocity distribution function must be used

$$f(\beta) = \left[2\pi K_2\left(\frac{m_e c^2}{\kappa T_e}\right)\right]^{-1} \frac{m_e c^2}{2\kappa T_e} \frac{\exp(m_e c^2/\kappa T_e(1-\beta^2)^{1/2})}{(1-\beta^2)^{5/2}}, \tag{4.1.2}$$

where K_2 is a modified Bessel function of the second kind of order two. If we expand this in powers of β, we recover to first order in β the Maxwellian distribution function. Remember that $m_e c^2/2\kappa T_e = c^2/a^2 \simeq \beta^{-2}$.

The fully relativistic expression for the scattered power is given in Section 3.5,

$$P_s(\mathbf{k}, \omega) \frac{NP_i r_e^2}{2\pi A} NS(\mathbf{k}, \omega), \tag{4.1.3}$$

where, for uncorrelated electrons, the scattered power may be calculated for each electron before performing the integral over velocity, and the relativistic fluctuation spectrum for noncollective scattering becomes

$$S(\mathbf{k}, \omega) = \lim_{\gamma\to 0, V\to\infty} \frac{2\gamma}{V} \left\langle \frac{1}{n_{e0}} \int d\mathbf{v} \left[\frac{\hat{s}\times\{(\hat{s}-\boldsymbol{\beta})\times\dot{\boldsymbol{\beta}}\}}{(1-\beta_s)^2}\right]^2 f(\mathbf{v})\delta(\mathbf{k}\cdot\mathbf{v}-\omega)\right\rangle. \tag{4.1.4}$$

As discussed in Section 1.7.2, the conventional approach is to consider a plane wave whose electric field $\mathbf{E_{i0}}$ is perpendicular to the scattering plane: consequently, $\hat{s}\cdot\hat{E}_{i0} = 0$. In addition, it is often convenient experimentally to apply a polarizer operating on the scattered radiation to select $\mathbf{E_s}$ parallel to $\mathbf{E_{i0}}$, which simplifies the expression

$$\left[\frac{\hat{s}\times\{(\hat{s}-\boldsymbol{\beta})\times\dot{\boldsymbol{\beta}}\}}{(1-\beta_s)^2}\right]^2 = \left|\frac{1-\beta_i}{1-\beta_s}\right|^2 |1-\boldsymbol{\beta}^2| \left|1-\frac{(1-\beta_E^2\cos\theta)}{(1-\beta_i)(1-\beta_s)}\right|^2. \tag{4.1.5}$$

Recognizing that $|(1-\beta_i)/(1-\beta_s)|^2$ may be taken outside the velocity integral, the total scattered power in the noncollective regime becomes

$$P_s(\mathbf{R},\omega)d\Omega d\omega_s = \frac{NP_i}{A^2} r_o^2 d\Omega d\omega_s \left|\frac{1-\beta_i}{1-\beta_s}\right|^2 \tag{4.1.6}$$

$$\times \int_{-c}^{+c} |1-\beta^2| \left|1 - \frac{1-\beta_E^2\cos\theta}{(1-\beta_i)(1-\beta_s)}\right|^2 f(\mathbf{v})\delta(\mathbf{k}\cdot\mathbf{v}-\omega)d\mathbf{v}.$$

Note that for comparison with previous work, we have set $n_i = 1$. This result is the simplified result of the full power spectrum derived in Section 3.5 when noting that both χ_e and χ_i are proportional to $\alpha^2 = 1/(k\lambda_{De})^2$; Eqn (3.5.15) reduces to the noncollective result in the limit $\alpha \to 0$. We note that if either f_{e0} or f_{i0} are non-Maxwellian or if $T_e \neq T_i$ or if the system as a whole is not in equilibrium with the surroundings, then f_{e0}, and consequently $P(\mathbf{k},\omega)$, will be time varying.

For $\alpha \geq 1$, that is, $\lambda \geq \lambda_{De}$, the incident wave interacts with the electrons that are a shield on each electron and on each ion, and the scattered spectrum, therefore, depends on the collective behavior of groups of charges; this is called "collective scattering" (see Chapter 5), and the more rigorous development of the scattered power must be used. In both cases, it must be remembered that the scattered power is inversely proportional to the square of the mass of the charge, and it is, therefore, the scattering from electrons that dominates.

Spectra computed by Matoba *et al.* (1979), showing the increased distortion of the spectrum as the electron temperature is raised, is shown in Fig. 4.1. Various analytic solutions have been derived that give a good approximation to the computed solutions. Zhuravlev & Petrov (1972, 1979) derived a solution using the approximation that the depolarization term [second term under the integral in Eqn (4.1.6)] varied slowly with v. However, using this approach required a calibration for each temperature and density, as pointed out by Beausang & Prunty (2008). The later authors derived a solution for the most awkward case of LIDAR backscattering ($\theta = 180°$, $\cos\theta = -1$). They showed that the polarization term can introduce as much as a 25% effect at $T_e = 50$ keV (see Fig. 4.2). The effect is primarily on the density measurement.

Matoba *et al.* (1979) computed the spectra and also extended the work of Pogutse (1963) and Sheffield (1972a, 1975) to second order in beta for the scattered spectrum. Selden (1980) computed scattering at 20 keV and to ~1% at 100 keV; however, he also found a difference in the transverse term in the two formulas of the scattering integral employed in the two papers that became significant above 20 keV. Subsequently, Naito *et al.* (1993) derived an analytic formula for the depolarization term requiring no calibration and accurate with a relative error of <0.1% up to 100 keV. Palastro *et al.* (2010) extended this work to a general fully relativistic formula applicable for all scattering angles to both collective and noncollective scattering.

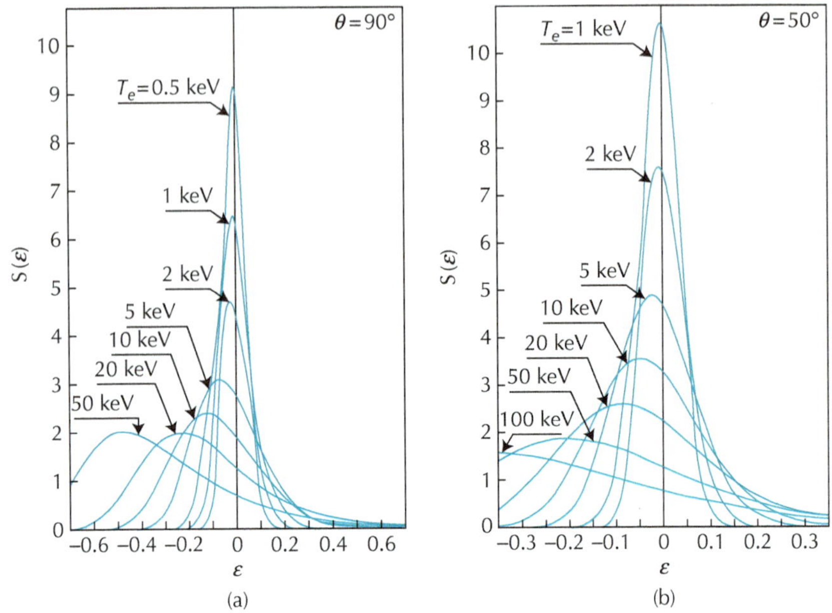

FIGURE 4.1 Spectral density function of the full relativistic Thomson scattering as a function of the normalized wavelength shift ($\epsilon = \Delta\lambda/\lambda_i$). (a) The scattering angle $\theta = 90°$ and (b) $\theta = 50°$. (Courtesy of T. Matoba (1979) and the Japanese Society of Applied Physics.)

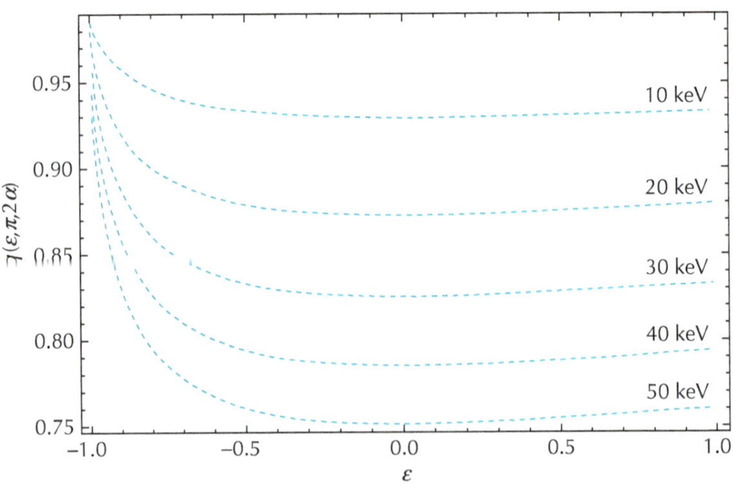

FIGURE 4.2 Depolarization term as a function of wavelength shift ($\epsilon = \Delta\lambda/\lambda_i$) for LIDAR scattering at electron temperature, in the range 10–50 keV. (Courtesy of K. Beausang (2008) and the Institute of Physics.)

4.2 THE ORIGIN OF THE DOPPLER SHIFT IN THE SCATTERED FREQUENCY

The scattered frequency is Doppler shifted from the incident frequency by an amount ω, which is proportional to the component of the electron velocity in the direction **k** (see Fig. 4.3).

The Doppler shift is in a sense the result of two effects. First, the electron "sees" the incident wave at the Doppler-shifted frequency $\omega' = \omega_i - \mathbf{k}_i \cdot \mathbf{v}_j$, because it is moving with respect to the source of radiation. Second, the electron has a velocity component in the direction of the observer, and this yields a second shift. From Eqns (3.2.3) and (3.2.4), we see that

$$t'_j = \frac{t}{1 - \frac{\hat{s}\cdot\mathbf{v}_j}{c}} = \frac{[R - \hat{s}\cdot\mathbf{r}_j(0)]}{c\left(1 - \frac{\hat{s}\cdot\mathbf{v}_j}{c}\right)}.$$

Differentiating, we obtain

$$\Delta t'_j = \frac{\Delta t}{1 - \frac{\hat{s}\cdot\mathbf{v}_j}{c}}.$$

Inverting this, we obtain the relationship between the frequency ω' radiated by the electron and the frequency ω_s received by the observer:

$$\omega_s = \frac{\omega'}{1 - \frac{\hat{s}\cdot\mathbf{v}_j}{c}} = \frac{\omega_i\left(1 - \frac{\hat{i}\cdot\mathbf{v}_j}{c}\right)}{\left(1 - \frac{\hat{s}\cdot\mathbf{v}_j}{c}\right)}.$$

Alternatively,

$$\omega_s = \omega_i + \mathbf{k}\cdot\mathbf{v}_j \quad \text{where} \quad \mathbf{k} = \mathbf{k}_s - \mathbf{k}_i.$$

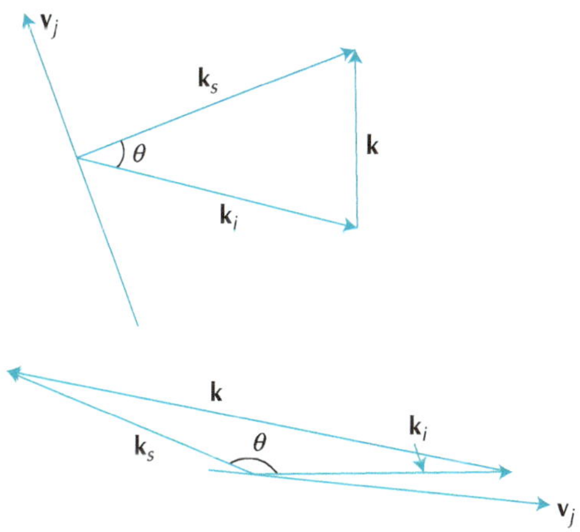

FIGURE 4.3 Wave vector diagrams for the scattering of radiation.

It is important to note that in a scattering experiment, the fixed quantities are ω_i, $\hat{i}$, $\mathbf{k}_i$, and $\hat{s}$, θ. Clearly, ω_s is a function of $\mathbf{v}_j$, and as a consequence, neither $\mathbf{k}_s$ nor $\mathbf{k}$ are constant in either magnitude or direction. It is only when $\beta = a/c \ll 1$, that is, for a low-temperature plasma, that we may reasonably set

$$|k| \simeq \text{constant} = 2|k_i|\sin(\theta/2).$$

The scattering is from N electrons within the volume V, the plasma is homogeneous in space, and the electron density is, therefore, $n_e = N/V$.

The level of the scattered power in the frequency range $\omega_s \to \omega_s + d\omega$ is proportional to the number of electrons that have the component of velocity $\mathbf{v}_k$ to $\mathbf{v}_k + d\mathbf{v}_k$ along $\mathbf{k}$ such that $\omega_s = \omega_i + \mathbf{v}_k \cdot \mathbf{k}$.

The number of electrons in the scattering volume with velocity in the range $\mathbf{v}$ to $\mathbf{v} + d\mathbf{v}_k$ is given in terms of the velocity distribution function as $Nf(\mathbf{v})d\mathbf{v}$ where $\int_{-\infty}^{+\infty} f(\mathbf{v})d\mathbf{v} = 1$. Strictly, the velocity limits should be $\pm c$, and one small point arises concerning v_k. Note first that for low densities where $\omega_{pe} \ll k_i c$,

$$k = \left(\omega_s^2 + \omega_i^2 - 2\omega_s\omega_i\cos\theta\right)^{1/2}/c,$$

and therefore,

$$\mathrm{v}_k = \frac{(\omega_s - \omega_i)c}{\left[(\omega_s - \omega_i)^2 + 2\omega_s\omega_i(1 - \cos\theta)\right]^{1/2}}.$$

Now $\mathrm{v}_k < c$, therefore, we must ignore terms in which ω_s and ω_i have different signs. The use of the infinite limits and a nonrelativistic distribution function is however unlikely to cause trouble until terms of order β^2 become important.

It is convenient to work in a coordinate system with three mutually perpendicular directions: $\hat{k}$, the perpendicular to $\hat{k}$ within the plane of $\hat{s}$ and $\hat{i}$, and the perpendicular to the plane of $\hat{s}$ and $\hat{i}$. The component velocities are denoted, respectively, by $\mathbf{v}_k$, $\mathbf{v}_{k\perp}$, and $\mathbf{v}_{kT}$ (see Fig. 4.4). The total scattered power in the solid angle $d\Omega$ and frequency range $\omega_s \to \omega_s + d\omega_s$, in the direction $\hat{s}$, is

$$P_s(\mathbf{R}, \omega_s)d\omega_s d\Omega = \frac{NP_i r_o^2}{A} d\Omega d\omega_s \left(1 + \frac{2\omega}{\omega_i}\right) |\hat{s} \times (\hat{s} \times \hat{E}_{io})|^2$$
$$\times \int\int\int_{-\infty}^{+\infty} d\mathrm{v}_x d\mathrm{v}_y d\mathrm{v}_z \cdot f(\mathbf{v}) \cdot \delta[(\omega_s - \omega_i) - \mathbf{k}\cdot\mathbf{v}], \quad (4.2.1)$$

where

$$r_o = e^2/m_o c^2 = 2.82 \times 10^{-13}\,\text{cm} \quad (4.2.2)$$

is the classical electron radius, and $d\Omega$ is the element of solid angle, centered on the angle θ in the plane at angle φ_0 to the incident electric vector $\mathbf{E}_{i0}$, over which the scattered light is collected. Here, A is the cross-sectional area of the incident

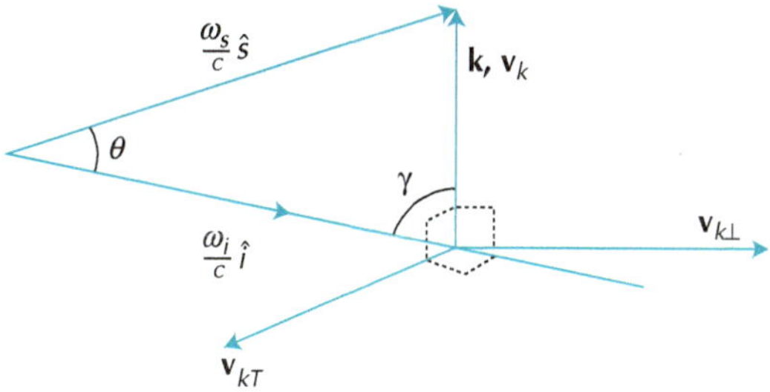

FIGURE 4.4 The wave vector diagram showing the velocity components.

beam of radiation, and L is the length of the scattering volume ($V = AL$). The factor

$$\left[\hat{s}\times(\hat{s}\times\hat{E}_{i0})\right]^2 = 1-\sin^2\theta\cos^2\varphi_0, \quad \text{polarized radiation}$$
$$= 1-(\sin^2\theta)/2, \quad \text{unpolarized, and averaged over } \varphi_0. \tag{4.2.3}$$

For $\beta \ll 1$, we may treat k as a constant, and in this approximation, when we transform to the new coordinate system, the Jacobian is unity:

$$d\mathrm{v}_x d\mathrm{v}_y d\mathrm{v}_z \to d\mathrm{v}_k d\mathrm{v}_{k\perp} d\mathrm{v}_{kT}.$$

The integrals over $\mathrm{v}_{k\perp}$ and v_{kT} simply give unity and

$$P_s(\mathbf{R},\omega_s)d\omega_s d\Omega = \frac{NP_i r_o^2}{A}d\Omega d\omega_s\left(1+\frac{2\omega}{\omega_i}\right)|\hat{s}\times(\hat{s}\times\hat{E}_{io})|^2 f(\omega/k)(d\omega_s/k)$$

if we restrict to cases where $2\omega/\omega_i \ll 1$ and assume that the distribution is Maxwellian and stationary, we may then make a best fit of the results to a one-dimensional Maxwellian distribution. This yields a value for the electron temperature T_e. In addition, if the apparatus is absolutely calibrated, it is possible to determine the electron density n_e from the signal integrated over all frequencies. Note that

$$\int_{-\infty}^{+\infty} f(\omega/k)(d\omega_s/k) = 1$$

and for a Maxwellian distribution,

$$\int_{-\infty}^{+\infty} \omega f(\omega/k)(d\omega_s/k) = 0,$$

and the total scattered power within the solid angle $d\Omega$ is, therefore,

$$P_s(\mathbf{R},\omega_s)d\Omega = P_i r_o^2 d\Omega L|\hat{s}\times(\hat{s}\times\hat{E}_{io})|^2 n_e = (P_i/A)r_o^2 d\Omega|\hat{s}\times(\hat{s}\times\hat{E}_{io})|^2 N. \tag{4.2.4}$$

When we integrate over $d\Omega$, the total power is simply N times the scattered power from one electron, as of course it must be for noncollective scattering.

We rearrange Eqn (4.2.4) and see that

$$\frac{\text{scattered power}}{\text{incident power}} = \frac{P_s}{P_i} \simeq r_o^2 n_e L d\Omega.$$

For a typical magnetic fusion plasma $n_e \simeq 10^{14}\,\text{cm}^{-3}$, $L \simeq 1\,\text{cm}$, $d\Omega \simeq 10^{-2}\,\text{sr}$, $r_o^2 = 7.95 \times 10^{-26}\,\text{cm}^2$. Substituting these values, we see that $P_s/P_i \simeq 10^{-13}$. This shows very clearly a major problem in applying scattering as a diagnostic, namely we require extremely powerful sources. This is discussed further in Section 4.5.

It is important to remember that the above results assumed that the distribution function does not change significantly during the integration time T. This is certainly true in the equilibrium case when we may use a Maxwellian distribution function.

It is not necessarily true if the distribution function is anisotropic, if the electrons and ions have different temperatures, or if the distribution is sufficiently extreme that the system is unstable. In any of these situations, there will be interactions (collisions) acting to restore equilibrium, and the scattering result will depend on the ratio T/τ_c, where τ_c is the appropriate collision time. The errors introduced by ignoring the factor $\left(1 + \frac{2\omega}{\omega_i}\right)$ are discussed below.

4.3 COMPARISON OF FIRST AND SECOND ORDER IN β

Matoba *et al.* (1979) calculated the fully relativistic noncollective spectral density function (S_R) and normalized it to the nonrelativistic one (S_N),

$$\begin{aligned}\frac{S_R\left(\frac{\Delta\lambda}{\lambda_i}\right)}{S_N\left(\frac{\Delta\lambda}{\lambda_i}\right)} &\simeq 1 - \frac{7}{2}\frac{\Delta\lambda}{\lambda_i} + \frac{c^2\Delta\lambda^3}{2a^2\lambda_i^3(1-\cos\theta)} - \frac{a^2}{8c^2}\left(\frac{39}{4} - 5\cos\theta\right) \\ &+ \frac{1}{8}\left(29 + \frac{5}{1-\cos\theta}\right)\frac{\Delta\lambda^2}{\lambda_i^2} - \frac{1}{16(1-\cos\theta)}\left(28 + \frac{1}{1-\cos\theta}\right)\frac{c}{a}\frac{\Delta\lambda^4}{\lambda_i^4} \\ &+ \frac{c^2}{8a^2(1-\cos\theta)^2}\frac{\Delta\lambda^6}{\lambda_i^6}.\end{aligned} \tag{4.3.1}$$

They compare this expression with Pogutse's (1963) use of only the first terms and Sheffield's (1972b, 1975) use of the first three terms [Eqn (3.4.1)] with the full relativistic solution as shown in Fig. 4.5.

4.4 THE FINITE TRANSIT TIME EFFECT

In the evaluation of the scattered intensity due to one electron (Section 1.7), we assume that this electron would remain within the scattering volume for the time of the measurement T. In fact, the incident beam is of finite size, and the

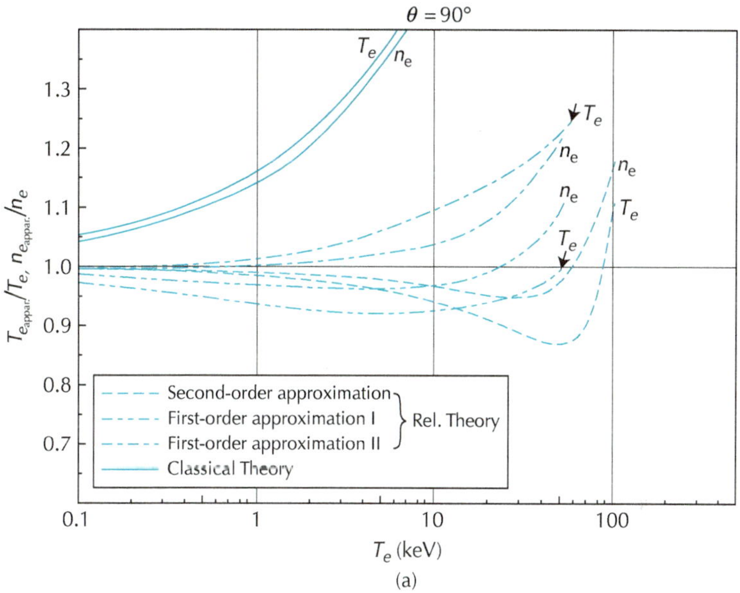

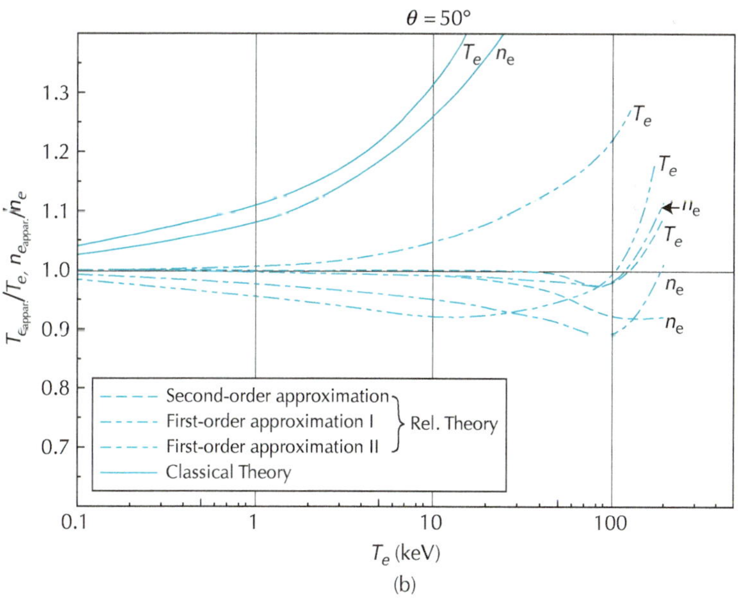

FIGURE 4.5 Comparisons among the apparent electron temperature and density derived from the classical theory, the first-order approximate theories, and the second-order approximate theory as a function of the true electron temperature. (a) $\theta = 90°$ and (b) $\theta = 50°$. (Courtesy of T. Matoba (1979) and the Japanese Society of Applied Physics.)

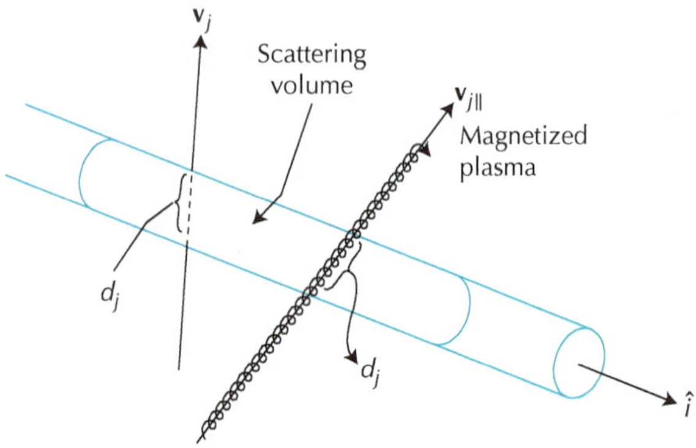

FIGURE 4.6 Electron orbits through scattering volume.

scattering volume is finite, and we will not be dealing over the whole time T with the same group of N electrons.

An electron of the type j has a velocity v_j and will spend a time $\Delta t' = d_j/\mathrm{v}_j$ within the scattering volume (see Fig. 4.6) where d_j is the dimension of the scattering volume in the direction $\hat{\mathrm{v}}_j$. Now, the mean density of any type of electron is constant; therefore, on average for every j-type electron that leaves the volume, another will enter. The result of this situation is that the wave packet from each electron will be curtailed leading to spread in the scattered frequency and wavelength. The wavelength spread is given by

$$\Delta\lambda_{FT} \sim \frac{\mathrm{v}_j \lambda_i}{c d_j} \lambda_i. \tag{4.4.1}$$

This may be compared to the wavelength shift owing to the temperature, which at half max is

$$\Delta\lambda_{1/2} \sim 2.34 a \lambda_i / c. \tag{4.4.2}$$

Thus,

$$\frac{\Delta\lambda_{FT}}{\Delta\lambda_{1/2}} \sim 0.43 \frac{\mathrm{v}_j \lambda_i}{a d_j}. \tag{4.4.3}$$

Salzmann (1986) gives the example of a 30 keV electron in a 3 keV plasma. With visible light and $d_j = 1\,\mathrm{mm}$, $\Delta_{FT} \sim 0.1\,\mu\mathrm{m}$ while $\Delta\lambda_{1/2} \sim 150\,\mathrm{nm}$.

We may extend the calculation to different wavelengths by setting d_j at the diffraction limit

$$d_j \simeq 2.4 f \lambda_i / D, \tag{4.4.4}$$

where f is the focal length of the input optic and D is its diameter.

$$\frac{\Delta\lambda_{FT}}{\Delta\lambda_{1/2}} \sim 0.18 \frac{\mathrm{v}_i}{a} \frac{D}{f}. \tag{4.4.5}$$

For the example above, assuming that $D/f \sim 10^{-2}$, the ratio is $\sim 0.6 \times 10^{-2}$.

4.5 NONCOLLECTIVE SCATTERING FROM A PLASMA IN THERMODYNAMIC EQUILIBRIUM

For a plasma in thermodynamic equilibrium, when $\beta \ll 1$, the electrons have a Maxwellian velocity distribution function

$$f(\mathbf{v}) = \frac{1}{(\pi a^2)^{3/2}} \exp\left\{-\left(\frac{\mathrm{v}_k^2 + \mathrm{v}_{k\perp}^2 + \mathrm{v}_{kT}^2}{a^2}\right)\right\},$$

where the mean velocity is $a = (2\kappa T_e/m_e)^{1/2}$, $\beta = a/c$. This is substituted in Eqn (4.2.1); the resultant scattered power spectrum may be presented in a number of ways.

4.5.1 The Frequency Spectrum of the Scattered Power

$$P_s(\mathbf{R}, \omega_s)d\Omega = P_i r_o^2 d\Omega n_e L\left(1 + \frac{2\omega}{\omega_i}\right)|\hat{s} \times (\hat{s} \times \hat{E}_{io})|^2 \frac{\exp[-(\omega/ka)^2]}{\pi^{1/2}ka} d\omega_s, \tag{4.5.1}$$

where $\omega = \omega_s - \omega_i$, $d\omega = d\omega_s$, and $k \simeq 2k_i \sin(\theta/2)$.

4.5.2 The Differential Cross Section

Some authors use the differential scattering cross section, which is defined as

$$\frac{d\sigma}{d\Omega} = \frac{P_s}{P_i}\frac{A}{d\Omega} = N r_o^2 \left(1 + \frac{2\omega}{\omega_i}\right)|\hat{s} \times (\hat{s} \times \hat{E}_{io})|^2 \frac{\exp[-(\omega/ka)^2]}{\pi^{1/2}ka}.$$

This may be integrated over the frequency ω_s and over the whole solid angle,

$$\sigma = N(8\pi/3)r_o^3.$$

The Thomson cross section is $\sigma_T = (8\pi/3)r_o^3$. It is the integrated cross section for a single electron.

4.5.3 The Wavelength Spectrum

In experimental work, the results are often expressed in terms of the wavelength shift of the scattered radiation, $\Delta\lambda$, where

$$\lambda_s = \lambda_i + \Delta\lambda.$$

To be consistent with our restriction to first order in β, we must only keep terms to first order in $\Delta\lambda/\lambda_i$. We note that at low densities ($\omega_{pe} \ll k_i c$)

$$\frac{1}{k} \simeq \frac{c}{2^{1/2}(1-\cos\theta)^{1/2}\omega_i}\left(1+\frac{\Delta\lambda}{2\lambda_i}\right)$$

$$\frac{\omega}{\omega_i} = -\frac{\Delta\lambda}{\lambda_i}; \quad d\omega_s \simeq \frac{2\pi c}{\lambda_i^2}\left(1-\frac{2\Delta\lambda}{\lambda_i}\right)d\lambda_s$$

$$\left(\frac{\omega_s-\omega_i}{ka}\right)^2 \simeq \frac{c^2\Delta\lambda^2}{4\lambda_i^2\sin^2\frac{\theta}{2}a^2\left(1+\frac{\Delta\lambda}{\lambda_i}\right)}. \tag{4.5.2}$$

We substitute in Eqn (4.5.1) and

$$P_s^0(\mathbf{R},\lambda_s)d\lambda_s \cdot d\Omega = \frac{P_i r_0^2 d\Omega n_e L}{2\pi^{1/2}\sin\frac{\theta}{2}}\left\{1-\frac{3.5\Delta\lambda}{\lambda_i}+\frac{c^2\Delta\lambda^3}{4a^2\lambda_i^3\sin^2\left(\frac{\theta}{2}\right)}\right\}$$
$$\times \frac{c\,d\lambda_s}{a\lambda_i}\exp\left\{-\frac{c^2\Delta\lambda^2}{4a^2\lambda_i^2\sin^2\left(\frac{\theta}{2}\right)}\right\}. \tag{4.5.3}$$

We retain the correction term with $\Delta\lambda^3/\lambda_i^3$ because it is multiplied by c^2/a^2. The correction to higher order in beta, derived by Matoba *et al.* (1979), is shown in Eqn (6.3.1).

4.5.4 The Application to Experiment

Spectral Shift

These extra terms shift the spectrum toward the short wavelength side of the incident wavelength. Differentiate Eqn (4.5.3) with respect to $\Delta\lambda$ and set the result equal to zero, and the solution of the equation gives the shift of the center of the spectrum $\Delta\lambda_m$. Now, $\Delta\lambda_m/\lambda_i \ll 1$, and we find

$$\frac{\Delta\lambda_m}{\lambda_i} = -\frac{7a^2\sin^2(\theta/2)}{c^2}. \tag{4.5.4}$$

The quantity

$$a^2/c^2 \simeq 4\times 10^{-6}T_e(\text{eV}),$$

therefore

$$\frac{\Delta\lambda_m}{\lambda_i} \simeq -2.8\times 10^{-5}T_e(\text{eV})\sin^2(\theta/2). \tag{4.5.5}$$

For $\theta = 83^\circ$, $T_e = 117$ eV, and $\lambda_i = 6943$ Å, $\Delta\lambda_m = -10.5$ Å. This value compares favorably with the value of 9 ± 3.5 Å obtained under these conditions

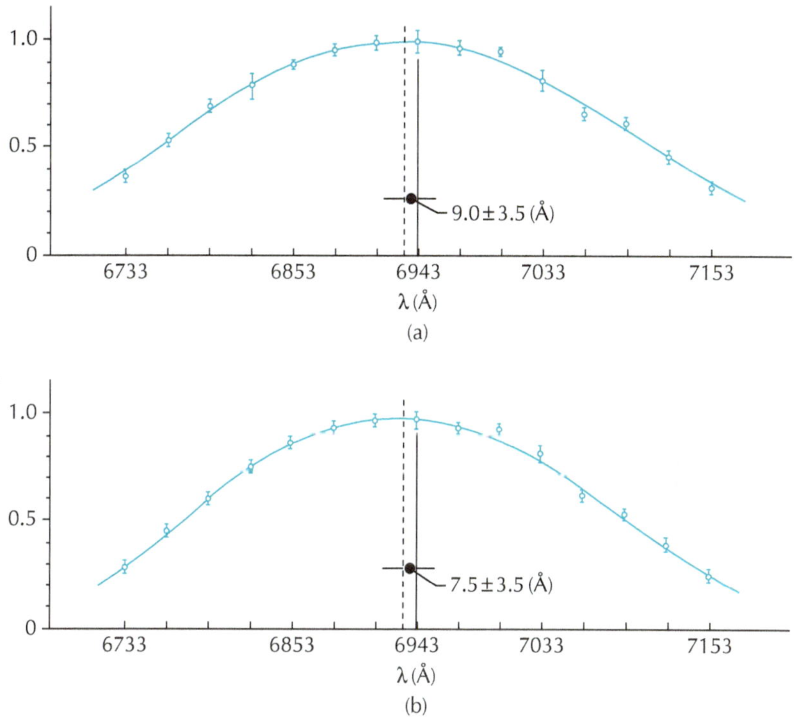

FIGURE 4.7 Scattered spectrum high-temperature shift to shorter wavelengths. (Courtesy Gondhalekar & Kronast (1971) and Nat. Res. Council of Can.) (a) Spectrum of scattering from plasma center with $\alpha_e = 0.2$, $T_e \cong 117$ eV and $\theta = 83°$; (b) spectrum of scattering from plasma sheath with $\alpha_e = 0.2$, $T_e \cong 95$ eV and $\theta = 83°$.

by Gondhalekar & Kronast (1973). The spectra obtained in this particularly experiment are shown in Fig. 4.7.

Electron Temperature Measurement

Figure 4.7 brings to our attention a major problem that occurs because of these high-temperature effects. Namely, that the temperature has sometimes been obtained from a plot of $\ln P_s^0$ against $\Delta\lambda^2$, with the assumption that the spectrum is Gaussian, and more important, symmetric about $\Delta\lambda = 0$. An estimate of the errors that will occur for a measurement on the short wavelength side of the spectrum is made below. This problem was discussed by Sheffield (1972a) and Zhuravlev & Petrov (1972).

Now, if we ignore the correction terms

$$1 - 3.5\frac{\Delta\lambda}{\lambda_i} + \frac{c^2\Delta\lambda^2}{4a^2\lambda_i^3\sin^2(\theta/2)}$$

and take the gradient, of the plot $\ln P_s^0$ versus $\Delta\lambda^2$, between $\Delta\lambda_1$ and $\Delta\lambda_2$, then we obtain (for simplicity, we take $\varphi_0 = 90°$)

$$\left(\frac{2\kappa T_e}{m_e}\right)_{\text{approx}} = \frac{c^2}{4\sin^2(\theta/2)}\left[\left(\frac{\Delta\lambda_2}{\lambda_i}\right)^2 - \left(\frac{\Delta\lambda_1}{\lambda_i}\right)^2\right]\cdot\frac{1}{G_0}, \tag{4.5.6}$$

where $G_0 = \ln[P_s^0(\Delta\lambda_1)] - \ln[P_s^0(\Delta\lambda_2)]$.

However, with the corrections terms included the result is

$$\left(\frac{2\kappa T_e}{m_e}\right)_{\text{corr}} = \frac{c^2}{4\sin^2(\theta/2)}\left[\left(\frac{\Delta\lambda^2}{\lambda_i}\right)^2 - \left(\frac{\Delta\lambda_1}{\lambda_i}\right)^2\right]$$

$$\times\left[G_0 + \ln\left(\frac{\left\{1 - 3.5\frac{\Delta\lambda_2}{\lambda_i} + \frac{c^2\Delta\lambda_2^3}{4a^2\lambda_i^3\sin^2(\theta/2)}\right\}}{\left\{1 - 3.5\frac{\Delta\lambda_1}{\lambda_i} + \frac{c^2\Delta\lambda_1^3}{4a^2\lambda_i^3\sin^2(\theta/2)}\right\}}\right)\right]^{-1} \tag{4.5.7}$$

For brevity, we set

$$\frac{c^2\Delta\lambda_1^2}{4\alpha^2\lambda_i^2\sin^2(\theta/2)} = \chi_1; \qquad \frac{c^2\Delta\lambda_2}{4\alpha^2\lambda_i^2\sin^2(\theta/2)} = \chi_2,$$

and then

$$G_0 = \chi_2 - \chi_1.$$

The result $\ln(1+\varepsilon) \cong \varepsilon$ for $\varepsilon < 1$ is used to simplify the formula, and rearranging, we obtain

$$T_{e_{\text{app}}} = T_{e_{\text{corr}}}\frac{\left\{\chi_2 - \chi_1 + \left[\chi_2^{1/2}\left(1 - \frac{\chi_2}{3.5}\right) - \chi_1^{1/2}\left(1 - \frac{\chi_1}{3.5}\right)\right]7\frac{a}{c}\sin(\theta/2)\right\}}{(\chi_2 - \chi_1)}. \tag{4.5.8}$$

The effect of the correction terms is indicated in the following examples where a comparison is made between the approximate and correct temperatures when the gradient is taken respectively between the points $e^{-\chi} = 0.9 - 0.3$ and $e^{-\chi} = 0.9 - 0.5$.

Case (1): $\chi_1 = 0.1$; $\chi_2 = 1.2$

$$T_{e_{\text{app}}} = T_{e_{\text{corr}}}(1 + 5.8\times10^{-3}\sin(\theta/2)[T_e\,(\text{eV})]^{1/2})$$

Case (2): $\chi_1 = 0.1$; $\chi_2 = 0.7$

$$T_{e_{\text{app}}} = T_{e_{\text{corr}}}(1 + 8.4\times10^{-2}\sin(\theta/2)[T_e\,(\text{eV})]^{1/2}).$$

TABLE 4.1 Temperature Values for the Short Wavelength Side

$T_{e_{corr}}$ (eV)	$T_{e_{app}}(0.90 \rightarrow 0.30)$ (eV)	$T_{e_{app}}(0.90 \rightarrow 0.50)$ (eV)
100	104	106
400	433	447
900	1011	1058
1600	1865	1980
2500	3000	3240

Some of these temperatures are evaluated for the short wavelength side in Table 4.1 for $\theta = 90°$.

For measurements on the long wavelength side, the correction terms take the opposite sign, and the temperature would be underestimated with the approximate formula Eqn (4.5.6). Obviously, if the whole spectrum is measured, there should be no problem, except that without the high-temperature corrections, the shift of the spectrum might be open to misinterpretation.

The formulas derived above are a good approximation for $T_e \lesssim 5\,\text{keV}$. For higher temperature, higher-order terms become important, and the spectra are quite drastically distorted as discussed in Section 4.2, where an analysis of various approximations is shown in Fig. 4.5 .

4.5.5 The Rate of Photon Scattering

One final point should be mentioned. We must remember to allow for the variation of the scattered photon energy with changing scattered frequency. We now divide the scattered power by $h\nu_s$; this is equivalent to multiplying the original result Eqn (4.5.3) by ω_i/ω_s (Williamson & Clarke, 1971). The extra terms are included and

$$\frac{\overline{dN_s}}{dt}(R, \lambda_s)d\lambda_s d\Omega = \frac{P_i r_0^2 d\Omega n_e L}{h\nu_i 2\pi^{1/2} \sin(\theta/2)} \left(1 - 2.5\frac{\Delta\lambda}{\lambda_i} + \frac{c^2 \Delta\lambda^2}{4a^2 \lambda_i^3 \sin(\theta/2)}\right)$$

$$\cdot \frac{c}{a} \exp\left(\frac{-c^2 \Delta\lambda^2}{4a^2 \lambda_i^2 \sin^2(\theta/2)}\right) \frac{d\lambda_s}{\lambda_i}. \tag{4.5.9}$$

Units

It is only necessary to be consistent in the evaluation of the quantities $P_i/h\nu_i, r_0^2 n_e L, c/a, \Delta\lambda^2/\lambda_i^2$, and $d\lambda_s/\lambda_i$. Thus, in the most commonly used practical units,

$$P_i(\mathrm{W})\quad h\nu_i(\mathrm{J}),\ h = 6.626\times 10^{-34}\,(\mathrm{J\cdot s}),\ \nu_s\nu_i\,(\mathrm{s}^{-1}),$$
$$r_0^2 = 7.95\times 10^{-26}\,\mathrm{cm}^2,\quad \text{with}\quad n_e(\mathrm{cm}^{-3}),\quad L(\mathrm{cm}),$$
$$c/a = 5\times 10^2/[T_e\,(\mathrm{eV})]^{1/2};\quad \Delta\lambda,\lambda_i,\lambda_s\,\text{Å (or }\mu\text{m)},$$
$$d\Omega \text{ in sr.}$$

4.5.6 Asymmetric Distribution Functions

An example is a distribution with a net drift in the scatter plane. For the case of a Maxwellian velocity distribution (Rose & Clark, 1961, p. 246),[1]

$$f_{e0}(\mathbf{v}) = \frac{\exp -(\mathbf{v}+\mathbf{v}_d/a)^2}{(\pi a^2)^{3/2}},\quad f_{i0}(\mathbf{v}) = \frac{\exp -(\mathbf{v}/b)^2}{(\pi b^2)^{3/2}}, \tag{4.5.10}$$

where

$$(\mathbf{v}+\mathbf{v}_d)^2 = (\mathrm{v}_k + \mathrm{v}_{kd})^2 + (\mathrm{v}_{k\perp} + \mathrm{v}_{\perp d})^2 + (\mathrm{v}_{kT})^2.$$

Note that to first order in β, we may ignore the change in direction of k with v while computing v_{kd} and $\mathrm{v}_{\perp d}$, and these may be taken as the drift components in the direction **k**. We will treat the simplest case, in this the scattering plane is perpendicular to $\mathbf{E}_{i0}$ so that $\cos\eta = \hat{\imath}\cdot\hat{0} = 0$ and $\hat{E}_{i0}\cdot\hat{0} = 1$. We also take the non-collective case ($\varepsilon = 1$), as this is important for the measurement of the electron drift velocity.

$$S(\mathbf{k},\omega) = \frac{2\pi}{k}\left(1+\frac{2\omega}{\omega_i}\right)\exp\left(-\left(\frac{\omega}{ka}+\frac{\mathrm{v}_{kd}}{a}\right)^2\right). \tag{4.5.11}$$

We differentiate with respect to ω, set the result to zero, and find the frequency shift of the center of the spectrum ω_m

$$\omega_m \simeq \left(-\frac{2\mathrm{v}_{kd}}{c}\sin\frac{\theta}{2} + 4\frac{a^2}{c^2}\sin^2\frac{\theta}{2}\right)\omega_i. \tag{4.5.12}$$

---

[1] The Maxwell-Juttner function is

$$f(\beta) = \left[2\pi K_2\left(\frac{m_e c^2}{\kappa T_e}\right)\right]^{-1}\cdot\frac{m_e c^2}{2\kappa T_e}\,\frac{\exp(m_e c^2/\kappa T_e(1-\beta^2)^{1/2})}{(1-\beta^2)^{5/2}}$$

where K_2 is a modified Bessel function of the second kind of order two. If we expand this in powers of β, we recover to first order in β the nonrelativistic Maxwellian. Remember that $m_e c^2/2\kappa T_e = c^2/a^2 \cong \beta^{-2}$.

4.5.7 The Measurement of Electron Temperature

The electron temperature may be obtained from a measured power spectrum by plotting the logarithm of the scattered power versus $(\Delta\lambda)^2$ [see Eqn (1.8.8)].

The gradient of this plot is

$$G = -\frac{c^2}{4a^2\lambda_i^2\sin^2(\theta/2)} = -\frac{6.4\times10^4}{\lambda_i^2\sin^2(\theta/2)}\frac{1}{T_e\,(\text{eV})}, \qquad (4.5.13)$$

where λ_i and $\Delta\lambda$ have the same units.

4.5.8 Electron Temperature Measurements on Magnetic Fusion Plasmas

An important application of noncollective scattering is the measurement of the electron temperature (distribution function) and density across a minor radius of magnetic fusion plasmas. This was first demonstrated on the Tokamak T-3 plasma containment device (Peacock *et al.*, 1969; Artsimovich *et al.*, 1969). A schematic diagram of the apparatus is given in Fig. 4.8. The incident radiation source was a Q-switched laser, $\lambda_i = 6943$ Å; $\Delta\lambda_i = 0.1$ Å giving 6 J in 25 ns. The scattered radiation was collected at $\theta = 90°$ by a simple lens and prism system. The radiation was dispersed and analyzed by a diffraction grating spectrometer combined with a ten-channel light guide output slit array; each channel was monitored by a photomultiplier. The main purpose of the experiment was to demonstrate that the high mean energy electron energy, ~ 1 keV, inferred from other measurements, was in fact associated with a high-temperature thermal distribution of electrons and not with a minority of very high-energy electrons riding on the tail of a low-temperature distribution. The electron density was determined from the absolute value of the total scattered power and was in good agrement with the results of microwave interferometry. This first scattering result showed that the bulk of the electrons had a thermal distribution at a high temperature ($T_e = 640$ eV).

A schematic diagram of a modern Thomson scattering apparatus (van der Meiden *et al.*, 2006) used to study the time evolution of the electron temperature and density distributions is given in Figs. 4.9 and 4.10. The intracavity ruby laser was switched by a Pockel cell and could produce a train of 40 pulses of 10–15 J each, with a repetition rate of 10 Hz, during the 10 ms flash lamp duration, thereby showing the temporal evolution of the electrons. As shown in Fig. 4.10, a Littrow grating spectrometer configuration was employed where the dispersed spectrum was directed onto the image intensifier (9) following which the light was imaged onto two CMOS cameras (12). The authors report the system was capable of measuring 120 spatial points (7.5 mm resolution)

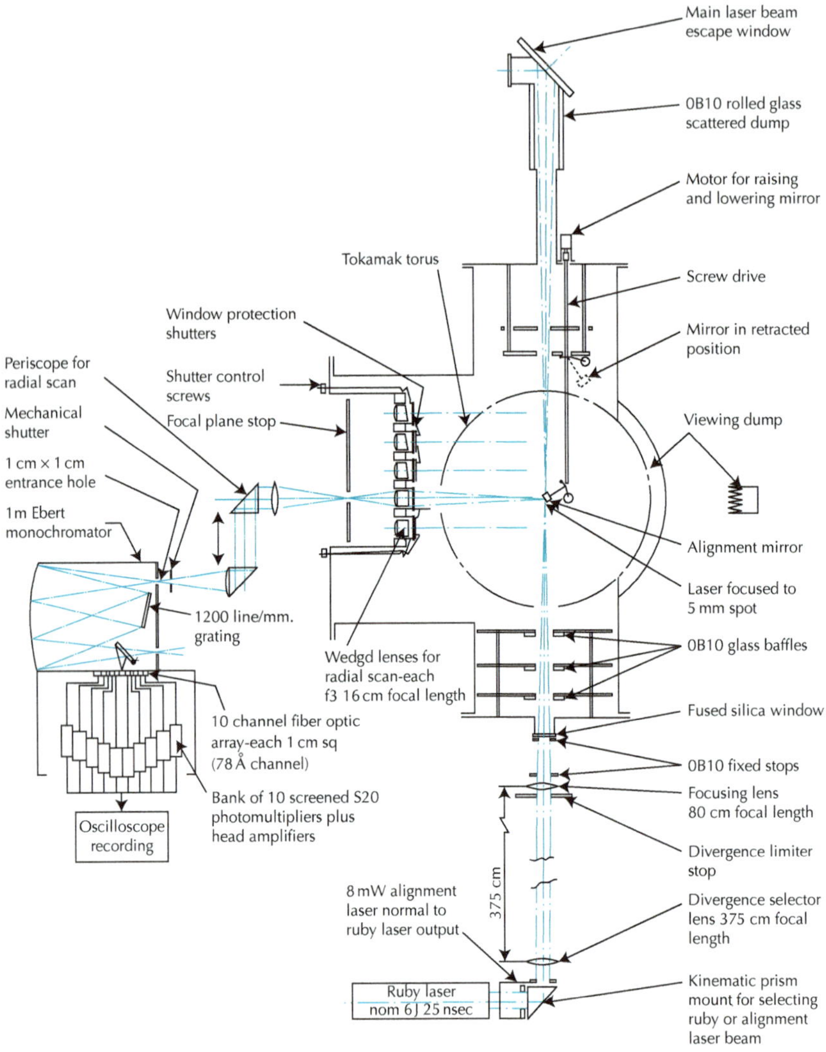

FIGURE 4.10 Schematic diagram of the laser scattering on Tokamak T-3. (Courtesy Culham Lab., UKAEA, U.K. and Peacock *et al.*, Rep. CLM-R-107.)

along a chord with a temperature error of 8% (over 50 eV–keV) and a density error of 4% at $n_e = 2.5 \times 10^{13}\,cm^{-3}$. Figure 4.11 displays a sequence of ten T_e and n_e profiles with a time separation of 200 ms obtained with this system. This experiment is discussed further in Chapter 7, where it is used as an example in the study of experimental techniques and problems.

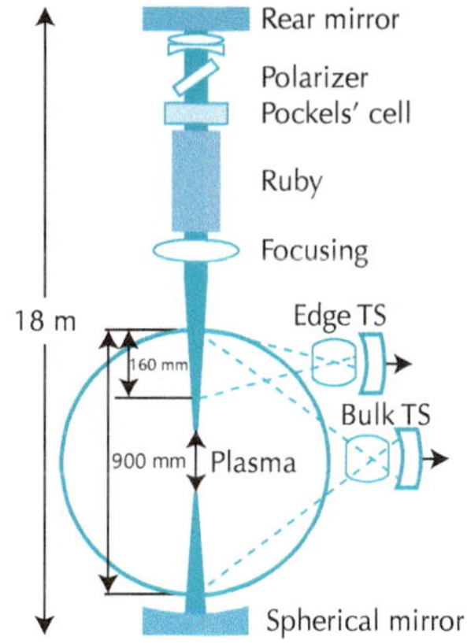

FIGURE 4.9 Schematic view of intercavity laser, the TEXTOR vessel is part of it. Both viewing systems for edge and full chord Thomson scattering are also shown. (Courtesy of van der Meiden).

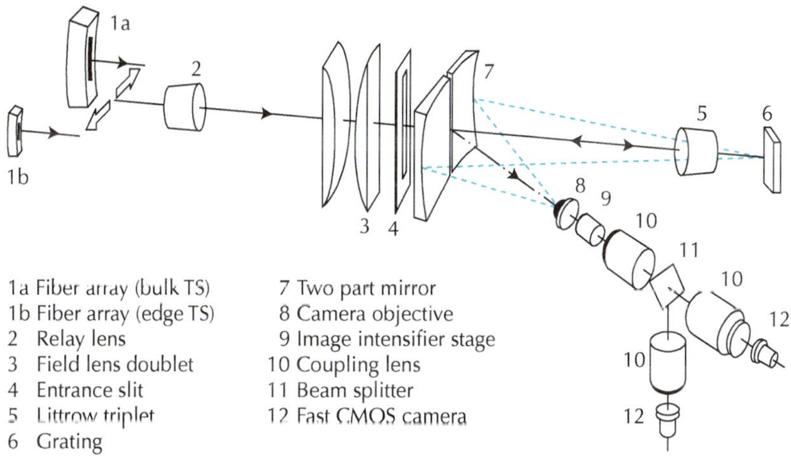

FIGURE 4.10 Spectrometer layout of the multipulse Thomson scattering system. (Courtesy of van der Meiden.)

4.6 NONCOLLECTIVE SCATTERING FROM A MAGNETIZED PLASMA

In the presence of a magnetic field, the orbits of the charges take on a helical form. Provided the incident beam does not interfere with the plasma, and the incident frequency exceeds Ω_e and ω_{pe}, this is the only additional effect we need consider in extending the noncollective theory to a low-temperature magnetized plasma.

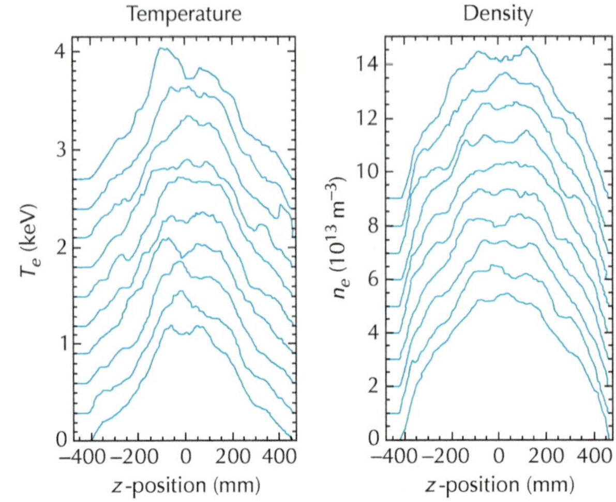

FIGURE 4.11 First ten of 18 T_e and n_e profiles, recorded every 200 ms, smoothed over four spatial points. The corresponding laser energy ranged from 1.2 to 8 J. The profiles of T_e and n_e are plotted over each other with an equidistance of 0.3 keV and $1 \times 10^{13}\,\text{cm}^{-3}$, respectively. (Courtesy of van der Meiden.)

The scattering geometry and electron orbit are shown in Fig. 4.12. The electron orbit is

$$\mathbf{r}\left(t_j'\right) = \mathbf{r}(0) + \rho_e \cos\phi\hat{x} + \rho_e \sin\phi\hat{y} + \mathrm{v}_{||}t_j', \tag{4.6.1}$$

where $\mathbf{r}(0)$ is the position of the guiding center at $t' = 0$, and $\mathrm{v}_{||}$ and $\mathrm{v}_{\perp}(t')$ are, respectively, the electron velocity parallel and perpendicular to the magnetic field $\mathbf{B}_o$. This field is taken in the $\hat{z}$ direction. The cyclotron radius is $\rho_e = \mathrm{v}_{\perp}/\Omega_e$ and

$$\phi\left(t_j'\right) = \Omega_e t_j' + \phi(0). \tag{4.6.2}$$

The velocity is

$$\mathbf{v}\left(t_j'\right) = -\mathrm{v}_{\perp}\sin\phi\hat{x} + \mathrm{v}_{\perp}\cos\phi\hat{y} + \mathrm{v}_{||}. \tag{4.6.3}$$

The retarded time for a low-temperature plasma ($\beta \ll 1$) is

$$t_j' \simeq t - \frac{R}{c} + \frac{\hat{s}\cdot\mathbf{r}(0)}{c}. \tag{4.6.4}$$

Now, with the aid of [see Eqn (10.1.5)]

$$e^{ia\sin\phi} = \sum_{l=-\infty}^{+\infty} J_l(a)e^{-il\phi}, \tag{4.6.5}$$

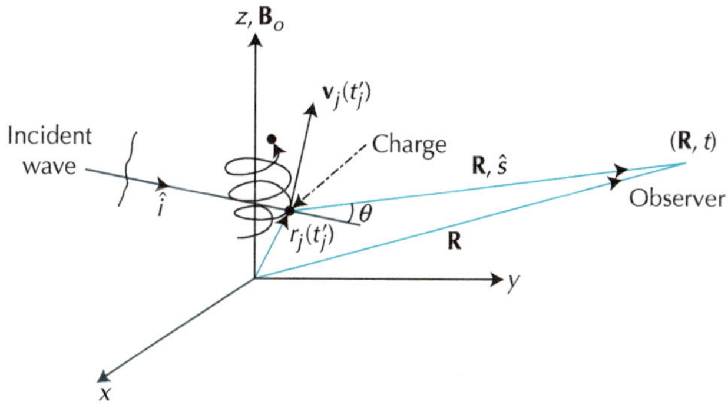

FIGURE 4.12 The scattering coordinate system for a magnetized plasma.

it can be shown that for low densities ($\omega_{pe} \ll kc$)

$$\cos\left[\mathbf{k}_i \cdot \mathbf{r}\left(t_j'\right) - \omega_j t_j'\right] = \sum_{l=-\infty}^{+\infty} J_l(k_\perp \rho_e) \cos[k_s R - \omega_s t - \mathbf{k}\cdot\mathbf{r}(0) - l(\phi(0)+\delta)] \tag{4.6.6}$$

$$\mathbf{k}_\mathrm{s} = \frac{\omega_\mathrm{s}}{c}\hat{s} \quad \text{and} \quad \omega_\mathrm{s} = \omega_\mathrm{i}\frac{(1-\hat{\imath}\cdot\mathbf{v}_{||}/c)}{(1-\hat{s}\cdot\mathbf{v}_{||}/c)} + l\Omega_\mathrm{e} = \omega_0 + l\Omega_\mathrm{e}, \tag{4.6.7}$$

where l is an integer and

$$\mathbf{k} = \mathbf{k}_s - \mathbf{k}_i, \quad \tan\delta = \frac{\mathbf{k}\cdot\hat{x}}{\mathbf{k}\cdot\hat{y}}, \quad k_\perp = [(\mathbf{k}\cdot\hat{x})^2 + (\mathbf{k}\cdot\hat{y})^2]^{1/2}, \quad k_{||} - \mathbf{k}\cdot\hat{z},$$

and using Eqn (4.6.6), we may rewrite Eqn (3.2.2) as

$$\mathbf{E}_{sj}(\mathbf{R}_j, t) = \left(\frac{e^2}{c^2 m_e R}\right)[\hat{s}\times(\hat{s}\times\mathbf{E}_{io})] \sum_{l=-\infty}^{+\infty} J_l(k_\perp \rho_e) \times \cos[k_s R - \omega_s t - \mathbf{k}\cdot\mathbf{r}_o - l(\phi(0)+\delta)]. \tag{4.6.8}$$

It can be seen from Eqns (4.6.7) and (4.6.8) that the scattered field contains an infinite series of components with frequencies that are multiples of the electron cyclotron frequency. The component with maximum amplitude occurs when $l \simeq k_\perp \rho_e$.

4.6.1 Scattered Power from One Electron

The scattered intensity is

$$I_{sj}(\mathbf{R},\omega_s)=\frac{c}{4\pi}\lim_{T\to\infty}\frac{1}{T}\int\limits_{-T/2}^{+T/2}\frac{e^4}{c^4m_e^2R^2}[\hat{s}\times(\hat{s}\times\mathbf{E}_{io})]^2$$
$$\times\sum_{l=-\infty}^{+\infty}J_l(k_\perp\rho_e)\sum_{m=-\infty}^{+\infty}J_m(k_\perp\rho_e)$$
$$\times\cos\left[(\omega_o+l\Omega_e)\left(\frac{R}{c}-t\right)\right]\cos\left[(\omega_o+m\Omega_e)\left(\frac{R}{c}-t\right)\right]. \quad (4.6.9)$$

The initial phase factor has been dropped for brevity; it does not enter into a discussion of noncollective scattering.

The only nonzero terms are those with $l=m$, and we obtain for the scattered power at the frequency ω_s, within the solid angle $d\Omega$,

$$P_{sj}(\mathbf{R},\omega_s)d\Omega=\frac{P_i}{A}r_o^2 d\Omega[\hat{s}\times(\hat{s}\times\mathbf{E}_{io})]^2\sum_{l=-\infty}^{+\infty}J_l^2(k_\perp\rho_e),$$

where $(\omega_s=\omega_o+l\Omega_e, \mathbf{k}=\mathbf{k}_s-\mathbf{k}_j)$.

4.6.2 Noncollective Scattering from a Magnetized Plasma in Thermodynamic Equilibrium

The total scattered power within the solid angle $d\Omega$ and in the frequency range $\omega_s\to\omega_s+d\omega_s$ is[2]

$$P_s(\mathbf{R},\omega_s)d\Omega d\omega_s=P_i r_o^2 n_e L d\Omega\frac{d\omega_s}{k_{||}}\left(1+\frac{2\omega}{\omega_i}\right)[\hat{s}\times(\hat{s}\times\hat{E}_{io})]^2$$
$$\times\int\limits_0^{2\pi}\int\limits_{-\infty}^{+\infty}\int\limits_0^{+\infty}d\phi d\mathrm{v}_{||}\mathrm{v}_\perp d\mathrm{v}_\perp$$
$$\times\sum_{l=-\infty}^{+\infty}J_l^2(k_\perp\rho_e)\left(\frac{m_e}{2\pi\kappa T_e}\right)^{3/2}\exp\left(\frac{\left(\mathrm{v}_\perp^2+\mathrm{v}_{||}^2\right)^2}{a^2}\right)$$
$$\times\delta\left\{\mathrm{v}_{||}-\left(\frac{\omega_s-\omega_i-l\Omega_e}{k_{||}}\right)\right\},$$

[2] Again we note that the limits should strictly be from $-c\to+c$, but that under our chosen conditions, we may set them at $\pm\infty$ (see Section 3.3).

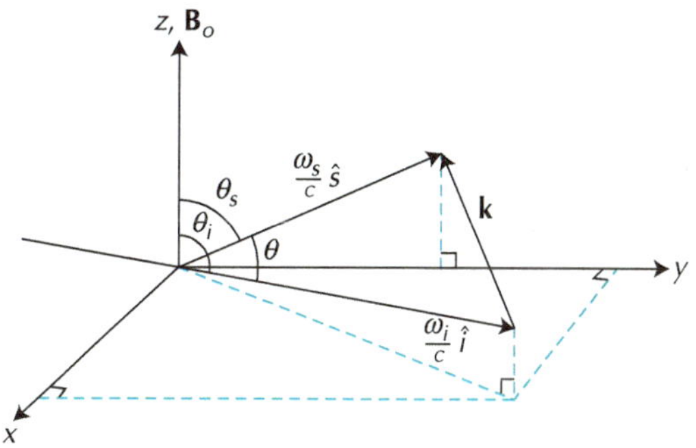

FIGURE 4.13 The wave vector diagram for a magnetized plasma.

where $a^2 = 2\kappa T_e/m_e$. This is obtained by the same procedure as that used in Section 4.6, but now, the coordinate system (see Fig. 4.13) is that appropriate to a magnetized plasma. We have, from Eqn (4.6.7),

$$v_{||} = (\omega_s - \omega_i - l\Omega_e)/k_{||} \quad \text{where} \quad k_{||} = (\omega_s/c)\cos\theta_s - (\omega_i/c)\cos\theta_i. \tag{4.6.10}$$

Now, see Eqns (10.1.9) and (10.1.8):

$$\int_0^\infty J_l^2(bt)\exp(-p^2t^2)tdt = \frac{1}{2p^2}\exp\left(-\frac{b^2}{2p^2}\right)I_t\left(\frac{b^2}{2p^2}\right); \tag{4.6.11}$$

$$J_{-l}(z) = (-)^l J_l(z) \quad \text{and} \quad I_{-l}(z) = I_l(z).$$

With the aid of these identities, we obtain

$$P_s(\mathbf{R}, \omega_s)d\omega_s d\Omega = P_i r_o^2 n_e L d\Omega \left(1 + \frac{2\omega}{\omega_i}\right)[\hat{s}\times(\hat{s}\times\hat{E}_{io})]^2 \exp\left(-\frac{k_\perp^2 a^2}{2\Omega_e^2}\right)$$
$$\times \sum_{l=-\infty}^{+\infty} I_l\left(\frac{k_\perp^2 a^2}{2\Omega_e^2}\right)\exp\left[-\left(\frac{\omega_s - \omega_i - l\Omega_e}{k_{||}a}\right)^2\right]\frac{d\omega_s}{\pi^{1/2}k_{||}a}. \tag{4.6.12}$$

This result was derived by Stewart (1972). The relativistic case was discussed by Nee *et al.* (1969). There are a number of features of this result that are worth discussing, the most obvious being the modulation at the cyclotron frequency Ω_e.

4.7 COMMENTS ON THE SCATTERED SPECTRUM

4.7.1 Fine Structure

The spectrum consists of an infinite series of peaks centered on frequencies separated from the incident frequency by multiples of the cyclotron frequency. The

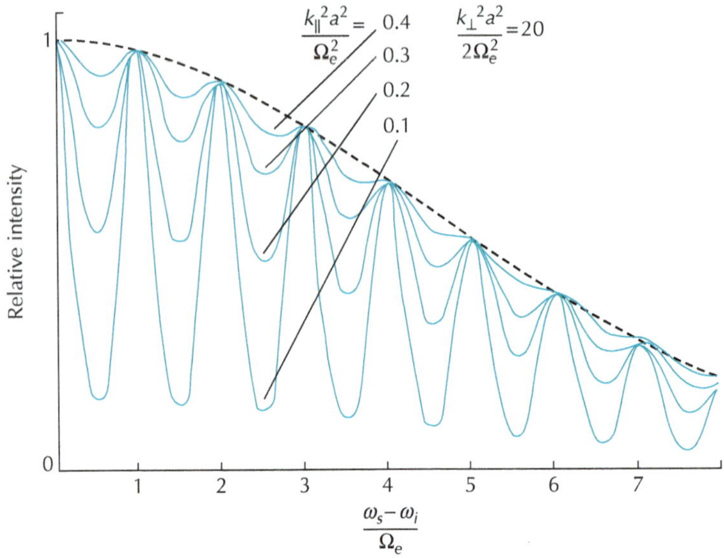

FIGURE 4.14 The noncollective scattered spectrum for a magnetized plasma as a function of $k_{||}a/\Omega_e$. (Courtesy of G. Lehner & F. Pohl (1970) *Z Phys.* 232, 405.)

frequency width of each peak is approximately $k_{||}a$. The modulation is therefore only significant when

$$|k_{||}a| \leq \Omega_e. \tag{4.7.1}$$

In the limit $k_{||} \to 0$, the peaks become narrower and narrower, and eventually at $k_{||} = 0$, we obtain a succession of spikes at the frequencies $\omega_i + l\Omega_e$. This is illustrated in Fig. 4.14.

This result is important because it gives us the possibility of measuring the magnitude of $\mathbf{B}_0$ from the period of the modulation and the direction of $\mathbf{B}_0$ from the degree, or in some cases, the mere presence of modulation (see Section 4.8). The modulation has been observed by Kellerer (1970); Evans & Carolan (1970); Lehner & Pohl (1970).

4.7.2 Resolution of Fine Structure

Two points have been raised regarding the resolution of this fine structure (Kellerer, 1970; Theimer & Theimer, 1970).

(a) First from Eqn (4.6.10) we see that for a fixed scattering angle θ_s, the parallel component of the differential scattering vector $k_{||}$ is a function of ω_s. We denote by ω_{os}, the frequency at which $\theta_B = 90^\circ$, i.e., $k_{||} = 0$; then from Eqn (4.6.10),

$$\omega_{os} \cos\theta_s = \omega_i \cos\theta_i. \tag{4.7.2}$$

We substitute in Eqn (4.7.1) and see that the modulation is resolvable in the frequency range $\omega_{os} \pm \Delta\omega_s$, where

$$|\Delta\omega_s|\cos\theta(a/c) \simeq \Omega_e. \qquad (4.7.3)$$

The full width of the scattered frequency spectrum is approximately $2ka = 4k_i a\sin(\theta/2)$ [see Eqn (4.5.1)]. This is substituted in Eqn (4.7.3), and we see that the fine structure is detectable over the whole spectrum for

$$4\omega_i(a/c)^2 \sin(\theta/2)\cos\theta_s \lesssim \Omega_e. \qquad (4.7.4)$$

(b) Second, in an experiment, radiation is collected over a finite angular range $\Delta\theta_s (\simeq \Delta\theta)$; the corresponding range of $k_{||}$ is approximately $\Delta k_{||} \simeq (\omega_i/c)\Delta\theta_s \sin\theta_s$. It might appear that in order to resolve the fine structure, we would require that

$$|\Delta k_{||}|a \simeq \omega_i(a/c)\Delta\theta_s \sin\theta_s \leq \Omega_e; \qquad (4.7.5)$$

in fact, this criterion is unnecessarily restrictive, as has been demonstrated by Carolan & Evans (1971). They also showed that it is the alignment of the collection cone axis with the perpendicular to $\mathbf{B}_o$ that matters rather than the angular range $\Delta\theta_s$ of the cone.

The magnitude of the magnetic field is determined from the period of the fine structure of the scattered spectrum when $k_{||} \to 0$. It can be seen from Eqn (4.6.12) that in this respect, there is no difference between the high- and low-temperature spectra to first order in beta.

Modification of the period occurs only when terms of order β^2 are important, because then the gyrofrequency $\Omega_e = |eB/m_0 c|(1-\beta^2)^{1/2}$ changes. $\beta^2 = 4\times 10^{-6}[T_e(\text{eV})]^{1/2}$, therefore, when $T_e < 10\,\text{keV}$; there should be no serious problems encountered in the measurement of the magnitude of the magnetic field. Equally there will he no additional problems in determining the direction of the magnetic field either by measuring the degree of the modulation or by detecting the modulation using Fourier transform spectroscopy.

4.7.3 The Envelope of the Spectrum

The envelope of the spectrum at high temperatures is shifted toward high frequency (small wavelengths) in a similar manner to the unmagnetized case.

For $k_\perp \to 0$, we find that since

$$I_l\left(k_\perp^2 a^2/2\Omega_e^2\right) \cong \frac{(k_\perp^2 a^2/4\Omega_e^2)_l}{l!},$$

only the term with $l=0$ remains. Further $k_{||} \to k$. Thus, for very small $k_\perp$ (note that it cannot in general be zero for all frequencies) because the direction of $\mathbf{k}$ is a function of ω,

$$P_{\mathrm{s}}^{0}(R, \omega_{\mathrm{s}}) d\omega_{\mathrm{s}} d\Omega \cong \frac{P_{\mathrm{i}} r_0^2 n_{\mathrm{e}} L d\Omega}{2\pi^{1/2} \sin(\theta/2)} \frac{c}{a} \exp\left\{-\frac{c^2 \Delta\lambda^2}{4a^2 \lambda_{\mathrm{i}}^2 \sin^2(\theta/2)}\right\} \frac{d\lambda_{\mathrm{s}}}{\lambda_{\mathrm{i}}}$$
$$\times \left\{1 - 3.5\frac{\Delta\lambda}{\lambda_{\mathrm{i}}} + \frac{c^2 \Delta\lambda^2}{4a^2 \lambda_{\mathrm{i}}^3 \sin^2(\theta/2)}\right\}. \qquad (4.7.6)$$

This is the same as the unmagnetized case Eqn (4.5.3) for $\chi = 90°$ and $\varphi_0 = 90°$. It was obtained for $\lambda_{\mathrm{s}} = \lambda_{\mathrm{i}} + \Delta\lambda$ by retaining terms to first order in $\Delta\lambda/\lambda_{\mathrm{i}}$.

4.7.4 The Measurement of the Magnitude of the Magnetic Field

The practical conditions that must be met for measuring the magnetic modulation of the scattered spectra have been discussed by Kellerer (1970). When the requirement that there must be a significant number of scattered photons within each peak of the modulation is included, the optimum working condition is for $\alpha = 1/k\lambda_{\mathrm{De}} \simeq 0.5$. From the general theory (see Chapter 10), we find that the spectrum then contains a collective contribution, which is concentrated near the incident line. This collective component is also modulated at the electron cyclotron frequency, but now, the source is the collective plasma behavior known as "Bernstein modes" (Bernstein, 1958) (see Appendix C). Nevertheless the modulation measurement is still possible even outside the strictly noncollective regime.

The apparatus used by Evans & Carolan (1970) for this measurement ($\alpha \cong 0.3$) is shown in Fig. 4.15. The plasma was produced by passing a current around a cylindrical thetatron coil. This device has the virtue that the direction of the magnetic field near the coil axis is well defined, i.e., with reasonable accuracy we may assume that it lies along the coil axis.

The incident radiation from a ruby laser ($\lambda_{\mathrm{i}} = 6943$ Å; 1.5 J in 30 ns) was directed at 15° to one side of the coil axis at the coil center, and the scattered radiation was collected at 15° to the other side. The mean scattering vector **k** was therefore close to the perpendicular to the field.

The electron temperature $T_{\mathrm{e}} \simeq 20$ eV and density $n_{\mathrm{e}} \cong 2 \times 10^{15}\,\mathrm{cm}^{-3}$ were obtained from the full spectrum profile. The modulation was measured over the central region of the spectrum, using a pressure-scanned Fabry–Perot etalon. Measured spectra are shown in Fig. 4.16. The period of the modulation corresponded in one case to a field of 14 kG, on the other to 5.5 kG. These results were in fair agreement with the values 16 and 8 kG, respectively, which were obtained independently by Faraday rotation measurements.[3] This experiment is discussed further in Chapter 7, where it is used to illustrate experimental problems and techniques.

---

[3]The general scattered spectrum from a magnetized plasma is calculated in Chapter 10.

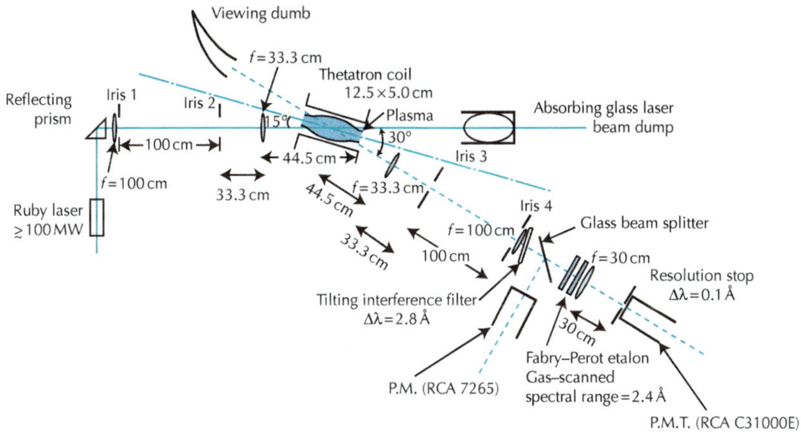

FIGURE 4.15 Layout of the experimental apparatus for the detection of magnetic modulation. (Courtesy of D. E. Evans & P. G. Carolan (1970).)

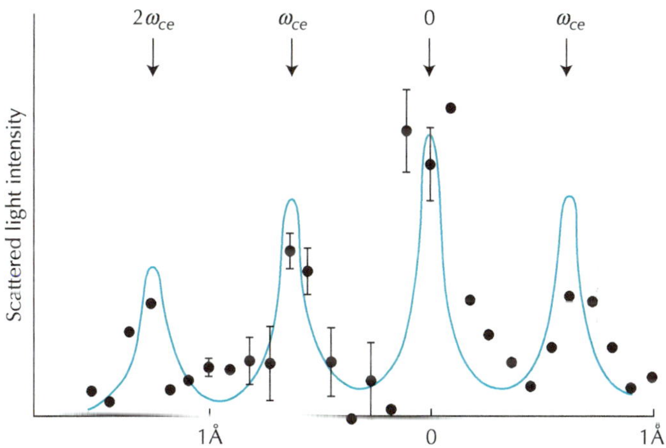

FIGURE 4.16 Scattered radiation spectrum from a magnetized plasma, $\lambda_i = 6943$ Å, $\theta = 30° \pm 0.85°$, $\alpha = 0.4$, $T_e = 20$ eV, $B = 16$ kG. (Courtesy of Carolan & Evans (1972).)

4.8 MEASUREMENT OF THE DIRECTION OF THE MAGNETIC FIELD IN A PLASMA

Numerous methods have been suggested whereby the detection of the magnetic modulation might be used as an indicator of the direction of a magnetic field in a plasma. Unfortunately, for the plasmas in devices, such as tokamaks, for which this measurement is of interest, there would be insufficient scattered photons to permit resolution of the modulation by conventional means for a ruby laser source.

(a) Consider a representative case, $n_e = 2 \times 10^{13}\,\text{cm}^{-3}$, $T_e = 900\,\text{eV}$, $B = 25\,\text{kG}$, $\lambda_i = 6943\,\text{Å}$, $W_i = 10\,\text{J}$, $h\nu_i = 1.65 \times 10^{-19}\,\text{J}$, $\theta = 9°$, $\eta = 6 \times 10^{-2}$, $L = 1\,\text{cm}$, and $T = 10^{-1}$.

From Eqn (4.7.2), we have to meet the condition $\Delta\theta_s \lesssim c\Omega_e/a\omega_i \sin(\theta/2)$, and $d\Omega \cong (\Delta\theta_s)^2$. For our case, $\Delta\theta_s \cong \pm 1°$ and $d\Omega \cong 10^{-3}\,\text{sr}$. Thus, mere detection of the modulation would fix the field direction within 1°. Now, the total number of detected photoelectrons, integrated over the whole spectrum, is $N_{pe} = (W_i/h\nu_i)r_0^2 n_e L d\Omega T \eta$. These photoelectrons are shared among the peaks, and very approximately these peaks number $2ka/\Omega_e$.

For this example, $N_{pe} \cong 360$ pe, and the number per peak is ~ 3 pe; thus, resolution of the modulation by conventional means is not possible.

(b) Murakami *et al.* (1970); Perkins (1970) suggested the use of a CO_2 laser. This source has the advantage that a larger value of $\Delta\theta_s$ may be used without smearing of the modulation, and there are less peaks, therefore more photoelectrons per peak. Further, for a comparable energy to a ruby laser, there are 14 times more photons. However, the limiting noise is from background radiation rather than from quantum statistics, and a heterodyne technique must be used to overcome the background noise.

Finally, note that the large value of $\Delta\theta_s$ for which there is modulation of the spectrum means that it is necessary to measure the degree of modulation and determine the angle at which it is a maximum rather than simply detect the presence of modulation which is sufficient for a ruby laser source.

(c) Multiplexing techniques have been proposed to take advantage of the narrow modulation angle seen with a ruby laser, i.e., techniques have been found whereby all the peaks and all the dips are superimposed, giving in effect one large peak and one large dip with a sufficient concentration of photoelectrons to allow detection of the modulation. Perkins (1970) suggested inserting a masking grid to select the peaks from the output spectrum of a grating spectrometer. Sheffield (1972b) suggested the use of a Fabry–Perot etalon of free spectral range equal to the cyclotron frequency. For this device, the ring pattern output spectra of frequencies separated by multiples of Ω_e are superimposed (see Sections 7.6.1 and 7.9). A similar suggestion using a Michelson interferometer was made by Katzenstein (1971).

There are three main problems in this technique; first the free spectral range must match the cyclotron frequency within $\sim \pm 1\%$ or the modulation is washed out. This restricts the use of the technique to situations for which the basic magnetic field is accurately known (a condition met in many tokamaks); second, if $k_\parallel = 0$ for $\omega_{0s} = \omega_i$, then we must restrict the transmission to $\pm\Delta\omega_s$, where $\Delta\omega_s = c\Omega_e/(a\cos\theta)$; this reduces the number of available photons. For our example, this is not a serious restriction. Third, the stray (parasitic) radiation is also transmitted in the peaks of the ring pattern, and for these low densities, this radiation can swamp the scattered signal. This problem can be overcome in two ways: either we can detect the dip in the transmitted signal by looking at radiation from between the cyclotron peaks or we can use an etalon with a free spectral range $\Delta\omega_F = 2\Omega_e$ and look at the alternate peaks that do not include the central and parasitic peaks Burgess (1971).

This measurement was performed on the DITE tokamak by Forrest *et al.* (1978). Their measurement of the poloidal field is shown in Fig. 4.17.

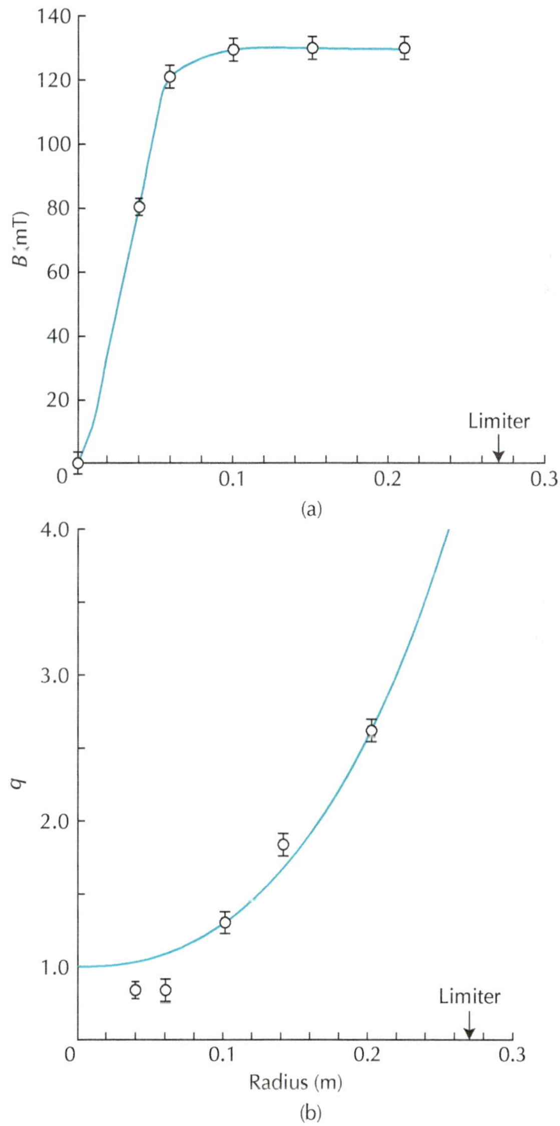

FIGURE 4.17 The measured poloidal field $B_\theta(r)$ distribution. (b) Comparison of 'q' distributions (circles) from poloidal field measurements and (line) electron temperature profiles assuming $j \propto T^{3/2}$ and constant Z_{eff} in similar plasma conditions. Courtesy of P. Carolan for Forrest *et al.* (1978).

Furthermore, pulse polarimetry, which combines the scattering of radiation with the Faraday effect to measure the magnetic field in a plasma, has been proposed by Smith (2008) (see Section 8.2).

PROBLEMS

4.1. From Eqn (1.7.19), derive $\mathbf{E}_s(R,t) = \dfrac{-e^2 E_{i0}(1-\beta^2)^{1/2}}{\left\{(m_e c^2 R)(1-\hat{s}\cdot\boldsymbol{\beta})^3\right\}}\cdot$

$$[(1-\beta_i)(1-\beta_s)-\{(1-\beta_i)\cos\eta+(\cos\theta-\beta_s)\beta_E\}\hat{s}+\beta_E(1-\beta_s)\hat{\imath} + \{(1-\beta_i)\cos\eta-(1-\cos\theta)\beta_E\}\boldsymbol{\beta}]_{ret}\times\cos(\mathbf{k}_i\cdot\mathbf{r}(t')-\omega_i t').$$

4.2. Show that the full velocity dependent term in the scattered power, without the restrictions of input E-field orientation or an output polarizer in the direction $\hat{e}_i$, is given by

$$(1-\beta^2)\left\{\frac{(1-\beta_i)^2}{(1-\beta_s)^2}\right\}\cdot\left[\frac{1-\cos^2\eta(1-\beta^2)}{(1-\beta_s)^2}+\left\{\frac{2\cos\eta\beta_E(1-\beta^2)}{(1-\beta_i)(1-\beta_s)^2}\right\}\times(1-\cos\theta)-\left\{\frac{\beta_E^2\cdot(1-\beta^2)}{(1-\beta_i)^2(1-\beta_s)^2}\right\}(1-\cos\theta)^2\right].$$

It is interesting that even in the general case, the factor $\left\{\frac{(1-\beta_i)^2}{(1-\beta_s)^2}\right\}$ may be extracted.

4.3. Comment on the dependence of the full depolarization term on the scattering angle θ. Specifically, is there any condition under which the term will have the same value at $\theta=0°$ and $180°$? Comment on the dependence on scattering angle for the case in which the incident electric field is perpendicular to the scattering plan and $\cos\eta=0$.

4.4. For circularly polarized input radiation, the incident wave may be written as $\mathbf{E}(\mathbf{R},t)=E_{io}(\hat{e}_1\pm i\hat{e}_2)\exp(ik\cdot r-i\omega t)$, where the $\pm$ refer to clockwise and anticlockwise rotation of the electric vector. $\hat{e}_1\cdot\hat{e}_2=0$, because the two components are 90° out of phase.

Show for 90° scattering that the velocity dependent terms under the integral, where η is the angle of one of the electric field components to the scattering vector, that

$$(1-\beta^2)\left\{\frac{(1-\beta_i)^2}{(1-\beta_s)^2}\right\}\cdot\left\{\left[\left\{\frac{1+2\cos\eta\beta_{E1}(1-\beta^2)}{(1-\beta_i)(1-\beta_s)^2}\right\}-\left\{\frac{\beta_{E1}^2\cdot(1-\beta^2)}{(1-\beta_i)^2(1-\beta_s)^2}\right\}\right]+\left[\left\{\frac{1+2\sin\eta\beta_{E2}(1-\beta^2)}{(1-\beta_i)(1-\beta_s)^2}\right\}-\left\{\frac{\beta_{E2}^2\cdot(1-\beta^2)}{(1-\beta_i)^2(1-\beta_s)^2}\right\}\right]\right\}.$$

4.5. Calculate the wavelength shift in the scattered spectrum owing to an electron temperature of 10 keV for scattering at 694.3 nm. Compare this to the scattered line broadening owing to the finite transit time for a 100 keV electron passing through the 1-cm scale scattering volume.

4.6. For the fully relativistic case, what is the formula for the rate of photon scattering equivalent to Eqn (4.2.1).

Show that the first order in beta correction becomes $\left(1+\frac{\omega}{\omega_i}\right)$.

4.7. Derive Eqn (4.5.12) for the shift of the spectrum center. Assume that $\left(\frac{v_{kd}}{c}\right)^2$ and $\left(\frac{a}{c}\right)^2$ are $\ll 1$ and that $\frac{v_{kd}}{c}$ is significant less than $\frac{a}{c}$.

4.8. On a graph of T_e versus n_e, plot the curves where $\alpha = 1$ for representative radiation sources, e.g., X-ray, 2ω and 3ω-YAG, ruby, CO_2, FEL, and D_2O for 10, 90, and 180 degree scattering.

4.9. An example of the difficulty with using a ruby laser-based system to measure the cyclotron peaks is given in the text. What would be the advantages and disadvantages of using a comparable energy, pulsed CO_2 laser system. Use the example lasers in Table 6.1 (ruby 25 J and 15 ns pulse, CO_2 17 J and 1 μs pulse, and assume the same scattering angle).

4.10. One approach to measuring the magnetic field direction in a plasma is to enhance the signal at right angles to the field with a Fabry–Perot etalon tuned to the cyclotron frequency. The number of photons scattered into the solid angle $d\theta d\varepsilon$ and frequency range $d\omega$ is given by

$$dN_{pe}(\varepsilon) = \left(\frac{W_i}{h\nu_i}\right) r_0^2 n_e L d\Omega \cdot \left(\frac{d\omega}{2\pi}\right) \exp\left(-\frac{k_\perp^2 a^2}{2\Omega_e^2}\right) \frac{2}{\pi}^{1/2}$$

$$\left[\sum_{1=-\infty}^{1=+\infty} I_1\left(\frac{k_\perp^2 a^2}{2\Omega_e^2}\right) \exp\left[-\frac{(\omega - 1\Omega_e)}{(k_\parallel a)}\right] / (k_\parallel a)\right].$$

Calculate the variation in the photons transmitted by the etalon as a function of the angle (ε) of the scattering vector to the perpendicular to the magnetic field angle to the magnetic field. Show for the case in which light is collected only between the cyclotron resonance peaks $\frac{s}{4} \to \frac{3s}{4}$ (where $s = \frac{\Omega_e}{k_\parallel a}$) is given by

$$N_{pe}(\varepsilon) = \left(\frac{W_i}{h\nu_i}\right) r_0^2 n_e T L d\theta d\varepsilon \cdot \frac{2}{\pi}^{1/2}$$

$$\cdot \left[\int_{s/4}^{3s/4} e^{-y^2} dy + \sum_{1=1}^{\infty} \int_{[s/4+1]}^{[3s/4+1]} e^{-y^2} dy\right].$$

What are the advantages and disadvantages of collecting scattered light at the resonances rather than between them.

SOLUTIONS TO ODD PROBLEMS

4.1. In Eqn (1.7.19), $\mathbf{E}_s(R,t) = \frac{e^2}{m_e c^2 R} \frac{(1-\beta^2)^{1/2}}{(1-\hat{s}\cdot\boldsymbol{\beta})^3} \left\{\hat{s} \times \left[(\hat{s}-\boldsymbol{\beta}) \times \left(\mathbf{E}_{i0} + \boldsymbol{\beta} \times \left\{n(\hat{i} \times \mathbf{E}_{i0})\right\} - \boldsymbol{\beta}(\boldsymbol{\beta}\cdot\mathbf{E}_{i0})\right)\right]\right\} \times \cos\left(k_i \cdot r(t') - \omega_i t'\right)$.

Using $A \times B \times C = (A \cdot C)B - (A \cdot B)C$.

$$\begin{aligned}
&[\hat{s} \times (\hat{s} - \boldsymbol{\beta}) \times \{\mathbf{E}_{i0} + \boldsymbol{\beta} \times \hat{i} \times \mathbf{E}_{i0} - \boldsymbol{\beta}(\boldsymbol{\beta} \cdot \mathbf{E}_{i0})] \\
&\quad = E_{i0}[\{-\hat{s}\cdot\hat{s} + \hat{s}\cdot\boldsymbol{\beta} + (\hat{s}\cdot\hat{s})(\boldsymbol{\beta}\cdot\hat{i}) - (\hat{s}\cdot\boldsymbol{\beta})(\boldsymbol{\beta}\cdot\hat{i})\}\hat{e}_i \\
&\qquad + \{\hat{s}\cdot\hat{e}_i + (\hat{s}\cdot\hat{i})((\boldsymbol{\beta}\cdot\hat{e}_i) - \hat{s}\cdot\hat{e}_i(\boldsymbol{\beta}\cdot\hat{i}) - (\hat{s}\cdot\boldsymbol{\beta})(\boldsymbol{\beta}\cdot\hat{e}_i)\}\hat{s} \\
&\qquad + \{-(\hat{s}\cdot\hat{s})(\boldsymbol{\beta}\cdot\hat{e}_i) + (\hat{s}\cdot\boldsymbol{\beta})(\boldsymbol{\beta}\cdot\hat{e}_i)\}\hat{i} \\
&\qquad + \{-(\hat{s}\cdot\hat{i})((\boldsymbol{\beta}\cdot\hat{e}_i) - \hat{s}\cdot\hat{e}_i\boldsymbol{\beta} + \hat{s}\cdot\hat{e}_i(\boldsymbol{\beta}\cdot\hat{i}) + (\hat{s}\cdot\hat{s})(\boldsymbol{\beta}\cdot\hat{e}_i)\}\boldsymbol{\beta}].
\end{aligned}$$

Substituting $\hat{s}\cdot\hat{i} = \cos\theta, \hat{s}\cdot\hat{e}_i = \cos\eta, \hat{e}_i = \hat{E}_{i0}, \boldsymbol{\beta}\cdot\hat{i} = \beta_i, \hat{s}\cdot\boldsymbol{\beta} = \beta_s$, and $\boldsymbol{\beta}\cdot\hat{e}_i = \boldsymbol{\beta}_E$ and rearranging yields Eqn (10.1.1).

$$\begin{aligned}
\mathbf{E}_s(R,t) = &-\frac{e^2 E_{i0}(1-\beta^2)^{1/2}}{\{(m_e c^2 R)(1-\hat{s}\cdot\boldsymbol{\beta})^3\}} \cdot \big[(1-\beta_i)(1-\beta_s)\hat{E}_{i0} \\
&- \{(1-\beta_i)\cos\eta + (\cos\theta - \beta_s)\beta_E\}\hat{s} + \beta_E(1-\beta_s)\hat{i} \\
&+ \{(1-\boldsymbol{\beta}_i)\cos\eta - (1-\cos\theta)\beta_E\}\boldsymbol{\beta}\big]_{ret} \times \cos(k_i \cdot r(t') - \omega_i t').
\end{aligned}$$

4.3. See problem 4.1.

$$(1-\beta^2)\left\{\frac{(1-\beta_i)^2}{(1-\beta_s)^2}\right\} \cdot \left[\frac{1-\cos^2\eta(1-\beta^2)}{(1-\beta_s)^2} + \left\{\frac{2\cos\eta\beta_E(1-\beta^2)}{(1-\beta_i)(1-\beta_s)^2}\right\}\right.$$
$$\left.\times(1-\cos\theta) - \left\{\frac{\beta_E^2\cdot(1-\beta^2)}{(1-\beta_i)^2(1-\beta_s)^2}\right\}(1-\cos\theta)^2\right].$$

For $\theta = 0°$, the term in square brackets becomes

$$\left[\frac{1-\cos^2\eta(1-\beta^2)}{(1-\beta_s)^2}\right].$$

For $\theta = 180°$,

$$\left[\frac{1-\cos^2\eta(1-\beta^2)}{(1-\beta_s)^2} + \frac{4\cos\eta\beta_E(1-\beta^2)}{(1-\beta_i)(1-\beta_s)^2} - \frac{4\beta_E^2\cdot(1-\beta^2)}{(1-\beta_i)^2(1-\beta_s)^2}\right].$$

The only condition under which the depolarization term would be the same at the extreme angles is when $\frac{4\cos\eta\beta_E(1-\beta^2)}{(1-\beta_i)(1-\beta_s)^2} = \frac{4\beta_E^2\cdot(1-\beta^2)}{(1-\beta_i)^2(1-\beta_s)^2}$ or $\cos\eta = \frac{\beta_E}{(1-\beta_i)}$.

When $\eta = 0$, the term becomes

$$(1-\beta^2)\left\{\frac{(1-\beta_i)^2}{(1-\beta_s)^2}\right\} \cdot \left[1 - \left\{\frac{\beta_E^2\cdot(1-\beta^2)}{(1-\beta_i)^2 \times (1-\beta_s)^2}\right\}(1-\cos\theta)^2\right].$$

For a given electron velocity distribution, the integral over velocity will yield some value $H(f(\mathbf{v}))$, and the scattered power will be proportional to

$$\left\{\frac{(1-\beta_i)^2}{(1-\beta_s)^2}\right\} \cdot [1 - H(f(\mathbf{v}))(1-\cos\theta)^2].$$

The depolarization term will decrease steadily from unity as the scattering angle is increased from 0° to180°.

4.5. The wavelength shift at the mean thermal speed of the electrons is given by $\Delta\lambda_{1/2} \sim 2.34 a \frac{\lambda_i}{c}$, where $a = 6 \times 10^5 [T_e\,(\mathrm{eV})]^{1/2}\,\mathrm{m\,s^{-1}}$.

For a temperature of 10 keV, $a = 6 \times 10^7\,\mathrm{m\,s^{-1}}$ and $\Delta\lambda_{1/2} = 325$ nm.

In contrast, the 100 keV electron will have a velocity of $1.9 \times 10^8\,\mathrm{m\,s^{-1}}$ and will traverse the scattering volume in 53 ps.

$\Delta\lambda_{FT} \sim \frac{(\mathrm{v}\lambda_i)}{(cd)} \cdot \lambda_i$ and the corresponding line broadening will be only 0.03 nm.

An electron has a velocity $\mathbf{v}$ and will spend a time $\Delta t' = \frac{d}{\mathrm{v}}$ within the scattering volume, where d is the dimension of the scattering volume in the direction $\mathbf{v}$.

The result of this situation is that the wave packet from each electron will be curtailed leading to a spread in the scattered frequency and wavelength. The wavelength spread is given by $\Delta\lambda_{FT} \sim \frac{(\mathrm{v}\lambda_i)}{(cd)} \cdot \lambda_i$.

This may be compared to the wavelength shift owing to the temperature, which at the mean thermal speed of the electrons is $\Delta\lambda_{1/2} \sim 2.34 a \frac{\lambda_i}{c}$.

4.7. To find the spectrum center, we equate $\frac{d}{d\omega}\left[\left(1+\frac{2\omega}{\omega_i}\right)\exp\left(-\left(\frac{\omega}{ka}+\frac{\mathrm{v_{kd}}}{a}\right)^2\right)\right]$ to zero and obtain $\frac{2}{\omega_i}+\left(1+\frac{2\omega_m}{\omega i}\right)\left(-\frac{2\omega_m}{ka}-\frac{\mathrm{v_{kd}}}{ka^2}\right)=0$, where ω_m is the frequency at the spectrum peak.

Now, $k = 2k_i \sin\frac{\theta}{2}$ and $k_i = \frac{c}{\omega_i}$. Substituting and rearranging leads to

$$4\omega_m^2+\omega_i\left(2+8\left(\frac{\mathrm{v_{kd}}}{c}\right)\sin\frac{\theta}{2}\right)\omega_m+\omega_i^2\left(4\left(\frac{\mathrm{v_{kd}}}{c}\right)\sin\frac{\theta}{2}\quad 8\sin^2\left(\frac{\theta}{2}\right)\left(\frac{a}{c}\right)^2\right).$$

The solution is

$$\omega_m=\frac{\omega_i}{4}\left[-\left(1+4\left(\frac{\mathrm{v_{kd}}}{c}\right)\sin\frac{\theta}{2}\right)\pm\left\{\left(1+4\left(\frac{\mathrm{v_{kd}}}{c}\right)\sin\frac{\theta}{2}\right)^2\right.\right.$$
$$\left.\left.-4\left(4\left(\frac{\mathrm{v_{kd}}}{c}\right)\sin\frac{\theta}{2}-8\sin^2\left(\frac{\theta}{2}\right)\left(\frac{a}{c}\right)^2\right)\right\}^{1/2}\right].$$

$$\omega_m\approx\left[-2\left(\frac{\mathrm{v_{kd}}}{c}\right)\sin\frac{\theta}{2}+4\sin^2\left(\frac{\theta}{2}\right)\left(\frac{a}{c}\right)^2\right]\omega_i.$$

4.8. On a graph of T_e versus n_e, plot the curves where $\alpha = 1$ for representative radiation sources, e.g, x-ray, 2ω and 3ω-YAG, ruby, CO_2, and HCN for 10, 90, and 180° scattering. (Ask for something different from figure in book.)

4.9. To resolve the peaks, we require $\Delta\Omega_s \le \frac{c\Omega_e}{a\omega_i}\sin\left(\frac{\theta}{2}\right)$ and $d\Omega \approx (\Delta\Omega_s)^2$.

The number of detected photoelectrons will be $N_{pe} = \left(\frac{W_i}{h\nu_i}\right) r_0^2 n_e L d\Omega T \eta$ photoelectrons. These will be shared among the approximately $\frac{2ka}{\Omega_e}$ cyclotron peaks.

Denoting the ruby laser result by subscript R and the CO_2 laser result by subscript C, we see that $\frac{(N_{pe})_C}{(N_{pe})_R} = \frac{(W_i/)_C}{(W_i/)_R} \cdot \frac{(\nu_i^3)_R}{(\nu_i^3)_C} \cdot \frac{\eta_C}{\eta_R}$.

For the example given, $\frac{(N_{pe})_C}{(N_{pe})_R} = 2.4 \times 10^3 \frac{\eta_C}{\eta_R}$.

Provided there is not too big a penalty for the lower detector efficiency (CHECK actual values) at 10.59 μm, the CO_2 system will measure many more photoelectrons than the ruby system per cyclotron peak. However, there will be a penalty in terms of the increased plasma light owing to the 67 times longer pulse length.

Chapter 5

Collective Scattering from a Plasma

Chapter Outline Head

DOI: 10.1016/B978-0-12-374877-5.00005-1

5.1 INTRODUCTION

Collective scattering is from electrons whose motion is correlated. It is seen when we are observing on a scale greater than the Debye length ($\lambda \geq \lambda_{De}$). In this regime, the scattering reflects the collective motion of electrons as they provide shielding with the ions and other electrons. In general, the thermal fluctuations are incoherent (i.e., reflect a broad frequency range), but the light scattered from these fluctuations can be coherent or at least consist of narrow frequency features on top of a broad incoherent feature.

A general expression for the spectrum of radiation scattered from an unmagnetized, collisionless, stable, low-temperature plasma is derived in Chapter 3. The scattered power, to first order in v/c, in the solid angle $d\Omega$, about the scattering angle θ, at a distance $R \gg L, \lambda_i$, and in the frequency range $\omega_s \rightarrow \omega_s + d\omega$, is

$$P_s(\mathbf{R}, \omega_s)d\Omega d\omega_s = P_i r_o^2 L d\Omega d\omega_s / 2\pi \left(1 + \frac{2\omega}{\omega_i}\right) |\hat{s} \times (\hat{s} \times \hat{E}_{io})|^2 n_e S(\mathbf{k}, \omega), \tag{5.1.1}$$

where L is the length of the scattering volume in the direction $\mathbf{k}_i, \mathbf{k} = \mathbf{k}_s - \mathbf{k}_i, \omega = \omega_s - \omega_i$, and P_i is the average incident power.

The scattered power is a function of the time through $P_i(t)$ and $S(\mathbf{k}, \omega)$, but note that the derivation was made for time variations slow compared to the microscopic times: ω_{pe}^{-1}, ν^{-1}, ω_i^{-1}. For arbitrary distribution functions, the spectral density function Eqn (3.4.3) is

$$S(\mathbf{k}, \omega) = \underbrace{\frac{2\pi}{k}\left|1 - \frac{\chi_e}{\epsilon}\right|^2 f_{eo}\left(\frac{\omega}{k}\right)}_{\text{Electron component}} + \underbrace{\frac{2\pi Z}{k}\left|\frac{\chi_e}{\epsilon}\right|^2 f_{io}\left(\frac{\omega}{k}\right)}_{\text{Ion component}}, \tag{5.1.2}$$

where f_{eo} and f_{io} are the normalized one-dimensional electron and ion velocity distribution functions, respectively, and Z is the charge on an ion. For plasmas with multiple ion species, the second term in Eqn (5.1.2) becomes

$$\sum_j \frac{2\pi}{k} \frac{Z_j^2 N_j}{N} \left|\frac{\chi_e}{\epsilon}\right|^2 f_{i0,j}\left(\frac{\omega}{k}\right),$$

where $N = \sum_j N_j Z_j$ and N_j is the number of jth ion species per unit volume. The longitudinal dielectric function becomes

$$\epsilon = 1 + \chi_e + \sum_j \chi_j,$$

where j is the sum over the ion species (Evans & Carolan, 1970).

In the collective region, the incident light wave "sees" the bunches of electrons that are a shield on each electron [given by the "electron component" which is the first term in Eqn (5.1.2)] and on each ion [given by the "ion component" which is the second term in Eqn (5.1.2)]. These shielding effects are contained in the dielectric function ϵ, and significant contributions to the scattering spectrum are obtained when $|\epsilon|^2$ is small; at the natural resonances of the plasma, i.e., in the unmagnetized case at the electron plasma frequency resonance $\omega \simeq \omega_{pe}$, and ion-acoustic resonance,

$$\omega \simeq \omega_{ac} \simeq k\sqrt{\frac{Z\kappa T_e + 3\kappa T_i}{m_\mathrm{i}(1+k^2\lambda_\mathrm{De}^2)}}.$$

For a stable plasma, Landau damping and collisions keep the resonant wave amplitude at a low level. The precise level of this damping depends critically on the form of the distribution function, and we must therefore specify f_{eo} and f_{io} in order to obtain a detailed spectrum.

In the following sections, we treat the case of Maxwellian distributions where the full nonrelativistic spectral function is assumed. A series of experimental applications are presented in Section 5.4 that demonstrate the power of collective Thomson scattering in measuring a range of plasma parameters. The effect of collisions on various features of the spectrum is investigated, and the effect of a small relative drift of the electrons and ions is discussed. Finally, the dependence of the total scattering cross section $S(\mathbf{k}) = S_e(\mathbf{k}) + S_i(\mathbf{k})$ on α and T_e/T_i is determined.

5.2 $S(\mathbf{k}, \omega)$, MAXWELLIAN DISTRIBUTION FUNCTIONS

In thermodynamic equilibrium, the one-dimensional velocity distributions are Maxwellian

$$f_{eo} = \left(\frac{1}{\pi a^2}\right)^{1/2} \exp(-\mathrm{v}^2/a^2) \quad \text{and} \quad f_{io} = \left(\frac{1}{\pi b^2}\right)^{1/2} \exp(-\mathrm{v}^2/b^2),$$

where the mean thermal speeds are

$$a = (2\kappa T_e/m_\mathrm{e})^{1/2}, \quad b = (2\kappa T_i/m_\mathrm{i})^{1/2}. \tag{5.2.1}$$

From Eqn (5.1.2), we can then write

$$S(\mathbf{k}, \omega) = \frac{2\pi^{1/2}}{ka}\left\{\frac{A_e}{|\epsilon|^2} + \frac{A_i}{|\epsilon|^2}\right\}, \tag{5.2.2}$$

where

$$A_e = \exp(-x_e^2)\left[\left(1+\alpha^2\frac{ZT_e}{T_i}\mathrm{Rw}(x_i)\right)^2 + \left(\alpha^2\frac{ZT_e}{T_i}\mathrm{Iw}(x_i)\right)^2\right]$$

$$A_i = Z\left(\frac{m_\mathrm{i}T_e}{m_\mathrm{e}T_i}\right)^{1/2}\exp(-x_i^2)\left[\left(\alpha^2\mathrm{Rw}(x_e)\right)^2 + \left(\alpha^2\mathrm{Iw}(x_e)\right)^2\right] \tag{5.2.3}$$

and

$$|\epsilon|^2 = \left[1+\alpha^2\left(\mathrm{Rw}(x_e) + \frac{ZT_e}{T_i}\mathrm{Rw}(x_i)\right)\right]^2 + \left[\alpha^2\mathrm{Iw}(x_e) + \alpha^2\frac{ZT_e}{T_i}\mathrm{Iw}(x_i)\right]^2 \tag{5.2.4}$$

$$x_e = \omega/ka, \qquad x_i = \omega/kb, \qquad \alpha = 1/k\lambda_\mathrm{De}.$$

For Maxwellian distribution functions, the electron and ion susceptibilities [see Eqns (3.3.9) and (3.3.10)] can be written as

$$\chi_e(\mathbf{k},\omega) = -\frac{\alpha^2}{2} Z'(x_e) = \alpha^2\,[\mathrm{Rw}(x_e) + i\mathrm{Iw}(x_e)] \tag{5.2.5}$$

$$\chi_i(\mathbf{k},\omega) = -\frac{\alpha^2}{2}\frac{ZT_e}{T_i} Z'(x_i) = \alpha^2 \frac{ZT_e}{T_i}\,[\mathrm{Rw}(x_i) + i\mathrm{Iw}(x_i)], \tag{5.2.6}$$

where Rw(x) and Iw(x) are, respectively, the real and imaginary parts of the derivative of the plasma dispersion function tabulated by Fried & Conte (1961). This expression can easily be computed using modern computers (see Appendix D). As discussed in the following section, it is possible to simplify this messy expression and derive useful analytic results.

Figure 5.1 shows, for real values of x, the real part,

$$\mathrm{Rw}(x) \equiv -\frac{1}{2}\mathrm{Re}[Z'(x)] = 1 - 2x\exp(-x^2)\int_0^x \exp(p^2)dp, \tag{5.2.7}$$

and the imaginary part of the susceptibility (Landau damping term),

$$\mathrm{Iw}(x) \equiv -\frac{1}{2}\mathrm{Im}[Z'(x)] = \pi^{1/2} x\exp(-x^2). \tag{5.2.8}$$

For $x<1$, i.e., $\mathrm{v}_{ph} = \omega/k < \mathrm{v}_{th}$ (ph is phase and th is thermal)

$$\mathrm{Rw}(x) \simeq 1 - 2x^2\left[1 - \frac{2x^2}{3} + \frac{4x^4}{15} - \cdots\right], \tag{5.2.9}$$

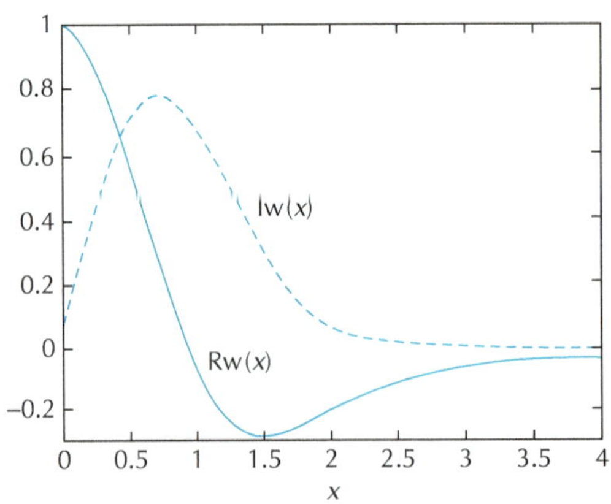

FIGURE 5.1 (solid) The real part of the plasma dispersion function, Eqn (5.2.7), and (dashed) the imaginary part (Landau damping term), Eqn (5.2.8), are shown.

while for $x \gg 1$, i.e., $v_{ph} \gg v_{th}$,

$$\mathrm{Rw}(x) \simeq -\frac{1}{2x^2}\left[1+\frac{3}{2x^2}+\frac{15}{4x^4}+\cdots\right]. \tag{5.2.10}$$

5.2.1 High Frequencies ($\omega \gg \omega_{pi}$)

The collective scattering spectrum can be broken into two regimes that are separated by the ability of the ions to respond to the electron motion; in the high-frequency regime ($\omega \gg \omega_{pi}$), the ions cannot respond ($x_i \gg 1$), and the ion component is negligible ($A_i/|\epsilon|^2 \to 0$) while the electron component $A_e/|\epsilon|^2$ dominates. Figure 5.2 illustrates the resonance observed in the high-frequency regime ("electron feature") when $\alpha > 1$ (see Section 5.4 for a discussion on the electron plasma waves).

5.2.2 Low Frequencies ($\omega \ll \omega_{pi}$)

In the low-frequency regime, the ion component dominates when $ZT_e/T_i < 1$ (strong ion wave damping). Figure 5.2a shows that under these conditions, the scattering spectrum results primarily from the Doppler shift due to scattering from electrons that are screening the ions and therefore reflects the ion thermal spread (i.e., the ion velocity distribution). For $ZT_e/T_i \simeq 1$, the phase velocity of the waves $v_{ph} \simeq (Z\kappa T_e/m_i)^{1/2} \simeq (\kappa T_i/m_i)^{1/2}$ sits in a region where heavy ion Landau damping occurs; in addition, the resonance is weak. As ZT_e/T_i increases, the ion-acoustic resonance becomes more marked, the phase velocity increases, and the waves propagate further into the tail of the ion distribution

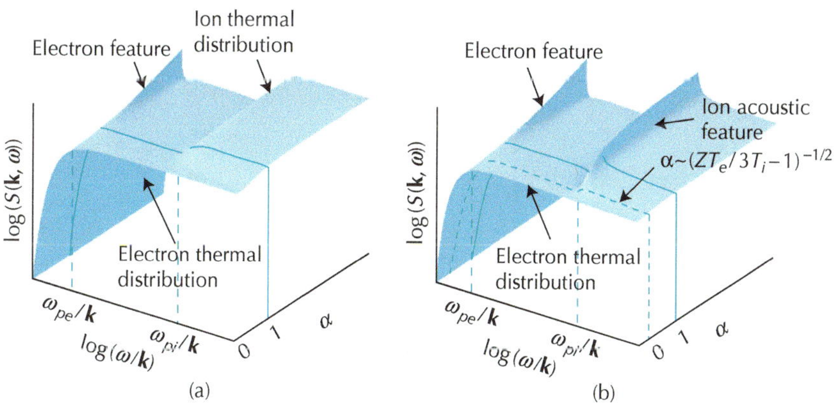

FIGURE 5.2 The features observed in the scattering spectrum [Eqn (5.2.2)] as a function of the scattering parameter, α. The electron density is varied to scale the scattering parameter while holding (a) $Z=1$, $T_e/T_i=0.1$ (strongly damped ion-acoustic waves) and (b) $Z=10$, $T_e/T_i=1$ (weakly damped ion-acoustic waves).

function where the damping is reduced. Here ($ZT_e/T_i \gg 1$), the ion component is reduced and electron component dominates. The reduction in the ion component occurs because of the factor $\exp(-x_i^2)$, as in this region $x_i^2 = ZT_e/2T_i \gg 1$. Although the condition for observing the electron feature is $\alpha > 1$, Fig. 5.2b shows that the ion-acoustic feature is observable in the spectrum for $\alpha < 1$ but greater than $\alpha \gtrsim \left(\frac{ZT_e}{3T_i} - 1\right)^{-1/2}$. This condition is derived and discussed further in the following section.

5.3 $S(\mathbf{k}, \omega), T_e/T_i \simeq 1$, THE SALPETER APPROXIMATION

It was first demonstrated by Salpeter (1960) that in this region, the spectrum takes a relatively simple form. From Eqn (5.2.1), we see that for $T_e/T_i \simeq 1$, since $m_e \ll m_i$, the quantity $x_e/x_i = (m_e T_i/m_i T_e)^{1/2} \ll 1$.

5.3.1 The Electron Component–Salpeter Approximation

We recollect that the first term in Eqn (5.2.2) includes the contribution to the spectrum from the free electrons when $\alpha \ll 1$ and from the electrons plus their electron Debye shields when $\alpha > 1$. Now, each shield is actually obtained by the repulsion of electrons and attraction of ions so that the form of the shield does involve the ions, hence the factor $(ZT_e/T_i)\alpha^2 \mathrm{Rw}(x_i)$ in this term. However, the majority of electrons are moving so fast that the ions cannot respond to their motion; the shielding is primarily achieved by the repulsion of other electrons. For the high-frequency part of the spectrum $x_e \simeq 1$, where $x_i \gg 1$, and from (5.2.10), $\mathrm{Rw}(x_i) \to 0$. Equally $\mathrm{Iw}(x_i) = 0$, because the ion Landau damping at these high phase velocities is negligible. It is only at low frequencies that the ions play a role, but as discussed above, for $ZT_e \simeq T_i$, the ion-acoustic resonance is weak, therefore to a good approximation

$$S_e(\mathbf{k}, \omega) \simeq \frac{2\pi^{1/2} \exp(-x_e^2)}{ka[(1+\alpha^2 \mathrm{Rw}(x_e))^2 + (\alpha^2 \mathrm{Iw}(x_e))^2]}. \tag{5.3.1}$$

5.3.2 The Ion Component–Salpeter Approximation

The second term in Eqn (5.2.2) gives the contribution to the scattering from electrons that have bunched to shield each ion. The shielding of the ion is by the attraction of electrons and repulsion of other ions. The term only contributes when we look on a scale comparable to or greater than a Debye length; this is seen in the dependence on α. Because of the factor $\exp(-x_i^2)$, the term is important only when $0 < x_i < 2$. A typical frequency for this part of the spectrum is

$$\omega \simeq k\left(\frac{2\kappa T_i}{m_i}\right)^{1/2} = \left(\frac{2T_i}{\alpha^2 ZT_e}\right)^{1/2} \omega_{pi}.$$

When $2T_i/\alpha^2 ZT_e \gg 1$ (i.e., $\omega/\omega_{pi} \gg 1$), the ions cannot respond collectively, and the spectrum reflects the ion thermal spread through scattering from the electrons shielding the ions. When $2T_i/\alpha^2 ZT_e \ll 1$, the ions play a major role in the shielding, and the spectrum is a combination of the ion thermal speed and the vestigial ion-acoustic waves.

For $\frac{T_e m_i}{T_i m_e} \gg 1$, we are probing in the flat center ($x_e \ll 1$) of the electron distribution, consequently $\mathrm{Rw}(x_e) \simeq 1$ and $\mathrm{Iw}(x_e) \simeq 0$ and the ion feature becomes

$$S_i(\mathbf{k},\omega) \simeq \frac{2\pi^{1/2}}{kb} \frac{Z\alpha^4 \exp(-x_i^2)}{[1+\alpha^2+\alpha^2(ZT_e/T_i)\mathrm{Rw}(x_i)]^2+[\alpha^2(ZT_e/T_i)\mathrm{Iw}(x_i)]^2}. \quad (5.3.2)$$

Note: This approximation is not valid either when ZT_e/T_i is significantly greater than one or when there is a relative drift of the ions and electrons. In both cases, the electron Landau damping becomes important (i.e., $\mathrm{Iw}(x_e) > 0$).

5.3.3 Complete Spectrum–Salpeter Approximation

Combining Eqns (5.3.1) and (5.3.2), we obtain

$$S(\mathbf{k},\omega) \simeq \frac{2\pi^{1/2}}{ka}\Gamma_\alpha(x_e) + \frac{2\pi^{1/2}}{kb} Z\left(\frac{\alpha^2}{1+\alpha^2}\right)^2 \Gamma_\beta(x_i), \quad (5.3.3)$$

where

$$\Gamma_\alpha(x) = \frac{\exp(-x^2)}{[1+\alpha^2\mathrm{Rw}(x)]^2+[\alpha^2\mathrm{Iw}(x)]^2} \quad \text{and} \quad \beta^2 = \frac{\alpha^2}{1+\alpha^2}\frac{ZT_e}{T_i}. \quad (5.3.4)$$

This result is accurate in the case where $\frac{T_e m_i}{T_i m_e} \gg 1$ and $ZT_e/T_i \leq 1$.

5.3.4 Electron Plasma Resonance

As we have discussed, there is a resonance in the spectrum at high frequencies, $\omega \simeq \omega_{pe}$, due to the plasma wave fluctuations. The ions cannot respond to such frequencies ($x_i \gg 1$), and we are therefore justified in dropping the ion component, which leaves us with the Salpeter approximation to the spectrum, Eqn (5.3.1).

The resonance occurs approximately where $\epsilon_R = 1+\alpha^2\mathrm{Rw}(x_e) = 0$, and inspection of Fig. 5.1 shows that for $\alpha > 1.86$, there are two roots. However, at the lower frequency root, the Landau damping term is large, and the resonance is strongly damped. We are therefore left with the high-frequency root, which we will denote by $x_{e\upsilon} = \omega_\upsilon/ka \gg 1$. The phase velocity is much greater than the electron mean thermal speed; thus, the waves propagate up in the tail of the electron velocity distribution where $(\partial f_{eo}/\partial \mathrm{v})$ is small, i.e., where the Landau damping is small. If we neglect the effect of damping on the position of the resonance, then with the aid of Eqn (5.2.10) we find

$$\frac{1}{\alpha^2} \simeq \frac{1}{2x_{eo}^2} + \frac{3}{4x_{eo}^4}.$$

This may be rearranged to give the dispersion relation

$$\omega_{\text{epw}}^2 \simeq \omega_{pe}^2 + \frac{3\kappa T_e}{m_e} k^2, \tag{5.3.5}$$

which is the Bohm and Gross relation for longitudinal electron plasma waves. This may be substituted in Eqn (5.3.1), and noting that this is a sharp resonance, we obtain

$$S_e(\mathbf{k},\omega) \simeq \frac{\pi^{1/2}}{2ka} \frac{x_{eo}^2 \exp(-x_{eo}^2)}{[\{x_e - x_{eo}\}^2 + \{(\pi^{1/2}/2)\alpha^2 x_{eo}^2 \exp(-x_{eo}^2)\}^2]}. \tag{5.3.6}$$

5.3.5 Ion-Acoustic Resonance

When $\alpha \gtrsim \left(\frac{ZT_e}{3T_i} - 1\right)^{-1/2}$ the scattering spectrum can reveal a low-frequency feature associated with the ion-acoustic waves. This resonance occurs at approximately $\omega_{ac} \approx k[(Z\kappa T_e + 3\kappa T_i)/m_i]^{1/2}$. In the vicinity of this frequency, $x_e(\omega_{ac}) \simeq \omega_{ac}/ka \ll 1$, and as a consequence, $\mathrm{Rw}(x_e) \simeq 1$ and $\exp(-x_e^2) \simeq 1$; thus, the scattering spectrum contains contributions from both ion and electron components,

$$S_{ac}(\mathbf{k},\omega) \simeq \frac{2\pi^{1/2}}{ka} \left\{ \frac{[1 + \alpha^2(ZT_e/T_i)\mathrm{Rw}(x_i)]^2}{|\epsilon|^2} + Z\left(\frac{m_i T_e}{m_e T_i}\right)^{1/2} \frac{\alpha^4 \exp(-x_i^2)}{|\epsilon|^2} \right\}, \tag{5.3.7}$$

where

$$|\epsilon|^2 \simeq \left\{ [1 + \alpha^2 + \alpha^2(ZT_e/T_i)\mathrm{Rw}(x_i)]^2 + [\alpha^2 \mathrm{Iw}(x_e) + \alpha^2(ZT_e/T_i)\mathrm{Iw}(x_i)]^2 \right\}. \tag{5.3.8}$$

The dispersion relation for ion-acoustic waves is obtained by setting $\epsilon_{Re} = 0$. Inspection of Fig. 5.1 shows that this condition is satisfied only when $(T_i/ZT_e)[(1+\alpha^2)/\alpha^2] < 0.29$. This leads to the condition for observing the ion feature $\alpha \gtrsim \left(\frac{ZT_e}{3T_i} - 1\right)^{-1/2}$ which is only strictly satisfied for $ZT_e/T_i > 3$. For smaller values of ZT_e/T_i, the resonance is strongly damped. For $ZT_e \gg T_i$, we may use the approximation Eqn (5.2.10) for $\mathrm{Rw}(x_i)$ and neglect the imaginary part of the resonance $\mathrm{Iw}(x_i) \simeq 0$ to find ion-acoustic dispersion relation

$$\omega_{ac} \simeq \pm k \left(\frac{\alpha^2 Z\kappa T_e}{(1+\alpha^2) m_i} + \frac{3\kappa T_i}{m_i} \right)^{1/2}. \tag{5.3.9}$$

5.3.6 Relative Drift of Electrons and Ions

We have seen in the preceding sections that the development of resonant oscillations in a plasma is limited by electron and ion Landau damping. The level of damping is proportional to the gradient of the velocity distribution functions, taken at the phase velocity of the waves. The resonances are strongly affected by a resonant drift of the ions and electrons because this shifts the phase velocity of the waves to a different part of the distribution function (see Fig. 5.3).

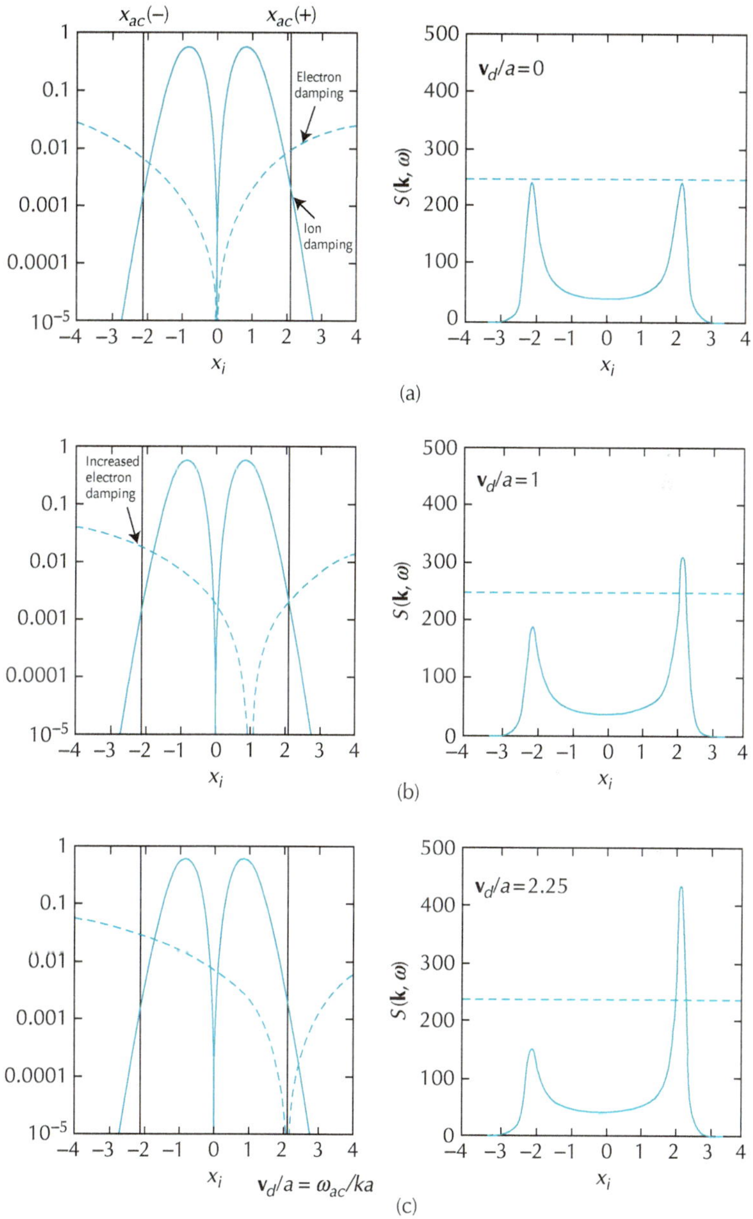

FIGURE 5.3 The electron (dashed) and ion (solid) contributions to the Landau damping are shown when there is (a) no relative drift between the electrons and ions ($x_d=0$), (b) a moderate drift ($x_d=1$), and (c) a drift equivalent to the ion-acoustic resonance ($x_d=x_{ac}\simeq 2.25$). The corresponding scattering spectrum is shown for each case where the average charge state is chosen to be $Z=6$ and $T_e/T_i=1$.

The effect of small electron drift velocities relative to the ions can be studied by shifting the electron velocity distribution function relative to the ion frame of reference. In this case, the Maxwellian distribution functions become

$$f_{e0}(\mathrm{v}) = \frac{\exp\left[-((\mathrm{v}-\mathrm{v_d})/a)^2\right]}{(\pi a^2)^{3/2}}, \quad f_{i0}(\mathrm{v}) = \frac{\exp\left[-(\mathrm{v}/b)^2\right]}{(\pi b^2)^{3/2}}. \tag{5.3.10}$$

Substituting into Eqn (5.1.2) yields

$$S(\mathbf{k},\omega) = \frac{2\pi^{1/2}}{ka}\left\{\frac{A_e}{|\epsilon|^2} + \frac{A_i}{|\epsilon|^2}\right\}, \tag{5.3.11}$$

where

$$A_e = \left[\left(1+\alpha^2\frac{ZT_e}{T_i}\mathrm{Rw}(x_i)\right)^2 + \left(1+\alpha^2\frac{ZT_e}{T_i}\mathrm{Iw}(x_i)\right)^2\right]\exp(-(x_e-x_d)^2)$$

$$A_i = Z\left(\frac{m_i T_e}{m_e T_i}\right)^{1/2}\left[\left(\alpha^2\mathrm{Rw}(x_e-x_d)\right)^2 + \left(\alpha^2\mathrm{Iw}(x_e-x_d)\right)^2\right]\exp(-x_i^2), \tag{5.3.12}$$

where $x_e = \omega/ka$, $x_d = \mathrm{v}_d/a\cos\chi$, and χ is the angle between $\mathbf{v}$ and the drift velocity $\mathbf{v_d}$.

$$|\epsilon|^2 = \left\{\left[1+\alpha^2\left(\mathrm{Rw}(x_e-x_d)+\frac{ZT_e}{T_i}\mathrm{Rw}(x_i)\right)\right]^2 + \left[\alpha^2\mathrm{Iw}(x_e-x_d)+\alpha^2\frac{ZT_e}{T_i}\mathrm{Iw}(x_i)\right]^2\right\}. \tag{5.3.13}$$

There are resonances when

$$1+\alpha^2[\mathrm{Rw}(x_e-x_d)+\frac{ZT_e}{T_i}\mathrm{Rw}(x_i)] \Rightarrow 0. \tag{5.3.14}$$

Figure 5.3 shows conditions where, in the absence of a drift, the resonant fluctuations at the ion-acoustic frequency are symmetric in amplitude indicating that the Landau damping at both the positive and negative roots are equal. For these conditions ($ZT_e/T_i = 6$), the ion acoustic Landau damping is small relative to the contribution from the electron Landau damping. Adding a drift reduces the damping on the positive root and increases the damping on the negative root by moving the resonant waves onto a shallower or steeper part of the distribution functions, respectively. For a normalized drift of $x_d = 1$, the electron and ion Landau damping on the positive root are nearly equal, and the electron Landau damping on the negative root has been significantly increased. For drift velocities larger than the ion-acoustic speed, $\mathrm{v}_d > (Z\kappa T_e/m_i)^{1/2}$, the electron damping on the positive root goes to zero, and if the growth of the electron wave is larger than the residual ion damping, the system will become unstable.

It must be emphasized here that our calculation is only valid for small drifts for which $\epsilon_{Im}(\omega_{Re})$ remains finite, and the level of the resonant fluctuations is small enough to justify our neglect of the nonlinear terms in the kinetic equations. These terms act not only on the fluctuations but also on the ensemble average conditions, and we have assumed these to be stationary.

It must also be remarked that Eqn (5.3.11) yields the spectrum in the **k** direction, but instabilities will occur whenever $\mathrm{Im}(\omega) < 0$, and this may not be in the direction of or at the particular **k** value we are studying (see Infeld *et al.*, 1972). Therefore, the use of these equations is valid only when the restriction to a small fluctuation level is satisfied in all directions. As a simple illustration of this, consider the case where **k** is perpendicular to v_d. Since $\cos\chi = 0$, Eqns (5.3.11) and (5.3.13) reduce to the standard drift free result, but of course, if v_d is large enough to cause instabilities, one would expect some enhanced fluctuations even in the perpendicular direction.

5.4 EXPERIMENTAL APPLICATIONS

A typical modern setup used for collective Thomson scattering measurements is discussed further in Section 7.8. Here, a typical Thomson scattering diagnostic focuses a probe beam through a plasma. The scattered light is collected and transported to a spectrometer that is coupled to either a streak camera, for temporally resolved spectra, or a charge-coupled device (CCD), for spatially resolved spectra. A spectrometer is chosen to have a wavelength range suitable for either measuring the electron plasma or the ion-acoustic waves. The following sections outline the possible measurements that can be made by Thomson scattering.

5.4.1 Experimental Geometry

The angle between the incident probe beam and the collection lens (θ) defines the direction of the fluctuations that are probed through the equations of momentum and energy conservation.

$$\mathbf{k} = \mathbf{k_s} - \mathbf{k_i},$$

$$\omega = \omega_s - \omega_i.$$

As the scattering takes place in the plasma, the magnitude of the wave vector being probed requires knowledge of the plasma parameters to solve the dispersion relation for the electromagnetic wave numbers

$$k_i = \left(\frac{\omega_i^2 - \omega_{pe}^2}{c^2}\right)^{1/2} \tag{5.4.1}$$

$$k_s = \left(\frac{\omega_s^2 - \omega_{pe}^2}{c^2}\right)^{1/2}. \tag{5.4.2}$$

In general, these equations can be used in concert with the full power spectrum to solve for a complete spectrum, but it is often convenient to obtain an estimate of the expected wavelength separation between the collective features by using the appropriate dispersion relation. For electron plasma waves, the dispersion relation when $\alpha \gtrsim 2$ is given by Eqn (5.3.5), and for ion-acoustic waves, the dispersion relation is given by Eqn (5.3.9).

When scattering from ion-acoustic waves, the scattered wavelength is approximately equal to the incident wavelength ($k_i \simeq k_s$) allowing the ion-acoustic wave vector to be given by the experimental geometry and wave number of the probe laser

$$k_{ia} \simeq 2k_i \sin\left(\frac{\theta}{2}\right). \tag{5.4.3}$$

To get an estimate of the expected scattered wavelength separation between the two ion-acoustic features ($\Delta\lambda_{ia}$) by substituting Eqn (5.4.3) into Eqn (5.3.9) and assuming $\omega_i \gg \omega_{pe}$,

$$\frac{\Delta\lambda_{ia}}{\lambda_i} \simeq \frac{4}{c}\sin\left(\frac{\theta}{2}\right)\sqrt{\frac{\kappa T_e}{m_{\mathrm{i}}}\left[\frac{Z}{\left(1+k_{ia}^2\lambda_{\mathrm{De}}^2\right)}+\frac{3T_i}{T_e}\right]}. \tag{5.4.4}$$

For 90° scattering, this leads to typical normalized peak-to-peak ion-acoustic separations of order $\Delta\lambda_{ia}/\lambda_i \sim 10^{-3}$.

For electron plasma waves, one can substitute Eqns (5.4.1) and (5.4.2) in the Bohm–Gross dispersion relation Eqn (5.3.5)

$$\begin{aligned}&\left(\frac{k_{\mathrm{epw}}}{k_i}\right)^4 - 4\cos\Phi\left(\frac{k_{\mathrm{epw}}}{k_i}\right)^3\\ &+4\left(\frac{k_{\mathrm{epw}}}{k_i}\right)^2\cos^2\Phi\left(1-\frac{1}{2\cos^2\Phi}\frac{\omega_p^2}{c^2k_i^2}-\frac{3}{\cos^2\Phi}\frac{a^2}{c^2}\right)\\ &+4\left(\frac{k_{\mathrm{epw}}}{k_i}\right)\frac{\omega_p^2}{c^2k_i^2}\cos\Phi - 4\frac{\omega_p^2}{c^2k_i^2}\left[1+\frac{3}{4}\frac{\omega_p^2}{c^2k_i^2}\right]=0,\end{aligned} \tag{5.4.5}$$

where Φ is the angle between $\mathbf{k}_{\mathrm{epw}}$ and $\mathbf{k_i}$. For a scattering angle $\theta = 90°$, $\cos\Phi = -k_i/k_{\mathrm{epw}} = -\lambda_i/\Delta\lambda_{\mathrm{epw}}$, and assuming $\frac{\omega_p^2}{c^2k_i^2} \sim \frac{n}{n_{cr}}$, the wavelength shift between the two scattered features from electron plasma waves can be approximated as

$$\frac{\Delta\lambda_{\mathrm{epw}}}{\lambda_i} \approx 2\left[\frac{n}{n_{cr}}+3\frac{a^2}{c^2}\right]^{1/2}\left(1+\frac{3}{2}\frac{n}{n_{cr}}\right). \tag{5.4.6}$$

This is in good agreement with the scattering results shown in Fig. 5.4a, where a $\frac{\Delta\lambda_{\mathrm{epw}}}{\lambda_i} \sim 0.18$ is estimated for a density of $n/n_{cr} = 0.005$.

5.4.2 Electron Plasma Waves

The electron feature contains the ability to accurately measure the electron density and in some regimes the electron and ion temperatures:

1. The wavelength shift $\frac{\Delta\lambda_{\mathrm{epw}}}{\lambda_i} \propto \omega_{pe}$ [see Eqn (5.4.6)], which depends, to first order, directly on n_{e}.
2. The absolute amplitude and width of the satellites is governed by the damping of the waves. In the collisionless equilibrium case, this is Landau damping, which depends on n_{e} and T_e.
3. For $ZT_e \simeq T_i$, the central part of the spectrum comes from the ion term, and the ratio of the cross section of one satellite to the central feature is from Eqn (5.6.3) given by

$$\frac{S_e(\mathbf{k})}{S_i(\mathbf{k})} = \frac{1+\alpha^2[1+(ZT_e)/T_i]}{2Z\alpha^4}, \tag{5.4.7}$$

where

$$\alpha^2 = \frac{1.17\times 10^{-8}[\lambda_i(\mathrm{cm})]^2 n_{\mathrm{e}}(\mathrm{cm}^{-3})}{\sin^2(\theta/2)T_e(\mathrm{eV})}. \tag{5.4.8}$$

Therefore, if T_e is known, we can estimate T_i. For a review of scattering measurements from the electron plasma waves, see Peacock *et al.*, 1969; Alladio *et al.*, 1977; Landen *et al.*, 1985; Behn *et al.*, 1989; Snyder *et al.*, 1994; Glenzer *et al.*, 1997; Pasqualotto *et al.*, 1999; Wrubel *et al.*, 2000; Gregori *et al.*, 2002; Dzierzega *et al.*, 2006; and Ross *et al.*, 2010.

Figure 5.4 demonstrates Thomson scattering spectra that span the noncollective (see Chapter 4) to collective regimes. Here, it is evident that when $\alpha < 1$, the scattering spectrum begins to reveal the electron distribution function. Increasing the scattering parameter above unity reveals the aforementioned electron plasma wave features where an accurate measure of both the electron temperature and density was made (Ross *et al.*, 2010).

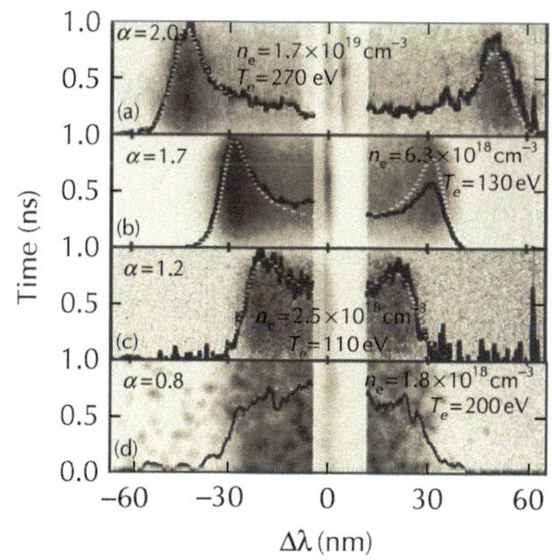

FIGURE 5.4 The measured time resolved scattering spectrum is shown as a function of the scattering parameter ($\lambda_i = 0.53\,\mu\mathrm{m}, \theta = 90°$). The spectra at 0.5 ns are plotted (black lines) with the corresponding fit using Eqn (5.3.7). (a) $\alpha = 2$, (b) $\alpha = 1.7$, (c) $\alpha = 1.2$, (d) $\alpha = 0.8$. (Courtesy of J. S. Ross.)

Electron Density and Temperature Measurement from the Electron Plasmas Resonance

The sensitivity of the electron feature to the electron temperature and density is shown in Fig. 5.5. Increasing the electron density increases the separation between the shifted peaks due to the increase in the electron plasma wave frequency, while increasing the electron temperature tends to increase the width of the resultant peaks due to the increase in Landau damping. The uncertainty in the measurement of electron density and temperature is determined by the extent in which the electron density and temperature can be varied before the fit falls outside the noise. The uncertainty in the temperature and density measurements are better than 15% in the collective regime for the measured spectra shown in Fig. 5.5. Furthermore, the calculated spectrum is unique; changing the temperature and the density simultaneously does not reproduce a given spectrum.

Asymmetric Electron Plasma Features

Figure 5.6 shows temporally resolved collective Thomson scattering from electron plasma waves for various phase velocities. The scattered power calculated to zero order in beta (nonrelativistic), see Section 2.3.3, does not fit the measurements, while including first order in beta relativistic corrections to the scattered power Eqn (5.1.1) agrees well with the measured spectra over a range of phase

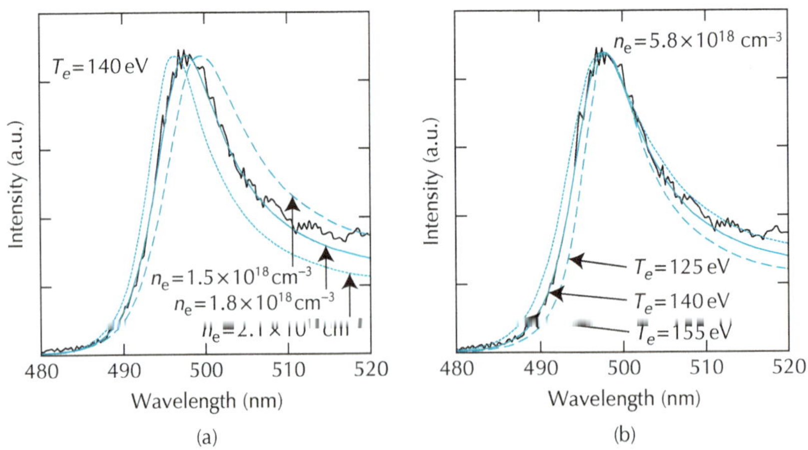

FIGURE 5.5 By varying the electron temperature and density, the sensitivity of the fit can be determined. The best fit to the experimental data (solid blue curves) is shown in solid black (T_e = 140 eV, n_e = 5.8 × 10^{18} cm^{-3}). In plot (a), n_e is increased by 15% (dotted curve) and decreased by 15% (dashed curve), and the electron temperature is held constant. In plot (b), the density is held constant, and the electron temperature is increased by 10% (dashed curve) and decreased by 10% (dashed curve). (Courtesy of J. S. Ross.)

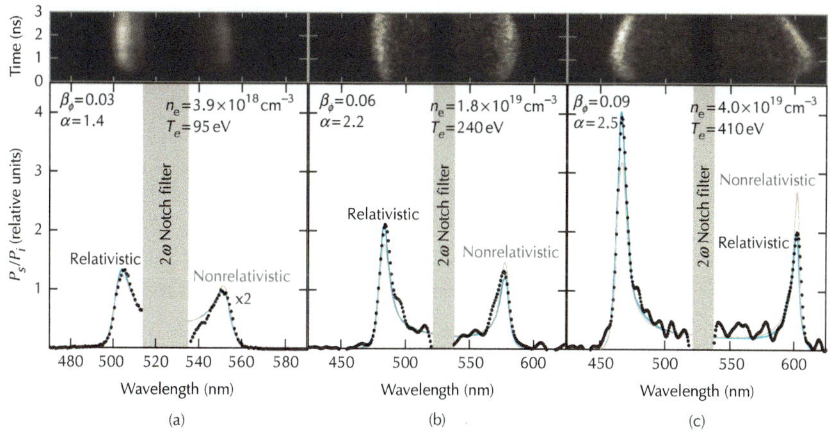

FIGURE 5.6 Top panels: raw streak data. Bottom panels: experimental data (dots) at 2 ns normalized to the blue-shifted feature and compared to the nonrelativistic (gray line) and relativistic (blue line) form factors. As the temperature and density increase, the phase velocity of the electron plasma wave increases (a) $\beta_\phi = 0.03$, (b) $\beta_\phi = 0.06$, (c) $\beta_\phi = 0.09$, and the difference between the nonrelativistic and relativistic form factors becomes more pronounced. (Courtesy of J. S. Ross (2010).)

velocities even at these low-temperature conditions ($v_{osc}/c = eE_0/m_e^2 \ll 1$ and $T_e < 1$ keV). When calculating this spectrum, first order v/c corrections must be included as a consequence of scattering from the fast electrons that make up the electron plasma wave which propagate with a velocity near the wave's phase velocity (Ross *et al.*, 2010).

In low-temperature plasmas, relativistic effects are observed in the asymmetric scattering spectrum and are attributed to two effects, corresponding to first order in beta corrections to the calculated scattered spectrum.[1] The effects of these corrections can be estimated by taking the ratio of the peak power in the blue- and red-shifted electron plasma wave resonances

$$\frac{P^{\text{blue}}}{P^{\text{red}}} \approx \underbrace{\frac{P_{NR}^{\text{blue}}}{P_{NR}^{\text{red}}}}_{A} \underbrace{\left(\frac{1+\beta_\phi \cos\phi}{1-\beta_\phi \cos\phi}\right)^2}_{B} \underbrace{\left(\frac{1+\beta_\phi \cos\Phi}{1-\beta_\phi \cos\Phi}\right)^2}_{C}, \tag{5.4.9}$$

where β_ϕ is the normalized phase velocity, ϕ is the angle between $\hat{k}$ and $\hat{k}_s$, Φ is the angle between $\hat{k}$ and $\hat{k}_i$, and $P_{NR}^{\text{blue}}/P_{NR}^{\text{red}}$ is the ratio of the peak powers in the blue- and red-shifted peaks, respectively, calculated using zero order in beta (nonrelativistic) the power scattered given by Eqn (2.3.13).

[1]Note that these are the same effects that are discussed for Section 2.3.3 and applied to the high-temperature noncollective experiments in Section 4.5.4.

The asymmetry in term A is a result of the difference in magnitude of the red- and blue-shifted resonant wave vectors, which is approximately $2\sqrt{\frac{n}{n_{cr}}} > 10\%$. When Landau damping is small ($\alpha \gg 1$), modest differences in the plasma wave vectors result in large relative changes in the damping of the plasma waves. Term B is a result of relativistic aberration, also referred to as the relativistic headlight effect, where light is preferentially directed in the emitter's direction of propagation. Term C is a result of the initial electron motion interacting with the magnetic field.

The electrons moving in the direction of the plasma wave and near the phase velocity of the plasma wave ($\mathbf{v}_\phi$) interact with the magnetic field of the Thomson scattering probe laser. The resulting $\mathbf{v}_\phi \times \mathbf{B}_i$ force is parallel or anti-parallel to the force induced by the electric field of the laser depending on the direction of $\mathbf{v}_\phi$ with respect to the incident laser beam ($\mathbf{k}_i$). Therefore, the scattered power into the blue-shifted feature is enhanced and the scattered power into the red-shifted feature is reduced. Terms B and C can be reduced to $(\omega_i + 2\omega_{\text{red}})/(\omega_i + 2\omega_{\text{blue}})$, which is the first-order correction to the non-relativistic scattered power derived in Section 2.3.3. This result highlights the importance of including relativistic corrections to the scattered power in the collective regime, even in low-temperature plasmas, as the electrons contributing to the collective features are moving near the phase velocity of the electron plasma waves, which can be a significant fraction of the speed of light.

For small Landau damping ($\alpha \gtrsim 2$), Term A has a small effect $\frac{P_{NR}^{\text{blue}}}{P_{NR}^{\text{red}}} \approx 1$, and one can estimate the correction to the intensities by substituting the Bohm–Gross frequency given by Eqn (5.3.5) into Eqn (5.1.1),

$$\frac{P^{\text{blue}}}{P^{\text{red}}} \approx \frac{\left(1+\frac{2\omega_{\text{epw}}}{\omega_i}\right)}{\left(1-\frac{2\omega_{\text{epw}}}{\omega_i}\right)} \approx 1+4\frac{\left(\omega_{pe}^2+(2\kappa T_e/m_e)k^2\right)^{1/2}}{\omega_i}. \tag{5.4.10}$$

Figure 5.7 compares the various approximations and shows that the first order in beta correction, Eqn (5.4.9), is in good agreement with experiments, and Eqn (5.4.10) reproduces the results to within 10%.

Effect of Non-Maxwellian Velocity Distribution Function

Non-Maxwellian electron velocity distribution functions primarily affect the damping of the electron plasma waves and to a lesser degree the ion-acoustic waves through a variation in the sound speed of the plasma (Afeyan *et al.*, 1998). The effects on the amplitude of the electron plasma wave are evident when writing Eqn (5.3.6) for an arbitrary, but stable, distribution

$$S_e(\mathbf{k},\omega) \simeq \frac{\pi}{2k} \frac{\omega_o^2 f_{eo}(\omega_o/k)}{\left[\{\omega-\omega_o\}^2+\{(\pi\omega_o\omega_{pe}^2/2k)(\partial f_{eo}(\mathrm{v})/\partial \mathrm{v}|_{\mathrm{v}=\omega/k})\}^2\right]}. \tag{5.4.11}$$

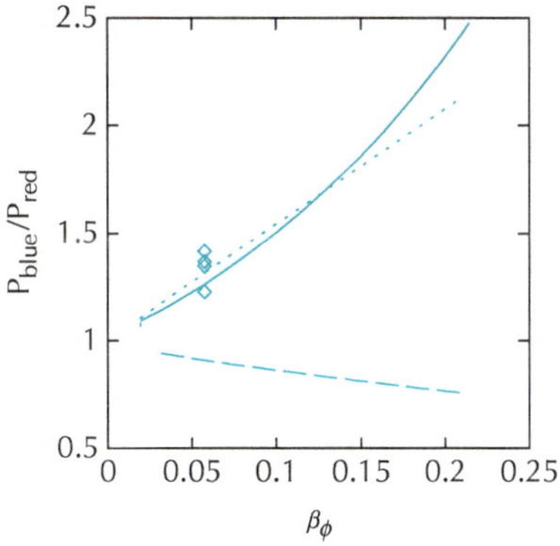

FIGURE 5.7 The ratio of the peak power scattered into the blue-shifted peak over the red-shifted peak is plotted as a function of the phase velocity normalized to the speed of light, β_ϕ. The scattering parameter is held constant ($\alpha = 2$), while the density and temperature are scaled from 2 to $150 \times 10^{18}\,\mathrm{cm}^{-3}$ and 30 to 3000 eV, respectively. The zero order in beta calculation [Eqn (2.3.13)] is shown as a dashed line, the first order in beta calculation [Eqn (5.4.9)] is shown as a solid line, and the approximate calculation [Eqn (5.4.10)] is shown as a dotted line. These calculated results are compared to experimental measurements (diamonds). (Courtesy of Ross *et al.* (2010).)

This shows how the level of Landau damping depends on the gradient of the distribution function taken at the phase velocity of the waves. A more complete derivation and discussion is made by Zheng *et al.* (1997) where they point out that if the velocity distribution function deviates from a Maxwellian and is unknown, the plasma parameters obtained from Thomson scattering by fitting the data are not unique.

Experiments by Glenzer *et al.* (2000b) have used simultaneous measurements of the electron plasma and ion-acoustic waves (see Fig. 5.8) to show that, for their plasma conditions, a Maxwellian electron velocity distribution function is a best fit to the data.

5.4.3 Ion-Acoustic Waves

Experimentally, the ion-acoustic features contain the ability to accurately measure the electron and ion temperatures, electron density, plasma flow, and the relative electron ion drive velocity.

1. The wavelength shift $\frac{\Delta\lambda}{\lambda_i} \propto \sqrt{\frac{Z\kappa T_e}{m_i}}$ [see Eqn (5.4.4)] provides an accurate measure of the electron temperature in plasmas where the ionization state is known and the ion temperature is either small ($3T_i < ZT_e$) or known.
2. An expanding plasma Doppler shifts the scattered frequency, $\Delta\omega = \omega_a + \mathbf{k} \cdot \mathbf{v}_f$.
3. In a collisionless plasma, the width of the resonance is given by Landau damping and is therefore a function of ZT_e/T_i. Therefore, when both the upshifted and downshifted resonances are resolved, ZT_e is obtained from the frequency separation, and T_i is given by the width of the resonances.

4. In plasmas that have multiple ion species, multiple resonances appear in the scattered spectra as multiple features that depend uniquely on ZT_e/T_i. Therefore, the ratio of the power scattered into these features can provide an accurate measure of the ion temperature.
5. The electron density can be determined by absolute calibration of the scattered power $P(\omega, \mathbf{k}) \propto n_e$, and when absolute calibration is not possible, multiple Thomson scattering diagnostics can be used (see Section 8.7).
6. The relative electron ion drift can be measured, see Section 5.3.6, which can be related to heat transport, see e.g., Hawreliak *et al.*, 2004; Froula *et al.*, 2007a; Moody *et al.*, 2003; Gregori *et al.*, 2004a.

Electron Temperature

Figure 5.8 shows Thomson scattering data from a Au coronal plasma using a $\lambda_i = 0.53\,\mu$m probe laser (Glenzer *et al.*, 2000b). For these experiments, the probe beam was focused to a 170 μm diameter focal spot and aligned parallel to a Au disk target. In addition to a high-resolution 1-m spectrometer, which was used to measure the ion feature, a 1/4-m spectrometer was used to observe the electron feature in the wavelength range of 550–800 nm. Optical streak cameras were employed to record the scattering spectra. Both spectrometers collected light from the same volume in the plasma.

The simultaneous observation of the ion-acoustic wave and the electron plasma wave spectra offers a unique way to obtain the electron temperature,

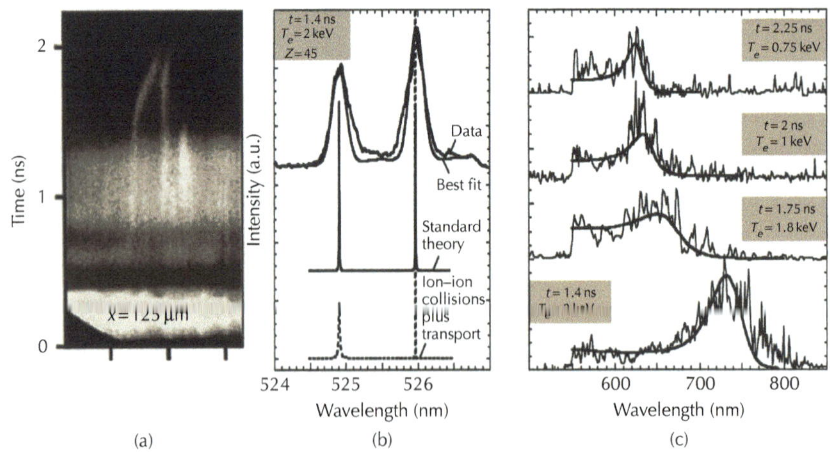

FIGURE 5.8 (a) The ion-acoustic spectrum measured from a Au coronal plasma is shown as a function of time. Laser beams (0.351 μm) with 3.8 kJ of total energy are used to heat the solid target for 1.5 ns. (b) The scattering spectrum at 1.4 ns is fit using the form factor [Eqn (5.1.2)] to determine the average ionization state ($Z = 45$). (c) The electron temperature and density are determined by fitting the form factor to the simulations measurement of the electron feature. (Courtesy of S. H. Glenzer *et al.* (2000b).)

electron density, and average ionization state. Here, they first determine the electron temperature and density from the electron plasma wave spectrum, as discussed in Section 5.4.2, and then use these values to determine the average ionization state from the ion-acoustic spectrum.

The full Thomson scattering spectrum is self-consistently fit using Eqn (5.1.2) assuming Maxwellian distribution functions. The ion-acoustic peaks are significantly broadened due to the inhomogeneity of the plasma in the scattering volume. The asymmetry in the ion-acoustic features is likely a result of the heat flux and the corresponding shift of the electron distribution function (see Section 5.3.6). Accounting for this asymmetry, the broadening from the instrument (0.05 nm), and the spatial gradients in temperature and plasma flow allows a good fit of the entire spectrum. The electron temperature was given by the electron feature with an error of about 15% determined by varying the calculated spectra within the noise of the data. Therefore, the average ionization state deduced from the ion feature was determined with an error of 20%.

Ion Temperature

In a collisionless plasma, the width of the ion-acoustic resonances is given by Landau damping, and a measurement of the ion-acoustic feature can be used to infer the ion temperature. But when the plasma is either collisional or the plasma gradients (T_e, v_f) within the Thomson scattering volume are significant, a multiple ion-species plasma can be used.

Ion-Acoustic Spectral Width

The ion temperature was first measured in a theta pinch by Ascoli-Bartoli *et al.* (1964), and more recently, (Behn *et al.*, 1989) use a D_2O laser, emitting 0.5 J in 1.4 μs at 385 μm to measure ion temperature in the TCA tokamak. Schottky-barrier-diode mixers and heterodyne detection were used to obtain a noise equivalent power down to 10^{-19} W/Hz and an IF bandwidth in the gigahertz range. The experimental arrangement is shown in Fig. 5.9, and the measured spectrum for a helium plasma is compared with theory in Fig. 5.10.

Multiple-Ion Species Plasmas

The effect of adding multiple species to the plasma dynamics has been studied in detail by several people over the last 30 years (see, e.g., Fried *et al.*, 1971; Williams *et al.*, 1995) and more recently used to obtain accurate ion-temperature measurements (see, e.g., Glenzer *et al.*, 1996; Froula, 2002b; Froula *et al.*, 2006b). Adding a second ion species to the plasma with a significantly different charge state introduces a second mode in the solution to the kinetic dispersion relation. The damping of these modes react differently when the ion temperature is changed because of their different phase velocities which probe different points of the distribution functions. Therefore, the relative damping of the modes provides an accurate ion temperature measurement, which is observed in the scattering amplitude of the ion-acoustic features.

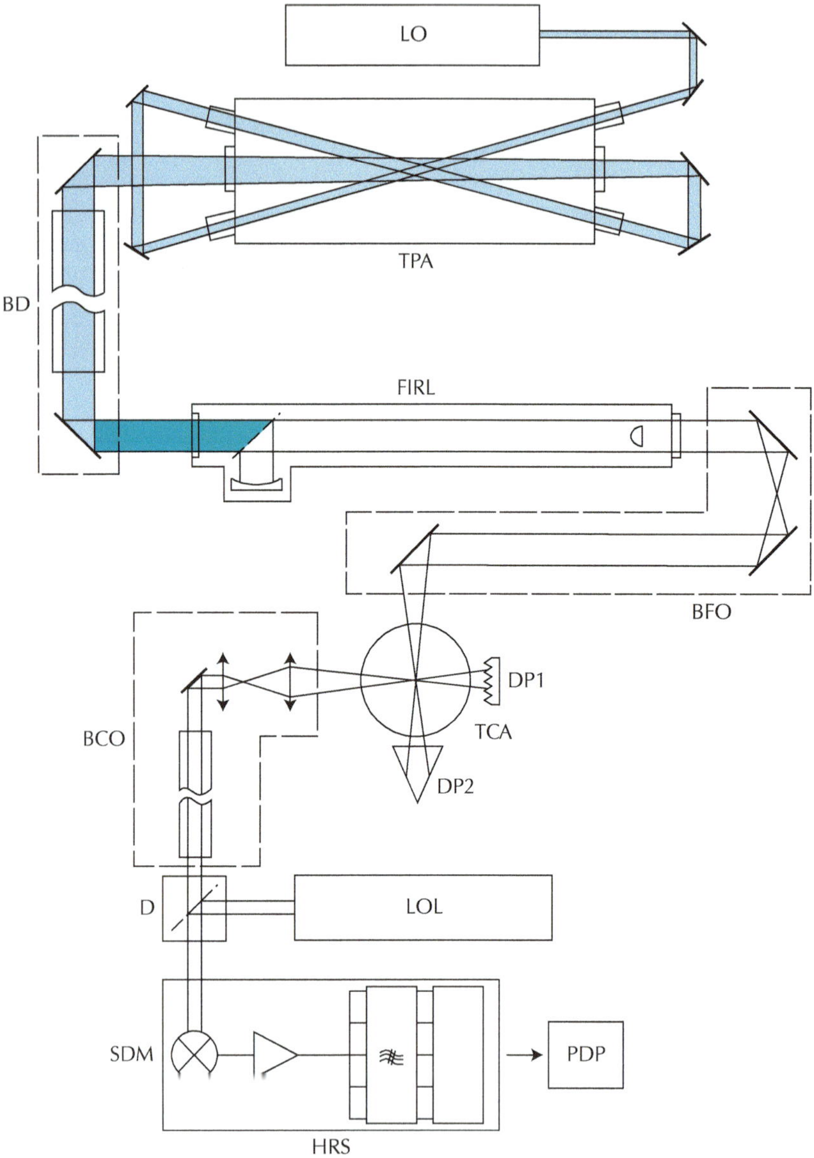

FIGURE 5.9 Configuration of the experimental system. LO: hybrid TEA CO_2 laser oscillator; TPA: triple-pass CO_2 laser amplifier, e-beam preionized; BD: beam duct, 70 m long; FIRL: D_2O FIR laser; BFO: beam-focusing optics; TCA: tockamak; DP1: viewing dump, Macor ceramic; DP2: beam dump, Pyrex cone; BCO: beam collection optics with 7-m long, dry-nitrogen-filled beam duct; D: diplexer; LOL: local oscillator laser; SDM: Schottky-diode mixer; HRS: heterodyne receiver system. (Courtesy of R. Behn (1989) and the American Physical Society.)

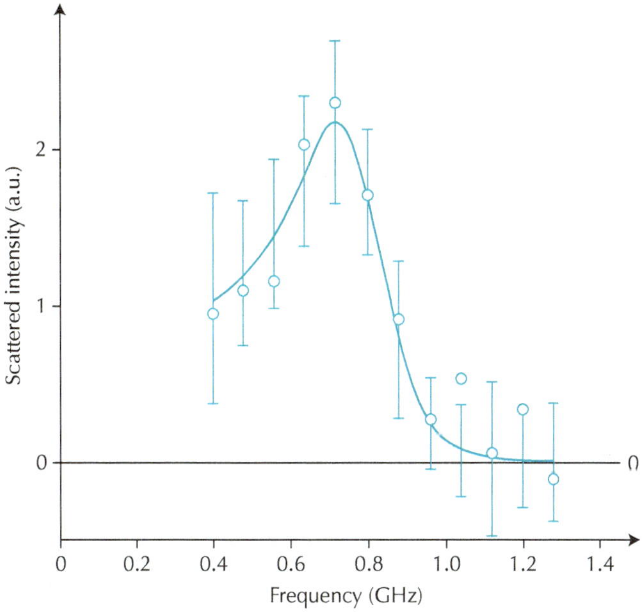

FIGURE 5.10 Measured spectrum for a He plasma in TCA. The solid curve is a least-squares-fit resulting in $n_e = 7 \times 10^{13}\,\text{cm}^{-3}$, $T_e = 670\,\text{eV}$, angle $(\mathbf{k}, \mathbf{B}) = 86°$, and $Z_{eff} = 4.4$. The fit yields $T_i = 250\,\text{eV}$. (Courtesy of R. Behn (1989) and the American Physical Society.)

The scattering spectrum shown in Fig. 5.11 demonstrates the change in relative damping of the two modes in a multi-ion carbon–hydrogen plasma. Early in time, the hydrogen-like mode is dominant; in the series of spectra that is shown in Figs 5.11b–5.11f, the carbon-like mode grows in time indicating an increase in the ratio between the ion and electron temperatures. As the ion temperature equilibrates with the laser-heated electrons, the carbon-like mode becomes evident and is dominant by the end of the scattering experiment; with increasing ion temperature, the width of the ion distribution begins to widen and effectively damp the fast (hydrogen-like) mode, while the distribution flattens at low velocities so that the slow (carbon-like) mode is less damped. Fitting the data with the Thomson scattering form factor for a two-ion-species plasma accurately measures the electron temperature and ion temperature in the carbon-hydrogen plasmas.

It should be noted that Liu *et al.* (2002) pointed out that the influence of super-Gaussian velocity distributions can affect the damping of the two ion-acoustic waves. In this case, the inferred plasma parameters such as electron and ion temperatures may suffer some errors. They point out that if an independent measure of the plasma parameters is made, a plasma with two-ion species could be used to infer super-Gaussian velocity distributions.

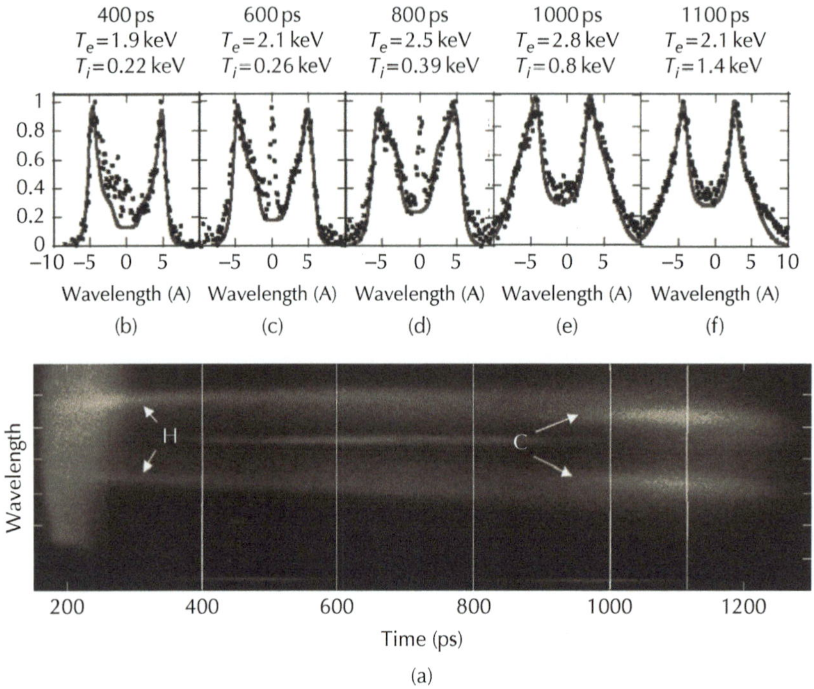

FIGURE 5.11 The time-resolved Thomson scattering spectrum shows a heavily damped carbon-like mode and a weakly damped hydrogen-like mode early in time. The relative damping of the modes is reversed at ~ 1 ns. The spectra are fit to determine T_e, T_i, and n_e at selected times. (b) 400 ps, (c) 600 ps, (d) 800 ps, (e) 1000 ps, (f) 1100 ps. (Courtesy of Froula *et al.* (2006b).)

5.5 COLLISIONAL PLASMAS

5.5.1 Electron Plasma Resonance

Again the Salpeter approximation may be used, because our interest still centers on the high-frequency regime in which the ions cannot respond, and from Eqns (3.7.11), (3.7.12), (3.7.15), and (3.7.16), we see that

$$S_e(\mathbf{k}, \omega, \nu_e) = \frac{2}{ka} \frac{\mathrm{Im}\left[i\pi^{1/2} \exp(-y_e^2) + 2\exp(-y_e^2) \int_0^{y_e} \exp(p^2)\, dp \right] - \dfrac{2ka}{\nu_e} |D_e|^2}{\left[1 + D_e + \alpha^2(\mathrm{Rw}(y_e) - i\,\mathrm{Iw}(y_e))\right]^2}, \tag{5.5.1}$$

where $y_e = (\omega - i\nu_e)/ka$, and since $\omega \cong \omega_{pe}$, we can expect to have $\nu_e \ll \omega$ for most plasmas of interest.

For $y_e = (\omega - i\nu_e)/ka \gg 1, \alpha^2 \gg 1$, and $\psi_e = \nu_e/ka \ll 1$, we may then use Eqns (3.7.17) and (3.7.18) with the expansion of Eqn (5.2.10) for Rw(y) and

$$D_e \cong \frac{i\nu_e}{\omega}\left(1 + \frac{1}{2y_e^2}\right), \quad \mathrm{Rw}(y_e) \cong -\frac{1}{2y_e^2} - \frac{i\nu_e}{y_e^2\omega} - \frac{3}{4y_e^4}$$

$$S_e(\mathbf{k}, \omega, \nu_e) \cong \frac{x_{e0}^2\left(\pi^{1/2}\exp(-x_{e0}^2) + \frac{\psi_e}{2x_{e0}^4}\right)}{2ka\left\{(y_e - x_{e0})^2 + \frac{x_{e0}^2\alpha^4}{4}\left(\frac{\psi_e}{2x_{e0}^3} + \pi^{1/2}x_{e0}\exp(-x_{e0}^2)\right)^2\right\}}. \tag{5.5.2}$$

where $x_{e0} = \omega_0/ka \gg 1$ is the high-frequency root (see Section 5.3.4). In terms of an arbitrary distribution function,

$$S_e(\mathbf{k}, \omega, \nu_e) \cong \frac{\pi}{2k}\frac{\omega_0^2\left\{f_{e0}\left(\frac{\omega}{k}\right) + \frac{k^3\kappa T_e\nu}{\pi m_e\omega_0^4}\right\}}{\left\{(\omega - \omega_0)^2 + \frac{\omega_{pe}^4}{\omega_0^4}\left(\frac{\nu_e}{\nu} - \frac{\pi\omega_0^3}{2k^2}\left[\frac{\partial}{\partial v}f_{e0}(\mathbf{v})\right]_{v=\omega_0/k}\right)^2\right\}}. \tag{5.5.3}$$

For these electron plasma frequency features, we may take ν_e to represent either electron–ion collision or electron–neutral collisions, and we may use for ν_{ei} and ν_{en} the values given by Eqns (2.2.7) and (2.2.9). Spectra with this form were obtained for a fully ionized plasma by Ron & Tzoar (1963), DuBois & Gilinsky (1964), and Perkins & Salpeter (1965). Their results differ by small numerical factors owing to the use of different approximations in the collision term (see Boyd, 1965). For a partially ionized plasma, the form of spectrum has been obtained for $T_e \cong T_i$ by Dougherty & Farley (1960).

The resonance occurs at $x_{e0}^2 \cong (\alpha^2/2) + \frac{3}{2}$ for $\alpha \gtrsim 4$, and if we denote the Landau damping coefficient by γ_L, then for a Maxwellian distribution

$$\gamma_L \cong (\pi/2)^{1/2}\omega_{pe}\,\alpha x_{e0}^2 \exp\left(-x_{e0}^2\right). \tag{5.5.4}$$

This may be compared with the collision frequencies (see Section 2.2.2)

$$\nu_{ei} = (2/\pi)^{1/2}\omega_{pe}\ln\Lambda/\Lambda \quad \text{where} \quad \Lambda = 12\pi n\lambda_{De}^3 \tag{5.5.5}$$

and

$$\nu_{en} = 2.8 \times 10^8[r_n(\mathrm{cm})]^2 n_n(\mathrm{cm}^{-3})[T_e(\mathrm{eV})]^{1/2}\,\mathrm{s}^{-1}. \tag{5.5.6}$$

For large α, i.e., $\lambda \gg \lambda_{De}$, we may move into the region where the wavelength we are probing is on the order of the distance between collisions ($\lambda \to \lambda_{coll}$) and the collisional damping becomes comparable with Landau damping in a fully ionized plasma.

5.5.2 Ion-Acoustic Resonance

The case of charged particle collisions is discussed by DuBois & Gilinsky (1964). Simple analytic results are obtained only in the limiting cases of

$$1 \gg \omega_{\mathrm{iac}}/\nu_{\mathrm{ii}} \gg (m_{\mathrm{e}}/m_{\mathrm{i}})^{1/2} \quad \text{and} \quad \omega_{\mathrm{ac}}/\nu_{\mathrm{ii}} \ll (m_{\mathrm{e}}/m_{\mathrm{i}})^{1/2}.$$

For the case of a partially ionized gas, when charge–neutral collisions are important, the spectrum is given by Eqns (3.7.11) and (3.7.12) and for the case of Maxwellian velocity distributions by the functions in Section 3.7.2. The case of $T_{\mathrm{e}} \cong T_{\mathrm{i}}$ has been treated by Dougherty & Farley (1963); in this region, the second term in Eqn (3.7.11) predominates, and the electron Landau damping is small. If we also neglect the electron collision terms, the spectrum is given by

$$S_{\mathrm{i}}(\mathbf{k}, \omega, \nu_{\mathrm{i}}) \cong \frac{2Z}{kb} \frac{\alpha^4 \left[\pi^{1/2} \exp(-y_{\mathrm{i}}^2) + (k^3 b^3 \nu_{\mathrm{i}}/2\omega^4)\right]}{\left[(1+\alpha^2)(1+D_{\mathrm{i}}) + \alpha^2 (ZT_{\mathrm{e}}/T_{\mathrm{i}})(\mathrm{Rw}(y_{\mathrm{i}}) - i\,\mathrm{Iw}(y_{\mathrm{i}}))\right]^2}, \quad (5.5.7)$$

$$y_{\mathrm{i}} = (\omega - i\nu_{\mathrm{in}})/kb.$$

As the ion–neutral collision frequency ν_{in} is increased, the ion-acoustic peaks become smaller and then disappear, and we then see the Doppler-shifted frequency resulting from the ion thermal motion. The scattering is from the electron Debye shields surrounding each ion; as the collision frequency is increased, the ions change direction more and more often, and the result is a narrowing of the spectrum. Eventually, the incident wave "sees" the ions only at their diffusion velocity; this occurs when $x_{\mathrm{i}} \gg 1$ or $\lambda_{\mathrm{imfp}} < 2\pi/k$. This is illustrated in Fig. 5.12, where spectra computed from the full expression for the spectrum are shown. At very high collision frequencies, the spectrum is similar to the collisional electron feature. The case $T_{\mathrm{e}} \gg T_{\mathrm{i}}$ is rather complicated, but one may note that for large x_{i}, when $x_{\mathrm{e}} \ll 1$ where electron Landau damping may be neglected, the result (5.5.7) still holds. Thus, for this region, we have from Section 3.8

$$S_{\mathrm{i}}(\mathbf{k}, \omega) \cong -\frac{2T_{\mathrm{i}}}{T_{\mathrm{e}} \alpha^2 \omega} \mathrm{Im}\left(\frac{1}{1+\alpha^2+C_{\mathrm{i}}}\right) \quad (5.5.8)$$

and

$$\left|1+\alpha^2+C_{\mathrm{i}}\right|^2 \cong \left[1+\alpha^2\left(1-\frac{ZT_{\mathrm{e}}}{2x_{\mathrm{i}}^2 T_{\mathrm{i}}}\right)\right]^2 + \left[\frac{\psi_{\mathrm{i}}}{2x_{\mathrm{i}}^3}\alpha^2 \frac{ZT_{\mathrm{e}}}{T_{\mathrm{i}}}\right]^2,$$

$$\psi_{\mathrm{i}} = \nu_{\mathrm{in}}/kb.$$

$S(\mathbf{k}, \omega)$ from the fluctuation-dissipation theorem is given in Eqn (3.8.1). Zheng *et al.* (1999) have derived the ion and electron susceptibilities [C_i and C_e in Eqn (3.8.1)] for the case of the collisional ion-acoustic region, using the Chang–Callen 13-moment approach. In the collisional limit, where the ion-ion mean free path (λ_{ii}) is much smaller than the scale that is being probed ($k\lambda_{ii} \ll 1$), and two low-frequency modes may exist in plasmas: ion-acoustic waves and

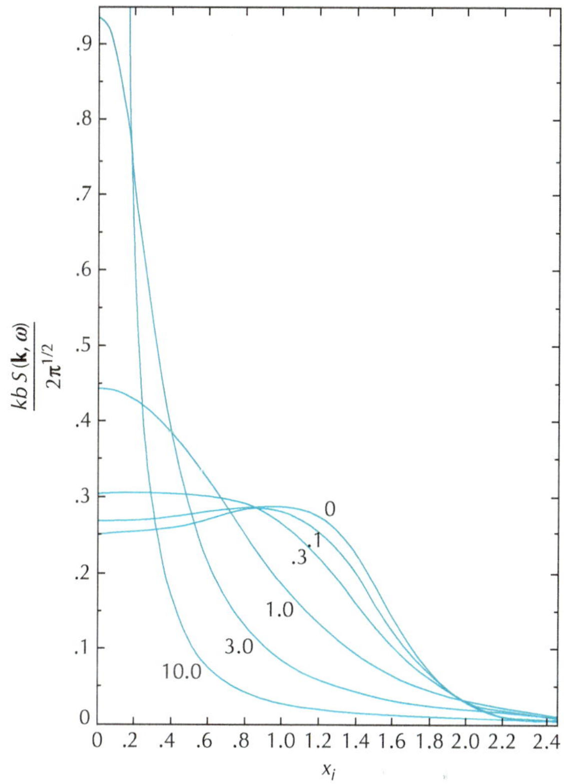

FIGURE 5.12 The effect of increasing collision frequency ψ_i, $\psi_e = \psi_i/10$, on the noncollective spectrum for $T_e = T_i$. (Courtesy of J. P. Dougherty & D. T. Farley (1963)).

entropy waves. In their solution, electron collisions are neglected, and the effect of ion–ion collisions is considered where C_e is given by Eqn (5.2.5), and C_i takes the following values, where $x_i - -i\nu/kb$ where ν is the total damping.

Collisional Limit, $k\lambda_{ii} \ll 1$

Solving for the ion susceptibility of the entropy mode in the collisional limit gives

$$C_i = \frac{ZT_e}{T_i}\alpha^2\left[1 + \frac{2}{5}\frac{s\frac{\nu/kv_i}{k\lambda_{ii}}}{1 - s\frac{\nu/kv_i}{k\lambda_{ii}}}\right], \tag{5.5.9}$$

where $s = 12[5\ln(1+\sqrt{2}) - 3\sqrt{2}]/\sqrt{2}$ and

$$\frac{\nu_{ent}}{kb} = \frac{5}{2s}k\lambda_{ii}\frac{2}{3}\frac{1+ZT_e/T_i}{5/3+ZT_e/T_i}. \tag{5.5.10}$$

In the case of the ion-acoustic mode, the ion susceptibility is given by

$$C_i = \frac{ZT_e}{T_i}\alpha^2\left(-\frac{1}{x_i^2} - \frac{5}{3x_i^4} + i\frac{10}{9}\frac{k\lambda_{ii}}{x_i^3}\right). \tag{5.5.11}$$

Intermediate Regime, $k\lambda_{ii}x_i \sim 1$

$$C_i \simeq \frac{ZT_i}{T_i}\alpha^2\left(-\frac{1}{x_i^2} - \frac{11}{5}\frac{1}{x_i^4} + i\frac{16}{25}\frac{1}{k\lambda_{ii}x_i^5} + i\sqrt{\frac{\pi}{2}}x_i e^{-x_i^2/2}\right), \tag{5.5.12}$$

where

$$\frac{\nu_{ia}}{kb} = \sqrt{\frac{\pi}{8}\frac{Zm_e}{m_i}}\left[1+\Gamma\frac{T_i}{ZT_e}\right]^{-3/2}\frac{\omega_{ia}}{kb} + \frac{8}{25}\frac{1}{k\lambda_{ii}(\omega_{ia}/kb)^2} + \sqrt{\frac{\pi}{8}}\left(\frac{\omega_{ia}}{kb}\right)^4$$
$$\times \exp\left[-\left(\frac{\omega_{ia}}{\sqrt{2}kb}\right)^2\right]. \tag{5.5.13}$$

Collisionless Limit, $k\lambda_{ii} \gg 1$

$$C_i \simeq \frac{ZT_i}{T_i}\alpha^2\left(-\frac{1}{x_i^2} - \frac{3}{x_i^4} + i\frac{8}{5}\frac{1}{k\lambda_{ii}x_i^5} + i\sqrt{\frac{\pi}{2}}x_i e^{-x_i^2/2}\right). \tag{5.5.14}$$

5.6 TOTAL CROSS SECTION

The total cross section is

$$S_T(\mathbf{k}) = \int_{-\infty}^{\infty} d\omega S(\mathbf{k},\omega). \tag{5.6.1}$$

If the full expression Eqn (5.2.2) for $S(\mathbf{k},\omega)$ is used, we must integrate Eqn (5.6.1) numerically. This has been done by Moorcroft (1963). We discuss below the situation where approximate analytic expressions may be obtained.

5.6.1 $T_e \simeq T_i$, Salpeter Approximation

From Eqn (5.3.4),

$$\int_{-\infty}^{\infty} \Gamma_\alpha(x)dx = -\text{Im}\left[\int_{-\infty}^{\infty} \frac{dx}{\pi^{1/2}x\alpha^{1/2}\epsilon(\alpha,x)}\right] = \frac{\pi^{1/2}}{(1+\alpha^2)}. \tag{5.6.2}$$

This integral is evaluated by a contour integration in the upper half of the complex plane (see Appendix A) using the properties of a stable plasma, namely that $\epsilon(z)|_{z\to\infty} = 1$, and $|\epsilon|^2$ is nowhere zero in the upper half-plane.

This is now used in conjunction with Eqn (5.3.3) and

$$S_T(\mathbf{k}) \simeq \underbrace{\frac{2\pi}{(1+\alpha^2)}}_{\text{Electron component}} + \underbrace{\frac{2\pi}{(1+\alpha^2)} \frac{Z\alpha^4}{[1+\alpha^2+\alpha^2(ZT_e/T_i)]}}_{\text{Ion component}}. \tag{5.6.3}$$

We define the first term on the right-hand side as the "electron component, $S_e(\mathbf{k})$" and the second term as the "ion component, $S_i(\mathbf{k})$."

Note: This result is only valid in the region where the ion-acoustic resonance is weak, i.e., $ZT_e/T_i < 3$. Two interesting features may be noted:

$$S_T(\mathbf{k}) \Rightarrow S_e(\mathbf{k}) = 1 \quad \text{for} \quad \alpha \to 0$$

$$S_T(\mathbf{k}) \Rightarrow S_i(\mathbf{k}) = \frac{Z}{1+Z} \quad \text{for} \quad \alpha \gg 1 \text{ and } T_e = T_i.$$

The latter result was first observed by Bowles (1958), and in fact, his observations stimulated much of the theoretical work in this field. Insight into these collective effects can be obtained from a simple fluid model derived from the total cross section for $\alpha \gg 1$ (see Section 5.6)

5.6.2 Arbitrary T_e/T_i

The spectrum (Eqn (5.1.2)) may be written as

$$S(\mathbf{k},\omega) = -\underbrace{\frac{2}{\alpha^2\omega}\left|\frac{1+\chi_i}{\varepsilon}\right|^2 \mathrm{Im}(\chi_e)}_{\text{electron component}} - \underbrace{\frac{2T_i}{T_e\alpha^2\omega}\left|\frac{\chi_e}{\varepsilon}\right|^2 \mathrm{Im}(\chi_i)}_{\text{ion component}}. \tag{5.6.4}$$

We note that the quantity

$$-\frac{2}{\alpha^2\omega}\mathrm{Im}\left\{\frac{(1+\chi_i)\chi_e}{1+\chi_e+\chi_i}\right\} = -\frac{2}{\alpha^2\omega}\left|\frac{1+\chi_i}{\epsilon}\right|^2 \mathrm{Im}(\chi_e) - \frac{2}{\alpha^2\omega}\left|\frac{\chi_e}{\epsilon}\right|^2 \mathrm{Im}(\chi_i). \tag{5.6.5}$$

Equation (6.7.6) differs from Eqn (6.7.5) by the factor T_i/T_e in the second term. If we can evaluate the ion component separately, then we can obtain the correct electron component from Eqn (6.7.6) by subtracting from it T_e/T_i times the ion feature contribution.

Now,

$$-\int_{-\infty}^{+\infty} d\omega \frac{2}{\alpha^2\omega}\mathrm{Im}\left\{\frac{(1+\chi_i)\chi_e}{1+\chi_e+\chi_i}\right\} = \frac{2\pi[1+\alpha^2(ZT_e/T_i)]}{1+\alpha^2+\alpha^2(ZT_e/T_i)}. \tag{5.6.6}$$

Again this integral is obtained by contour integration in the upper plane using the result that for complex $\omega = \infty$, $\chi_e = \chi_i = 0$; for complex $\omega = 0$, $\chi_e = \alpha^2$, $\chi_i = \alpha^2 ZT_e/T_i$.

$T_e \simeq T_i$

To check this approach, we subtract from Eqn (5.6.6), T_e/T_i times the ion feature of Eqn (5.6.3), and we obtain the correct electron feature $S_e(\mathbf{k}) = 2\pi/(1+\alpha^2)$.

$T_e/T_i \leq 1$

In this region where the ion-acoustic resonance is weak, we have $\mathrm{Iw}(x_e) \ll (ZT_e/T_i)\mathrm{Iw}(x_i)$ and recover the original result, while for $T_e/T_i > 1$, the resonance is sharp and the damping is the dominant part of the denominator. Thus,

$$S_i(\mathbf{k})_{\mathrm{corr}} \simeq \frac{2\pi\left(\frac{ZT_e}{T_i}\right)^3 \exp\left(\frac{-ZT_e}{T_i}\right) Z\alpha^4}{\left\{\left(\frac{Zm_e}{m_\mathrm{i}}\right)^{1/2} + \left(\frac{ZT_e}{T_i}\right)^{3/2} \exp\left(\frac{-ZT_e}{2T_i}\right)\right\}^2 (1+\alpha^2)\left(1+\alpha^2+\alpha^2\frac{ZT_e}{T_i}\right)}, \tag{5.6.7}$$

where we have set $\omega_{ac} \simeq k(Z\kappa T_e/m_\mathrm{i})^{1/2}$. Now,

$$S_e(\mathbf{k}) \simeq \frac{2\pi[1+\alpha^2(ZT_e/T_i)]}{[1+\alpha^2+\alpha^2(ZT_e/T_i)]} - \frac{T_e}{T_i} S_i(\mathbf{k})_{\mathrm{corr}}. \tag{5.6.8}$$

This result is shown plotted in Fig. 5.13 for large α; for hydrogen, it compares favorably with the computed results of Moorcroft (1963). Note that the dependence upon Z and m_i appears in (5.6.8).

Correction to Salpeter for $\frac{T_e}{T_i} > \left(\frac{2Zm_e}{m_\mathrm{i}}\right)^{1/2}$

In this region, the cross section given by Eqn (5.6.3) is incorrect only because we have neglected the electron Landau damping term. A good approximation to the correct value is obtained by multiplying Eqn (5.6.3) by

$$\left.\left(\frac{Z(T_e/T_i)\mathrm{Iw}(x_i)}{\mathrm{Iw}(x_e)+(ZT_e/T_i)\mathrm{Iw}(x_i)}\right)^2\right|_{\omega/k=\omega_{ac}/k}.$$

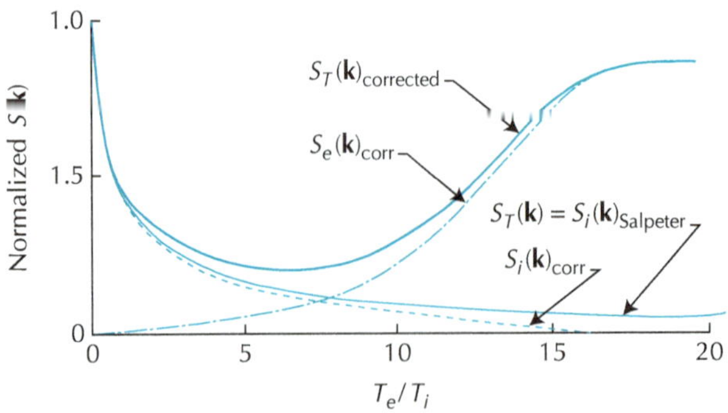

FIGURE 5.13 The normalized total cross section $S_T(\mathbf{k}) = S_e(\mathbf{k}) + S_i(\mathbf{k})$ as a function of T_e/T_i from Eqn (5.6.8) with $\alpha \gg 1$ and $Z = 1$.

5.6.3 Fluid Model

Insight into the collective effects can be obtained by considering the following simple model for the case $\alpha \gg 1$. At these wavelengths, the incident wave "sees" not the individual charge but the shielded or "dressed" charge. We can determine the form of the dressing by using a simple fluid description of the background plasma. The potential around a charge q is $\phi(r) = q/r$. In thermodynamic equilibrium and for a stationary charge, the electron and ion densities adjust within the potential to

$$n_e(r) = n_{e0} \exp(e\phi/\kappa T_e), \qquad n_i(r) = n_{i0} \exp(e\phi/\kappa T_i). \tag{5.6.9}$$

We may assume overall charge neutrality, therefore $n_{e0} \simeq n_{i0} \equiv n$. For a moving charge, the inertia of the ions will slow their response. To a good approximation, we may ignore the ion interaction if the charge velocity exceeds the ion mean thermal speed b. The closest separation of the charges is approximately $n^{1/3}$, and $|eq/\kappa T_e n^{1/3}|$ is typically much less than one; therefore, we may approximate Eqn (5.6.9) by

$$n_e(r) = n + \left(\frac{eqn}{\kappa T_e r}\right), \qquad n_i(r) = n + \left(\frac{eqn}{\kappa T_i r}\right). \tag{5.6.10}$$

The total perturbation in the charge density is obtained by integrating over all space. Strictly, we should use a consistent potential, which allows for the movement of the background charge. Here, we will keep $\phi = q/r$ and limit the integration to a distance D from the charge. We determine D from the condition that there is no net charge within a sphere of radius D. The net perturbations, that is, the number of electrons and ions, respectively, that provide the dressing are

$$\Delta N_e = \left(\frac{2\pi eq}{\kappa T_e D}\right) nD^3 \quad \text{and} \quad \Delta N_i = -\left(\frac{2\pi eq}{\kappa T_i D}\right) nD^3.$$

There are three cases to consider:

(a) *Fast electrons*, i.e., $v_e > b$: The ions cannot respond, and the dressing is obtained by the removal of the equivalent of one electron from the vicinity of the test charge. Both the electron and "hole" have the same mass and can respond to the beam; the dressed electron is neutral, and there is no scattering.

(b) *Slow electrons*, i.e., $v_e \leq b$: The shielding is a result of an increase in the ion density and a decrease in the electron density. The more massive ions do not react to the incident wave, and the scattering is from that "fraction of the electron test charge" that is neutralized by the ions. Unless $T_i \gg T_e$, there are very few electrons in this category, and as a result, little scattering occurs.

(c) *Ions*: Both the electrons and other ions can respond. The form of the shielding is obtained from the requirement that the net charge including the test ion must be zero, that is, $q = +e$, and

$$(+e) + 2\pi \frac{e^2 n D^3}{\kappa T_e D}(-e) - 2\pi \frac{e^2 n D^3}{\kappa T_i D}(+e) = 0$$

Rearranging we have $D^2 = 2\lambda_{\mathrm{De}}^2 T_i/(T_e + T_i)$. We see that the shielding distance D is of the order of the Debye length $\lambda_{\mathrm{De}}^2 = \kappa T_e / 4\pi e^2 n$.

The dressed ion consists of

$$(+e) + \left(\frac{T_i}{T_e + T_i}\right)(-e) - \frac{T_e}{T_e + T_i}(+e) = 0.$$

The dressing consists of a fraction of an electron $T_i/(T_e + T_i)$, which has been attracted into the Debye sphere, and the loss of a fraction of an ion $T_e/(T_e + T_i)$, which has been repelled out. For $T_e = T_i$, the dressing is half an electron and half the lack of an ion. To the incident wave, each ion acts like half an electron. The total cross section is half that for $\alpha \ll 1$, and since the dressing follows the ion, the frequency spectrum reflects the Doppler shift resulting from the ion motion.

The frequency spectrum ($\alpha > 1$) is in fact more complicated than this even in thermodynamic equilibrium, because as we have seen, electron plasma waves are excited. Nevertheless, this fluid limit model provides a simple physical explanation of the results in Eqn (5.6.3).

PROBLEMS

5.1 (a) Show that for Maxwellian distributions,

$$\chi_e(\mathbf{k}, \omega) = -\frac{\alpha^2}{2} Z'(x_e) = \alpha^2\,[\mathrm{Rw}(x_e) + i\mathrm{Iw}(x_e)].$$

Start with the following definition of the electron susceptibility,

$$\chi_e(\mathbf{k}, \omega) = \int_{-\infty}^{+\infty} d\mathbf{v} \frac{4\pi e^2 n_{eo}}{m_e k^2} \frac{\mathbf{k} \cdot \partial f_{eo}/\partial \mathbf{v}}{\omega - \mathbf{k} \cdot \mathbf{v} - i\gamma}.$$

(b) Using the solutions from above show that Eqn (5.1.2) can be written as

$$S(\mathbf{k}, \omega) = \frac{2\pi^{1/2}}{ka} \left\{ \frac{A_e}{|\epsilon|^2} + \frac{A_i}{|\epsilon|^2} \right\},$$

where

$$A_e = \exp\left(-x_e^2\right)\left[\left(1 + \alpha^2 \frac{ZT_e}{T_i} \mathrm{Rw}(x_i)\right)^2 + \left(\alpha^2 \frac{ZT_e}{T_i} \mathrm{Iw}(x_i)\right)^2\right]$$

$$A_i = Z\left(\frac{m_i T_e}{m_e T_i}\right)^{1/2} \exp\left(-x_i^2\right)\left[\left(\alpha^2 \mathrm{Rw}(x_e)\right)^2 + \left(\alpha^2 \mathrm{Iw}(x_e)\right)^2\right]$$

and

$$|\epsilon|^2 = \left\{\left[1+\alpha^2\left(\mathrm{Rw}(x_e)+\frac{ZT_e}{T_i}\mathrm{Rw}(x_i)\right)\right]^2 + \left[\alpha^2\mathrm{Iw}(x_e)+\alpha^2\frac{ZT_e}{T_i}\mathrm{Iw}(x_i)\right]^2\right\}.$$

5.2. What approximations are made in the Salpeter approximation? Explain the approximations in terms of where the phase velocity resides within the relevant distribution function.

(a) Show

$$S_e(\mathbf{k},\omega) \simeq \frac{2\pi^{1/2}\exp(-x_e^2)}{ka\left[\left(1+\alpha^2\mathrm{Rw}(x_e)\right)^2+(\alpha^2\mathrm{Iw}(x_e))^2\right]}$$

(b) Show

$$S_i(\mathbf{k},\omega) \simeq \frac{2\pi^{1/2}}{kb}\frac{Z\alpha^4\exp(-x_i^2)}{\left[1+\alpha^2+\alpha^2(ZT_e/T_i)\mathrm{Rw}(x_i)\right]^2+\left[\alpha^2(ZT_e/T_i)\mathrm{Iw}(x_i)\right]^2}.$$

5.3. Under what conditions will

(a) the electron plasma features be evident in the scattering spectrum?

(b) the ion thermal distribution be evident in the scattering spectrum?

(c) the ion-acoustic features be evident in the scattering spectrum?

5.4. **(a)** Derive the ion-acoustic dispersion relation, and show ion-acoustic waves exist for $ZT_e/T_i > 3$. Explain.

(b) When is this approximate dispersion relation valid?

5.5. Derive the wavelength separation between the electron plasma features.

(a) Starting with the Bohm–Gross dispersion relation for electron plasma waves, show

$$\left(\frac{k_{\mathrm{epw}}}{k_i}\right)^4 - 4\cos\Phi\left(\frac{k_{\mathrm{epw}}}{k_i}\right)^3 + 4\left(\frac{k_{\mathrm{epw}}}{k_i}\right)^2\cos^2\Phi\left(1-\frac{1}{2\cos^2\Phi}\frac{\omega_p^2}{c^2k_i^2}-\frac{3}{\cos^2\Phi}\frac{a^2}{c^2}\right)$$

$$+4\left(\frac{k_{\mathrm{epw}}}{k_i}\right)\frac{\omega_p^2}{c^2k_i^2}\cos\Phi - 4\frac{\omega_p^2}{c^2k_i^2}\left[1+\frac{3}{4}\frac{\omega_p^2}{c^2k_i^2}\right]=0$$

(b) Show that for $\theta = 90°$, the above equation reduces to

$$\frac{k_{\mathrm{epw}}}{k_i} = \left(2+\frac{\omega_{pe}^2}{c^2k_i^2}+6\frac{a^2}{c^2}\right)$$

$$\pm 2\left[\frac{\omega_{pe}^2}{c^2k_i^2}+\left(\frac{\omega_{pe}^2}{c^2k_i^2}\right)^2+6\frac{a^2}{c^2}\left(1+\frac{1}{2}\frac{\omega_{pe}^2}{c^2k_i^2}+\frac{3a^2}{2c^2}\right)\right]^{1/2}$$

(c) From the above equation, show that when assuming $\frac{\omega_{pe}^2}{c^2 k_i^2} \sim \frac{n}{n_{cr}}$, the wavelength separation is

$$\frac{\Delta\lambda_{\text{epw}}}{\lambda_i} \approx 2\left[\frac{n}{n_{cr}} + 3\frac{a^2}{c^2}\right]^{1/2}\left(1 + \frac{3}{2}\frac{n}{n_{cr}}\right).$$

Start with the fact that

$$\Delta\lambda \equiv \lambda^{\text{red}} - \lambda^{\text{blue}} = 2\pi\left(\frac{1}{k^{\text{red}}} - \frac{1}{k^{\text{blue}}}\right),$$

where λ^{red} and λ^{blue} are the wavelengths for the resonant up- and down-shifted electron plasma features, respectively.

5.6. Derive the wavelength separation between ion-acoustic features.

(a) Show that when scattering from ion-acoustic fluctuations, the wave vector is given by

$$k_{ia} \simeq 2k_0 \sin\left(\frac{\theta}{2}\right).$$

(b) Now, show

$$\frac{\Delta\lambda_{ia}}{\lambda_i} \simeq \frac{4}{c}\sin\left(\frac{\theta}{2}\right)\sqrt{\frac{\kappa T_e}{m_{\text{i}}}\left[\frac{Z}{(1 + k_{ia}^2\lambda_{\text{De}}^2)} + \frac{3T_i}{T_e}\right]}.$$

5.7. Relativistic corrections to the scattered power in the collective regime.

(a) Show that Terms B and C reduce to the correction Term D, which is derived when calculating the scattered power to first order in v/c (see Section 2.3.3).

$$\frac{\text{P}^{\text{blue}}}{\text{P}^{\text{red}}} \approx \underbrace{\frac{\text{P}_{NR}^{\text{blue}}}{\text{P}_{NR}^{\text{red}}}}_{A}\underbrace{\left(\frac{1+\beta_\phi\cos\phi}{1-\beta_\phi\cos\phi}\right)^2}_{B}\underbrace{\left(\frac{1+\beta_\phi\cos\Phi}{1-\beta_\phi\cos\Phi}\right)^2}_{C} = \underbrace{\frac{\text{P}_{NR}^{\text{blue}}}{\text{P}_{NR}^{\text{red}}}\left(\frac{\omega_i + 2\omega_{\text{epw}}}{\omega_i - 2\omega_{\text{epw}}}\right)}_{D}$$

Hint: Assume a scattering angle of $\theta = 90°$.

(b) Use the Bohm–Gross dispersion relation to show that when $2\omega_{\text{epw}}/\omega_i$ is small,

$$\frac{\text{P}^{\text{blue}}}{\text{P}^{\text{red}}} \approx \frac{\left(1 + \frac{2\omega_{\text{epw}}}{\omega_i}\right)}{\left(1 - \frac{2\omega_{\text{epw}}}{\omega_i}\right)} \approx 1 + 4\frac{\left(\omega_{pe}^2 + 3(\kappa T_e/m_e)k^2\right)^{1/2}}{\omega_i}.$$

(c) When is Eqn (5.4.10) a good approximation to the ratio in power scattering into the blue- and red-shifted electron plasma resonance?

(d) Calculate the asymmetry in the peaks of light scattered from the electron plasma frequency resonances for the case of $\theta = 90°$, $\lambda_i = 532$ nm, $n_e = 4 \times 10^{19}$ cm^{-3}, and $T_e = 410$ eV. Assume $\text{P}_{NR}^{\text{blue}}/\text{P}_{NR}^{\text{red}} = 1$ using Eqns (5.4.9) and (5.4.10) and compare them to the results

presented in Fig. 5.6c. How does the phase velocity compare to the mean electron thermal velocity and the velocity of light?
Hint: Use the solution to 6(b) to determine the scattering wave vector.

5.8. What set of Thomson scattering measurements could be used to measure:
(a) The electron temperature;
(b) The ion temperature;
(c) The electron density; and what else must be known or assumed in each case?

5.9. Under what conditions (n_e and T_e) will ion–electron collisions affect ion-acoustic waves? For simplicity, look for approximate answers, and consider a single ion species with $Z=1$, $A=1, 10$, and 50, $T_i=T_e$, and $\alpha=2$. Comment on the effects of other situations.

5.10. Using the fluid approximation, derive the dispersion relation for electro static waves in a cold, unmagnetized plasma in which the ions are treated as a fixed, uniform, neutralizing background, but with the electron fluid drifting with a constant drift velocity v_0 in the $\hat{z}$ direction. Plot ω vs. k and identify the "fast" and "slow" space charge waves. Find the group velocity for each. Derive an expression for the dielectric constant. Hint: Put Poisson's equation into the form $\nabla\cdot(\epsilon E_1)$. Using the expression for the time-average energy in a wave (kinetic plus electrostatic)

$$W_E=\frac{E_p^2}{8\pi}\frac{\partial(\epsilon\omega)}{\partial\omega}.$$

SOLUTIONS TO ODD PROBLEMS

5.1. (a) Using the usual definitions for the Debye length and the electron thermal speed, the electron susceptibility becomes

$$\chi_e(\mathbf{k},\omega)=\frac{\alpha^2}{2}a^2\int_{-\infty}^{\infty}d\mathrm{v}\frac{\mathbf{k}\cdot\frac{\delta f}{\delta \mathrm{v}}}{\omega-\mathbf{k}\cdot\mathbf{v}}.$$

Assuming a Maxwellian velocity distribution and that k is parallel to v and $\frac{\delta f}{\delta \mathrm{v}}$,

$$\chi_e(\mathbf{k},\omega)-\frac{\alpha^2}{(\pi a^2)^{1/2}}\int_{-\infty}^{\infty}d\mathrm{v}\frac{(\mathrm{v}/a)e^{-\mathrm{v}^2/a^2}}{(\mathrm{v}/a)-\omega/(ka)}.$$

Changing variables from v/a to p and defining $x_e=\omega/(ka)$,

$$\chi_e(\mathbf{k},\omega)=\frac{\alpha^2}{(\pi)^{1/2}}\int_{-\infty}^{\infty}dp\frac{pe^{-p^2}}{p-x_e}.$$

Breaking the integral into two parts by replacing the p in the numerator with $(p+x_e-x_e)$ transforms the integral into

$$\chi_e(\mathbf{k},\omega)=\alpha^2\left[\left(\pi^{-1/2}\int_{-\infty}^{\infty}dpe^{-p^2}\right)+(x_eZ(x_e))\right]$$

with the plasma dispersion function defined as $Z(x_e)=\pi^{-1/2}\int_{-\infty}^{\infty}dp\frac{e^{-p^2}}{p-x_e}$. Since the first term in the brackets is equal to 1, the electron susceptibility can be recast now as

$$\chi_e(\mathbf{k},\omega)=\alpha^2\left[-\frac{1}{2}Z'(x_e)\right],$$

where $Z'=\frac{dZ}{dx_e}=-2(1+x_eZ(x_e))$. Assuming now that x_e is purely real,

$$Z'(x_e)=-2\left[1+xe^{-x^2}\left(i\pi^{1/2}-2\int_0^{x_e}dte^{t^2}\right)\right],$$

and with the definition of the functions $\mathrm{Rw}(x_e)=-\left(\frac{1}{2}\right)\mathrm{Re}(Z'(x_e))$ and $\mathrm{Iw}(x_e)=-\left(\frac{1}{2}\right)\mathrm{Im}(Z'(x_e))$, the susceptibility takes on the familiar form of

$$\chi_e(\mathbf{k},\omega)=\alpha^2\left[\mathrm{Rw}(x_e)+i\mathrm{Iw}(x_e)\right].$$

(b) Start with Eqn (5.1.2),

$$S(\mathbf{k},\omega)=\frac{2\pi}{k}\left|1-\frac{\chi_e}{\epsilon}\right|^2 f_{e0}\left(\frac{\omega}{k}\right)+\frac{2\pi Z}{k}\left|\frac{\chi_e}{\epsilon}\right|^2 f_{i0}\left(\frac{\omega}{k}\right).$$

Using 1D Maxwellian distribution functions evaluated at ω/k, with the definitions of the electron and ion thermal speeds, the above equation becomes

$$S(\mathbf{k},\omega)=\frac{2\pi}{k}\left|1-\frac{\chi_e}{\epsilon}\right|^2\left(\pi a^2\right)^{-1/2}e^{-x_e^2}+\frac{2\pi Z}{k}\left|\frac{\chi_e}{\epsilon}\right|^2\left(\pi b^2\right)^{-1/2}e^{-x_i^2},$$

where $x_e=\omega/ka$ and $x_i=\omega/kb$. Using the susceptibilities above and realizing that $1-\frac{\chi_e}{\epsilon}=\frac{1+\chi_i}{\epsilon}$,

$$|1+\chi_i|^2=\left(1+\alpha^2\frac{ZT_e}{T_i}\mathrm{Rw}(x_i)\right)^2+\left(\alpha^2\frac{ZT_e}{T_i}\mathrm{Iw}(x_i)\right)^2$$

$$|\chi_e|^2=(\alpha^2\mathrm{Rw}(x_e))^2+(\alpha^2\mathrm{Iw}(x_e))^2.$$

Factoring out the π's and substituting $a/b = \left(\frac{m_i T_e}{m_e T_i}\right)^{1/2}$ results in

$$S(\mathbf{k}, \omega) = \frac{2\pi^{1/2}}{ka} \left\{ \frac{A_e}{|\epsilon|^2} + \frac{A_i}{|\epsilon|^2} \right\},$$

where

$$A_e = \exp(-x_e^2)\left[\left(1+\alpha^2 \frac{ZT_e}{T_i}\mathrm{Rw}(x_i)\right)^2 + \left(\alpha^2 \frac{ZT_e}{T_i}\mathrm{Iw}(x_i)\right)^2\right]$$

$$A_i = Z\left(\frac{m_i T_e}{m_e T_i}\right)^{1/2} \exp(-x_i^2)\left[(\alpha^2 \mathrm{Rw}(x_e))^2 + \left(\alpha^2 \mathrm{Iw}(x_e)\right)^2\right]$$

and

$$|\epsilon|^2 = \left\{\left[1+\alpha^2\left(\mathrm{Rw}(x_e) + \frac{ZT_e}{T_i}\mathrm{Rw}(x_i)\right)\right]^2 + \left[\alpha^2 \mathrm{Iw}(x_e) + \alpha^2 \frac{ZT_e}{T_i}\mathrm{Iw}(x_i)\right]^2\right\}.$$

5.3. **(a)** When the scattering parameter $\alpha = 1/k\lambda_{\mathrm{De}} \gtrsim 1$, collective wave motion is probed, and electron plasma features will be evident in the scattering spectrum.

(b) Scattering from the ion thermal distribution is present when probing frequencies less than the ion plasma frequency, i.e., $\alpha^2 \gg 2T_i/ZT_e$.

(c) Ion-acoustic features are observed with the condition of (b) is met and the waves are not strongly damped $\left(\left(\frac{T_i}{ZT_e}\right)\left[\left(1+\alpha^2\right)/\alpha^2\right] < 0.29\right)$.

5.5. Electron plasma fluctuations

(a) Start with,

$$\omega_s^2 = \omega_{pe}^2 + c^2 k_s^2$$
$$\omega_s = \omega + \omega_i$$
$$\mathbf{k}_s = \mathbf{k} + \mathbf{k}_i$$

Assume 1D,

$$(\omega + \omega_i)^2 = \omega_{pe}^2 + c^2(k + k_i)^2$$
$$\omega^2 + 2\omega\omega_i + \omega_i^2 = \omega_{pe}^2 + c^2k^2 + 2c^2kk_ic_1 + c^2k_i^2,$$

where $c_1 = \cos\Phi$ is the angle between k and k_i.

Use $\omega_i^2 - \omega_{pe}^2 + c^2k_i^2$

$$\omega^2 + 2\omega\omega_i = c^2k^2 + 2c^2kk_ic_1.$$

Use $\omega^2 = \omega_{pe}^2 + 3a^2k^2$

$$\omega_{pe}^2 + 2\left(\omega_{pe}^2 + 3a^2k^2\right)^{1/2}\omega_i = \left(c^2 - 3a^2\right)k^2 + 2c^2kk_ic_1.$$

Assume that $3a^2 \ll c^2$,

$$2\left(\omega_{pe}^2+3a^2k^2\right)^{1/2}\omega_i = c^2k^2+2c^2kk_ic_1-\omega_{pe}^2.$$

Square both sides,

$$k^4+4k^3k_ic_1+4k^2k_i^2c_1^2\left(1-\frac{\omega_{pe}^2}{2c^2k_i^2c_1^2}-\frac{3\omega_i^2a^2}{c^4k_i^2c_1^2}\right)$$

$$-4kk_ic_1\frac{\omega_{pe}^2}{c^2}+\frac{\omega_{pe}^4}{c^4}-4\frac{\omega_{pe}^2\omega_i^2}{c^4}=0.$$

Use $\omega_i^2=\omega_{pe}^2c^2k_i^2$,

$$k^4+4k^3k_ic_1+4k^2k_i^2c_1^2\left(1-\frac{\omega_{pe}^2}{2c^2k_i^2c_1^2}\frac{2\omega_{pe}^2a^2}{c^2c_1^2}\right)$$

$$+4kk_ic_1\frac{\omega_{pe}^2}{c^2}-4\frac{\omega_{pe}^2k_i^2}{c^2}\left(1+\frac{3}{4}\frac{\omega_{pe}^2}{c^2k_i^2}\right)=0.$$

Dividing through by k_i^4 reveals

$$\left(\frac{k_{\mathrm{epw}}}{k_i}\right)^4-4\cos\Phi\left(\frac{k_{\mathrm{epw}}}{k_i}\right)^3$$

$$+4\left(\frac{k_{\mathrm{epw}}}{k_i}\right)^2\cos^2\Phi\left(1-\frac{1}{2\cos^2\Phi}\frac{\omega_p^2}{c^2k_i^2}-\frac{3}{\cos^2\Phi}\frac{a^2}{c^2}\right)$$

$$+4\left(\frac{k_{\mathrm{epw}}}{k_i}\right)\frac{\omega_p^2}{c^2k_i^2}\cos\Phi-4\frac{\omega_p^2}{c^2k_i^2}\left[1+\frac{3}{4}\frac{\omega_p^2}{c^2k_i^2}\right]=0$$

(b) Define $\Delta\equiv k/k_i$, $c_2\equiv(a/c)^2$, and $R=\dfrac{\omega_{pe}^2}{c^2k_i^2}$

$$\Delta^4-4c_1\Delta^3+4\Delta^2c_1^2\left(1-\frac{R}{2c_1^2}-\frac{3c_2}{c_1^2}\right)+4\Delta c_1R-4R\left(1+\frac{3R}{4}\right)=0$$

For $\theta=90^\circ$ scattering, $k^2=k_i^2+k_s^2$ and $c_1=\cos\Phi=-\frac{k_i}{k}=-\frac{1}{\Delta}$ and

$$\Delta^4-4\left(1+\frac{R}{2}+3c_2\right)\Delta^2+4\left(1-\frac{3R^2}{4}\right)=0.$$

Using the quadratic formula,

$$\Delta^2=(2+R+6c_2)\pm 2[R+R^2+3c_2(2+R+3c_2)]^{1/2}.$$

(c) Assuming $R \sim \left(\frac{n}{n_{cr}}\right)$,

$$\Delta^2 \simeq 2 + \left(\frac{n}{n_{cr}}\right) + 6c_2^2 \pm 2\left[\frac{n}{n_{cr}} + 6c_2^2 + \left(\frac{n}{n_{cr}}\right)^2\right]^{1/2}.$$

Take the square root of both sides and assume c_2 is small compared to N/N_{cr},

$$\frac{k}{k_i} = \sqrt{2}\sqrt{1 + \frac{1}{2}\left(\frac{n}{n_{cr}}\right) \pm \left[\left(\frac{n}{n_{cr}}\right) + 6c_2^2 + \left(\frac{n}{n_{cr}}\right)^2\right]^{1/2}}$$

Expand around $\frac{n}{n_{cr}}$,

$$\frac{k}{k_i} = \sqrt{2}\left[1 + \frac{1}{4}\left(\frac{n}{n_{cr}}\right) \pm \frac{1}{2}\left[\left(\frac{n}{n_{cr}}\right) + 6c_2^2 + \left(\frac{n}{n_{cr}}\right)^2\right]^{1/2}\right]$$

Defining,

$$\Delta\lambda \equiv \lambda^{\text{red}} - \lambda^{\text{blue}} = 2\pi\left(\frac{1}{k^{\text{red}}} - \frac{1}{k^{\text{blue}}}\right)$$

$$\frac{\Delta\lambda}{\lambda_i} = \left(\frac{k_i}{k^{\text{red}}} - \frac{k_i}{k^{\text{blue}}}\right)$$

$$\frac{\Delta\lambda}{\lambda_i} = \left[\frac{1}{2}\left(\frac{n}{n_{cr}}\right) + 3c_2^2\right]^{1/2}.$$

In terms of the scattered wave,

$$\frac{\Delta\lambda_s}{\lambda_i} = \left(\frac{k_i}{k_s^{\text{red}}} - \frac{k_i}{k_s^{\text{blue}}}\right)$$

$$\frac{k_s}{k_i} = \sqrt{\Delta^2 - 1}$$

$$\frac{k_s}{k_i} = 1 \pm \left[\left(\frac{n}{n_{cr}}\right) + 6c_2^2 + \left(\frac{n}{n_{cr}}\right)^2\right]^{1/2}$$

substitution gives

$$\frac{\Delta\lambda_s}{\lambda_i} = 2\left[\left(\frac{n}{n_{cr}}\right) + 6c_2^2 + \left(\frac{n}{n_{cr}}\right)^2\right]^{1/2}\left[1 + \left(\frac{n}{n_{cr}}\right)\right],$$

which is reduced to

$$\frac{\Delta\lambda_{\text{epw}}}{\lambda_i} \approx 2\left[\frac{n}{n_{cr}} + 3\frac{a^2}{c^2}\right]^{1/2}\left(1 + \frac{3}{2}\frac{n}{n_{cr}}\right).$$

5.7. Relativistic corrections to the scattered power in the collective regime.

(a) Notice that $\cos\Phi = k_s/k$ and $\cos\phi = k_i/k$, so

$$\frac{P^{blue}}{P^{red}} \approx \frac{P^{blue}_{NR}}{P^{red}_{NR}} \left(\frac{1+\beta_\phi \cos\phi}{1-\beta_\phi \cos\phi}\right)^2 \left(\frac{1+\beta_\phi \cos\Phi}{1-\beta_\phi \cos\Phi}\right)^2$$
$$= \frac{P^{blue}_{NR}}{P^{red}_{NR}} \left(\frac{1+\beta_\phi \frac{k_i}{k}}{1-\beta_\phi \frac{k_i}{k}}\right)^2 \left(\frac{1+\beta_\phi \frac{k_s}{k}}{1-\beta_\phi \frac{k_s}{k}}\right)^2.$$

Expand keeping only terms of β and noting that $k^2 = k_i^2 + k_s^2$,

$$\frac{P^{blue}}{P^{red}} \approx \frac{P^{blue}_{NR}}{P^{red}_{NR}} \left(\frac{k^2+2\beta_\phi k_i k + 2\beta_\phi k_s k}{k^2-2\beta_\phi k_i k - 2\beta_\phi k_s k}\right) = \frac{P^{blue}_{NR}}{P^{red}_{NR}} \left(\frac{\omega_i + 2\omega}{\omega_i - 2\omega}\right).$$

(b) Use the geometric series to expand around small $\frac{2\omega_{epw}}{\omega_i}$,

$$\frac{P^{blue}}{P^{red}} \approx \frac{\left(1+\frac{2\omega_{epw}}{\omega_i}\right)}{\left(1-\frac{2\omega_{epw}}{\omega_i}\right)} \approx 1 + 4\frac{\omega}{\omega_i} + 4\left(\frac{\omega}{\omega_i}\right)^2.$$

Keep only first-order terms and substitute $\omega = \left(\omega_{pe}^2 + 3(\kappa T_e/m_e)k^2\right)^{1/2}$

$$\frac{P^{blue}}{P^{red}} \approx 1 + 4\frac{\omega}{\omega_i} + 4\left(\frac{\omega}{\omega_i}\right)^2 \approx 1 + 4\frac{\left(\omega_{pe}^2 + 3(\kappa T_e/m_e)k^2\right)^{1/2}}{\omega_i}.$$

(c) Equation (5.4.10) is a good approximation to the scattered power ratio when alpha is large and the phase velocity is small compared to c. When alpha is large, the Landau damping is small, and there is very little difference between the damping rate between the two resonances and therefore the scattered power between the two electron plasma resonances. Equation (5.4.10) only takes into account terms of order beta, the normalized phase velocity, so when beta is large, Eqn (5.4.10) is no longer a good approximation.

(d) The calculated asymmetry using Eqn (5.4.9) is 1.87, 1.61 using the approximate Eqn (5.4.10), and 2.0 from Fig. 5.6 where Term A is not assumed to be unity. $a = 0.028c$, $v_\phi = 0.066c$, and $v_\phi/a \simeq 2.3$.

5.9. The effects of collisions will become noticeable when the collision frequency approaches the ion-acoustic frequency, or

$$2.92\times 10^{-6}Z^2 n_i\,[\mathrm{cm}^{-3}](T_e\,[\mathrm{eV}])^{-3/2}\ln\Lambda \rightarrow$$

$$k\left(\frac{\alpha^2 Z}{1+\alpha^2}+3\frac{T_e}{T_i}\right)^{1/2}\left(\frac{m_e}{m_i}\right)^{1/2}\left(\frac{\kappa T_e}{m_e}\right)^{1/2},$$

where $\Lambda = 1.5\times 10^{10}\,(T_e\,[\mathrm{eV}])^{s/2}\,(n_e\,[\mathrm{cm}^{-3}])^{-1/2}$. Now,

$$k \approx \frac{1.36\times 10^{-3}}{\alpha}\left(\frac{n_e\,[\mathrm{cm}^{-3}]}{T_e\,[\mathrm{eV}]}\right)^{1/2}$$

A	T_e [eV]	n_e [cm^{-3}]
1	10	$\rightarrow 5.6\times 10^{18}$
1	100	$\rightarrow 5.6\times 10^{21}$
10	10	$\rightarrow 1.8\times 10^{18}$
10	100	$\rightarrow 1.8\times 10^{21}$
50	10	$\rightarrow 0.8\times 10^{18}$
50	100	$\rightarrow 0.8\times 10^{21}$

and

$$\left(\frac{\kappa T_e}{m_e}\right)^{1/2} - 4.243\times 10^{7}(T_e\,[\mathrm{eV}])^{1/2}.$$

Rearranging leads to

$$\frac{n_i\,[\mathrm{cm}^{-3}]\ln\Lambda}{(n_e\,[\mathrm{cm}^{-3}])^{1/2}(T_e\,[\mathrm{eV}])^{3/2}} \rightarrow 1.98\times 10^{10}\frac{1}{Z^2}\left(\frac{Z}{1+\alpha^2}+\frac{3}{\alpha^2}\frac{T_e}{T_i}\right)^{1/2}\left(\frac{m_e}{m_i}\right)^{1/2}.$$

For the conditions of this problem

$$\left(n_e\,[\mathrm{cm}^{-3}]\right)^{1/2}\frac{\ln\Lambda}{T_e\,[\mathrm{eV}]} \rightarrow 1.93\times 10^{10}\left(\frac{m_e}{m_i}\right)^{1/2}.$$

Alternatively,

$$\frac{(n_e\,[\mathrm{cm}^{-3}])^{1/2}}{(T_e\,[\mathrm{eV}])^{3/2}} \rightarrow \frac{1.93\times 10^{10}}{\ln\Lambda}\left(\frac{m_e}{m_i}\right)^{1/2}.$$

or

$$\ln\Lambda \rightarrow \ln\left[0.777\left(\frac{m_e}{m_i}\right)^{1/2}\ln\Lambda\right],$$

which is a slowly varying function of $\left(\frac{m_e}{m_i}\right)\left(\frac{m_e}{m_i}\right)^{1/2} = 1837A$, where A is the atomic mass number). The $\ln\Lambda$ will run from about 5 for hydrogen to about 7.2 for A = 50. Using the value $\ln\Lambda = 6$ leads to $(n_e\,[\mathrm{cm}^{-3}])^{1/2}/T_e\,[\mathrm{eV}]^{3/2} \rightarrow 7.5\times10^7[1/A]^{1/2}$ Comments: Clearly collisions can be important in the warm dense matter region and in inertial fusion energy situations. The Z^2 term will be important at the higher temperatures. Note that for $Z > 1$, $n_i < n_e$.

Chapter 6

Constraints on Scattering Experiments

Chapter Outline Head

DOI: 10.1016/B978-0-12-374877-5.00006-3

6.1 INTRODUCTION

A number of excellent articles were of great assistance to the authors in the preparation of this chapter and Chapter 7: Kunze (1968), Evans & Katzenstein (1969a), DeSilva & Goldenbaum (1970), Donné *et al.* (2008), and Luhmann *et al.* (2008).

6.2 CHOICE OF A SOURCE ($\lambda_i, \Delta\lambda_i$)

With the aid of sources listed in Tables 6.1 and 6.2, measurements have been made from $> 0.05\,\text{eV}$ at $\sim 10^{16}\,\text{m}^{-3}$ and to $9\,\text{keV}$ at densities of $10^{30}\,\text{m}^{-3}$. We require the following:

1. $\omega_i > \omega_{pe}$ for transmission of the incident radiation;
2. $\alpha = \lambda_i/[4\pi\lambda_{De}\sin(\theta/2)] \lesssim 0.1$ for noncollective scattering and $\alpha \gtrsim 1$ for collective scattering;
3. A very powerful source, because the scattering cross section is very small and $P_s/P_i \cong 10^{-13}$ [see Eqn (4.2.4)] for densities typically encountered in magnetic fusion devices. For high density plasmas produced in inertial confinement fusion research, higher scattering fractions are obtained but required higher photon energy probes. Even for scattered signals in the microwatt region, we typically require an input power for fast pulsed sources of $P_i \gtrsim 10\,\text{MW}$. Signal-to-noise considerations (these are discussed below) lead to an even greater power requirement. Of course, if the pulse duration is very long, we may relax this requirement;
4. The source beam divergence must be small enough for us to focus the incident beam to a reasonable diameter d in the plasma (e.g., $d \ll$ the plasma radius);
5. The incident radiation must be sensibly monochromatic, that is, $\Delta\lambda_i \ll \Delta\lambda_{s1/2}$ (the half-width of the scattered spectrum).

6.2.1 X-Ray Probe Requirements

In dense plasmas, where optical probes no longer penetrate, laser-produced x-ray probes (Glenzer *et al.*, 2003b, 2007) or short wavelength free-electron

TABLE 6.1 A Selection of Sources Suitable for Thomson Scattering

Source	λ_i	Pulse Duration	Peak Power	Energy	Reference
X-ray	3000–9000 eV	10–100 ps		0.1 J	Glenzer & Redmer (2009)
Free Electron Laser (FEL)	60–8000 eV	40–900 fs		60–3 mJ	Fäustlin *et al.* (2010b), Glenzer & Redmer (2009)
Nd-YAG	1064, 532, 261 nm	0.1–20 ns	~200 MW	> 2 J	Ross *et al.* (2006), Glenzer *et al.* (1999a), MacKinnon *et al.* (2004)
Ruby	694.3 nm	15 ns	1.7 GW	~25 J	Barth *et al.* (2001)
Ti:sapphire ($Ti:Al_2O_3$)	800 nm/(10–100 nm) (650–1000 nm)	10–200 fs	~1 PW	~30 J	
Alexandrite	750 nm	350 ps		1–2 J	Gowers *et al.* (1990)
CO_2	10.6 μm	CW	25–60 W		Lin *et al.* (2006)
CO_2	10.59 μm	1 μs	17 MW	17 J	Kondoh *et al.* (2007)
CO_2	10.6 μm	5 ps	20 TW		Haberberger *et al.* (2010)
CH_3OD	57.2 μm	CW	1.6 W		Nakayama *et al.* (2004)
D_2OH	118.8 μm	CW	1.6 MW		Kawahata *et al.* (2005)
D_2O	385 μm/100 MHz FWHM	1 μs	2.5 MW	~2.5 J	Semet *et al.* (1983)
HCOOH	393 μm	CW	150 mW		Lehecka *et al.* (1990)
HCOOH	743 μm	CW	12 mW		Lehecka *et al.* (1990)
$C^{13}H_3F$	1.22 mm	~1 μs	4 kW	0.5 mJ	Brower *et al.* (1981)
$C^{13}H_3F$	1.22 mm	CW	25 mW		Lehecka *et al.* (1990)

TABLE 6.2 A Selection of Nonlaser Sources Suitable for Thomson Scattering

Source	f_i	Pulse Duration	Peak Power	Reference
Gyrotron	50 GHz	CW-long pulse	1 MW	Thumm (2007)
GaAs MMIC	100 GHz	CW	> 250 mW	Mehdi *et al.* (2008)
SiC IMPATT	100 GHz	CW	~1 W	Shur (2008)
Diode Doubler	110 GHz	CW	~300 mW	Virginia Diodes, Inc.
FEM	130–220 GHz	CW	1 MW	Verhoeven *et al.* (1998)
EIK	140 GHz	CW	50 W	Communications and Power Industries
Gyrotron	170 GHz	CW-long pulse	1 MW	Thumm (2007)
InP GUNN	193 GHz	CW		Eisele & Kamoua (2004)
Orotron	200–370 GHz	CW	30–50 mW	Insight Product Co.
Diode Doubler	210–360 GHz	CW	~25 mW	Virginia Diodes, Inc.
EIK	220 GHz	CW	6 W	Communications and Power Industries
Carcinotron	280 GHz	CW	300 mW	Thomson-CSF data sheet
Gyrotron	345 GHz	CW	80 W	Ogawa *et al.* (2006)
Clinotron	345–390 GHz	CW	100 mW	Vavriv *et al.* (2007)
Carcinotron	400 GHz	CW	50 mW	Thomson-CSF data sheet
Clinotron	442–510 GHz	CW	50–100 mW	Vavriv *et al.* (2007)
Gyrotron	460 GHz	CW	8 W	Han *et al.* (2006)
Diode Doubler	473–490 GHz	CW	~6 mW	Virgina Diodes, Inc.
BWO	526–714 GHz	CW	4–15 mW	Insight Product Co.
BWO	789–968 GHz	CW	3–8 mW	Insight Product Co.
BWO	1034–1250 GHz	CW	0.5–2 mW	Insight Product Co.
VUV FEL	90 eV	40 fs	10^{13} photons	Ackermann (2007)
X-ray FEL	3000 eV	100 fs	10^{12} photons	Akre (2008)
K-alpha source	4500 eV	10 ps	10^{12} photons	Kritcher (2008)
He-alpha	4750–9000 eV	1 ns	10^{15} photons	Glenzer (2003a)
Ly-alpha	3000 eV	1 ns	10^{15} photons	Glenzer (2007)

lasers (Fäustlin *et al.*, 2010b) have been employed for scattering experiments. These sources must provide a narrow bandwidth of $\Delta E/E \simeq 0.01$ for noncollective scattering (Landen *et al.*, 2001b) and $\Delta E/E \simeq 0.002$ for collective scattering (Urry *et al.*, 2006), respectively. For example, probing solid-density beryllium plasmas at temperatures of 50 eV and electron density of $n_e = 3 \times 10^{23}\,\mathrm{cm}^{-3}$ in backscatter geometry, $\theta = 125°$ and with a moderate x-ray energy source such as titanium He$-\alpha$ at $E_0 = 4.75\,\mathrm{keV}$, will access a scattering vector of $|\mathbf{k}| = 4 \times 10^{-10}\,\mathrm{m}^{-1}$. In this case, with $\alpha = 0.3$, the configuration results in noncollective scattering. The Compton feature (see Chapter 11) will be shifted from E_0 by $\Delta E_C = \frac{\hbar^2 k^2}{2m_e} = 70\,\mathrm{eV}$, and the spectral broadening by thermal motion will result in a width of 200 eV. To resolve this feature, an x-ray probe bandwidth of 40 eV such as provided by the titanium He$-\alpha$ spectrum that can be produced with energetic long-pulse lasers is sufficient.

Plasmon measurements, on the other hand, require forward scattering measurements, moderate x ray probe energies, and high spectral resolution of about 7 eV. Energetic long pulse lasers have been employed to efficiently produce the chlorine Ly$-\alpha$ at 3 keV to meet these requirements. Subsequently, ultrashort pulse laser–produced K$-\alpha$ fluorescence radiation has been shown to be a viable alternative (Kritcher *et al.*, 2008) because dielectronic satellite features that may blend with the downshifted plasmon feature do not exist. These probes are typically produced with lower efficiency and lower laser energies than long pulse laser–produced x-rays, but they are short lived allowing the use of more efficient nongated detectors.

The x-ray yield, Y, of a laser source is defined as the number of photons per 1 J of incident laser energy emitted into the full sphere. With E_L being the total laser energy on target, the total number of photons produced is YE_L. The x-ray conversion efficiency is then defined as $\eta_x = Yh\nu$ and is a dimensionless parameter that yield the number of x-ray photons in a single x-ray spectral line produced in a single laser shot. For 4$-$10 keV He$-\alpha$ and Ly$-\alpha$ x-rays, a conversion efficiency of $0.0005 < \eta_x < 0.01$ has been demonstrated (see Kauffman, 1991; Park *et al.*, 2006; Glenzer & Redmer, 2009, for reviews). For example, the titanium He$-\alpha$ transition has been produced by irradiating solid disk targets with $\eta \approx 10^{-3}$ roughly corresponding to 10^{12} titanium He$-\alpha$ (4.75 keV) x-ray photons per 1 J of incident laser energy and into 4π sr. Conversion efficiencies scale linearly with laser energy and are weakly dependent on pulse length. More importantly, a strong dependence on laser frequency for low-energy x-rays ($< 4\,\mathrm{keV}$) has been observed with up to an order of magnitude higher conversion efficiency measured with 351 nm (3ω) compared to 1053 nm (1ω) laser irradiation.

In contrast to He$-\alpha$ and Ly$-\alpha$ sources, irradiating targets with $Z > 20$ with short-pulse lasers result in K$-\alpha$ conversion efficiencies that are almost constant or slightly increasing with increasing K$-\alpha$ x-ray energy. With increasing Z, radiation processes dominate over nonradiative Auger processes, and fluorescence x-rays are emitted by inner-shell transitions in relatively cold atoms where the electron holes are created by interactions with fast electrons from short-pulse laser interactions with the solid target. After optimization of laser energy, power, polarization, prepulse, angle of incidence and spot size on target, these sources yield $10^{-5} < \eta_x < 10^{-4}$.

X-ray free-electron laser have been developed to provide a peak brilliance of 10^{32} to 10^{34} photons per second mrad^2, mm^2, and 0.1% bandwidth for photon energies of 10^2 to 10^4 eV (Ackermann *et al.*, 2007). Although the laser energy can be tuned and highly focussed, the number of x-ray photons available for probing is smaller compared to laser sources. For present single shot x-ray scattering applications, a total of 10^{12} probe x-ray photons can be provided at the dense plasma within a bandwidth of $\Delta E/E \leq 0.5\%$ and in a 40 fs long pulse. At lower x-ray energies, the number of photons available is slightly higher with total energy deliverable on target being constant with a limit of 10^{12} photons at an energy of 8.5 keV (Akre, 2008).

6.3 CHOICE OF A SCATTERING ANGLE (θ, $\Delta\theta$)

This is a difficult compromise, which involves the minimization of parasitic radiation from the source and of background radiation from the plasma; a decision on a spectral width to be consistent with available spectrometers, which finally in laboratory experiments, often brings us to the realization that the plasma device itself is physically so restrictive, because of magnetic field coils, pumping ports, etc. that we have very little choice.

We may rearrange Eqn (1.8.8) to obtain

$$\sin\frac{\theta}{2} = \frac{1.08 \times 10^{-4} \lambda_{\mathrm{i}}\,(\mathrm{cm})}{\alpha} \left[\frac{n_{\mathrm{e}}(\mathrm{cm}^{-3})}{T_{\mathrm{e}}(\mathrm{eV})}\right]^{1/2}. \tag{6.3.1}$$

For a given plasma n_e, T_e, and source λ_i, the permissible range of θ is determined by the specified scattering regime. The demarcation line between collective and noncollective scattering may be set at $\alpha \cong 1$ to define the range of θ, and we see that collective scattering is obtained at smaller scattering angles than noncollective scattering. Numerous factors influence the choice of θ.

For example, to obtain $\alpha \geq 1$ with 90° scattering as T_e ranges from 1 eV to 100 keV, the density must be, respectively, 4.3×10^{15} cm^{-3} to 4.3×10^{20} cm^{-3} at $\lambda_i = 1\,\mu$m and 4.3×10^9 cm^{-3} to 4.3×10^{14} cm^{-3} at $\lambda_i = 1$ mm.

6.3.1 Parasitic Radiation

This is radiation from the source that reaches the observer by routes other than scattering from the plasma. As a general rule, parasitic radiation is greatest near 0°, because we are looking nearly into the source, and near 180°, because we collect radiation reflected by the beam dump. It is not possible to make general predictions about the level of parasitic (or stray) radiation, because

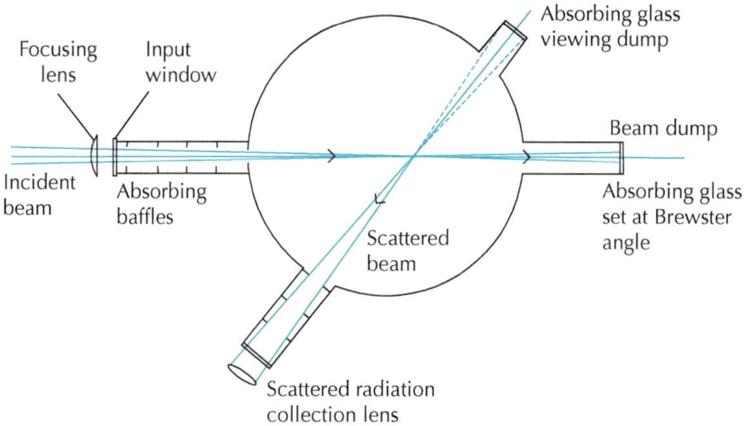

FIGURE 6.1 A typical arrangement for the reduction of parasitic (stray) radiation.

it obviously depends on the peculiarities of each experiment. It is, however, generally considered that the main sources even in a well-designed experiment are the final input window, or focusing lens for the incident beam, and the wall that is seen by the observer; this is illustrated in Fig. 6.1. An important development to significantly reduce parasitic radiation is the use of gated or streaked detectors that measure the Thomson scattered light and shut off before radiation from farther-away surfaces reach the detector.

Optical Components in the Beam

These should be of high optical quality, and they must be kept clean. The dust and imperfections contribute to the parasitic radiation, and they lead to strong absorption of the incident beam, which can cause severe damage to the optical components.

For measurements in the visible with a Q-switched laser, high-quality borosilicate crown glass may be used with low incident powers, say less than 20 MW/cm^2, and barite crown glass withstands more than 100 MW/cm^2 (Eidmann *et al.*, 1972).

At higher powers, the best quality fused silica[1] may be used; this can withstand more than 400 MW/cm^2. Another suitable material is Neodymium laser glass, for which the damage threshold at 1055 nm is in the region of 50 J cm^{-2} in 30 ns (Glass & Guenther, 1970). For high power ultraviolet ($\lambda_i = 266$ nm) probing with nonosecond pulses Eximer grade fused silica optics (corning 7980, grade A) has been applied with good performance up to 1 J/cm^2 (Ross, 2010; Glenzer, 1999; Mackinnon, 2004).

[1]For example, Corning 7940, General Electric 150, or Thermal Syndicate, Spectrosil A and B.

Among suitable materials for the near infrared are germanium, silicon, sodium chloride, and gallium arsenide.[2] There are certain problems in the use of some of these materials. For example, sodium chloride is hygroscopic and must be protected; germanium has a refractive index ~4 and therefore has a poor transmission coefficient (~22%). In addition, it is possible for electrons in germanium and gallium arsenide to be excited to higher levels by the beam, and this leads to absorption, which limits their ability to withstand high power levels.

Absorbing Baffles

Black anodized aluminum or blackened steel have been used and in the visible, colored glass (for example, Schott NG, KG, UG glass). For example, numerous types of blue glass[3] will readily absorb ruby laser light and may also be used for the beam dump. In the latter application, the glass is tilted at the Brewster angle. These baffles act to reduce the level of primary scattered radiation, from the input system and beam dump, at the viewing dump.

Ideally the viewing dump should be a deep cavity similar to the beam dump so that parasitic radiation would have to scatter and reflect many times before reaching the observer. This was possible on the magnetic modulation experiment (see Fig. 4.15), where a Rayleigh horn was used. This device directs the majority of the radiation away from the observer by successive reflections and absorptions. For the Tokamak T-3 experiment (Fig. 4.8), there was less room available, and the problem was overcome with a series of parallel polished steel knife edges. These had an included angle of about 20° and acted in a similar manner to a Rayleigh horn. At a minimum, the viewed surface should be roughened and if possible blackened. In the FIR and millimeter regions, people have looked at both graphite beam dump tiles with with grooves or holes and silicon carbide pyramidal tiles (Woskoboinikow *et al.*, 1983; Rhee *et al.*, 1992).

Dust and Deposited Material

The scattering cross section of micron-size particles is enormous, and if any of these are in the beam or scattering volume, they will certainly lead to a very high level of parasitic radiation. This problem is particularly acute when Rayleigh scattering from high pressures of nitrogen is used to calibrate the collection system. In this situation, it is common practice to leave the gas overnight to settle down. The problem can also occur when a plasma is created, and great care must be taken to ensure that the system is clean.

Another problem is deposition of material on windows. Yoshida *et al.* (1997) discussed an approach to make in situ transmission measurements and to use a laser to blow off deposits.

---

[2] Handbook of Military and I. R. Technology. Office of Naval Research, Dept. of the Navy, Washington, D.C.

[3] For example, Corning, C. S. series–Chance–Pilkington O. B. 10.

Note that Thomson scattering is also used to monitor dust in magnetic fusion plasmas.

6.3.2 The Extent of the Scattering Angle $\Delta\theta$

Radiation is collected in a finite solid angle $d\Omega$; this means that we look at a range of scattering angles $d\Omega \cong (d\theta)^2$. The spectrum is a function of θ; therefore, in any measurement, we obtain some average value for the spectrum. Further, the size of the scattering volume is proportional to $d\theta$. Thus, we gain in signal level, through increasing both $d\Omega$ and V. When we increase $d\theta$, however, we do this at the expense of spatial resolution and preciseness in our resolution of the spectrum. In general, we set $\Delta\theta$ as small as possible consistent with signal-to-noise and resolution requirements. This problem has been discussed for the magnetic modulation experiment by Carolan & Evans (1971). For x-ray scattering experiments, small $\Delta\theta$ is advantageous to resolve collective plasmon features (Gregori, 2006; Glenzer & Redmer, 2009).

6.4 SIGNAL-TO-NOISE RATIO

The minimum permissible values for beam energy W_i and beam duration $\delta\tau_i$ are determined by signal-to-noise requirements. At short wavelengths ($\lambda_s < 1\,\mu m$), where the photons are energetic, the main limitation is placed by quantum statistics, i.e., if our detector is 10% efficient and we bring up 1000 photons, in one detector response time, we will detect on the average 100 photons, but the precise number will vary from pulse to pulse leading to an uncertainty on a single measurement. In addition to this source of noise, there are contributions from plasma radiation, local thermal radiation from the detector surroundings, and circuit noise. The latter two problems, which are very important at long wavelengths ($\lambda_s > 1\,\mu m$), are discussed in Section 6.7.

6.4.1 Quantum Statistical Noise

Note: The collector system receives during the measurement a number of scattered photons N_p. Strictly, the numbers that concern us here are those that arrive within the integration time τ_I of the detector circuit; however, the integration time is normally set close to the pulse duration so that we may act as if all the scattered photons are counted (a good example is given in Section 9.1).

For a measurement of density, all the photons may be taken to one detector. More often the photons are dispersed and divided among a number of channels C in order that we may analyze the spectrum. Each channel feeds a separate detector, and these each receive approximately $N_{pc} = N_p/C = (P_s\tau_I)/h\nu_i C$ photons, where P_s is the scattered power.

Not all the photons are detected. For photomultipliers of the type used in the T-3 experiment, the quantum efficiency (η photoelectrons per photon) was approximately 2.5×10^{-2} at 7000 Å. As discussed in Section 6.7, today's detectors are more efficient.

For the situation in which the detection probability is small, statistical theory (Lambe, 1967; Oliver, 1965) tells us that the probability of obtaining (N_{pec}) photoelectrons is the Poisson distribution

$$P(N_{\text{pec}})=e^{-m}m^{-N_{\text{pec}}}/N_{\text{pec}}, \tag{6.4.1}$$

where $m=N_{\text{p}}\cdot\eta$ is the most probable number of photoelectrons. The variance of the measurement is $\sigma^2=m$. The standard deviation is $\sigma=m^{1/2}$.

6.4.2 Measurement on a Single Pulse

The signal from a given channel is therefore

$$N_{\text{pec}}=N_{\text{pc}}\cdot\eta\pm(N_{\text{pc}}\cdot\eta)^{1/2} \quad \text{photoelectrons.} \tag{6.4.2}$$

That is, if a channel detects 100 photoelectrons on a given pulse, the uncertainty in this result is ± 10 photoelectrons.

Put another way, our signal-to-noise ratio is

$$S/N=N_{\text{pec}}/(N_{\text{pec}})^{1/2}=(N_{\text{pec}})^{1/2}. \tag{6.4.3}$$

6.4.3 Average Value Measured over a Number of Pulses

Now, if we repeat the measurement s times, we improve on the accuracy of the mean value. The mean value is

$$\overline{N_{\text{pec}}}=\sum_s N^s_{\text{pec}}/s, \tag{6.4.4}$$

and the error of the mean is $\pm(N_{\text{pec}}/s)^{1/2}$. The signal-to-noise ratio is now

$$S/N=(s\overline{N_{\text{pec}}})^{1/2}. \tag{6.4.5}$$

Note that from pulse to pulse, the conditions of the plasma may change, and there may always be a definite spread in the measurement.

The standard deviation of the mean is

$$\overline{\sigma}=\left[\sum_s \frac{(N^s_{\text{pec}}-\overline{N_{\text{pec}}})^2}{s}\right]^{1/2}.$$

We must take care to differentiate between the accuracy of our average values and the standard deviation, which is a measure of the reproducibility of the experiment.

6.4.4 Plasma Noise

Unfortunately, the signal-to-noise ratio is made worse by the presence of plasma radiation. This radiation arises from free–free transitions (bremsstrahlung), free–bound transitions (recombination radiation), and bound–bound transitions (line radiation).

It has been argued that bremsstrahlung is the main cause of plasma radiation (Evans, 1969); however, this is at odds with the Tokamak T-3 measurements for which it was in fact line radiation from chromium and iron impurities and from hydrogen that caused problems. More to the point is the feeling that it is only for bremsstrahlung that we can make a general calculation to illuminate the problems, because the impurity level will vary drastically from one experiment to another. This calculation is presented in Section 6.5.

Let the number of plasma photons detected by a given channel be $N_{\mathrm{pl}} \cdot \eta = N_{\mathrm{plc}}$ photoelectrons. These additional photoelectrons must be added to the genuine signal in computing the signal-to-noise ratio, and this now becomes for a single pulse

$$S/N = N_{\mathrm{pec}}/(N_{\mathrm{pec}} + N_{\mathrm{plc}})^{1/2} = (N_{\mathrm{pec}})^{1/2}/[1 + (N_{\mathrm{plc}}/N_{\mathrm{pec}})]^{1/2}. \tag{6.4.6}$$

Again we must differentiate between the accuracy of a single measurement and the accuracy of the average of a number of measurements. For the latter case where there are m measurements,

$$S/N = (N_{\mathrm{pec}} m)^{1/2}/(N_{\mathrm{pec}} + N_{\mathrm{plc}})^{1/2}. \tag{6.4.7}$$

There will still be a spread in the measurements owing to the irreproducibility of the plasma, and the standard deviation of the mean is $\overline{\sigma}$ where

$$(\overline{\sigma})^2 = s^{-1}\left\{\sum_s [N^s_{\mathrm{pec}} + N^s_{\mathrm{plc}} - (\overline{N_{\mathrm{pec}} + N_{\mathrm{plc}}})]^2\right\} \tag{6.4.8}$$

In most circumstances, there will be no correlation between the scattered and plasma photons, and therefore, $(\overline{\sigma})^2 = (\overline{\sigma}_{\mathrm{pec}})^2 + (\overline{\sigma}_{\mathrm{plc}})^2$.

Three final points should be made in this general discussion.

1. It is assumed in the above that allowance is made for changes in the incident power, i.e., the incident power is monitored and each measurement is normalized to it.
2. From a practical point of view, we gain by working with polarized incident radiation, with the scattering plane perpendicular to the incident **E** vector (i.e., $\varphi_0 = 90°$, see Sections 1.7.2 and 2.3.2). The scattered radiation is all polarized perpendicular to the scattering plane (for high-temperature plasmas, see Chapter 4), and a polarizer set in this direction will discriminate against the plasma radiation. The latter is assumed to be unpolarized. For a single pulse with a polarizer,

$$S/N \cong N_{\mathrm{pec}}/(N_{\mathrm{pec}} + N_{\mathrm{plc}}/2)^{1/2}.$$

3. It is not only the noise on the plasma radiation that can cause problems.

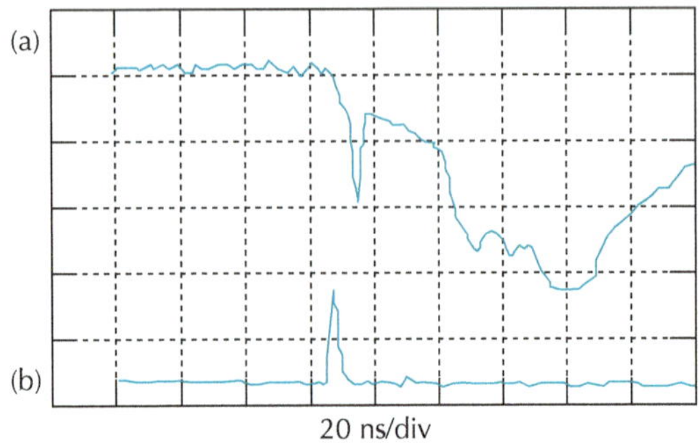

FIGURE 6.2 (a) This trace shows how rapidly varying plasma radiation can interfere with the resolution of the scattered signal. (b) Monitor pulse from laser.

(a) If the mean level of the radiation is varying on the same timescale as the incident beam, this can make it difficult to isolate the scattered signal (see Fig. 6.2).

(b) If the mean radiation level is high, there is sometimes a danger of draining the supplies of pulsed detectors prior to the scattering measurement. For this reason, it is common practice either to operate the detector only while scattering is occurring or to have a fast shutter in front of the detector.

6.5 RATIO OF SCATTERED POWER TO BREMSSTRAHLUNG RADIATION POWER

6.5.1 Bremsstrahlung Radiation

The electron–ion bremsstrahlung power emitted by the volume of plasma V_p into the solid angle $d\Omega$, and in the wavelength interval $\lambda_s \rightarrow \lambda_s + d\lambda_s$, is (Glasstone & Lovberg, 1960; Finkelburg & Peters, 1957; Gabriel *et al.*, 1962)

$$P_B d\Omega d\lambda_s = 2.09 \times 10^{-36} g Z^2 \left(\frac{n_e n_i}{\lambda_s^2 T_e^{1/2}} \right) \exp\left[-\left(\frac{1.24 \times 10^{-4}}{\lambda_s T_e} \right) \right] V_p \frac{d\Omega}{4\pi} d\lambda_s \text{ (W)}, \tag{6.5.1}$$

where Z is the charge on the ion. Generally, the Gaunt factor g is temperature and density dependent (Fortmann *et al.* 2006). For typical laboratory plasmas it is near unity for $\lambda_s \lesssim 10^{-5}$ cm and increases to ~5 in the far infrared (~10^{-2} cm).

In general, there will be numerous ions in the plasma and $P_B \propto \sum_Z \frac{Z^2 n_i}{n_e}$. Naito and Hatae (2003) show that it is possible to measure the effective ionization state Z_{eff} as a function of position using the background radiation signal provided that the line radiation is less than 10% of the background. For hydrogen (H_I) in the visible, this becomes

$$P_B d\Omega d\lambda_s = 2.09 \times 10^{-36} \left\{ \frac{n_e n_i}{\lambda_s^2 T_e^{1/2}} \right\} \exp\left[-\left(\frac{1.24 \times 10^{-4}}{\lambda_s T_e} \right) \right] V_p \frac{d\Omega}{4\pi} d\lambda_s \text{ (W)} \tag{6.5.2}$$

(n_e and n_i, cm^{-3}; V_p, cm^3; λ_s, cm; T_e, eV).

In order to determine the signal-to-noise ratio, we must calculate the ratio N_{pec} to N^B_{plc}. Strictly, we should start by integrating the equations for scattered and plasma power over the wavelength range covered by each channel, $\lambda_s \rightarrow \lambda_s + \Delta\lambda_c$, making due allowance for the wavelength response of all the components of the system. More realistically, because this calculation only sets a best possible signal-to-noise level, we make a number of practical approximations.

6.5.2 The Ratio of Scattered to Bremsstrahlung Power

To illustrate the problems caused by plasma radiation and so that we may compare our calculations with the Tokamak T-3 results, we consider the case of a hydrogen plasma $T_e > 10\,\text{eV}$ with scattering from a ruby laser source, $\lambda_i = 6943$ Å.

1. To a good approximation, the exponential term in Eqn (6.5.2) is unity, and P_B varies only slowly with wavelength.
2. In general, we work with $[\hat{s} \times (\hat{s} \times \hat{E}_{i0})]^2 \cong 1$.
3. $\Delta\lambda_{s(1/e)} \ll \lambda_i$, therefore $\lambda_s \approx \lambda_i$.
4. For simplicity, we divide the scattered spectrum equally among C channels. Note that when we integrate over the spectrum Eqn (1.8.8),

$$\int_{-\infty}^{+\infty} \left[b\,\text{cexp}(-b^2 x^2)/\pi^{1/2} \right] dx = 1.$$

5. Both the scattered and plasma photons follow the same optical path, and radiation is collected over the same solid angle and for the same length of time.

Now, we use Eqns (4.5.1) $\left[\text{ignoring the correction terms in } \frac{\Delta\lambda}{\lambda_i}\right]$ and Eqn (6.5.2) and obtain

$$\frac{N_{pec}}{N^B_{plc}} = \frac{\int^{\tau_I} dt P_i(t) r_0^2 \lambda_i^2 T_e^{1/2} 4\pi}{\tau_I\, CA 2.09 \times 10^{-36} n_e \Delta\lambda_c} \left(\frac{V}{V_p} \right), \tag{6.5.3}$$

where τ_I is the integration time of the detector circuit, V is the scattering volume, $r_0^2 = 7.95 \times 10^{-26}\,\text{cm}^2$, and A is the area of the incident beam.

6.5.3 Interesting Features of the Scattered to Bremsstrahlung Power Ratio

1. The scattering volume V and the plasma volume observed V_p are not the same (see Fig. 6.3). Obviously, in designing an experiment, if we have a choice, we must try and keep V_p to a minimum (see Section 6.3). The plasma volume does not radiate into the same solid angle as that subtended by the observer at the scattering volume. A calculation by DeSilva & Goldenbaum (1970) indicates that the equivalent plasma volume for scattering into $d\Omega$ is typically the image of the viewing aperture in the plasma times the thickness of the plasma slab along the line of sight.
2. The ratio is inversely proportional to the width of the scattered spectrum $\Delta\lambda \cong C\Delta\lambda_c$. For noncollective scattering from a thermal plasma, $\alpha \ll 1$; $\Delta\lambda_{1/e} = (2a/c)\lambda_i \sin(\theta/2)$; therefore, we gain by going to smaller scattering angles (see Section 6.3).
For collective scattering, $\alpha \gtrsim 1$ (this is discussed in Chapter 5), the spectral width is typically $(m_e/m_i)^{1/2}$ smaller, and the bremsstrahlung noise is less of a problem.
3. The value we find for Eqn (6.5.3) clearly depends on the integration time τ_I relative to the incident pulse duration $\delta\tau_i$.

   (a) It is desirable to work with $\tau_I \le \delta\tau_i$, though there is little virtue in working with $\tau_I \ll \delta\tau_i$ unless there is a sufficient number of scattered photons within the integration time, that we can measure the time variation of the scattered radiation.

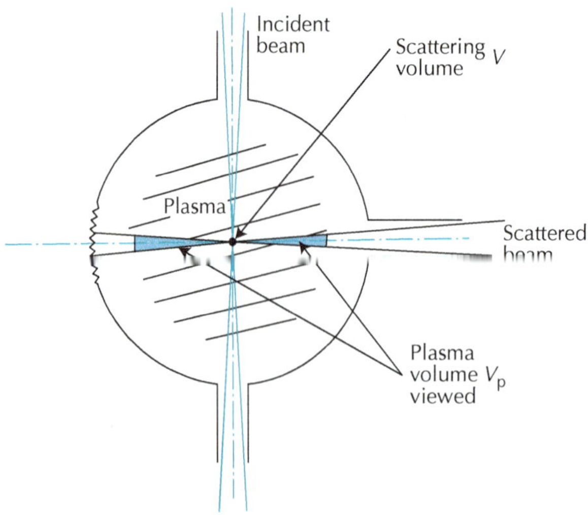

FIGURE 6.3 An illustration of the difference between the scattering volume and the plasma volume viewed by the collection optics.

(b) It is not to our benefit to work with $\tau_I > \delta\tau_i$ because we have reached the limit of scattered photons at $\tau_I = \delta\tau_i$ and are simply adding to the number of plasma photons. However, streak cameras may work in this limit where τ_i denotes the streak window. This set up allows one to subtract bremsstrahlung and background radiation from the scattering data by using measurements immediately before and after the laser pulse. Now, $\tau = \frac{1}{2}\Delta f$, where Δf is the frequency bandwidth of the detection system, and this shows that it is necessary for the detector system to have a large enough bandwidth to match the duration of the incident pulse.

(c) $\tau_I \cong \delta\tau_i$: We integrate Eqn (6.5.3) and find

$$\frac{N_{pe}}{N_{pl}^B} = \frac{W_i r_0^2 \lambda_i^2 T_e^{1/2} 4\pi}{\delta\tau_i 2.09 \times 10^{-36} Z_{eff} n_e 2\,\Delta\lambda_{1/e}} \frac{L}{V_p}, \tag{6.5.4}$$

where $C\Delta\lambda_c$ has been replaced by $2\Delta\lambda_{1/e}$ and W_i is the total energy in the incident beam in Joules and $L = V/A$ cm.

Now, it is in general true because of thermal boundary problems that the source power is greatest for the shortest duration pulses, and therefore, we gain on signal-to-plasma noise by working with a fast pulse (see Fig. 6.5).

This result must, however, be qualified by the requirement that W_i be large enough to give us an adequate number of photoelectrons for a significant measurement.

$$N_{pe} = (W_i/h\nu_i) r_0^2\, d\Omega\, n_e L T \eta, \tag{6.5.5}$$

where T is the transmission coefficient of the collection and detection system and η is the quantum efficiency of the detector.

6.5.4 Line Radiation

Line radiation power density is given by Jensen *et al.* (1977),

$$P_{LR} = \sum_Z n_e n_Z f(Z)\ (\mathrm{W\,m^{-3}}), \tag{6.5.6}$$

where $f(Z)$ is shown for some representative elements in Fig. 6.4.

6.5.5 Synchronous Detection

One approach to deal with an unfavorable signal-to-plasma radiation level, and this is particularly important at low plasma densities $n_e \lesssim 10^{14}\,\mathrm{cm^{-3}}$, is to use a modulated (probably CW) source and then detect the scattered radiation with an amplifier tuned to the modulation frequency. This of course does not discriminate against parasitic radiation. The technique has been applied by Koons & Fiocco (1968) to an arc discharge, $n \cong 5 \times 10^{12}\,\mathrm{cm^{-3}}$, $T_e \cong 1.2\,\mathrm{eV}$. They used an argon-ion laser 0.5 W, modulated at 385 Hz. In addition, for this demonstration, they could modulate the arc current (70 Hz), and this enabled them

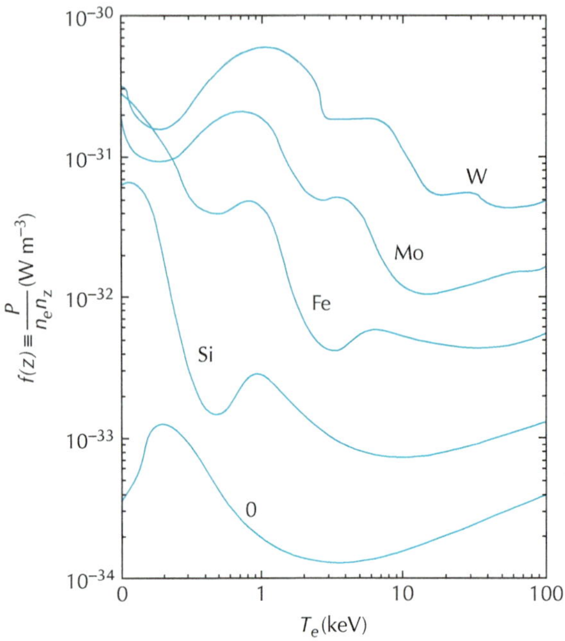

FIGURE 6.4 Line radiation factor $f(Z)$ as a function of electron temperature for representative impurities (Courtesy of R. V. Jensen (1977) and Nuclear Fusion.)

to discriminate against the parasitic radiation by detecting at the sum frequency (455 Hz). A schematic diagram of the apparatus is shown in Fig. 6.6. The results are impressive, and the rates of the photoelectron count of the background plasma radiation, parasitic radiation, and scattered radiation were, respectively, 10^5, 5×10^3, and 64 photoelectrons/s.

In general, it is not possible to modulate the plasma, and parasitic radiation must be removed by other means. In this connection, the homodyne (heterodyne) technique is valuable; the signal is mixed with radiation from another source, possibly the incidence source itself, and the signal at the difference frequency is detected. This is discussed in Section 6.7.5. Yokoyama *et al.* (1971) have used a CO_2 laser modulated at 2 kHz for a similar measurement of electron densities, $n_e \cong 10^{13} \rightarrow 10^{14}\,\text{cm}^{-3}$ in an arc discharge. The measured densities were in good agreement with interferometric measurements.

6.6 EFFECT OF THE INCIDENT BEAM ON THE PLASMA

6.6.1 Introduction

In Section 1.8.1, we looked briefly at the macroscopic interaction of an electromagnetic wave with a nonmagnetized plasma in the collisionless case. We found

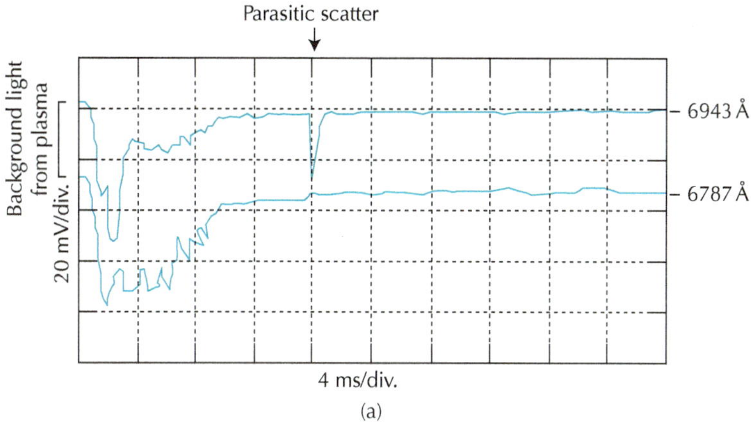

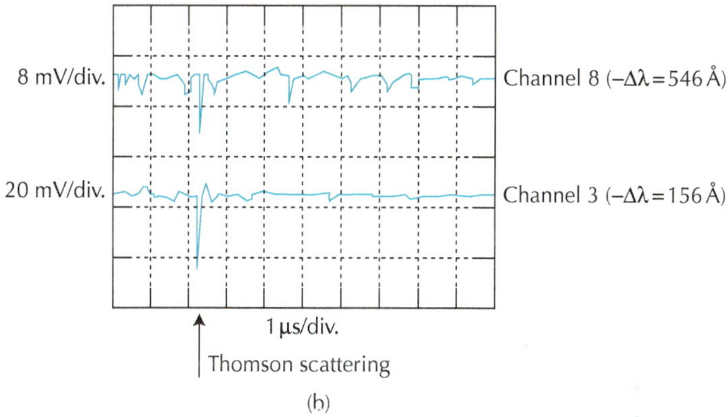

FIGURE 6.5 (a) In the relaxation mode of operation, the scattered signal is swamped by the background radiation. The parasitic signal at 6943 Å indicates where the scattered signal is buried. (b) In the fast pulsed mode, the scattered signal exceeds the background radiation. (Courtesy of M. J. Forrest *et al.* (1970), Culham Lab., U.K.)

that for $\omega_i > \omega_{pe}$, the wave is all transmitted, while for $\omega_i < \omega_{pe}$, it is all reflected. In real life, a given charge in the plasma (i.e., electrons at the higher frequencies $\omega_i \gtrsim \omega_{pe}$) cannot oscillate freely in the wave; it will interact with other charges and with neutrals, and some of the incident energy will be dissipated in the plasma. The order of magnitude of the dissipation can be determined by a straightforward classical calculation using the Langevin equation to represent the motion of the electrons. A simple collision frequency ν is used to represent the given dissipational process.

For a plasma in equilibrium, in the linear approximation we have

1. Direct heating via electron–ion collisions. In the quantum mechanical treatment, this is termed "inverse bremsstrahlung." In the vicinity of natural

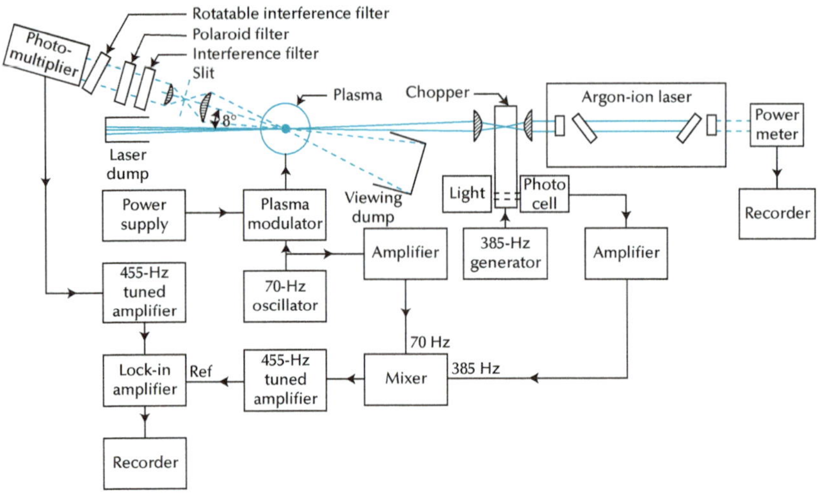

FIGURE 6.6 Schematic diagram of apparatus used for the synchronous detection of scattered radiation from a low-density arc. (Courtesy of H. C. Koons & G. Fiocco (1968).)

plasma resonances, ω_{pe}, Ω_e, we must correct the simple collision term to allow for coupling of the radiation to waves in the plasma.[4]

2. Direct heating via electron–neutral collisions and ionization of the background neutrals.

   At a very high level of incident intensity, we must include nonlinear effects; therefore, in addition, we have

3. A modification of the inverse bremsstrahlung absorption cross section by multiphoton processes.
4. An increase in the effective collision frequency because the electrons excite plasma waves, and these may be Landau damped or decay via wave-coupling processes. Not only does this lead to the transfer of more energy to the plasma but in addition the scattered spectrum is enhanced.
5. Finally, if we drive the plasma hard enough, particularly if ω_i or its harmonics and subharmonics come close to a natural resonance, we may cause instabilities (see Section 12.7). Estimates of the absorption of radiation when ω_i is near a natural resonance have been made by, for example, Kaw *et al.* (1970), Kruer *et al.* (1970), Porkolab (1971), Freidberg *et al.* (1972), Yamanaka *et al.* (1972), and Martineau & Pepin (1972).

This latter regime is sometimes not suitable for scattering measurements, and our main interest lies in scattering when $\omega_i \gg \omega_{pe}$, Ω_e. However, scattering is undertaken at UHR resonances, e.g., Gusakov *et al.* (2006).

---

[4]The nonmagnetized case is discussed by Dawson & Oberman (1962) and the magnetized case by Aliev *et al.* (1966)

6.6.2 Classical Collisional Dissipation

The electrons are accelerated by the incident wave, and some of the directed energy they gain is randomized by collisions.

$$\frac{d\mathbf{u}}{dt} = -\frac{e}{m_e} E_{i0} \cos(\mathbf{k}_i \cdot \mathbf{r} - \omega_i t) - \nu \mathbf{u}, \tag{6.6.1}$$

where

$$\mathbf{r} = \int_0^t \mathbf{u}\, dt + \mathbf{r}(0).$$

Now for an electromagnetic wave, $\mathbf{k}_i$ and $\mathbf{E}_{i0}$ are orthogonal, and if we ignore the magnetic field of the wave (i.e., this is a nonrelativistic treatment, $\omega_i/k_i = c \gg u$), we see that any change in position $\mathbf{r}$ induced by the wave will be perpendicular to $\mathbf{k}_i$. We set $\mathbf{u} = \mathbf{u}_k + \mathbf{u}_E + \mathbf{u}_\perp$. Equation (4.6.1) is integrated and

$$\mathbf{u} = -\frac{e\mathbf{E}_{i0}}{m_e[\nu^2 + (\omega_i - k_i u_k)]^2} \{\nu \cos(\mathbf{k}_i \cdot \mathbf{r} - \omega_i t) - (\omega_i - k_i u_k) \sin(k_i \cdot \mathbf{r} - \omega_i t)\}. \tag{6.6.2}$$

The average power dissipated per unit volume is

$$W_D = \frac{1}{2T} \int_{-T}^{+T} dt (\mathbf{J} \cdot \mathbf{E}) = \frac{n_e e^2 \nu}{m_e c \varepsilon_0 \omega_i^2} \frac{P_i}{A} (\mathrm{W m^{-3}}), \tag{6.6.3}$$

where $\mathbf{J} = n_e e\mathbf{u}$ is the current density induced by the incident field. The incident power per unit area is

$$P_i/A = c\varepsilon_0 E_{i0}^2/2 \ (\mathrm{W\,m^{-2}})$$

with

$$\varepsilon_0 = 8.85 \times 10^{-12}\ \mathrm{F\,m^{-1}}, \quad n_e\ (\mathrm{m^{-3}}), \quad c\ (\mathrm{m\,s^{-1}}),$$
$$\omega_i\ (\mathrm{rad\,s^{-1}}), \quad \nu\ (\mathrm{s^{-1}}), \quad e = 1.6 \times 10^{-19}\ \mathrm{C}, \quad m_e = 0.911 \times 10^{-30}\ \mathrm{kg}.$$

6.6.3 Coulomb Collisions

It is the electrons that gain the random energy in the collision. Kunze (1968) has compared the energy deposited per unit volume by the incident waves with the electron thermal energy density $U_0 = \frac{3}{2}\kappa T_e n_e$. The relative change in electron temperature is

$$\frac{\Delta T_e}{T_e} = \frac{\int_0^\tau \overline{W_D}\, dt}{\frac{3}{2}\kappa T_e n_e}; \tag{6.6.4}$$

τ is the incident pulse duration.

For $\tau > \tau_{ie}$ the equilibration time with the ions, the energy is shared with the ions in the scattering volume.

For $\tau > \tau_{Tc}$ the thermal conduction time, the heat is conducted away to the neighboring plasma.

The maximum temperature rise occurs for $\tau < \tau_{ie}, \tau_{Tc}$. Now, for singly charged ions,

$$\nu_{ei} = 2.9 \times 10^{-2} n_i(m^{-3})[T_e(\mathrm{eV})]^{-3/2} \ln\Lambda\,\mathrm{s}^{-1}, \tag{6.6.5}$$

where $\Lambda = 12 n_e \lambda_{De}$, usually $\ln\Lambda \sim 10$. We substitute and obtain

$$\frac{\Delta T_e}{T_e} = 1.28 \times 10^2 \frac{n_i \ln\Lambda}{\omega_i^2 A (T_e)^{5/2}} \int_0^{\tau} P_i\, dt. \tag{6.6.6}$$

Inspection of the equation shows that the problems will be most severe at high density, and low temperature, and for the smaller incident frequencies.

Example

$$n_e = 10^{21}\,\mathrm{m}^{-3}, \qquad T_e = 1\,\mathrm{eV} \qquad (\ln\Lambda = 6),$$
$$\omega_i = 2.7 \times 10^{15}\,\mathrm{rad\,s^{-1}} \qquad (\text{ruby laser}),$$

and the total incident energy $U_T = 6\,\mathrm{J}$. For $A = 10^{-6}\,\mathrm{m}^2$, we obtain

$$\Delta T_e / T_e \cong 1.$$

For a CO_2 laser,

$$\omega_i = 1.76 \times 10^{14}\,\mathrm{rad\,s^{-1}}$$

of the same energy.

$$\Delta T_e / T_i \gtrsim 1 \quad \text{for} \quad n_i = 10^{21}\,\mathrm{m}^{-3} \quad \text{when} \quad T_e \leq 10\,\mathrm{eV}.$$

6.6.4 Quantum Mechanical Effects

In quantum mechanical terms, this collision process is inverse bremsstrahlung (Spitzer, 1962; Bekefi, 1966). Radiation is *absorbed* by an electron as it has a Coulomb interaction with an ion. There is in addition induced bremsstrahlung emission, this reduces the net absorption, and the quantum correction factor is

$$(h\omega_i/\kappa T_e)(1 - \exp(-h\omega_i/\kappa T_e)).$$

This factor tends to unity as $h\omega_i/\kappa T_e \to 0$; the correction is therefore negligible for most scattering cases with the exception of x-ray probes (see Chapter 11). Note that even for the energetic photons of a ruby laser, $h\omega_i$ is only 1.79 eV.

For completeness, we note that for very intense incident beams, there are corrections (Rand, 1964; Nicholson-Florence, 1971) to this result. Multiple photon processes are important when the energy gained by the electron in the incident field $\left(\frac{1}{2}m_e u_0^2 \text{ where } u_0 = \frac{eE_{i0}}{m_e \omega_i}\right)$ becomes comparable with the photon energy ($h\omega_i$).

Example

$$\frac{m_e u_0^2}{2h\omega_e} = \frac{e^2 E_{i0}^2}{2m_e \omega_i^3} = \frac{10^{29}}{[\omega_i \, \mathrm{rad\,s^{-1}}]^3}\left[\frac{P_i}{A}\mathrm{W\,m^{-2}}\right] \quad \text{for} \quad P_i = 10^9\,\mathrm{W}.$$

$$A = 10^{-6}\,\mathrm{m}^2 \quad \text{for} \quad \omega_i = 2.7\times 10^{15}\,\mathrm{rad\,s^{-1}},$$

$$m_e u_0^2 / 2h\omega_i \cong 5\times 10^{-3}.$$

Thus, there is no significant correction for a typical ruby laser, but a correction might be necessary for a CO_2 laser at comparable power density.

6.6.5 Collisions with Neutrals and the Ionization of Neutrals

The electron–neutral collision frequency is (see Section 2.2.2)

$$\nu_{en} = 2.8\times 10^6 [r_n(\mathrm{m})]^2 n_n(\mathrm{m}^{-3})[T_e(\mathrm{eV})]^{1/2}\mathrm{s}^{-1}. \qquad (6.6.7)$$

The radius of the neutral r_n is typically 10^{-10} m.

Again it is in low-temperature plasmas that the effect will be important. For hydrogen, at most densities, essentially complete ionization is obtained (Tanenbaum, 1967, pp. 12, 348) for $T_e > 1$ eV.

Let

$$\ln\Lambda = 10, \quad r_n = 10^{-10}\,\mathrm{m},$$

then,

$$\nu_{en}/\nu_{ei} \cong 10^{-3}(n_n/n_i)[T_e(\mathrm{eV})]^2.$$

Clearly, neutrals play a role in randomizing the electron motion only in very lightly ionized plasmas, and in general, this effect is not significant.

Ionization can occur either by direct electron impact or by photoionization. In the former case, we require

$$\frac{1}{2}m_e \mathrm{v}_{max}^2 > eV_i \quad \text{or} \quad \frac{e}{m_e c\varepsilon_0 \omega_i^2}\frac{P_i}{A} > V_{i.}$$

$\omega_i \gg \omega_{pe}, \nu$

6.6.6 Collective Effects

$\omega_i \gg \omega_{pe}, \nu$

If the electrons are accelerated to velocities comparable with their thermal speed, they can excite electron plasma waves. The growth of these waves is balanced by Landau damping, and in principle, we can also have decay into ion-acoustic waves. This is a mechanism for transferring energy from the incident field to the plasma, but of course, it only occurs for very high-incident power densities. We require

$$v_{max} = \frac{eE_{i0}}{m_e\omega_i} = \frac{e}{m_e\omega_i}\left(\frac{2P_i}{c\varepsilon_0 A}\right)^{1/2} \cong \left(\frac{\kappa T_e}{m_e}\right)^{1/2}$$

or

$$\frac{1.14\times 10^7}{\omega_i(\text{rad s}^{-1})}\left(\frac{P_i}{A}\frac{W}{m^2}\right)^{1/2} \cong [T_e(\text{eV})]^{1/2}. \qquad (6.6.8)$$

Example

$$P_i = 10^9\,\text{W}, \quad A = 10^{-6}\,\text{m}^2, \quad \omega_i = 2.7\times 10^5\,\text{rad s}^{-1}$$

(ruby laser). The effect may be important for $T_e \lesssim 2\,\text{eV}$.

For a CO_2 laser, $\omega_i = 1.76\times 10^{14}\,\text{rad s}^{-1}$, giving the same power density the effect may be important for $T_e \lesssim 40\,\text{eV}$.

This nonlinear effect, where Landau damping balances the growth of the waves, has been treated by Albini & Rand (1965). They calculate that the power dissipated per unit volume by this process is

$$W_{DNL} = 2.3\times 10^{-31}\frac{n_i(\text{m}^{-3})n_e(\text{m}^{-3})}{[T_e(\text{eV})]^{1/2}}F(B,\lambda)\,\frac{\text{W}}{\text{m}^3}, \qquad (6.6.9)$$

where

$$B \equiv 0.21\times 10^{30}[T_e(\text{eV})]^3/[\omega_i\,(\text{rad s}^{-1})]^2$$

and

$$\lambda = \frac{0.64\times 10^{14}}{T_e\,(\text{eV})[\omega_i\,(\text{rad s}^{-1})]^2}\left(\frac{P_i}{A}\frac{\text{W}}{\text{m}^2}\right).$$

For $B\lambda \ll 1$,

$$F(B,\lambda) = \lambda\int_{1/B}^{\infty} e^{-y}dy/y.$$

It must be remembered that this is a calculation for $\omega_i \gg \omega_{pe}$, and therefore, ω_i values must be consistent with this at the given plasma density.

6.7 DETECTORS

6.7.1 Photoelectric Detectors

In these devices, incident radiation releases electrons, which are subsequently detected, either into vacuum or into the conduction band. At high photon energies of ~1000 to 15,000 Å (1.5 μm), the process is photoemission. In a photomultiplier, photoelectrons first emerge when the incident photon energy exceeds the work function of the phocathode which has motivated the development of low work function cathodes for long wavelength light detection, as well as high quantum efficiency. To provide measurable current levels, the electrons are "multiplied." In the traditional photomultiplier tube (PMT), the photoelectrons are accelerated toward an electrode denoted as a "dynode" with the potential difference adjusted to maximize the secondary emission yield. The secondary electrons in turn impinge upon another dynode with the process continued until a final gain of $\sim 10^5 - 10^7$ is achieved. Secondary electrons from the final dynode are collected by an anode, which provides an external current through a resistor ($\sim 100\,\Omega$) or a high impedance follower circuit, and the voltage is subsequently measured. A review of photomultiplier technology may be found in Boutot *et al.* (1987) and Hungerford & Birch (1996).

Another "multiplication" process (Dhawan & Majka, 1977) involves the use of microchannel plates (MCPs), which can be thought of as arrays of PMTs since they consist of many (millions) of microscopic channels/pores (4 to 25 μm in diameter) in a thin disk of Pb oxide glass with the channels running parallel to each other from one face to the other. Each channel functions like a tiny image intensifier: electrons striking the walls eject additional secondary electrons resulting in a cascade of electrons with the channels providing electron confinement and thus improved resolution. As discussed later, charge-coupled devices (CCDs) dominate in the optical region, with MCPs finding greater use in UV and x-ray detection.

Due to the differences in the multiplication processes, photocathodes are configured as opaque or semitransparent, where the former refers to the case where the light is incident on a thick photoemissive material with the electrons emitted from the same side. In the second type, a semitransparent photocathode is made by the deposition of photoemissive material on a transparent medium so that the electrons are emitted from the back of the photocathode opposite the incident light. Depending upon the spectral region of interest, a variety of cathode materials are employed including CsI, CsTe, Cs_3Sb, bialkali, trialkali, GaAs(Cs), InGaAs(CS), and InP/InGaAsP(Cs). Figure 6.7 illustrates the spectral response in the visible for GaAs and InGaAs MCP/PMTs. Tables 6.3 and 6.4 summarize the properties of common photocathode materials. In an image converter, the same process is used, but the final electrons impinge on a photosensitive surface, and an enhanced picture of the incident radiation is received.

Typical photochathode materials in the 50 Å to 2000 Å UV region have been of alkali halides such as CsI and KBr (Siegmund, 1998; Tremsin & Siegmund, 2000). Development activities on other materials (GaN and diamond) have yielded encouraging results; GaN phocathodes have achieved detection $\eta > 50\%$ in the UV, while diamond photocathodes have been made with detection quantum efficiency of 40% and sensitivity up to 200 nm (Siegmund, 2004). A variety of techniques have been employed in the x-ray region. For example, one may employ a "hybrid" PMT where a scintillating window precedes the PMT (D'Ambrosio & Leutz, 2003). Researchers have also reported results of

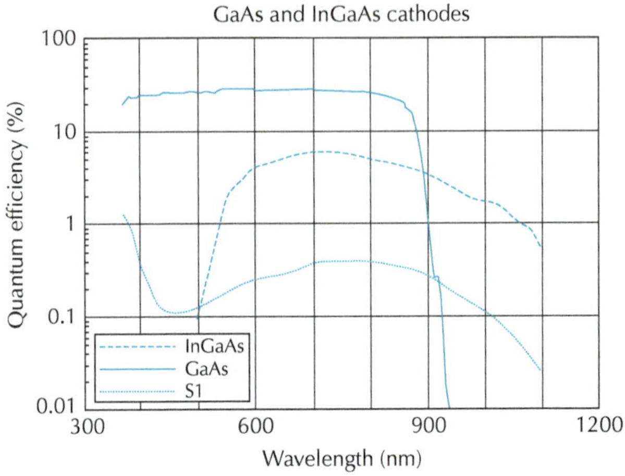

FIGURE 6.7 Spectral response in the visible for GaAs and InGaAs MCP/PMTs and comparison with S1 response. (Courtesy of Burle Industries Inc.)

studies regarding the application of large area avalanche photodiodes as soft x-ray detectors (Yatsu *et al.*, 2006).

At lower energies, the photons cannot excite electrons over the surface barrier, and the lower energy transition between the valence and conduction band is used for detection. In a pure (intrinsic) material, this band can be as low 0.1 eV, and this allows detection out to 10 μm.[5] To extend the operation to longer wavelengths, compound semiconductors are employed. A frequently employed material is HgCdTe with a band gap which can be varied ($1.00-0.09$ eV) so that the critical wavelength can in turn be varied ($1.24-14$ μm). These are operated cooled ($20-80$ K), thereby providing low noise and high efficiency (Lerner, 2009).

Beyond this region, excitation of impurity levels in the forbidden gap of an extrinsic semiconductor permits detection in the range $10-200$ μm corresponding to the binding energies (a few meV to 10's of meV) of the shallow levels in Si, Ge, and GaAs. A frequently employed detector material is p-type gallium-doped germanium (Ge:Ga), which results in detectors with a cutoff wavelength of ~ 115 μm. Operated at 1.8 K, they exhibit a responsivity of 7 A/W, $\eta = 20\%$, and a dark current $< 200 e^-/s$ (Young, 2000). By applying uniaxial stress to the detector,[6] the response can be extended to ~ 200 μm.

A blocked-impurity-band (BIB) detector configuration consists of a doped absorbing layer coupled with a high-purity blocking layer to suppress the contribution of the hopping conduction to the dark current. The dopant concentration may be increased leading to both an enhanced linear optical absorption coefficient and longer wavelength response. The use of lightly doped GaAs:Se

[5] In the case of intrinsic Si and Ge, the long wavelength cutoff is 1.1 μm and 1.8 μm, respectively.

[6] Thereby splitting the degeracy of the Γ_4 valence band edge and consequently reducing the energy difference between the bound acceptor states and the top of the valence band.

TABLE 6.3 Quick Reference for Typical Spectral Response Characteristics

Curve Code (S Number)	Photocathode Material	Window Material	Luminous Sensitivity (Typ.) (μA/1m)	Spectral Response: Spectral Range (nm)	Peak Wavelength: Radiant Sensitivity (mA/W)	Radiant Sensitivity (nm)	Quantum Efficiency (%)	Quantum Efficiency (nm)
100 M	Cs-1	MgF_2	—	115 to 200	14	140	13	130
200 S	Cs-Te	Quartz	—	160 to 320	29	240	14	210
200 M	Cs-Te	MgF_2	—	115 to 320	29	240	14	200
400 K	Bialkali	Borosilicate	95	300 to 650	88	420	27	390
400 U	Bialkali	UV	95	185 to 650	88	420	27	390
400 S	Bialkali	Quartz	95	160 to 650	88	420	27	390
401 K	High temp. bialkali	Borosilicate	40	300 to 650	51	375	17	375
500 K (S-20)	Multialkali	Borosilicate	150	300 to 850	64	420	20	375
500 U	Multialkali	UV	150	185 to 850	64	420	25	280
500 S	Multialkali	Quartz	150	160 to 850	64	420	25	280
501 K (S-25)	Multialkali	Borosilicate	200	300 to 900	40	600	8	580
502 K	Multialkali	Borosilicate (prism)	230	300 to 900	69	420	20	390
700 K (S-1)	Ag-O-Cs	Borosilicate	20	400 to 1200	2.2	800	0.36	740
—	InP/InGaAsP(Cs)	—	—	950 to 1400	10	1250	1.0	1000 to 1200
—	InP/InGaAs(Cs)	—	—	950 to 1700	10	1550	1.0	1000 to 1200

Courtesy of Hammamatsu.

TABLE 6.4 Quick Reference for Typical Spectral Response characteristics

Curve Code (S Number)	Photocathode Material	Window Material	Luminous Sensitivity (Typ.) (μA/1 m)	Spectral Response				
				Spectral Range (nm)	Peak Wavelength			
					Radiant Sensitivity		Quantum Efficiency	
					(mA/W)	(nm)	(%)	(nm)
150 M	Cs-1	MgF2	—	115 to 200	25.5	135	26	125
250 S	Cs-Te	Quartz	—	160 to 320	62	240	37	210
250 M	Cs-Te	MgF2	—	115 to 320	63	220	35	220
350 K (S-4)	Sb-Cs	Borosilicate	40	300 to 650	48	400	15	350
350 U (S-5)	Sb-Cs	UV	40	185 to 650	48	340	20	280
351 U (Extd S-5)	Sb-Cs	UV	70	185 to 750	70	410	25	280
452 U	Bialkali	UV	120	185 to 750	90	420	30	260
456 U	Low dark bialkali	UV	60	185 to 680	60	400	19	300
552 U	Multialkali	UV	200	185 to 900	68	400	26	260
555 S	Multialkali	UV	525	185 to 900	90	450	30	260
650 U	GaAs(Cs)	UV	550	185 to 930	62	300 to 800	23	300
650 S	GaAs(Cs)	Quartz	550	160 to 930	62	300 to 800	23	300
851 K	InGaAs(Cs)	Borosilicate	150	300 to 1040	50	400	16	370
—	InP/InGaAsP(Cs)	Borosilicate	—	300 to 1400	10	1250	1.0	1000 to 1200
—	InP/InGaAs(Cs)	Borosilicate	—	300 to 1700	10	1550	1.0	1000 to 1200

Courtesy of Hammamatsu.

and GaAs:Te has made possible detectors operating in the $170-320\,\mu m$ region (Watanabe *et al.*, 2008). Their respective responsivities were $1.46\,A/W$ and $1.86\,A/W$ at a wavelength of $287\,\mu m$ with the GaAs:Te sample having an noise-equivalent-power (NEP) of $8.52\times 10^{-15}\,WHz^{-1/2}$. Longer wavelength responses can be obtained with InAs and InSb extrinsic semiconductors, which possess electron (hole) binding energies lower than those in GaAs. Specifically, the electron binding energies are $\sim 1.4\,meV$ for InAs and $\sim 0.7\,meV$ for InSb, which lead to photoconductive onset of $886\,\mu m$ and $1772\,\mu m$, respectively (Haller & Beeman, 2002).

Another important IR detector is the quantum-well IR photoconductor (QWIP), which employs IR excitation of carriers in quantum wells formed by growing layered structures of wide-band-gap materials (Rogalski, 2003). Here, band-gap "engineering" is employed to tailor the material such that the energy difference between selected states in the structure matches the energy of the infrared photons to be detected.

At the very lowest energies, direct absorption by electrons in the conduction band is used. At very low temperatures ($\sim 4\,K$), the resultant change in mobility is detectable out to $1000\,\mu m$. It is necessary to cool these long wavelength detectors, because with the small energy gaps, thermal agitation at room temperature can cause significant excitation of electrons, and this leads to electrical noise. The various regions are indicated in Fig. 6.7 along with some typical quantum efficiencies for photomultipliers. For detectors in the infrared, the quantum efficiency varies from $0.15 < \eta < 1.0$, but is typically 0.05.

6.7.2 Thermal Detectors

In these devices, incident radiation is absorbed by the body of the detector, and some temperature-dependent quantity, such as resistivity, is monitored. They have the advantage that they may be used over a whole spectral range. However, poor frequency response and sensitivity have limited their usefulness. The use of micromachining and thin-film lithographic fabrication techniques combined with new materials and antenna coupling schemes has resulted in sensitive, high-speed microbolometers, which have found widespread application in the millimeter wave and THz region (Dietlein *et al.*, 2007; Miller *et al.*, 2004; Grossman *et al.*, 2004; Middleton & Boreman, 2006; Rebeiz *et al.*, 1987; Neikirk *et al.*, 1984).

A readily fabricated detector element is the bolometer whose resistance depends upon temperature which is increased by radiation. A limitation associated with conventional bolometers is their relatively large thermal mass which limits their speed of response, as well as their NEP.[7] This problem led to efforts to reduce the size of the element and hence the thermal mass (typical

[7] In the case of a phonon-noise-limited bolometer, the NEP is given by $(4\kappa T^2 C/\tau_0)^{1/2}$ where C is the heat capacity, G is the thermal conductance between the bolometer and the heat sink, and $\tau_0 = C/G$ is the thermal time constant.

microbolometer dimensions are $4\,\mu m^2$ and 1000 Å thick). However, the reduced element size degraded the collection efficiency; a problem ameliorated by the development of substrate antennas for radiation collection with the temperature change confined to the bolometer element. The response time and NEP were further improved by reducing the thermal coupling to the substrate by techniques such as detector fabrication on thin membranes[8] (Rebeiz *et al.*, 1987). The goal was to achieve a large temperature, coefficient of resistance α, as defined $\alpha = \frac{1}{\rho}\frac{d\rho}{dT}$. This factor determines the responsively, which is given by $\Re_V = \frac{i_{bias} R \alpha \eta}{G\sqrt{1+\omega^2\tau^2}}$ where R is the resistance of the bolometer and i_{bias} is the current bias through the bolometer. The original microbolometers of Neikirk *et al.* (1984) employed bismuth, which has an α of only $0.003\,K^{-1}$. By employing tellurium, this figure was improved by a factor ~ 3 (Wentworth & Neikirk, 1989). Miller *et al.* (2004) have reported an un-cooled THz Nb microbolometer (~ 35 nm film thickness), which exhibited a responsivity in excess of 85 V/W at 95 GHz with an NEP of $2.5 \times 10^{-11}\,WHz^{-1/2}$.

Yet higher values of temperature coefficient of resistance ($> 0.50\,K^{-1}$) are possible by operating a superconductor near its transition temperature. Such bolometers, fabricated with a superconducting thin film and operated near the transition edge, exhibit high sensitivity and are referred to as transition-edge microbolometers or transition edge superconducting (TES) microbolometers (Kenyon *et al.*, 2006; Wentworth & Neikirk, 1990).

Hot-electron direct detectors (HEDD) utilizing micron-size hot-electron sensors made from thin Ti and Nb films have been shown capable of detecting single quanta of THz radiation due to the small heat capacity of electrons (Karasik *et al.*, 2003). At 0.3 K, an NEP of $\sim 10^{-19}\,WHz^{-1/2}$ was obtained and decreased to 10^{-20} at 0.1 K. One micron long devices made from a 12 nm thick Nb film on a Si substrate with a planar spiral antenna exhibited a flat response over the range 250 − 900 GHz. Nanobolometers using titanium and niobium flims on silicon substrates covered with a layer of SiO_2 or Si_3N_4 have yielded an NEP of $9 \times 10^{-21}\,WHz^{-1/2}$ at ~ 65 mK (Wei *et al.*, 2008). References to other THz thermal detectors may be found in Siegel (2002).

6.7.3 X-Ray Detectors

For scattering experiments with x-ray probes charged-coupled device (CCD) cameras are routinely used for detection, primarily because of their high detection efficiency and versatility. The quantum efficiency (η) approaches values close to unity for x-rays with energies $E < 4$ keV decreasing to 0.1 for $E =$ 10 keV and becomes negligible for energies $E > 20$ keV. The CCD linearity is $\pm 0.25\%$ for signals between $< 0.5\%$ and $> 80\%$ of the full dynamic range with a readout noise of < 5 electrons per pixel.

[8] The use of thin membranes also reduces substrate modes.

For higher x-ray energies and also in high *emp*, neutron and high gamma flux environments, film has traditionally been the alternative to CCD cameras and is now being replaced by image plates (IP) (see Izumi *et al.* (2006)). These x-ray detectors are sensitive up to x-ray energies of about 100 keV with $\eta > 0.03$ and less sensitive than CCD cameras for $E < 4$ keV. For energies of $2\,\text{keV} < E < 20\,\text{keV}$, the quantum efficiency of IP detectors has been estimated by Amemiya & Miyahara (1988) to be 60 – 80%. The high values are explained by high absorption efficiency of the phosphor. The dynamic range of $1:10^5$ with a high linearity and the spatial resolution of 150 μm (full width at half maximum) match the requirements for measuring the x-ray spectra from scattering experiments in dense plasmas.

For long duration x-ray probes produced by energetic nanosecond laser beams, scattering spectra have been measured with temporal resolution of ~ 100 ps by using microchannel plate detectors. Quantum efficiency values (back illuminated) are in the range of 0.01 – 0.1 for energies up to 100 keV Bateman (1977). Higher efficiencies of CsI photo cathodes by one order of magnitude have been reported by Lowney *et al.* (2004) for use with grazing incidence x-rays with energies $E \leq 1$ keV. For pulsed microchannel plate detectors, the dependence on the angle of incidence has been investigated by Landen *et al.* (2001b). A factor of about three improvements can be achieved when applying grazing incidence detection at $\sim 3°$ as opposed to standard configurations that use $\sim 8°$.

In the future, CMOS (complementary metal oxide semiconductor) image sensors may be also applied. The quantum efficiency is comparable to front illuminated CCD detectors but with a fill factor of 65%. Their read out noise as well as gain non-uniformity is significantly larger than for CCD cameras; the noise is 200 electrons per pixel with 3.6 eV per electron hole pair. The CMOS development is directed toward smaller features and lower voltage both affecting imaging performance, but also implies less sensitivity to neutron-induced noise. For neutron energies of 3 to 10 MeV, CMOS detectors have $< 2 \times 10^{-10}$ damage sites per pixel and per incident neutron. CCDs have 4×10^{-8} damage sites per pixel and per incident neutron, two orders of magnitude more than for CMOS. These features indicate that strong shielding is required in a neutron environment and will likely limit routine scattering experiments with these detectors to neutron flux environments of $< 10^{14}/4\pi$ neutrons.

Photon (x-ray) Estimates

Estimating that the thermal He$-\alpha$ foil sources provide a conversion efficiency of about $\eta_x \simeq 0.004$ at 4.75 keV into a bandwidth of $\Delta E/E = 0.5\%$ provides a total photon number of 5×10^{12} photons per incident laser energy in joules and into a solid angle of 4π sr. Assuming laser pulses with energy of 1 kJ per 100 ps and a dense plasma subtending a solid angle of $d\Omega = 1$ sr will yield a peak x-ray brilliance of $B(4.75\,\text{keV}) = 5 \times 10^{12}$ photons $\times 10^3 \times (4\pi)^{-1} = 4 \times 10^{14}$ photons/100 ps/0.5% bw/sr at the dense plasma in a single shot.

In x-ray scattering experiments, the intense x-ray source and the closely coupled geometry results in a substantial fraction of scattered photons. For $n_e = 3 \times 10^{23}\,\text{cm}^{-3}$, the Thomson scattering cross section, and a path length of $L = 0.1\,\text{cm}$, the scattering fraction is $n_e \sigma_T L = 0.02$, close to the maximum desirable for avoiding multiple scattering. Coupled with a source solid angle of 1 sr, the scattered efficiency is 10^{-3}, which is substantially larger than that available for optical Thomson scattering experiments allowing single-shot experiments with 10^{12} photons at the plasma sample.

The total number of the detected photons, N_p, is estimated from Eqn (11.2.6)

$$N_p = \left(\frac{E_L}{h\nu}\eta_x\right)\left(\frac{\Omega_{\text{plasma}}}{4\pi}\eta_{\text{att}}\right) \times \left(\frac{n_e \sigma_T L}{1+\alpha^2}\right)\left(\frac{\phi_c R_{\text{int}}}{4\pi}\eta_d\right), \tag{6.7.1}$$

where E_L is the laser energy, η_x is the conversion efficiency from the laser energy into the probe x-rays, η_{att} is the attenuation of the probe x-rays through the dense plasma, Ω_{plasma} is the solid angle that the plasma subtends with respect to the x-ray source, ϕ_c is the acceptance angle determined by the crystal or detector width and distance from the plasma, R_{int} is the integrated reflectivity of the crystal, and η_d is the efficiency of the detector including the MCP efficiency and filter transmission. The factor $(1+\alpha^2)^{-1}$ applies for estimates of the inelastic scattering signal (see Eqn (5.6.3)).

With $\eta_{\text{att}} \approx 1/e$, a killojoule laser produced He$-\alpha$ source at 4.75 keV effectively creates

$$\frac{E_L}{h\nu}\eta_x \frac{\Omega_{\text{plasma}}}{4\pi}\eta_{\text{att}} \approx 2 \times 10^{14}$$

photons at the target. The collection efficiency of $0.1\,\text{rad} \times 3\,\text{mrad}/4\pi \simeq 3 \times 10^{-5}$ (Section 7.4.2), and a detection efficiency for a MCP detector plus filter of $\eta_d \simeq 0.01$, results in a collection fraction of 3×10^{-7}. Combined with a scattering fraction of 0.02, the total number of collected photons is $\sim 10^6$.

The detection efficiency has been substantially improved for experiments with K$-\alpha$ sources. A curved crystal improved the collection efficiency by a factor of 2 (Urry *et al.*, 2006) and use of a IP detector improved the detection efficiency by a factor of about 50. At 4.5 keV, the K$-\alpha$ source provides 2×10^{12} photons into 1 sr. For the experiments of Kritcher *et al.* (2008), the target subtends 0.25 sr and $L = 0.01\,\text{cm}$ yielding $N_{\text{photons}} \simeq 3 \times 10^3$ photons collected in a single shot.

For use with a CCD camera with a gain of $g_{CCD} = 3.5\,\text{eh/count}$ (electron holes per count) and $g_{eh} = 3.6\,\text{eV/eh}$ is the energy required to liberate an electron hole (eh) pair per incident photon of energy E, and a total number of counts of $N_{\text{photons}} \times E/(g_{CCD} g_{eh})$ is expected. Experiments by Kritcher *et al.* (2007) have employed curved HOPG crystal spectrometers illuminating about 50×20 pixels. Thus, the collected photons are distributed over 10^3 pixels resulting in a scattering signal of about $\sim$1000 counts/pixel.

Photometrics sets the practical limits on the temporal and spatial resolution of x-ray Thomson scattering measurements. Scattering experiments with x-ray free-electron lasers will provide single-shot detection signals comparable to Kritcher *et al.* (2008) with the added benefit of 20 fs temporal resolution and possible spatial resolution of order 1 μm. In addition, high repetition rate capability that will provide spectra with signal-to-noise ratios of > 100 compared to $10-20$ in single shot experiments. For future experiments on the National Ignition Facility, a suitable x-ray probe at energies of, e.g., 18 keV (where x-rays penetrate through the dense capsule with $\eta_{att} \approx 1/e$), must overcome bremsstrahlung emission of $10^{-7}\,\mathrm{J}(\mathrm{eV\ sr})^{-1}$ to achieve 10^4 detected photons per eV. In this case, more than two orders of magnitude higher electron densities than the experiment by Kritcher *et al.* (2008) will result in acceptable signal-to-noise ratios.

6.7.4 Noise Equivalent Power

The incident signal power that will produce a signal equal to the rms background noise power is called the "noise equivalent power." We will establish here the order of magnitude of this quantity for an idealized detector.

For uniform irradiation, the black body power incident on an area A situated in a cavity at temperature T in the frequency range $\nu \to \nu + d\nu$ is

$$P_B d\nu = 2\pi(\nu^2/c^2)[Ah\nu/(e^{h\nu/\kappa T-1})]d\nu \text{ (W)}. \tag{6.7.2}$$

The average number of detected photons incident in the detector response time ($\tau_1 = 1/2\Delta f$, where Δf is the detector bandwidth) and integrated over all frequencies is

$$\overline{N_{Be}} = \int_0^\infty \frac{2\pi(\kappa T)^3}{c^2 h^3} A\eta 2\Delta f \frac{x^2 e^{-x}}{(e^x-1)^2} dx \text{ photoelectrons,} \tag{6.7.3}$$

where $x = h\nu/\kappa T$. Since we are interested in the noise on the detected signal, we have introduced η, the quantum efficiency for the photoelectric detector. The discussion is in fact restricted now to photoelectric detectors because normally we require a good frequency response in scattering measurements. For a thermal detector, η would represent the absorption coefficient. A is the sensitive area of the detector.

The root mean square noise level from statistical theory (see Section 6.4) is given by

$$\sqrt{\overline{N_{Be}^2}} = \sqrt{\overline{N_{Be}}}.$$

1. Now, our detectors have a threshold of $h\nu$ below which photons are not detected; therefore, we approximate $\eta = \eta_0, x \geq x_0, \eta = 0, x < x_0$.

2. For a photoemittive detector, the noise equivalent power at some wavelength λ is

$$P_{\mathrm{N}} \cong \frac{(2\pi)^{1/2}(\kappa T)^{3/2}}{\lambda h^{1/2}} \left(\frac{A\Delta f}{\eta_0}\right)^{1/2} \left[\int_{x_0}^{\infty} \frac{x^2 e^{-x} dx}{(e^x-1)^2}\right]^{1/2} \text{(W)}. \tag{6.7.4}$$

(Note that the amount of detected power is $\eta_0 P_{\mathrm{N}}$.)

3. For a photoconductive detector, the total noise power includes an essentially equal contribution to allow for the recombination of the excited photons. Thus, P_{N}(photoconductive)$\cong 2P_{\mathrm{N}}$(photoemittive).

Detectivity

It is common practice to use the quantity detectivity to characterize a detector. We define

$$D \equiv P_{\mathrm{N}}^{-1}. \tag{6.7.5}$$

Further for the comparison of detectors, we introduce the specific detectivity

$$D^* \equiv D(A\Delta f)^{1/2}. \tag{6.7.6}$$

In Fig. 6.8, a plot is given of the specific detectivity of an ideal photoelectric device working in a room-temperature environment (290° K) and operating at its threshold wavelength ($\lambda = \lambda_0$) for each wavelength. Also shown are the characteristics of some typical infrared detectors. These detectors are in fact commonly operated at low temperatures, i.e., less than the equivalent of their threshold energy in order to reduce the background noise. The heterodyne (homodyne) technique that may sometimes be used to overcome the background noise problem is discussed below.

6.7.5 The Heterodyne and Homodyne Method for Overcoming Background Noise[9]

If we add the field of a local oscillator to the scattered signal, taking care to maintain a constant relative phase at the detector across the whole wavefront, then the total electric field at the detector, including the background radiation field, is

$$E_{\mathrm{t}} = E_{\mathrm{s}} \cos \omega_{\mathrm{s}} t + E_{\mathrm{B}}(t) + E_{\mathrm{L}} \cos(\omega_{\mathrm{L}} t + \varphi). \tag{6.7.7}$$

The intensity at the detector averaged over high frequencies such as $2\omega_{\mathrm{s}}, 2\omega_{\mathrm{L}}$, and $\omega_{\mathrm{s}} + \omega_{\mathrm{L}}$ is

$$I_{\mathrm{t}} = I_{\mathrm{s}} + I_{\mathrm{L}} + I_{\mathrm{B}} + 2(I_{\mathrm{s}} I_{\mathrm{L}})^{1/2} \cos[(\omega_{\mathrm{s}} - \omega_{\mathrm{L}})t], \tag{6.7.8}$$

---

[9] Teich (1970).

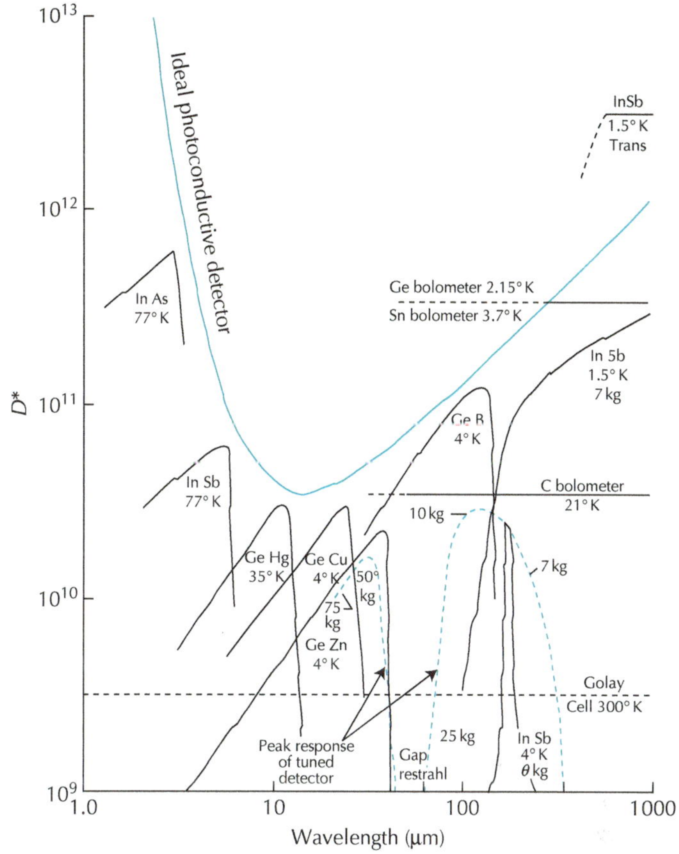

FIGURE 6.8 The performance of far infrared detectors. (Courtesy of E. H. Putley (1966) and The Institute of Physics.)

where the arbitrary phase factor has been dropped. There is no correlation between the background field E_B and either E_L or E_S; therefore, on the average, we obtain no cross-correlational contribution from this quarter.

The mean square of the number of detected photons in the detector response time, which are associated with the scattered signal, is

$$\left\langle N_{pes}^2 \right\rangle = \left(\frac{2\eta A \Delta f}{h\nu} \right)^2 I_s I_L + \left(\frac{2\eta A \Delta f I_s}{h\nu} \right)^2, \tag{6.7.9}$$

where A is the area of the detector and η is the quantum efficiency. The noise level involves the statistical fluctuation due to all the photons that arrive in a response time and is given by

$$\langle N_{pes}^2 \rangle = \frac{2\eta a \Delta f}{h\nu}(I_s + I_L + I_B). \tag{6.7.10}$$

The signal-to-noise ratio as we have defined it previously in Section 6.4 is

$$S/N = [\langle N_{\text{pes}}^2 \rangle / \langle N_{\text{pen}}^2 \rangle]^{1/2}. \tag{6.7.11}$$

Clearly, if we make the local oscillator powerful enough, both the signal and noise will be dominated by it, and we can overcome the background radiation. Thus, in the limit $|I_L| \gg I_s, I_B$,

$$S/N \Rightarrow (2\Delta f \eta P_s / h\nu)^{1/2} = (N_{\text{pes}})^{1/2}. \tag{6.7.12}$$

Thus, we have recovered the quantum noise limit previously discussed in Section 6.4 for photoemittive detectors.

For photoconductive detectors where approximately the noise is doubled by recombination, we find

$$S/N \Rightarrow (\Delta f \eta P_s / h\nu)^{1/2}. \tag{6.7.13}$$

It is conventional to describe this as a "heterodyne" method when the local oscillator is separate from the source of the incident radiation. When the local oscillator is radiation tapped from the incident source, it is termed a "homodyne" method. Possible application is to forward scattering at small angles where the parasitic radiation might serve as the local oscillator signal.

There are a number of terminologies associated with heterodyne mixer/receiver systems (particularly in the microwave to THz region) that are helpful in understanding the literature and which are illustrated in Fig. 6.9.

The noise figure is defined as the ratio of the signal-to-noise ratio at the input compared to the signal-to-noise at the output, measured at 290 K:

$$NF = \frac{(S/N)_i}{(S/N)_o} \geq 1. \tag{6.7.14}$$

The equivalent noise temperature is defined as the absolute temperature required to generate the same noise power[10]: $T_n = P_n/\kappa B$, where B is the bandwidth. For a cascaded circuit comprised of N blocks with noise figure NF_i and gain G_i, the overall noise figure is simply an expression of that for two blocks which is given by

$$NF = NF_1 + \frac{(NF_2 - 1)}{G_i}. \tag{6.7.15}$$

Conversion loss, L_c, is defined as the ratio of the wanted output signal level to the input and is normally expressed in dB. Therefore, for a simple mixer-receiver, we have $T_{\text{sys}} = T_{\text{mixer}} + L_C T_{IF}$.

Note that the mixing process produces down-converted (difference) signals at $f_{\text{rf}} - f_{\text{LO}}$ and $f_{\text{LO}} - f_{\text{im}}$, where f_{rf} is the desired signal frequency and f_{im} is the

---

[10] Note that this is not the physical temperature of the device. Also, the above utilizes the relation involving the power from a noisy resistor $P_n = \left(\frac{v_n}{2R}\right)^2 R = \frac{v_n^2}{4R} = \kappa TB$, which results in a maximum power delivered to a load given by $P_n = \kappa TB$.

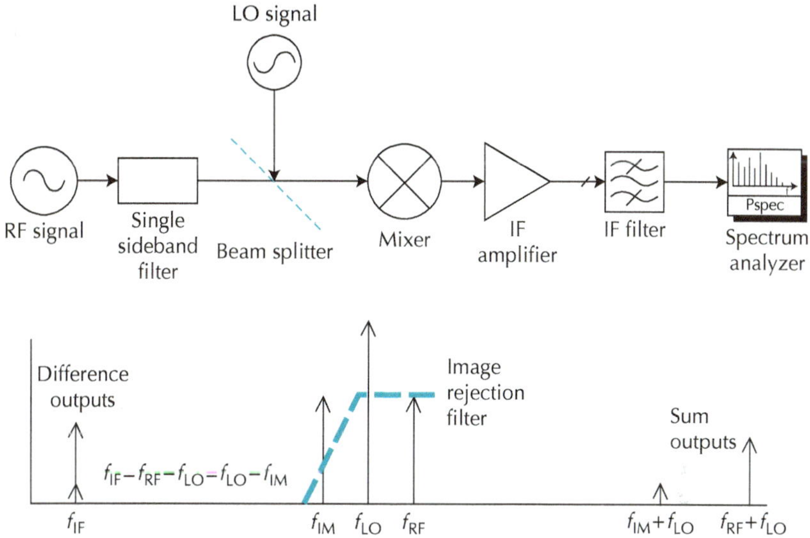

FIGURE 6.9 A schematic of at heterodyne mixer/receiver system is shown.

so-called image frequency. The Double Sideband (DSB) Noise Figure includes noise and signal contributions at both the RF and the image frequencies. In the case of the Single Sideband (SSB) Noise Figure, no image signal is included although image noise is included. Assuming the mixer performance is the same at the image and the desired frequencies, the SSB noise factor is twice the DSB noise factor. The term image-reject mixer refers to mixer configurations where there is cancelation of the image signal.

A common technique for mixer calibration and characterization is the so-called Y-factor or Hot-Cold load technique. This utilizes noise sources that are placed at the input of the mixer (or amplifier) where the two noise sources have different noise "temperatures" (equivalent temperature in the frequency range of interest).

Using these two terminations (hot and cold), two measurements of noise power are made at the output of the receiver or amplifier, and from these two measurements, the gain and noise figure of the device under test can be calculated. The output power in each case is given by

$$P_n(H) = (T_H + T_s)\kappa B G_{IF} \tag{6.7.16}$$

$$P_n(C) = (T_C + T_s)\kappa B G_{IF}, \tag{6.7.17}$$

where T_s is the system temperature, B is the IF bandwidth, and G_{IF} is the gain of the IF amplifier system. In the above, $P_n(H)$ and $P_n(C)$ are the system output power, when the input is terminated with a load at temperature $T_H(T_C)$. Combining Eqns (6.7.16) and (6.7.17), the system noise temperature can be calculated, yielding

$$T_s = \frac{P_c(C)T_H - P_n(H)T_C}{P_n(H) - P_n(C)}. \tag{6.7.18}$$

Often one is simply interested in the receiver noise temperature and defines a Y factor given by $Y \equiv P_H/P_C$ and one can write

$$T_R = \frac{T_H - YT_C}{Y-1} \tag{6.7.19}$$

Here, we note that thus far we have tacitly assumed that we are in a regime where the Rayleigh–Jeans noise formula $P^{R-J} = \kappa T\,\mathrm{W/Hz}$ is applicable. This, of course, is the small $(h\nu/\kappa T)$ limit of the Planck expression

$$P^{\text{Planck}} = \kappa T \left[\frac{h\nu/\kappa T}{\exp(h\nu/\kappa T) - 1} \right]. \tag{6.7.20}$$

As pointed out by Kerr *et al.* (1997) and Kerr (1999), this neglects the zero-point (quantum) noise $h\nu = 2\,\mathrm{W/Hz}$, which is nonnegligible for low-noise millimeter-wave and photonic devices. Therefore, when the quantum noise is prominent and the nonlinearity of the Planck law can no longer be neglected, Kerr notes that the correct expression is that of Callen & Welton (1951).

$$P(\text{C \& W}) = \kappa T \left[\frac{h\nu/\kappa T}{\exp(h\nu/\kappa T) - 1} \right] + \frac{h\nu}{2}. \tag{6.7.21}$$

Predictions of these three relations were plotted by Kerr (1999) for frequencies of 100 GHz (3 mm) and 200 THz (1.5 μm) and are shown in Fig. 6.10. It is seen that for the 1.5 μm case, the Rayleig–Jeans and Callen and Welton differ significantly with the zero-point noise dominating below 3000 K.

PROBLEMS

6.1. Calculate the related electron densities and probing beam pulse lengths at which the scattered signal exceeds bremsstrahlung by a factor of 10 for visible light. Assume a hydrogen plasma with $Z_{\text{eff}} = 1$, $W_{\text{i}} = 10\,\text{J}$, 90° scattering, and $L/V_{\text{p}} = 0.1$ and 0.01. Comment on the coupled requirements for input wavelength and pulse length for a magnetic fusion plasma at $n_{\text{e}} = 10^{14}\,\text{cm}^{-3}$.

Discuss why the simplified formula for visible light does not apply to a warm dense matter plasma at $10^{22}\,\text{cm}^{-3}$ and $T_{\text{e}} = 50\,\text{eV}$. Note that the input wavelength must be less than the critical wavelength $\lambda_{\text{icrit}} = 3.33 \times \frac{10^6}{(n_{\text{e}}(\text{cm}^{-3}))^{1/2}}\,\text{cm}$.

6.2. Line radiation, particularly from the plasma edge, often exceeds bremsstrahlung radiation in plasmas. Approximately $P_{\text{LR}}/P_{\text{B}} \approx \frac{\Sigma_z n_{\text{e}} n_z f(z)}{(2\times 10^{-38} n_{\text{e}}^2 Z_{\text{eff}}[T_{\text{e}}(\text{eV})])^{1/2}}$.

Oxygen from water trapped on surfaces is a common impurity, particularly in unbaked systems. Calculate the values of n_z/n_{e} and $P_{\text{LR}}/P_{\text{B}} = 1$ for the case of a hydrogen plasma having $T_{\text{e}} \sim 400\,\text{eV}$ and $Z_{\text{eff}} = 1.2$ and an

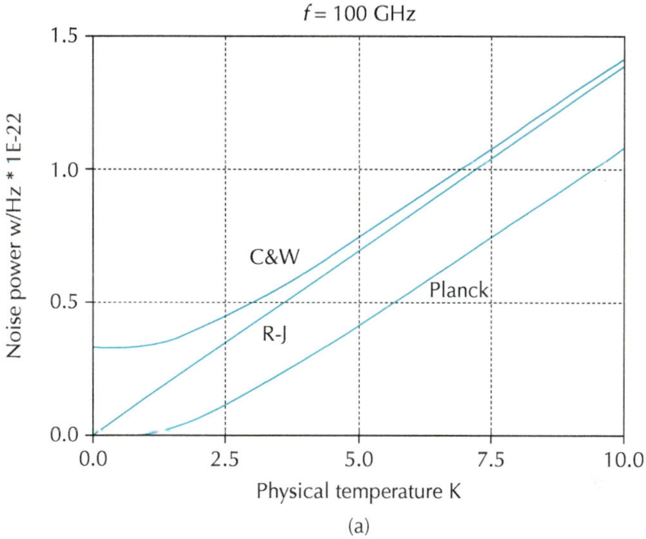

(a)

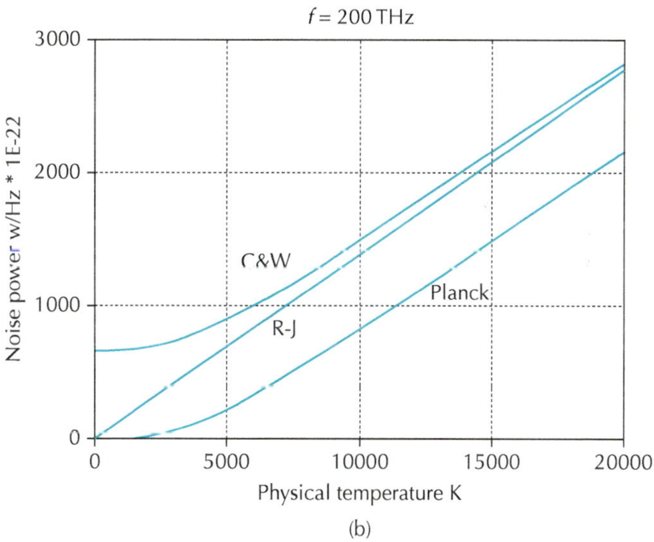

(b)

FIGURE 6.10 Noise power density versus physical temperature for a resistor, computed at (a) 100 GHz and (b) 200 THz ($\lambda = 1.5\,\mu$m) according to the Rayleigh–Jeans, Planck, and Callen and Welton laws.

oxygen impurity (take an average $Z = 4$ and $f_Z \sim 6 \times 10^{-34}$ Wm^3). Assume that $n_Z/n_e \sim$ constant. Comment on when line radiation will be important.

Note that $Z_{\text{eff}} = \Sigma_Z Z^2 n_Z/n_e$ and $n_i = n_e - Zn_Z$.

6.3. Discuss the consequences of successively reducing the pulse length of the probing beam.

6.4. Generally, bremsstrahlung in a reasonably clean hydrogenic plasma should not interfere much with a scattering measurement. An exception can occur when very long measurement times are used, even though the input beam energy will be increased (e.g., the difference between a 10 J, 20-ns duration and a 100 J, 1-ms duration.

Line radiation can pose a much greater problem, as discovered in the seminal T-3 tokamak electron temperature measurements.

Consider the S/N issues for a hydrogen plasma with $T_e = 1\,\text{keV}$, $n_e = 5\times 10^{13}\,\text{cm}^{-3}$, $Z_{eff} = 2$, and containing oxygen, iron, molybdenum, and tungsten as impurities. For simplicity, assume that each imputity has a charge state equal to half of its atomic number [O(4), Fe(13), Mo(21), W(37)]. In reality, the situation will be much more complicated with multiple impurities and n_z and Z varying across the line of sight, but this problem serves to illustrate the effects of impurities. Let the measurement be in the visible at 700 nm, and note that it is assumed that the line radiation is in region of the scattered spectrum.
Remember that $Z_{eff} = \Sigma_z Z^2 n_z/n_e$ and $n_i = n_e - Zn_z$.

$$P_{LR}/P_B \approx \Sigma_z n_e n_z f(z)/(2\times 10^{-38} n_e^2 Z_{eff}[T_e(\text{eV})])^{1/2},$$

where approximately $f(z) = \text{O}(2\times 10^{-34})$, $\text{Fe}(4\times 10^{-32})$, $\text{Mo}(2\times 10^{-31})$, $\text{W}(6\times 10^{-31})$.

Approximately $\frac{N_{pe}}{N_B} = \frac{(7.5\times 10^{38} W_i r_0^2 \lambda_i L)}{(\delta\tau_i n_e \sin(\theta/2) V_p)}$, all dimensions in centimeters.

W_i (J) is the energy in the input beam, $\delta\tau_i$(s) is the measurement duration, L is the length of the scattering volume, and V_p is the plasma volume viewed by the collection optics.

Let $L = 1$ cm and $V_p = 100\,\text{cm}^3$.

6.5. Estimate the number of photoelectrons per channel obtained in the tokamak T-3 scattering measurement described in Section 6.4: parameters $n_e = 2.5\times 10^{19}\,\text{m}^{-3}$, $\lambda_i = 694.3\,\text{nm}$, $L = 0.007\,\text{m}$, $V_p = 10^{-5}\,\text{m}^3$, $W_i = 6\,\text{J}$, transmission coefficient for scattered light to the detector $T = 0.1$, $d\Omega = 2.3\times 10^{-2}$ sr, $\eta = 2.5\times 10^{-2}$ photoelectrons/photon.

6.6. A phase contrast imaging (PCI) experiment employs a liquid nitrogen cooled HgCdTe semiconductor diode for detecting the infrared light. This produces a current which is proportional to the incident photon flux:

$$< i_0 > = Ae\eta < \Phi_0 >.$$

In the above, A is the detection area, e is the electron charge, η is the quantum efficiency, and Φ is the photon flux (from Weisen).

(a) What is the noise power $< i_n^2 >$ associated with $< i_0 >$ if it is only due to Poisson statistics of the photons ($\Delta N = N^{1/2}$)? You should consider that a bandwidth Δf corresponds to an effective integration time $\tau_{eff} = 1/(2\Delta f)$. This noise is called shot noise.

(b) What is the S/N ratio $(< i_n^2 >)^{1/2}/i_0$ for $\lambda = 10.6\,\mu\text{m}$, $A = 1\,\text{mm}^2$, $h\nu\Phi_0 = 2\,\text{mW/mm}^2$, $\eta = 0.5$ and $\Delta f = 0.5\,\text{MHz}$, which are typical for PCI?

(c) What is the smallest detectable phase shift using PCI. Note that $h\nu\Phi_0 = 2\,\mathrm{mW/mm}^2$ corresponds to a level close to saturation of the detector, so it should not be exceeded.

6.7. An electromagnetic beam is made to cross a turbulent refractive medium with mean amplitude of refractive index fluctuations Δn and coherence length l in the z direction and L being the length of the refractive medium (from Weisen).

(a) What is the mean amplitude (variance$^{1/2}$) of the phase fluctuations of the laser beam after traversing the medium? Assume geometrical optics are valid. Express the result in terms of an effective integration length.

(b) It is considered that the turbulence is isotropic. Show that then the autocorrelation function in the direction X of $\Delta\varphi$ is proportional to the Abel transform of the autocorrelation function of Δn expressed in the Δr variable.

6.8. You are involved with a microwave scattering experiment and are configuring a 3-stage amplifier chain with the following properties:

Stage	Power Gain	Noise Figure
1	10 dB	3 dB
2	13 dB	6 dB
3	14.8 dB	7.8 dB

What is the overall gain and noise figure?

6.9. Consider the general case when the signal wavefront strikes the mixer detector at an angle θ with respect to the perfectly collimated laser wavefront as shown in the figure below (Problem courtesy of G. H. Rieke (1994)).

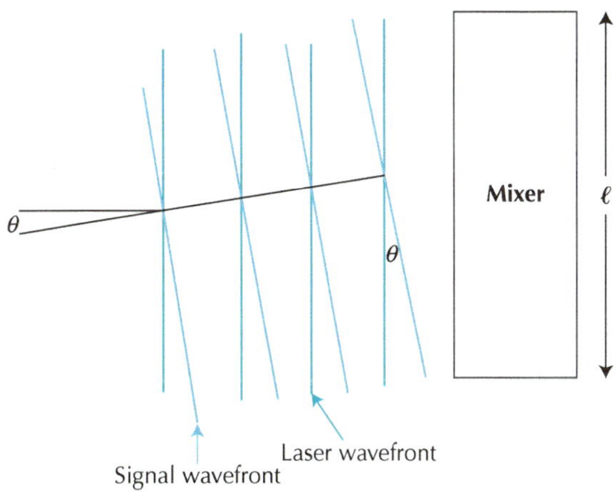

There will thus be a phase shift across the detector between the laser and the signal.

(a) Determine the condition under which the resulting IF signal will be fully out of phase.

(b) From this, show that $\Omega A - \lambda^2$.

SOLUTIONS TO ODD PROBLEMS

6.1. $\frac{N_{pe}}{N_B} = \frac{(4\pi W_i r_0^2 \lambda_i^2 T_e^{1/2} L)}{(2.09\times 10^{-36}\delta\tau_i n_e 2\Delta\lambda_{1/e} V_p)}$.

Now, $\Delta\lambda_{1/e} = \frac{2a\lambda_i \sin(\theta/2)}{c}$, where $a = 6\times 10^7 [T_e(\text{eV})]^{1/2}\,\text{cm}\,\text{s}^{-1}$.

Substitute $\frac{N_{pe}}{N_B} = \frac{(7.5\times 10^{38} W_i r_0^2 \lambda_i L)}{(\delta\tau_i Z_{eff} n_e \sin(\theta/2) V_p)}$.

Using the values in the example and with $r_0^2 = 7.95\times 10^{-26}\,\text{cm}^2$, $n_e\delta\tau_i/\lambda_i \le 8.41\times 10^{13}$ to $8.43\times 10^{12}\,\text{cm}^{-4}\,\text{s}$.

For the magnetic fusion plasma, we need $\delta\tau_i/\lambda_i \le 5.62\times 10^{-1} - 5.62\times 10^{-2}\,\text{s}\,\text{cm}^{-1}$.

$\lambda_{icrit} = 3.33\times 10^{-1}$ cm, e.g., let $\lambda_i = 6.943\times 10^{-5}$ cm and $\delta\tau_i \le 3.9\times 10^{-5} - 3.9\times 10^{-6}$ s.

For the WDM plasma and $\lambda_{icrit} = 3.33\times 10^{-5}$ cm, the term in the expression for bremsstrahlung power, $\exp[-1.24\times 10^{-4}/(\lambda_s T_e)]$, is not near unity.

Consider the case in which $\lambda_i \approx \lambda_s = 10^{-7}$ cm and the exponential term $= 1.7\times 10^{-11}$.

$\delta\tau_i/\lambda_i \le 497$ to $49.7\,\text{cm}^{-1}\,\text{s}$, e.g., $\lambda_i = 10^{-7}$ cm and $\delta\tau_i \le 4.97\times 10^{-5}$ to 4.97×10^{-6} s.

6.3. A typical pulse length for the ruby laser systems commonly used in scattering measurements has been around 10 ns. The laser frequency is $4.32\times 10^{14}\,\text{s}^{-1}$, and the period is 2.3×10^{-15} s. The inverse of the pulse length is $10^{-8}\,\text{s}^{-1}$, which is very small compared to either the input frequency or to a typical half width of the scattered spectrum $\sim 8\times 10^{12} \rightarrow 8\times 10^{13}\,\text{s}^{-1}$ (for T_e in the range 10–1000 eV).

However, powerful lasers have been made with pulse lengths well below 1 ns, e.g., 1 ps, and here, the inverse of the pulse length is $10^{-12}\,\text{s}^{-1}$, which may not be neglected.

A second effect is that, at very short pulse lengths, the length of the incident beam can be less than the scale of the scattering volume. For example, a typical scale length is 10^{-2} m, and at 10 ps, the incident beam's length is only 3×10^{-3} m. As a consequence, electrons moving with the beam will scatter for longer than those moving against it.

6.5. $N_{pe} = (W_i/h\nu_i) r_0^2 d\Omega n_e L T \eta$ photoelectrons $h = 6.626\times 10^{-34}$ J s, $\nu_i = 4.32\times 10^{14}\,\text{s}^{-1}$, $r_0^2 = 7.95\times 10^{-30}\,\text{m}^2$.

Substituting the values for the various terms leads to 1677 photoelectrons.

6.7. **(a)** The problem corresponds to that of a random walk with $m = L/l$ steps of amplitude $s = k_0 l \Delta n$. The average walk amplitude for random walk being $S = sm^{1/2}$ and the average phase perturbation is $\varphi_{\text{rms}} = k_0 (Ll)^{1/2} \Delta n$. $\sqrt{Ll}$ is an effective integration length.

(b) The answer made be found in Weisen *et al.* (1988b).

6.9. **(a)** The resulting IF signal will be fully out of phase when

$$\ell \sin(\theta_D) = \frac{\lambda}{2} \cong \ell \theta_D \leftarrow \theta_{\text{Destructive}}$$

(b) Efficient interference (in phase IF signals) will be maintained for ~ half the above angle: Recall that

$$\Omega = \Pi \sin^2(\theta)$$

Then,

$$\Rightarrow \Omega \approx \pi \, (\lambda/2\ell)^2$$

but, since $\ell^2 \sim A \Rightarrow \boxed{\Omega A \sim \lambda^2}$.

Chapter 7

Optical Systems

Chapter Outline Head

DOI: 10.1016/B978-0-12-374877-5.00007-5

7.1 INTRODUCTION

In the previous chapter, we discussed the problems that arise in obtaining a useful amount of scattered radiation from a plasma. In this chapter, we investigate the techniques that may be used for analyzing this scattered radiation.

For a given scattering experiment, the spectral width will lie within prescribed bounds. For example, for noncollective scattering, the half-width is

$$\Delta\lambda_{1/e} = 4\times 10^{-3}\lambda_{\mathrm{i}}[T_{\mathrm{e}}(\mathrm{eV})]^{1/2}\sin(\theta/2); \tag{7.1.1}$$

$\Delta\lambda_{1/e}$ and λ_{i} have the same units.

For the resolution of the magnetic modulation at the cyclotron frequency,

$$\Delta\lambda_{\mathrm{ce}} \cong \lambda_{\mathrm{i}}(\Omega_{\mathrm{e}}/\omega_{\mathrm{i}}). \tag{7.1.2}$$

It is clear that $\Delta\lambda/\lambda_{\mathrm{i}}$ can vary from ~ 1 down to certainly 10^{-5}.

For collective scattering, there are several considerations depending on what is required: resolution of the spectral features or merely the separation of these features. In general, resolving the width of the feature provides information on the damping of the wave. Resolving the separation provides a measure of the resonant frequency, which is related to the sound speed for ion-acoustic waves or the plasma frequency for electron plasma waves; see Section 5.4 for details on the associated plasma parameters that can be obtained.

A convenient estimate of the scattered wavelength separation between the two ion-acoustic features is given by Eqn (5.4.4),

$$\frac{\Delta\lambda_{ia}}{\lambda_i} \simeq \frac{2}{c}\sin\left(\frac{\theta}{2}\right)\sqrt{\frac{\kappa T_e}{m_i}\left[\frac{Z}{\left(1+k_{ia}^2\lambda_{De}^2\right)}+\frac{3T_i}{T_e}\right]}. \tag{7.1.3}$$

For 90° scattering, this leads to typical normalized peak-to-peak ion-acoustic separations of order $\Delta\lambda_{ia}/\lambda_i \sim 10^{-3}$.

For Thomson scattering with a 90° scattering angle, the wavelength shift between the two electron plasma wave features given by Eqn (5.4.6)

$$\frac{\Delta\lambda_{\mathrm{epw}}}{\lambda_i} \approx 2\left[\frac{n}{n_{cr}}+3\frac{a^2}{c^2}\right]^{1/2}\left(1+\frac{3}{2}\frac{n}{n_{cr}}\right). \tag{7.1.4}$$

Here, $\frac{\Delta\lambda_{\mathrm{epw}}}{\lambda_i} \sim 0.18$ is estimated for a density of $n/n_{cr} = 0.005\%$.

The first factor in the choice of a dispersing element is the resolving power since we need to resolve some small fraction, say $\frac{1}{10}$th of $\Delta\lambda$. The second factor is the efficiency of the instrument, and this includes the light-gathering power (etendue), as well as the transmissivity.

The most commonly used instruments in laboratory experiments are the diffraction grating spectrometer, the Fabry–Perot etalon spectrometer (the theory of these is discussed in detail), and interference filters. (The properties of some of the image dissectors and detectors, which are used with these instruments, are also reviewed.)

Finally, some interesting applications of the technique are analyzed.

7.2 GENERAL PROPERTIES OF SPECTROMETERS: INSTRUMENT FUNCTION

The instrument function $K(p,\theta_1,\theta_2,\lambda)$ describes how the incident radiation, in the wavelength interval[1] $\lambda \rightarrow \lambda + d\lambda$ and in the solid angle defined by $\theta_1 \rightarrow \theta_1 + d\theta_1, \theta_2 \rightarrow \theta_2 + d\theta_2$, is redirected (dispersed) by the spectrometer into the output parameter range $p \rightarrow p + dp$. Often p (it may depend on a number of parameters) is related to the angle φ between an incident and transmitted wavefront (see Fig. 7.1).

Let $I_i(\lambda,\theta_1,\theta_2)d\lambda\, d\theta_1\, d\theta_2$ be the incident power per unit area in the range $\lambda \rightarrow \lambda + d\lambda, \theta_1 \rightarrow \theta_1 + d\theta_1, \theta_2 \rightarrow \theta_2 + d\theta_2$ at the input plane. The transmitted power per unit area in the range $p \rightarrow p + \Delta p$ at the output plane is

$$\int_p^{p+\Delta p} I_t(p)\, dp = \int_p^{p+\Delta p} dp \int_{-\infty}^{+\infty} \lambda \int_0^{\pi/2} d\theta_1 \int_0^{2x} d\theta_2\, I_1(\lambda,\theta_1,\theta_2) K(p,\theta_1,\theta_2,\lambda_1). \qquad (7.2.1)$$

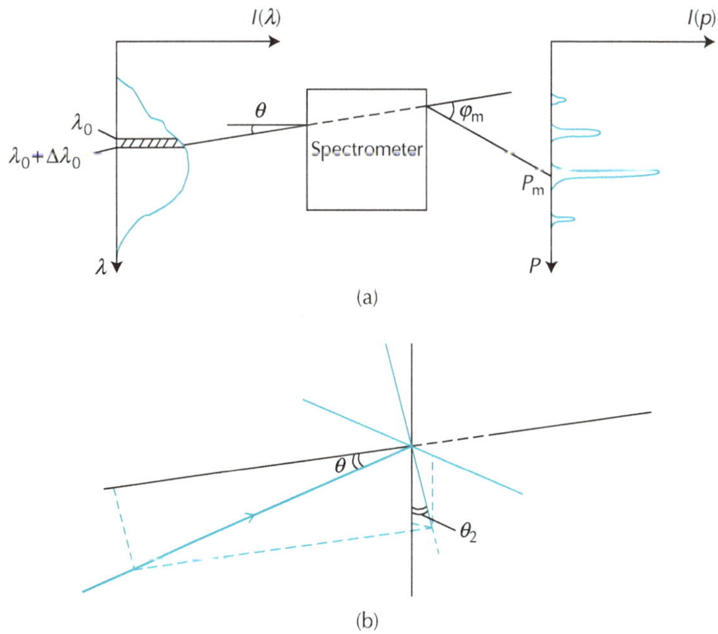

FIGURE 7.1 (a) Illustration of the response of a spectrometer to incident radiation. (b) Geometry of incident radiation.

[1] This function may equally well be defined in terms of the frequency ω.

7.2.1 Monochromatic Spectrum

The instrument function is simply the normalized transmitted spectrum in the limit $\Delta p \to 0$ when plane monochromatic radiation is incident, i.e., $I_i = I_i(\lambda, \theta_1, \theta_2)\delta(\lambda_0 - \lambda)\delta(\theta_{10} - \theta_1)\delta(\theta_{20} - \theta_2)$, when

$$K(p, \theta_{10}, \theta_{20}, \lambda_0) = I_t(p, \theta_{10}, \theta_{20}, \lambda_0)/I_i(\lambda_0, \theta_{10}, \theta_{21}). \tag{7.2.2}$$

The quality of the instrument is determined by this function (see Fig. 7.1) where we show that $K(\rho, \theta_{10}, \theta_{20}, \lambda_0)$ is typically sharply peaked at some angle φ_m, which corresponds to, say, an interference maximum for λ_0. In addition, there are usually other peaks (orders). Note that there is always some contribution at every value of p.

7.2.2 Resolving Power

When λ_0 is changed, the central value p_0 of each output peak changes. It is normally considered that two neighboring monochromatic lines λ_{01} and $\lambda_{02} = \lambda_{01} + \Delta\lambda_0$ are resolvable if the corresponding shift $p_{01} - p_{02} = \Delta p_0$ is greater than the width in p of the instrument function. This is illustrated in Fig. 7.2; the resolving power is defined as $\lambda_{01}/\Delta\lambda_0$.

7.2.3 Broad Input Spectrum

In an experiment, we measure the power in each range $p \to p + \Delta p$ and obtain the spectral function $P_t(p)\Delta p$. Now, this does not correspond exactly to the incident power spectrum $P_i(\lambda)d\lambda$ because, though the main part of $P_t(p)$ comes from a small range of incident wavelengths, there will be some contribution from all other wavelengths.

To obtain $P_i(\lambda)$, we must either unfold $P_t(p)$ using the instrument function or make a best fit of the type of spectrum we expect, with a correction for the instrument response, or if we choose to interpret $P_t(p)dp$ as a reflection of $P_i(\lambda)d\lambda$, we must be prepared to accept the errors which are involved. In some cases, we can overcome the problem by calibrating the instrument with a spectrum similar to the one being investigated.

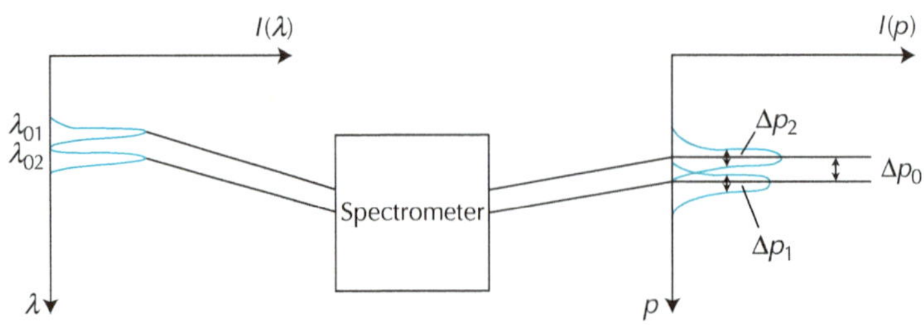

FIGURE 7.2 Illustration for the definition of spectral resolution.

7.3 DIFFRACTION GRATING SPECTROMETER: THEORY

7.3.1 Introductory Remarks

Calibration of a Grating

A schematic diagram of the apparatus that is used to calibrate the efficiency of a grating is shown in Fig. 7.3. A double monochromator is used as a filter for white light, and with it, one can select essentially monochromatic light over a wide range of wavelengths. The light illuminates a thin slit, which is at the focal point of a mirror (or lens), and the parallel light from this mirror is incident on the test grating. The detector first monitors the incident radiation, and it is then rotated to collect the diffracted radiation. The ratio of the signals is the efficiency of the grating. The grating may be rotated in order to bring different orders to the detector.

Blazed Reflection Grating

It is a common practice to cut the mirror (lines or rulings) at an angle φ_B to the back of the grating (see Fig. 7.4). The angle φ_B is chosen so that the

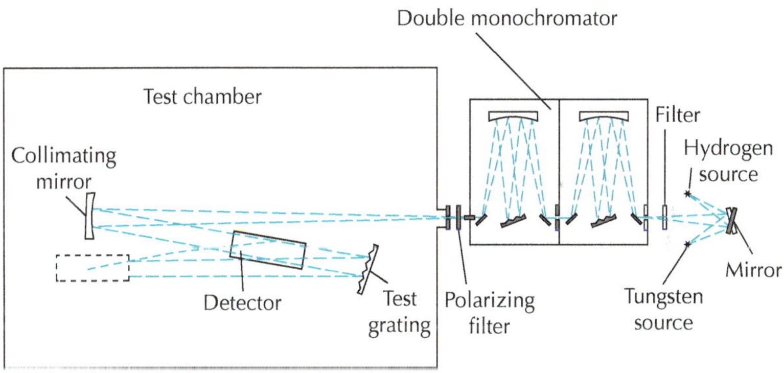

FIGURE 7.3 Schematic diagram of apparatus used for the calibration of a grating. (Courtesy of R. G. Schmitt, Jarrell-Ash Division of Fisher Scientific Company, Waltham, Massachusetts.]

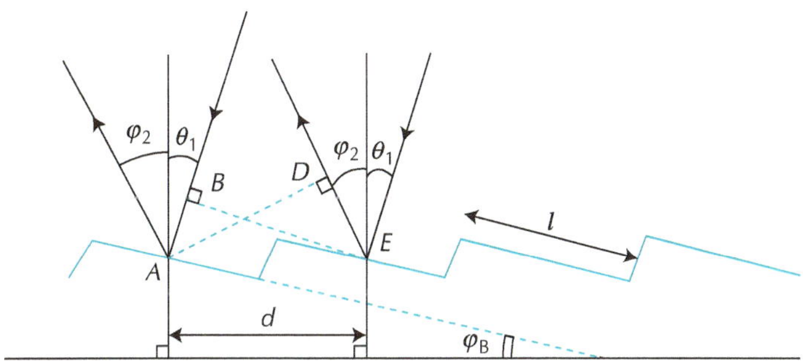

FIGURE 7.4 Geometry of the reflection of radiation by a blazed grating.

interference maxima for the wavelengths of interest occur when the incident and reflected radiation are nearly normal to the mirror surface. This gives the maximum transmission for the device. The line separation as shown is d; the line width $l \cong d$. The path difference between parallel rays which fall on neighboring mirrors is given by

$$AB - EB = d(\sin\theta_1 - \sin\varphi_2) = p. \tag{7.3.1}$$

Interference maxima occur when this path difference is an integral number of wavelengths

$$p = m\lambda_0, \quad m = 0, \pm 1, \pm 2, \text{ etc.} \tag{7.3.2}$$

Diffraction Effects

These maxima are superimposed on the Fraunhofer diffraction pattern, which is associated with each mirror (Born & Wolf, 1965, pp. 401–404). The blaze wavelength λ_B is that wavelength for which the interference maxima (often the first order) coincides with the maximum of the diffraction pattern.

For $\lambda_0 \ll l$, the diffraction pattern for a single mirror is given by

$$D(q, \lambda_0) = \left| \frac{\sin(\pi lq/\lambda_0)}{\pi lq/\lambda_0} \right|^2 = |U(q, \lambda_0)|^2, \tag{7.3.3}$$

where $q = \sin(\theta_1 - \varphi_B) - \sin(\varphi_2 + \varphi_B)$.

For $\lambda_0 \cong l$, the pattern broadens, tending in the limit when $\lambda_0 \gg l$ (all the electrons in the mirror surface then see the incident wave at the same phase) to the dipole-like pattern of a single radiator. Most systems work somewhere between these two regimes, and it is, therefore, necessary to calibrate the grating to determine the diffraction effects. In addition, the response depends strongly upon the polarization of the incident radiation, as can be seen in Fig. 7.5.

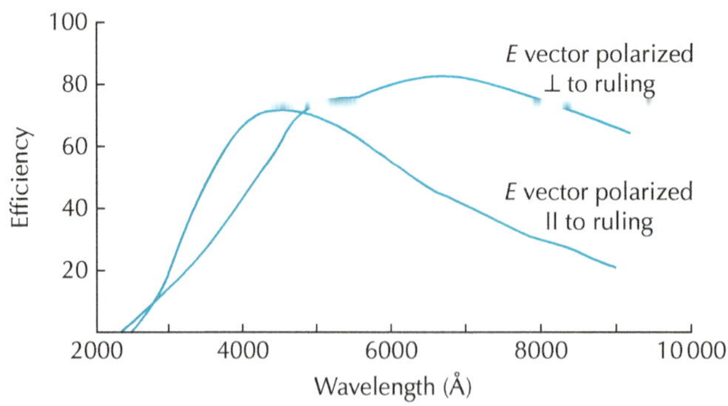

FIGURE 7.5 The measured efficiency of a blazed grating; 1180 lines mm, $\lambda_B = 5000$ Å. (Courtesy of R. G. Schmitt, Jarrell-Ash Division of Fisher Scientific Company, Waltham, Massachusetts.)

7.3.2 Instrument Function

Let the distance from an input plane (say the focusing mirror) to the nearest mirror on the grating and back to some output plane be X_0. Then, the distance to the second mirror is X_0+p, and the gth mirror is X_0+gp. Consider a monochromatic plane wave incident on the grating at the angle θ_1

$$E_\mathrm{i}=E_0\exp[i(2\pi X/\lambda_0)-i\omega_0 t]. \qquad (7.3.4)$$

The total reflected field at some angle φ_2 at the output plane, where we take an average of the M contributions because the radiation will eventually be combined by a lens or mirror, is

$$E_\mathrm{t}^\mathrm{T}=(aE_0U/M)U(q,\lambda_0)\sum_{g=0}^{M-1}\exp[i(2\pi/\lambda_0)(X_0+gp)-i\omega_0 t_1], \qquad (7.3.5)$$

where M is the number of mirrors (lines) on the grating. It is assumed that they are equally illuminated. The factor $A=a^2$ allows for simple absorption and scattering losses.

It is obvious that a maximum in the reflected field occurs whenever gp is an integral number of wavelengths. The incident intensity is

$$I_\mathrm{i}=(c/4\pi)\overline{|E_\mathrm{i}|^2}=(cE_0^2/8\pi).$$

The bar denotes a time average.

The reflected intensity is

$$I_\mathrm{t}=I_\mathrm{i}\frac{AD(q_\mathrm{i},\lambda_0)}{M^2}\sum_{g=0}^{M-1}\sum_{l=0}^{M-1}e^{i(2\pi p/\lambda_0)(g-l)}. \qquad (7.3.6)$$

However,

$$\sum_{g=0}^{M-1}e^{i(2\pi pg/\lambda_0)}=\left[1-e^{i(2\pi Mp/\lambda_0)}\right]/\left[1-e^{i(2\pi p/\lambda_0)}\right]$$

since

$$1/(1+b)=1-b+b^2-b^3\ldots, \quad -1<b<1.$$

Therefore, rearranging we have the instrument function of the grating

$$K(\lambda_0,p,q)=\frac{I_\mathrm{t}(\lambda_0,p,q)}{I_\mathrm{i}(\theta_1,\lambda_0)}=AD(q,\lambda_0)\left|\frac{\sin(\pi Mp/\lambda_0)}{M\sin(\pi p/\lambda_0)}\right|^2. \qquad (7.3.7)$$

The theoretical form of this function and the component functions are sketched in Fig. 7.6, where A represents absorption and $D(q,\lambda_0)$ is the single mirror (ruling) diffraction pattern, $q=\sin(\theta_1-\varphi_\mathrm{B})-\sin(\varphi_2+\varphi_\mathrm{B})$ (see Fig. 7.4). The final term is the interference pattern, and $p=d(\sin\theta_1-\sin\varphi_2)$. M is the

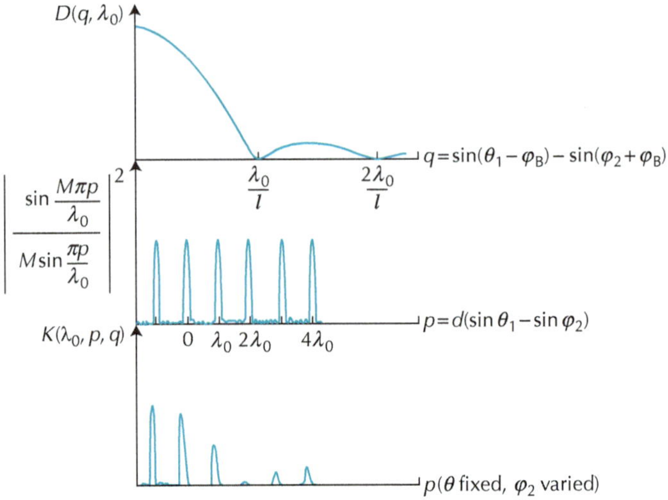

FIGURE 7.6 The diffraction pattern, interference pattern, and instrument function of a diffraction grating. (See text for definitions of variables.)

number of rulings, which are illuminated, d is the ruling spacing, θ_1 is the angle of incidence, φ_2 is the angle of reflection, and φ_B is the blaze angle.

The measured efficiency of a grating, i.e., $K(\lambda_0)$ versus λ_0 for $I(\lambda_0) =$ constant, is shown in Fig. 7.6; it is a measure of the factor $AD(q, \lambda_0)$. The grating is typical of those used to analyze scattered visible laser radiation. The ruling width is 8500 Å, the blaze wavelength for the first order is 5000 Å, and the effects discussed above when $\lambda_0 \cong d$ are very obvious; they were obtained using the apparatus of Fig. 7.3.

7.3.3 Temporal Instrument Function

The temporal broadening introduced by a spectrometer is a result of the path length difference light travels through the spectrometer (Visco *et al.*, 2008). In a standard optical spectrometer with reflective optics, the path difference is introduced by the effects of diffraction at the grating. This path difference for two parallel rays, as shown in Fig. 7.4, is given Eqn (7.3.1). Summing over all the illuminated grooves (M) leads to a total path length difference introduced by the grating,

$$p_{tot} = Md(\sin\theta_1 + \sin\psi_2), \text{ or}$$
$$p_{tot} = Mm\lambda_0. \tag{7.3.8}$$

Therefore, photons arriving at the detector from opposite ends of the grating are offset temporally by

$$\Delta t_g = \frac{Mm\lambda_0}{c}. \tag{7.3.9}$$

Notice that there is no broadening for the zero order case for which $\theta_1 = -\psi_2$. This is sensible as the zeroth-order case corresponds to a purely reflective imaging system, through which all optical paths are equal. For optical scattering, the temporal dispersion introduced by the spectrometer is typically on the order of 100 ps. It is standard practice to reduce the number of grooves illuminated by masking the grating to improve the temporal resolution. This reduces the theoretical resolving power, see Section 7.4.1, but the theoretical resolving power is often much better than the practical resolving power given by Eqn (7.4.4) and the input slit width.

7.4 SPECTROMETER: IMAGE DISSECTORS, APPLICATION

7.4.1 Grating Spectrometer

A typical spectrometer is shown in Fig. 7.7, and the instrument function is given by Eqn (7.3.7).

We will set $AD(q, \lambda_0) = 1$; the errors involved in this are discussed above. In this case,

1. Maxima occur when $p = \pm m\lambda_0, m = 0, 1, 2, \ldots.$
2. Minima occur when $Mp/\lambda_0 = \pm n, n = 1, 2, 3, \ldots.$
3. The separation of the first maximum from the minimum occurs when $\Delta p = \lambda_0 / M$.

If we change $\lambda_0 \rightarrow \lambda_0 + \Delta\lambda_0$, then the change in p at the mth maximum is $\Delta p' = |m|\Delta\lambda$. From the condition $\Delta p = \Delta p'$, we obtain, the following.

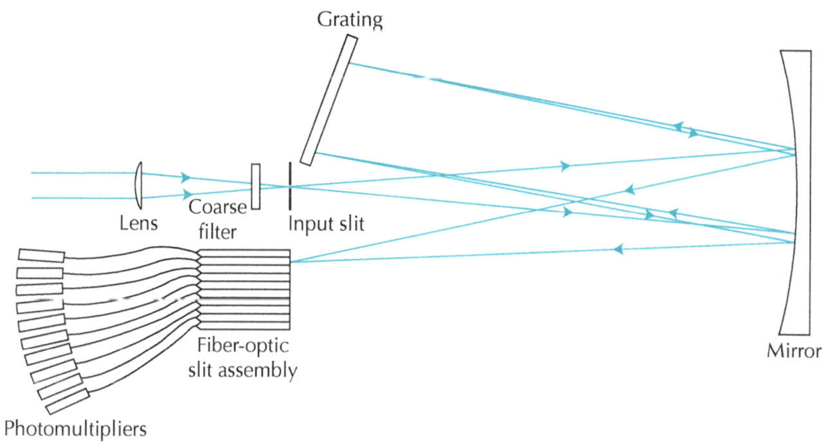

FIGURE 7.7 The layout of a typical multichannel diffraction grating spectrometer.

Theoretical Resolving Power

$$R^{\mathrm{T}} = \lambda_0/\Delta\lambda = |m|M = (Md/\lambda_0)(\sin\theta_1 - \sin\varphi_2); \quad (7.4.1)$$

the maximum possible value is $(\lambda_0/\Delta\lambda)_{\max} = 2Md/\lambda_0$.

Dispersion

We combine Eqns (7.3.1) and (7.3.2) and differentiate with θ_1 fixed to obtain

$$d\varphi_2/d\lambda_0 = m/(d\cos\varphi_2). \quad (7.4.2)$$

Linear Dispersion

If a lens or mirror of focal length f_2 is used to collect the output radiation, and dl represents the infinitessimal scale in the direction of dispersion in the focal plane, $d\varphi_2 = dl/f_2$ and

$$D_{\mathrm{G}} = d\lambda_0/dl = (d\cos\varphi_2)/f_2|m|. \quad (7.4.3)$$

Practical Resolving Power

Let the output slit width be Δl_0. The practical minimum resolvable wavelength change is $\Delta\lambda_{\min} = \Delta l_0 D_{\mathrm{G}}$. The practical resolving power is, therefore,

$$R_{\mathrm{p}} = \lambda_0/\Delta\lambda_{\min} = f_2|m|\lambda_0/(\Delta l_0\, d\cos\varphi_2). \quad (7.4.4)$$

Obviously, the input slit and mirror or lens must be consistent with this, i.e., $f_1 = f_2$, then $\Delta\lambda_{\mathrm{i}} \leq \Delta l_0$.

Etendue

The light gathering power is $E_{\mathrm{G}} = aA/f_1^2$, where a is the area of the input slit and A is the smaller of the area of the collection lens or mirrors or the area of the grating. The input slit is at the focal point f_1.

Efficiency

Light is scattered and absorbed in the region between the mirror surfaces of the rulings; nevertheless a high-quality grating can have an efficiency near the peak of the diffraction pattern as high as 80% (see Fig. 7.5). This level may be achieved over a wide wavelength band if $d \cong \lambda_0$ and the blaze wavelength $\lambda_{\mathrm{B}} < \lambda_{\min}$, and further the incident E vector should be perpendicular to the rulings. Other losses will occur because radiation from the edge of the diffraction pattern of the incident slit misses the collection mirror or the grating. Nevertheless overall efficiency greater than 50% is possible.

Image Dissection

In the past, a common approach was to use a multichannel fiber optic, or light guide, slit assembly, where each fiber bundle, or group of bundles, feeds a detector. A more modern approach is to use an intensified CCD camera.

Overall Transmission

In systems that use multiple lenses, cut-off filters, fiber-optics, and spectrometers, the transmission from the scattering volume to the detectors may only be of order 0.1. Note that each such optical component can lead to losses of 10% or more. Very carefully designed systems and systems that use mirrors and waveguides, such a microwave systems, can have transmission up to 0.3.

7.4.2 Crystal Spectrometers

Measuring the x-ray Thomson scattering signal requires highly efficient spectrometers, as well as moderate spectral resolution, to resolve the elastic and inelastic scattering components. For free-electron or soft x-ray laser experiments in the vacuum-ultraviolet spectral range, $6.7\,\mathrm{nm} < \lambda < 32\,\mathrm{nm}$, reflection grating spectrometers fielded by Beiersdorfer *et al.* (1999); Nakano *et al.* (1984); Harada *et al.* (1999) are available. They are advantageous compared to transmission grating spectrometers (Jasny *et al.*, 1994) because they are highly reflective and further do not show ghosts that occur due to the support structure of transmission gratings.

For x-ray wavelength below 6.5 nm, crystals with high reflectivity are available (Beiersdorfer *et al.*, 2004). In x-ray scattering experiments between $3\,\mathrm{keV} < E < 9\,\mathrm{keV}$, Bragg crystals in the mosaic focusing mode (Yaakobi *et al.*, 1983; Glenzer *et al.*, 2003a) have been employed to spectrally disperse the scattered photons onto an efficient detector (see Section 6.7). The crystals intersect a solid angle of $0.1\,\mathrm{rad} \times 3\,\mathrm{mrad}/4\pi$ to measure the spectrum for a specific **k**-vector (Pak *et al.*, 2004). Alternatively, a wide angle spectrometer with a highly oriented pyrolytic graphite (HOPG) crystal has been developed by Saiz *et al.* (2007) to measure the energy-resolved spectrum of scattered x-rays from a dense plasma over a wide range of angles in a single shot.

The ideal mosaic crystal consists of mosaic blocks with slightly different orientation of the reflecting planes characterized by the mosaic spread γ. These blocks must individually fulfill the Bragg condition $n\lambda = 2d\sin\theta_B$, where n is the diffraction order, λ is the x-ray wavelength, and θ_B is the Bragg angle. This property leads to a broad rocking curve. In addition, the mosaic crystal should be thick enough so that it is almost completely reflective in the middle of the mosaic distribution. For x-ray energies of $4.5\,\mathrm{keV} < E < 9\,\mathrm{keV}$, a reflectivity of 2–9 mrad has been reported by Marshall & Oertel (1997) and Pak *et al.* (2004) for HOPG crystals with a $\gamma = 0.4°$ (ZYA) and $\gamma = 3.5°$ (ZYH). Döppner *et al.* (2008) have shown a reflectivity of 4 mrad at 12.6 keV in first order. In second order, the authors report a reflectivity of 1 mrad at 12.6 keV and 0.5 mrad at 22.2 keV linearly decreasing by a factor of 10 when going to fourth order.

By differentiating the Bragg law, we obtain a simple estimate of the crystal dispersion, $\Delta\lambda/\lambda = \Delta E/E = \Delta\theta/\tan\theta_B$, where E is the x-ray energy and $\Delta\theta$ is the angular spread of the incident x-rays. On the other hand, the spatial distribution of energies (i.e., the dispersion) at a normal image plane is

$$\frac{\Delta E}{\Delta x} = \frac{E}{2f \tan \theta_B} \tag{7.4.5}$$

with $\Delta x = 2f\Delta\theta$ and f being the focal length (source to crystal distance). To obtain optimum signal collection, crystals curved along the sagittal (i.e., nondispersive) plane with a radius of curvature of $R = 115$ mm have been fielded by Urry *et al.* (2006). Best focusing in both the dispersive and sagittal directions is thus achieved when the source-to-crystal and crystal-to-image distances are given by $f = R/\sin\theta_B$.

Figure 7.8 shows the measured spectral resolution of the graphite crystal using a titanium K$-\alpha$ x-ray source at 4.5 keV. When the distance between the source and the crystal equals the distance between the crystal and the detector plane, a focussed image (mosaic focusing) with a spectral resolution of $\Delta\lambda/\lambda = 0.003$ has been observed. This value for the resolution is determined by volume (depth) broadening. The latter also induces an asymmetry enhancing the high-energy wing of the spectral feature. If the ratio between the distances deviates significantly from unity, the mosaic structure of the crystal is observed resulting in a rather poor spectral resolution. These tests have further confirmed that the graphite crystal has a high reflectivity, a factor of 3 larger than LiF, and more than an order of magnitude larger than PET or KAP crystals.

7.4.3 Examples

Example 1

Consider the situation shown in Fig. 7.7, with the following values. The collection lens has $f = 50$ cm, diameter $= 10$ cm, $A = 79$ cm^2. The scattered radiation

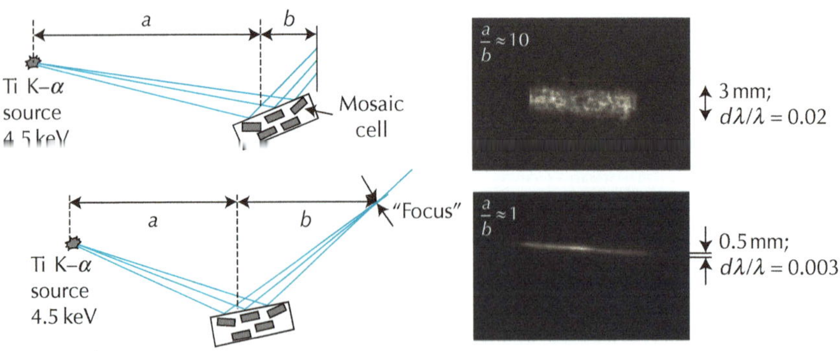

FIGURE 7.8 Demonstration of mosaic focusing mode using a HOPG Bragg crystal from Glenzer *et al.* (2003a). This configuration provides simultaneously a high reflectivity from the carbon crystal and a sufficient wavelength resolution of $\Delta\lambda/\lambda = 0.003$ for spectrally resolved x-ray Thomson scattering measurements.

is collected from 1-cm length of the incident beam, which has a diameter of 1 mm. The etendue of the scattered light collection system is

$$E_E = (0.1 \times 1.0 \times 79)/(50)^2 = 3.1 \times 10^{-3}\,\text{cm}^2;$$

the collection lens focuses the light on the input slit, with unit magnification. The slit is 1 cm long in the direction of the incident beam, and for simplicity, $\Delta l_\text{i} = \Delta l_0$.

The grating is $10 \times 10\,\text{cm}^2$, 1200 lines/mm ($d = 8400\,\text{Å}$) $M = 1.2 \times 10^5$ lines. The grating is mounted with the long side of the slit parallel to the lines. The focal length of the mirror is $f = 100$ cm. The incident spectrum is centered on $\lambda_0 = 7000\,\text{Å}$.

(a) In the first order ($m = 1$) for $\varphi_2 \cong 0$, we have $D_G = 8.4\,\text{Å/mm}$. For maximum transmission, we require $E_\text{G} \geq E_\text{i}$ so that all the scattered radiation is accepted.

$$E_\text{G} = \frac{\Delta l_0 \times 1.0 \times 100}{(100)^2} = \Delta l_0\,(\text{mm}) \times 10^{-3}\,\text{cm}^2.$$

Equating E_G to E_E gives us $\Delta l_0 \geq 3.1\,\text{mm}$. $R_T = 1.2 \times 10^5$, and for the minimum Δl_0, we have $R_\text{p} = 2.8 \times 10^2$, $\Delta\lambda_\text{min} = 25\,\text{Å}$.

(b) To obtain better resolution, we can go to fourth order ($m = 4$), and for $\varphi_2 \cong 0$ with $\Delta l_\text{i} = \Delta l_0 = 0.1\,\text{mm}$, we obtain

$$D_\text{G} = 2.1\,\text{Å/mm}, \quad R_\text{p} = 3.3 \times 10^4, \quad \text{and} \quad \Delta\lambda_\text{min} = 0.2\,\text{Å}.$$

However,

$$E_\text{G} = (0.01 \times 1.0 \times 100)/(100)^2 = 10^{-4}\,\text{cm}^2 \ll E_E,$$

and we see that the better resolution has been obtained at the expense of a considerable reduction in transmission.

This illustrates the point that for scattering measurements where good transmission is usually vital, the grating is best suited to the analysis of the broader spectra where it combines good light-gathering power with an adequate resolution.

Example 2

Consider the response of the grating spectrometer to a broad plane spectrum ($\Delta\lambda \gg \Delta\lambda_\text{min}$), in the limit $\Delta p \to 0$ and ignoring the absorption and diffraction response of the grating. For this situation, Eqn (7.3.7) becomes

$$I_\text{t}(p) = I_\text{i}(p) \int_{-\infty}^{\infty} d\lambda \left| \frac{\sin(\pi Mp/\lambda)}{M \sin(\pi p/\lambda)} \right|^2.$$

Now, $\sin(\pi Mp/\lambda)$ changes very rapidly with λ compared to $\sin(\pi p/\lambda)$ so that to a good approximation

$$I_t(p) = I_i(p) M\pi p \int_{-\infty}^{\infty} dx |(\sin X)/X|^2 = \pi^2 M[pI_i(p)]. \qquad (7.4.6)$$

It can be seen that the instrument function distorts the output spectrum, and the incident spectrum is given by $I_i(p) \propto I_t(p)/p$. This effect can be dealt with by calibrating with a white light source, as can the considerable corrections which arise from the absorption, diffraction characteristics, and detector response.

7.5 FABRY–PEROT ETALON: THEORY

The response to plane monochromatic incident radiation is discussed by Born & Wolf (1965, pp. 323–333). In order to illustrate the effect of the etalon on a spectrum, we will consider here the response to an arbitrary plane spectrum. Let the time-averaged incident intensity be $I_i(X_1)$, where X_1 is a plane in front of the etalon. Then,

$$I_i(X_1) = (c/4\pi) \lim_{T\to\infty} \frac{1}{T} \int_{-\infty}^{+\infty} dt |E_i(X_1, t)|^2. \qquad (7.5.1)$$

The incident field may be Fourier analyzed, and the component at the frequency ω may be given by

$$E_i(X_1, \omega) = \int_{-\infty}^{+\infty} dt E_i(X_1, t) e^{-i\omega t}. \qquad (7.5.2)$$

By Parseval's theorem (Appendix A),

$$I_i(X_1) = (c/4\pi) \lim_{T\to\infty} \frac{1}{T} \int_{-\infty}^{+\infty} (d\omega/2\pi) |E_i(\omega)|^2. \qquad (7.5.3)$$

The average intensity per unit frequency interval at ω is simply

$$I_i(\omega) = \frac{c}{4\pi} \lim_{T\to\infty} \frac{1}{T} |E(\omega)|^2. \qquad (7.5.4)$$

The effect of the etalon is illustrated in Fig. 7.9. The transmitted field is the sum of the multiply reflected fields. Here, r, a, and r', a' are the reflectivity and transmissivity of, respectively, the radiation entering and leaving the

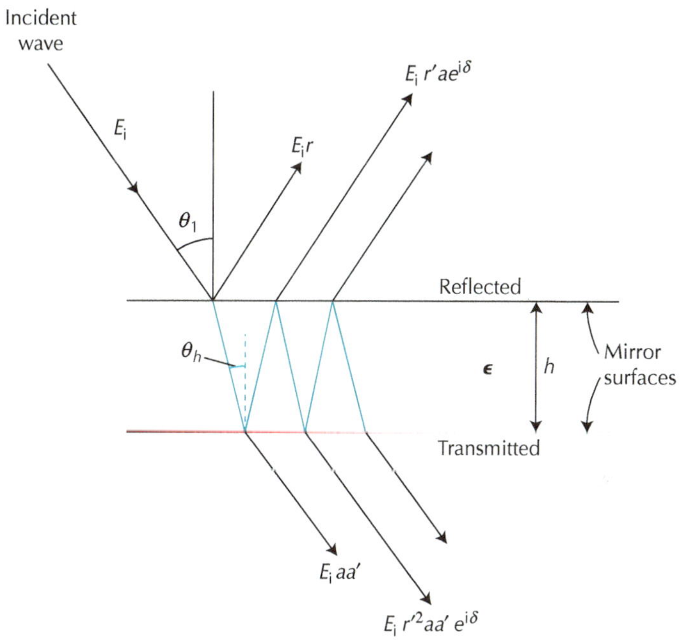

FIGURE 7.9 Illustration of the effect of a Fabry–Perot etalon on plane incident radiation.

etalon. The intensity reflectivity is $R=r^2=r'^2$ where $r^2=-r$. The transmission coefficient is $T=a^2=a'^2$ and $R+T=1$. Strictly, there is of course absorption and scattering of the radiation; we denote this by A and

$$R+T+A=1. \tag{7.5.5}$$

The optical path difference between successive transmitted beams is $2\varepsilon h\cos\theta_h$, where θ_k is the angle of incidence to the mirror faces of the etalon. The plate separation is h, and the refractive index between the plates is ε. Let X_0 be the path difference from the incident plane at X_1 to the transmitted plane at X_2, for the directly transmitted beam. All the other transmitted components will have traveled a greater distance, which is some multiple of $2\varepsilon h\cos\theta_h$. The total transmitted electric field at the frequency ω is

$$E_t(X_2,\omega)=E_\omega e^{-i\omega X_{0/c}}\sum_{g=0}^{\infty} AR^g \exp[-ig(2\omega/c)\varepsilon h\cos\theta_h]. \tag{7.5.6}$$

This is substituted into Eqn (7.5.3), and with the aid of Eqn (7.5.4), we obtain

$$I_t(X_2)=\int_{-\infty}^{+\infty}(d\omega/2\pi)\left\{\sum_{g=0}^{\infty}\sum_{j=0}^{\infty}A^2R^{g+j}\exp[i(g-j)(2\omega/c)\varepsilon h\cos\theta_h]\right\}I_i(\omega). \tag{7.5.7}$$

The quantity in braces is simply the instrument function of the etalon; it gives the response to monochromatic radiation.

Autocorrelation Function

The autocorrelation function of the intensity spectrum is defined by

$$F(\tau)=\int_{-\infty}^{+\infty}(d\omega/2\pi)e^{i\omega\tau}I_{\rm i}(\omega). \tag{7.5.8a}$$

Alternatively,

$$I_{\rm i}(\omega)=\int_{-\infty}^{+\infty}d\tau e^{-i\omega\tau}F(\tau). \tag{7.5.8b}$$

For our case,

$$i(g-j)[(2\varepsilon h\cos\theta_h)/c]\times\omega=i\omega\tau.$$

Thus,

$$\tau=(g-j)[(2\varepsilon h\cos\theta_h)/c],$$

and

$$I_{\rm t}(X_2)=\sum_{g=0}^{\infty}\sum_{j=0}^{\infty}A^2R^{g+j}F(\tau). \tag{7.5.9}$$

Example

Consider the response to a Lorentzian intensity function

$$I_{\rm i}(\omega)=\left\{\nu/\left[(\omega-\omega_0)^2+\nu^2\right]\right\}I_0 \tag{7.5.10a}$$

or

$$F(\tau)=e^{-\nu|\tau|}I_0. \tag{7.5.10b}$$

The constant ν is the half-width of the spectrum at half-height.

We substitute this in Eqn (7.5.9) and obtain

$$I_{\rm t}(\theta_h)=\frac{(1-R)}{(1+R)}\frac{\left[1-(Re^{-\nu\tau_1})^2\right]}{\left[(1-Re^{-\nu\tau_1})^2+4Re^{-\nu\tau_1}\sin^2(\omega_0\tau_1/2)\right]}, \tag{7.5.11}$$

where $\tau_1=(2\varepsilon h\cos\theta_h)/c$, h is the mirror separation, and ε is the refractive index between the plates.

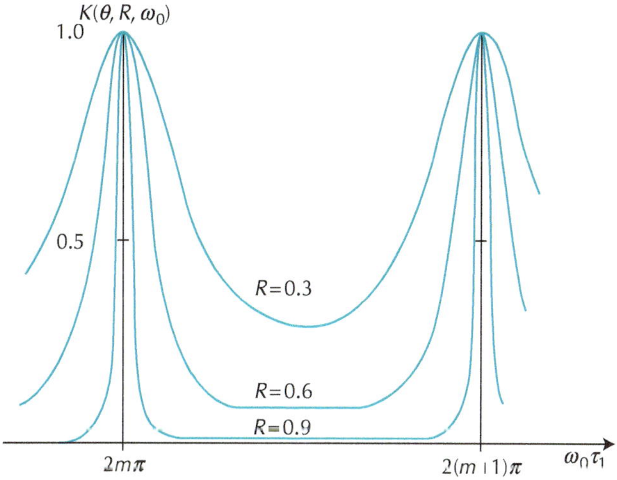

FIGURE 7.10 The instrument function for a Fabry–Perot etalon for different values of the reflection coefficient R.

Instrument Function

If we take $\nu \to 0$, we have monochromatic incident radiation at the frequency ω, and we find

$$K(\theta_h, \varepsilon h, R, \omega) = \frac{(1-R)^2}{(1-R)^2 + 4R\sin^2(\omega\tau_1/2)}, \tag{7.5.12}$$

where R is the reflectivity of the plates. This is shown in Fig. 7.10.

7.6 FABRY–PEROT ETALON SPECTROMETER: IMAGE DISSECTORS, APPLICATION

7.6.1 Typical Spectrometer

A typical spectrometer is shown in Fig. 7.11. Note that in a conventional Fabry–Perot, the medium between the mirrors is usually a gas and $\varepsilon \cong 1$, and further the same medium is often outside the etalon, as a consequence, $\theta_h \cong \theta_1$, the incident angle. In all the discussion, we therefore drop the subscript h. (For a filter, however (see Section 7.6.4), this is not the case, and we must allow for the difference in θ_h and θ_1 when we tilt the filter.)

1. Maxima occur when

$$\omega\tau_1/2 = [(2\varepsilon h\cos\theta_h)/\lambda]\pi = m\pi, \tag{7.6.1}$$

where m is an integer.

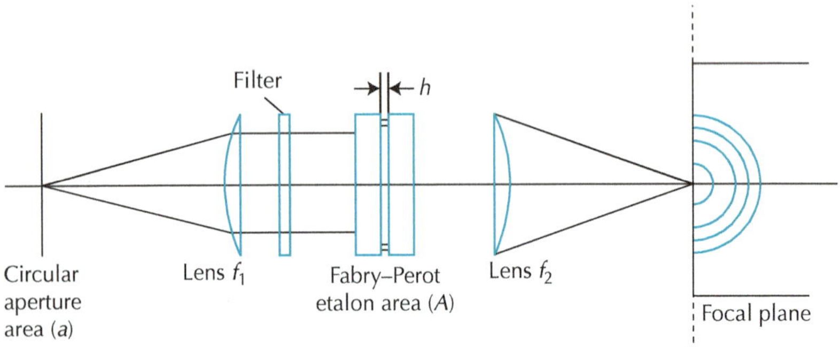

FIGURE 7.11 The layout of a typical Fabry–Perot etalon spectrometer.

2. For a fixed incident frequency, the focused output is a set of circular bright fringes. Each fringe corresponds to a particular angle of incidence.

   Note: Radiation at intermediate angles is primarily reflected and also a particular fringe or section of a fringe appears only if radiation is brought into the etalon at the appropriate angle.

3. *Free spectral range in frequency* $\Delta\omega_F$: A fringe appears for a certain angle for a frequency ω_0; the same fringe will appear for the neighboring frequencies $\omega_0 \pm n\Delta\omega_F$, where n is an integer (for different order numbers m of course) where

$$\Delta\omega_F = \pi c/(\varepsilon h \cos\theta). \tag{7.6.2}$$

4. *Fringe width at half-height* $\Delta\omega_{1/2}$: For good reflectivity, we have approximately that

$$\frac{(1-R)^2}{(1-R)^2 + 4R\left(\frac{1}{2}\Delta\omega_{1/2} \times \frac{1}{2}\tau_1\right)} = \frac{1}{2}$$

   from which we obtain

$$\Delta\omega_{1/2} = 2(1-R)/\tau_1 R^{1/2}. \tag{7.6.3}$$

5. *Finesse*: A parameter commonly used to describe the etalon:

$$F \equiv \frac{\Delta\omega_F}{\Delta\omega^{1/2}} = \frac{\pi R^{1/2}}{(1-R)} = \frac{\text{useful frequency interval}}{\text{resolvable frequency interval}}. \tag{7.6.4}$$

   Practically, this theoretical value will not be realized because it requires absolutely flat mirror surfaces. If the surfaces are flat to, say, λ/S, then as a rough estimate we will find $F \lesssim S/2$.

6. *Resolving power*:

$$R_T = \frac{\omega}{\Delta\omega_{1/2}} = \frac{\lambda}{\Delta\lambda_{1/2}} = \frac{\omega\varepsilon h \cos\theta}{\pi c} F. \tag{7.6.5}$$

7. *Dispersion* D_{FP}: We differentiate Eqn (7.6.1) and find that a given maximum is shifted by $\Delta\theta$ when we change the wavelength by $\Delta\lambda$, and $\Delta\lambda/\lambda = \frac{\sin\theta}{\cos\theta}\Delta\theta \cong \theta d\theta$, for small θ (θ in radians). The transmitted radiation is collected by a lens f_2, and in the focal plane radiation in the angles $\theta \to \theta + \Delta\theta$ is spread radially from $r \cong \theta f_2 \to r + \Delta r \cong \theta f_2 + \Delta\theta f_2$.

$$D_{FP} = \Delta\lambda/\Delta r = -\lambda r/f_2. \tag{7.6.6}$$

If an aperture of radius Δr_1 is used on the axis of the system, we have

$$r \cong \Delta r_1 \quad \text{and} \quad \Delta\lambda_{min} \cong \lambda \Delta r_1^2/f_2^2. \tag{7.6.7}$$

8. *Contrast*:

$$C = \frac{\text{maximum intensity}}{\text{minimum intensity}} = \frac{(1+R)^2}{(1-R)^2}. \tag{7.6.8}$$

9. *Transmission*: Monochromatic radiation incident at an angle corresponding to a maximum is all transmitted, barring a small percentage which is absorbed. Radiation incident at a minimum is mainly reflected. On the average if radiation covers a free spectral range, either because it is a broad spectrum or because it is an incident at a range of angles, then the majority is reflected. On the average, only a fraction $1-R$ is transmitted at the first mirror, and then, roughly half the radiation in the cavity is transmitted, and half is reflected. The average transmission is

$$\overline{T} \cong 2/F. \tag{7.6.9}$$

A method of improving the transmission by collecting reflected radiation and sending it back at a different angle has been suggested by Ascoli-Bartoli *et al.* (1967).

10. *Etendue*: Now, $E_{FP} = A d\Omega$, where A is the area of the Fabry–Perot and $d\Omega$ is the solid angle consistent with a given resolution. For an on-axis aperture $d\Omega \cong \theta\, d\theta - \Delta\lambda_{min}/\lambda$, where $\Delta\lambda_{min}$ is the required wavelength resolution; thus,

$$E_{FP} \cong (A\Delta\lambda_{min})/\lambda. \tag{7.6.10}$$

7.6.2 Image Dissectors, Multietalon Systems

The spectrum may be constructed from a series of measurements by varying the optical path and using an on-axis aperture in the output focal plane. A convenient method of changing the optical path is to use a gas-filled etalon and vary the gas pressure; SF_6 (Daehler & Ribe, 1967; Daehler *et al.*, 1969) and freon (Evans *et al.*, 1966) have been used. At the very high-order numbers commonly used, only a small adjustment in the refractive index ($\varepsilon \cong 1$) is required to cover a free spectral range. Alternatively, a piezoelectric crystal mount may be used for one of the plates. This is a useful approach if rapid variation of the optical path is required. The inertia of the plates sets a limit on the rapidity.

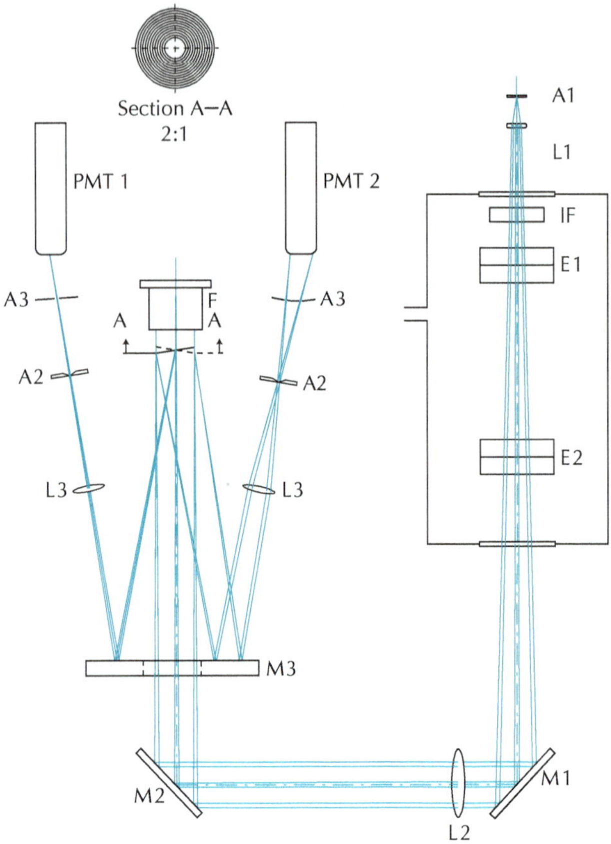

FIGURE 7.12 An illustration of the use of a Fafnir multimirror system *F* in a multichannel Fabry–Perot spectrometer. (Courtesy of M. Daehlcr, G. A. Sawyer, & K. S. Thomas (1969). Phys. Fluids 12, 225.)

For single pulse, spectrum measurements, some cunning is required because of the awkward form of the fringe pattern. A commonly used solution is the "Fafnir" mirror system (Hirschberg & Platz, 1965; Hirschberg, 1967). This is a set of concentric annular mirrors (typically 10) each tilted to throw radiation in a different direction (see Fig. 7.12). In addition, other methods have been tested, which use a concentric, offset Fresnel lens system (Hirschberg, 1967), a thin axicon (Katzenstein, 1965), and a fiber optic image dissector (Forrest, 1967).

Multietalon Systems

Considerable improvements in resolution and contrast are obtained when a series of etalons is used. A system using three etalons was investigated in detail by Mack *et al.* (1963) and has been applied to scattering measurements by Daehler & Ribe (1967).

7.6.3 Examples

Example 1

We want to resolve the form of a spectrum of width $\Delta\omega_{max} = 3.8 \times 10^{11}$ rad s^{-1} ($\Delta\lambda_{max} = 1$ Å) centered on $\omega_0 = 2.7 \times 10^{15}$ rad s^{-1}($\lambda_0 = 7000$ Å), using 10 channels. We will work close to normal incidence with an on-axis aperture, using a gas-filled etalon; thus, $\cos\theta \cong 1$, $\varepsilon \cong 1$.

First, we set $\Delta\omega_F > \Delta\omega_{max}$ so that different orders will not overlap; therefore, $\Delta\omega_F = 2\Delta\omega_{max} = 7.6 \times 10^{11}$ rad s^{-1}. This fixes the etalon plate separation Eqn (7.6.2) at $h = 0.124$ cm. Let the input and output lens be $f_1 = f_2 = 50$ cm. In order to be able to resolve $\Delta\omega_{max}/10 = \Delta\omega_F/20$, we require $F \geq 20$. From Eqn (7.6.4), this tells us that the reflectivity must be $R \geq 0.85$. The plate surface tolerance must be at least as good as $\lambda/40$. From Eqns (7.6.6) and (7.6.7), we can calculate the maximum radius of the output aperture for resolution of $\Delta\omega - \Delta\omega_{max}/10$

$$\Delta r_1 = (\Delta\omega/\omega_0)^{1/2} f_2 = 0.2\,\text{cm}.$$

By symmetry, we see that the input aperture must be the same radius. From Eqn (7.6.10), the etendue is $E_{FP} = (A\pi\,\Delta r_1^2)/f_2^2$.

A good etalon can have a working area of 20 cm^2, and we see that for a system which can resolve 0.1 Å, the etendue $E_{FP} \cong 10^{-3}$ cm^2 (compare this with the case of a grating Example 1(b), Section 7.4.3).

Example 2: Effect of the Etalon on a Broad Spectrum

Consider a Lorentzian spectrum of comparable width to the free spectral range. We traverse the free spectral range with a very small change in τ_1; therefore, we may ignore the effect of the change on the exponent in Eqn (7.5.11). We interpret the output in terms of a frequency change, that is, we set

$$\omega(\tau_1 + \Delta\tau_1) = (\omega + \Delta\omega)\tau_1, \quad \text{where} \quad \Delta\omega = \omega - \omega_0 \quad \text{and} \quad \omega\tau_1 = m2\pi,$$

m is an integer. Now from Eqn (7.6.2),

$$\Delta\omega_F = \pi c/(\varepsilon h \cos\theta) = 2\pi/\tau_1,$$

and from Eqn (7.6.3), the half-width of the instrument function is

$$\Delta\omega_{1/2} = \left[(1 - R)/\pi R^{1/2}\right]\omega_F.$$

We normally work with the spectral width much less than the free spectral range to avoid overlapping of orders; therefore,

$$\nu \ll \Delta\omega_F \quad \text{and} \quad \nu\tau_1 \ll 1. \tag{7.6.11}$$

Finally, for good reflectivity, $R^{1/2} \cong R \cong 1$, and with these substitutions,

$$I_t = \frac{(1-R)}{(1+R)} \frac{\omega_F}{2\pi} \left\{ \frac{\nu + \frac{1}{2}\Delta\omega_{1/2}}{(\omega-\omega_0)^2 + \left(\nu + \frac{1}{2}\Delta\omega_{1/2}\right)^2} \right\}. \tag{7.6.12}$$

Comparison with Eqn (5.5.10) shows that the etalon broadens the spectrum by an amount equal to the width of the instrument function. Probably of more importance is the more marked increase in the height of the spectral wings.

7.6.4 Interference Filters

These filters are Fabry–Perot etalons with a very small optical path length εh between the mirrors. In this circumstance, the order number $m = (2\varepsilon h/\lambda)\cos\theta_h$ Eqn (7.6.1) is small, and the free spectral range Eqn (7.6.2) is large. Thus, it is possible to isolate an interference maximum, i.e., a limited frequency region of high transmission, using simple gelatine cutoff filters coupled with the interference filter. Remember that θ_h is the angle of the radiation between the mirrors. This is related to the external incident angle θ_1 by $\varepsilon \sin\theta_h = \sin\theta_1$ (for a vacuum outside the filters). The fringe width at half-height is $\Delta\omega_{1/2} = c(1-R)/(R^{1/2}\varepsilon h \cos\theta_h)$ Eqn (7.6.3), and this is made small by working with high-reflectivity mirrors. These are usually dielectric mirrors with $R > 0.95$.

The filter is normally used at close to normal incidence ($\theta_h \cong 90°$), and the passband is varied by tilting the filter. The value of the wavelength at peak transmission is

$$\lambda \cong \lambda_0 \left\{1 - \frac{1}{2}\theta_h^2\right\} \cong \lambda_0 \left\{1 - \frac{1}{2}(\theta_1/\varepsilon)^2\right\}, \tag{7.6.13}$$

where λ_0 is the peak wavelength of normal incidence.

A modern filter polychrometer is discussed in Section 7.8 (Carlstrom *et al.*, 1990, 1992). The transmission curves are shown in Fig. 7.13.

7.7 CALIBRATION AND ALIGNMENT

There are three main aspects to a complete calibration. First, we need to know the relative transmission (this includes detector response) of the system over the wavelength range of the scattered spectrum. Second, for an absolute calibration, we must have a precise assessment of the scattering volume, solid angle, and transmission for, certainly, one wavelength. Third, the absolute calibration should ideally be made on a timescale comparable to that of the scattering measurement.

For the relative spectral response, we may use a chopped white light source or possibly the plasma continuum radiation (Kunze, 1968). Four approaches have been used for making the absolute calibration:

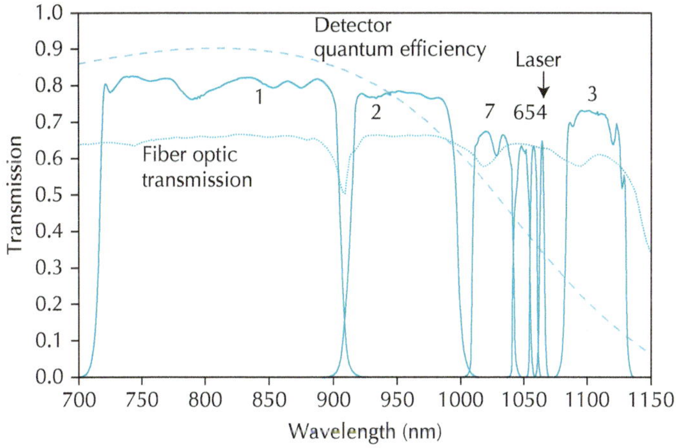

FIGURE 7.13 Measured transmission curves (solid) for the polychrometer. Also shown are the measured fiber-optic transmission (dotted) and the estimated quantum efficiency (dashed). (Courtesy of T. Carlstrom (1992) and the American Institute of Physics.)

1. Rayleigh scattering,
2. Raman scattering,
3. Electron cyclotron emission cut-off data, and
4. Calibration for microturbulence and collective wave studies.

7.7.1 Rayleigh Scattering

In the case of Rayleigh scattering, a gas target is substituted for the plasma (see, for example, DeSilva & Goldenbaum, 1970; van de Sande, 2002). As discussed by Van de Sande, the total Rayleigh scattered power is given by

$$P_R = P_i n_R \frac{d\sigma_R}{d\Omega} \Delta\Omega, \tag{7.7.1}$$

where n_R is the density of the gas providing the Rayleigh scattering.

The differential scattering cross section for Rayleigh scattering by an atomic gas is related to the polarizability α of the atoms,

$$\frac{d\sigma_R}{d\Omega} = \frac{\pi^2\alpha^2}{\epsilon_0^2\lambda_i^4}\left(1-\sin^2\theta\cos^2\phi\right). \tag{7.7.2}$$

The (microscopic) polarizability may also be written in terms of the (macroscopic) refractive index of the gas,

$$\alpha = \frac{3\epsilon_0}{n_\mu}\frac{\left(\mu^2-1\right)}{\left(\mu^2+2\right)} \approx \frac{2\epsilon_0(\mu-1)}{n_\mu}, \tag{7.7.3}$$

for $(\mu-1) << 1$.

TABLE 7.1 Examples of $d\sigma_R/d\Omega$ Are Given by van de Sande for $\lambda_i = 532$ nm and $\theta = 90°$

Particle:	Ar	Ar^+	He	N_2	O_2	CO_2
$d\sigma_R/d\Omega$ (10^{-12} m^2):	5.40	2.12	0.087	6.07	4.99	13.6
$d\sigma_{J\to J}/d\Omega$				0.038	0.10	0.38

TABLE 7.2 σ_T/σ_R given by DeSilva and Goldenbaum (1970)

Gas:	Dry Air	N_2	O_2	H_2
σ_T/σ_R:	401	380	462	1769

Table 7.1 shows examples of $d\sigma_R/d\Omega$ are given by van de Sande for $\lambda_i = 532$ nm and $\theta = 90°$.

For the case of N_2 and $\lambda_i = 532$ nm, practically all the Rayleigh scattered radiation is polarized like the Thomson scattered radiation.

Rayleigh scattering by molecules is also discussed by van de Sande. The total cross section is now given by a polarized contribution from the mean polarizability, plus a partly depolarized contribution ($\rho = 3/4$) of the polarizability anisotropy γ. This latter contribution is the sum of all rotational transitions $J \to J$.

$$\frac{d\sigma_R}{d\Omega} = \frac{d\sigma_{R,\alpha}}{d\Omega} + \sum_J \frac{n_J}{n} \frac{d\sigma_{J\to J}}{d\Omega}, \tag{7.7.4}$$

where (n_J/n) is the fraction of molecules in the state J.

For Thomson scattering, $\frac{d\sigma_T}{d\Omega} = r_0^2\left(1 - \sin^2\theta\cos^2\phi\right)$. The ratio of the Thomson to Rayleigh scattering cross section (σ_T/σ_R) given by DeSilva and Goldenbaum for a number of commonly used gases is summarized in Table 7.2.

The total integrated intensities in the given scattering solid angle must be compared, as the spectral width due to Doppler shifts is much less than for the Rayleigh case. If we assume that both spectra are Gaussian, then the relative scattered powers are

$$\frac{P_R}{P_T} = \frac{\sigma_R n_R}{\sigma_T n_T} = \frac{\left(\Delta\lambda_R I_{Rp}\right)}{\left(\Delta\lambda_T I_{Tp}\right)}, \tag{7.7.5}$$

where $\Delta\lambda$ is the spectral half-width and I_p is the peak scattered intensity. The advantages of this method are that the calibration signal traverses the complete optical path and starts from the correct scattering volume, and the time variation is the same as for the Thomson scattered signal. Further, we can use the signal to optimize the alignment of the collection system with the incident system.

The disadvantages are that we need relatively high gas pressures (1 Torr < pressure < 760 Torr) to obtain a signal comparable with the Thomson scattered signal. Dust is stirred up at these pressures, and we must leave the system for a long time (ideally overnight) for the dust to settle. Further, the components of the vacuum system may move as the pressure is changed, and this can alter alignment. As a precaution, it is sensible to check that the scattered signal varies linearly with filling pressure.

For initial alignment of the beam, it is common practice to determine the path of the incident beam by using it to burn marks on developed film. When the film is at the collection volume, the beam may be illuminated by, for example, a He–Ne laser to provide an object for the alignment of the collection optics. Alternatively, a small mirror or ground class can be used as the object (see Fig. 4.8). Other alignment techniques are discussed in Section 7.8.2.

7.7.2 Raman Scattering

The use of Raman-scattered radiation for calibration was proposed by Röhr (1977, 1981). Examples of applications are given by Howard *et al.* (1979); Flora & Giudicotti (1987); Muraoka *et al.* (1998); van de Sande (2002); LeBlanc (2008).

An advantage of Raman scattering over Rayleigh scattering is that the scattered radiation is at a different wavelength than the incident radiation. Consequently, the calibration system does not have to use a different arrangement from the Thomson scattering setup, which may need filters to deal with stray radiation and Rayleigh scattered radiation.

The differential cross section for a Raman transition $J \to J'$ given by Penney *et al.* (1974) is

$$\frac{d\sigma_{J\to J'}}{d\Omega} = \frac{64\pi^4}{45c_0^2}\frac{b_{J\to J'}\gamma^2}{\lambda^4_{J\to J'}}\left|(1-\rho)\cos^2[\xi\,(1-\sin(2\theta)\cos(2\phi))]+\rho\right], \tag{7.7.6}$$

where $b_{J\to J+2} = \frac{3(J+1)(J+2)}{2(2J+1)(2J+3)}$, $b_{J\to J-2} = \frac{3J(J-1)}{2(2J+1)(2J-1)}$, γ is the anisotropy of the molecular polarizability tensor, and ρ is the depolarization ratio, which equals 3/4 for simple linear molecules. In Table 7.3, estimates for γ^2 for the case of $\lambda_i = 532$ nm incident radiation are given for a variety of molecules.

In a test of Raman scattering, LeBlanc found good agreement between the calibration factors using Rayleigh scattering from N_2 at 1064 nm and Raman scattering at 1048 nm with a Nd:YAG laser in measurements made across a major radius in National Spherical Torus Experiment (NSTX).

Bassan *et al.* (1993) offer a caution that at sufficiently high filling density, depending on incident power, that stimulated Raman scattering can affect the calibration.

TABLE 7.3 The Values for $\lambda_i = 532$ nm Incident Radiation are Given by van de Sande and for the Other Incident Wavelengths by LeBlanc

$\gamma^2(10^{-82}\ F^2m^4)$	532 nm	694.3 nm	1064 nm
N_2	$0.395 \pm 8\%$	0.353%	$0.400 \pm 5\%$
O_2	$1.02 \pm 10\%$		
CO_2	$4.08 \pm 13\%$		

7.7.3 Electron Cyclotron Emission Cutoff Data

A disadvantage of using high-pressure gases to calibrate the scattering system is that some gas molecules end up in the vacuum vessel walls. Their release during subsequent discharges interferes with the plasma. An alternative calibration technique employed on Alcator C-Mod, Zhurovich *et al.* (2005), uses measurements of electron cyclotron emission (ECE) from the plasma to detect the position of the critical density layer at which the frequency is cut off. Specifically, the technique involves monitoring the second harmonic ECE perpendicular to the magnetic field in the extraordinary mode. The ECE signal is cut off if the second harmonic ECE frequency is less than the right-hand cut off frequency

$$\Omega_{ce} < \omega_R = \Omega_{ce}\frac{\left[1+\left(1+\frac{4\omega_{pe}^2}{\Omega_{ce}^2}\right)\right]^{1/2}}{2} \quad \text{or} \quad \Omega_{ce}^2 > \frac{\omega_{pe}^2}{2}. \tag{7.7.7}$$

In the above equation, the critical density $n_{\text{crit}} = 2\Omega_{ce}^2 m_e \epsilon_0 / e^2$.

In a tokamak, the applied toroidal field varies as $1/R^2$ and is known. Corrections to obtain the total field owing to the poloidal field and paramagnetic effects may be calculated using plasma modeling codes.

This measurement allows calibration of the density at selected points during a discharge. Calibration of the Thomson scattering system at its measurement radii comes from interpolating between the ECE points. A comparison between the integrated line density from the scattering measurements and that from an interferometer shows agreement to within about 5% (see Fig. 7.14). Overall, the absolute accuracy is estimated to be 10% or better.

7.7.4 Calibration of Collective Scattering Systems for Microturbulence and Coherent Wave Studies

Collective scattering is often employed in the study of microturbulence and coherent wave phenomena. However, referring to the simplified equation for the scattered power together with Fig. 7.15, it is seen that scattering system parameters such as scattering length and wave number resolution must be carefully

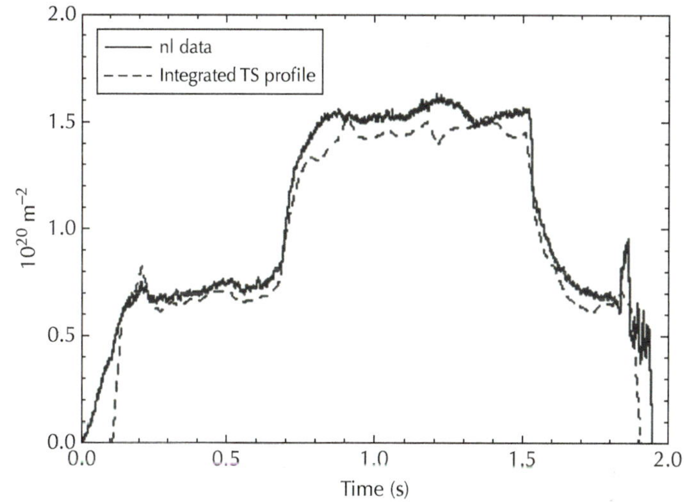

FIGURE 7.14 Comparison of the integral under the Thomson scattering profile (dashed line) with the line integrated density measurements from one of the two-color interferometer chords (solid line) as a function of time. (Courtesy of K. Zhurovich (2005) and The American Institute of Physics.)

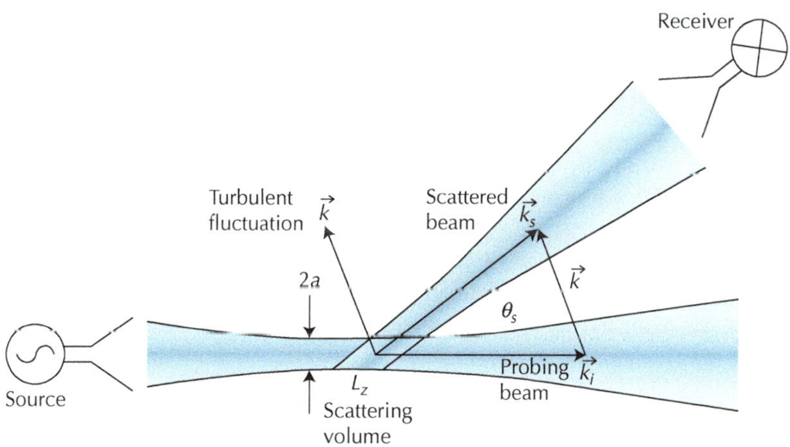

FIGURE 7.15 Collective scattering geometry for the study of microturbulence and coherent wave phenomena.

measured in order to obtain accurate wave dispersion relations and amplitudes via collective scattering. Furthermore, absolute density fluctuation and spectral density [$S(k)$] measurements require an absolute calibration of the entire optical system, as well as the receiver system. Here, we note that in the simplest case of isotropic turbulence and a simple geometry, the scattering volume is simply the beam overlap region. However, a nonuniform magnetic field (curvature and shear) detunes portions of the overlap region to the detector, thereby imposing an instrumental selectivity function which constricts the scattering length (Mazzucato, 2006). It is also to be noted that anisotropy of the plasma microturbulence with the fluctuation wave nearly perpendicular to the local magnetic field can improve the spatial resolution.

An extremely useful approach for millimeter wave and far-infrared scattering systems has been to employ an acoustic cell for optical system calibration (Saito *et al.*, 1981; Park *et al.*, 1982). Figure 7.15 illustrates the approach. Here, the acoustic cell consists of a main body possessing low loss transmission at the desired FIR operating frequency (e.g., TPX, polyethylene) and a piezoelectric transducer (PZT) to launch acoustic waves at a frequency consistent with that of the phenomena of interest (i.e., microturbulence for example). For an accurate calibration, the acoustic wave amplitude must be precisely controlled, which presents a challenge since even if the PZT driving voltage is kept constant, the amplitude of the acoustic waves is a sensitive function of the frequency since the PZT transducer exhibits a series of resonances related to its thickness, d, by

$$d=\frac{\lambda}{2},\frac{3\lambda}{2},\frac{5\lambda}{2}\ldots, \tag{7.7.8}$$

where λ is the acoustic wavelength in the PZT transducer. Because of this strong frequency dependence, the amplitude of the launched acoustic waves must be monitored within the acoustic cell at each frequency. This was accomplished by Park *et al.* (1982) through the use of small angle He–Ne laser scattering orthogonal to the FIR scattering plane, which was made possible through the use of TPX plastic for the Bragg cell since it has good transmission in the FIR, as well as satisfactory transmission in the visible. In order to determine the wavelength of the acoustic wave within the Bragg cell, a time-of-flight technique was employed to launch wave packets thus permitting the determination of the sound speed ($\sim 2.2\times 10^5$ cm/s for TPX) in the medium (and thus, k since ω is known). Referring to Fig. 7.16, it is seen that one must also correct for the bending of the FIR radiation at the dielectric-air interface for all angles other than exact normal incidence. One must, therefore, employ Snell's law and correct the data for this angular shift.

Using the above described techniques, Park *et al.* (1982) calibrated the optical system for a multichannel, far-infrared scattering instrument. By maintaining constant acoustic wave amplitude while varying the frequency, the wave number resolution for each channel was obtained as shown in Fig. 7.17. Using these data, they concluded that $\Delta k = 1.1\,\mathrm{cm}^{-1}$, which was in good agreement with theoretically predicted value $\Delta k = 2/a_0$ using the separately measured value $a_0 \approxeq 1.8\,\mathrm{cm}$ for the beam waist.

The scattering length may be determined by scanning the acoustic cell along the incident beam and monitoring the scattered power; experimental results from Park *et al.* are shown in Fig. 7.18. The scattering length defined as the

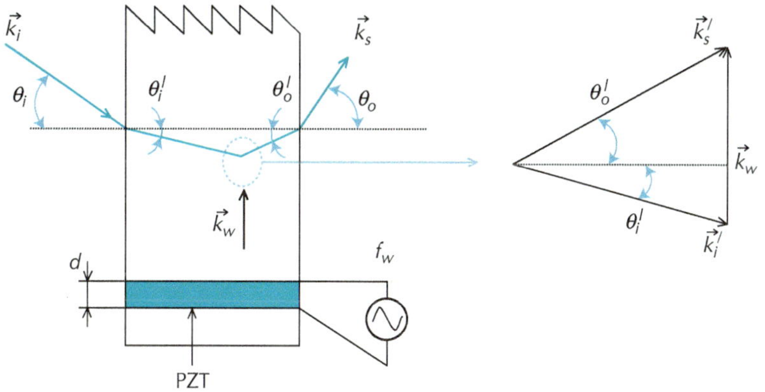

FIGURE 7.16 Schematic of the scattering geometry for system calibration using a Bragg cell.

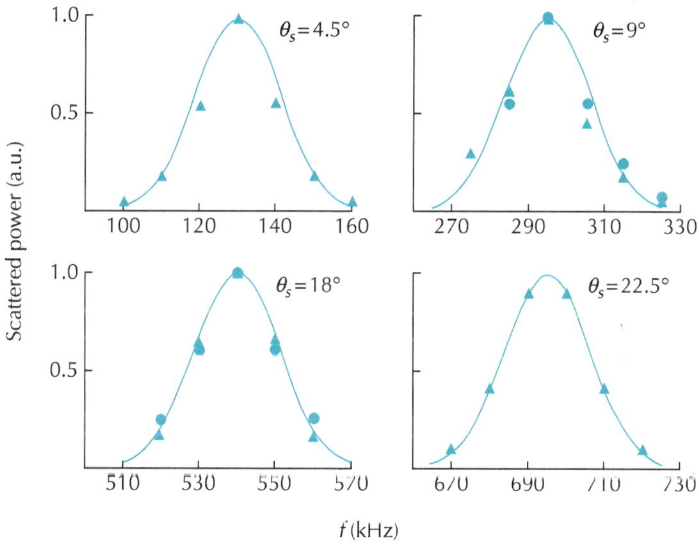

FIGURE 7.17 Acoustic cell calibration of scattering system wave number resolution. (Courtesy of Park *et al.* (1982).)

full width at half-maximum of these curves was in good agreement with the theoretical value $L_z = 2a_0/\sin\theta_s$. Note that the scattering lengths for the 4.5° and 9° channels exceed the plasma diameter.

The above approach is one of the techniques that have been employed in the calibration of the five-channel, 280-GHz collective scattering system used in the investigation of electron gyro-scale density fluctuations in National Spherical Torus Experiment (NSTX) plasmas (Lee *et al.*, 2008, 2009a,b; Smith *et al.*, 2004).

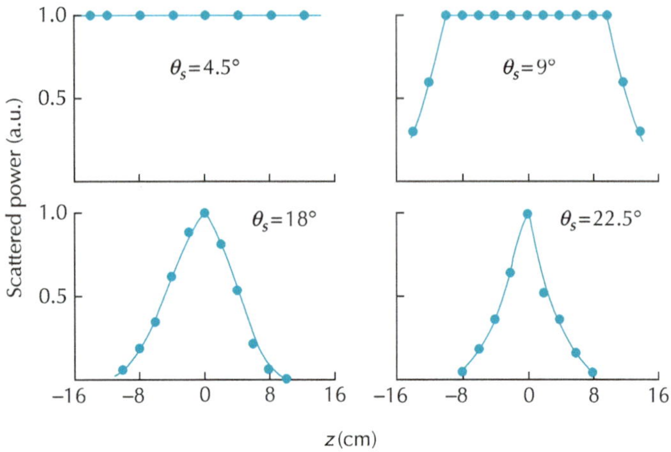

FIGURE 7.18 Acoustic cell calibration of the scattering length for four of the six system channels. (Courtesy of Park *et al.* (1982).)

Here, they accounted for the complicated magnetic field geometry, which significantly affects the longitudinal scattering volume length, and targeting wave number. They begin with the expression

$$P_s = \frac{1}{4} P_i n_e^2 r_0^2 \lambda_i^2 L_z^2 f\left(\mathbf{k}_w, \mathbf{k}_m\right), \tag{7.7.9}$$

where P_i is the probe beam power, n_e is the density fluctuation, $r_0 = e^2/(4\pi\epsilon_0 m_e c^2)$ is the classical electron radius, λ_i is the probe beam wavelength, and L_z is the interaction length. In the above, the k-matching function is defined as

$$f\left(\mathbf{k}_w, \mathbf{k}_m\right) = \exp\left[-2\frac{\left(\mathbf{k}_w - \mathbf{k}_m\right)_\perp^2}{\left(2/a_0\right)^2}\right] \sin\left(c^2\left[\frac{\left(\mathbf{k}_w - \mathbf{k}_m\right)_{||}^2 L_z}{2}\right]\right). \tag{7.7.10}$$

In employing the acoustic cell, they took account of the differences in the cases of uniform and toroidal curvature: the acoustic cell has a fixed angle in the uniform case and can be pointed to the radial direction along the probe beam path in the toroidal curvature case. Magnetic shear adds the complication that a three dimensional alignment of the acoustic cell angle is required at each probe beam path.

Receiver noise temperatures and conversion loss may be obtained from standard Hot–Cold load measurements and Y-factor analysis as described in Section 6.7.5. One may also employ a coherent source to obtain the overall power calibration of the receiver system as was used by Lee *et al.* (2009a). A schematic of their configuration is shown in Fig. 7.19 where the calibration source power was −4 dBm(=0.4 mW), which was subsequently attenuated to −54, −49, and −47 dBm by a stack of calibrated papers and standard ECCOSORB® microwave absorber. For their NSTX implementation, Faraday

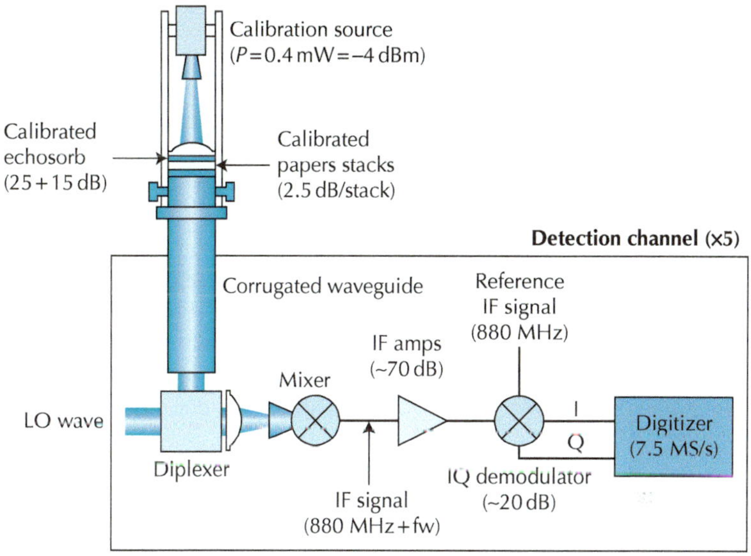

FIGURE 7.19 Experimental setup for the power calibration of the heterodyne detection system. (Courtesy of Lee *et al.* (2009b).)

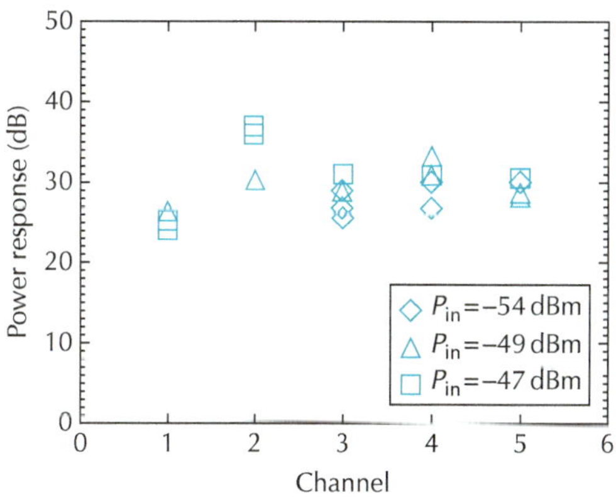

FIGURE 7.20 Measured power response. The difference in the response among channels is mainly due to the difference in the coupling efficiency to the mixer and the amplifier gain. The maximum value in each channel is the most probable value of the detected power. (Courtesy of Lee *et al.* (2009b).)

rotation of the linearly polarized probe and scattered wave is nonnegligible. Using an overall efficiency estimate of 0.5, they estimated the power responses of the five channels were 26, 36, 30, 33, and 30 dB, respectively, with the difference in the response among the channels attributed to various factors such as coupling efficiency to the mixer, transmission line losses, and amplifier gain variations. Figure 7.20 displays their measured channel responses.

As another note on acousto-optic cells, it is useful to mention the work of Dürr & Schmidt (1985). Here, they describe studies of the use of Ge instead of TPX and present measurements at both 119 μm and 10.6 μm. They contend that while Germanium exhibits considerable electromagnetic absorption, this is partially compensated for by low acoustic absorption and high figure of merit.

Finally, we note that Chang *et al.* (1990a) developed a dual laser system that worked with acoustic waves driven in air to calibrate a collective CO_2 laser scattering system. In this approach, they determine the sound pressure from heterodyne detection of the acousto-optical diffraction instead of directly detecting the sound pressure of the acoustic field or via detection of the power diffracted from the CO_2 laser beam. Here, acoustic waves were launched at 245 kHz via a piezoelectric transducer with the scattered He–Ne laser beam detected with a photomultiplier tube and the CO_2 beam with a HgCdTe detector.

7.7.5 Scattering Volume and Alignment

The scattering volume is approximately the intersection of two cylinders. The maximum scattered power is obtained when the axis of the receiving cylinder intersects the axis of the input radiation. A detailed discussion of this topic has been given by Bindslev *et al.* (2003) in regard to fast ion collective scattering on ITER. They calculate that the beam overlap O_b is given by

$$O_b = 2\frac{\sin^{-1}\theta}{2\pi\left[\left(\omega_n^i\right)^2+\left(\omega_n^s\right)^2\right]}\exp\left\{\frac{-2\Delta n}{\left(\omega_n^i\right)^2+\left(\omega_n^s\right)^2}\right\}, \tag{7.7.11}$$

where θ is the scattering angle, ω_n^i, ω_n^s are the Gaussian half widths of the input and receiver beams, and Δn is the shortest distance between the beam center lines. They define the scattering volume as spatial region from which 90% of the received scattered radiation comes and which has the smallest volume.

Yoshida *et al.* (1997) describe a quantitative method for precise quick and reliable alignment of the collection optics using special alignment fiber bundles; see also the TFTR system, Johnson (1992), and the D-IIID system, Carlstrom *et al.* (1992).

7.8 TYPICAL COLLECTIVE SCATTERING SETUP AND CONSIDERATIONS

7.8.1 Setup

Figure 7.21 shows a typical Thomson scattering system; a lens collects and collimates the light scattered by the plasma. The collimated light is transported to a lens that focuses the light into a spectrometer. In general, a high-resolution spectrometer is required to resolve the ion-acoustic features; for a typical fully ionized nitrogen plasma ($T_e = 1\,\text{keV}$, $T_e/T_i = 3$, $n_e = 10^{20}\,\text{cm}^{-3}$, and $Z = 7$) and

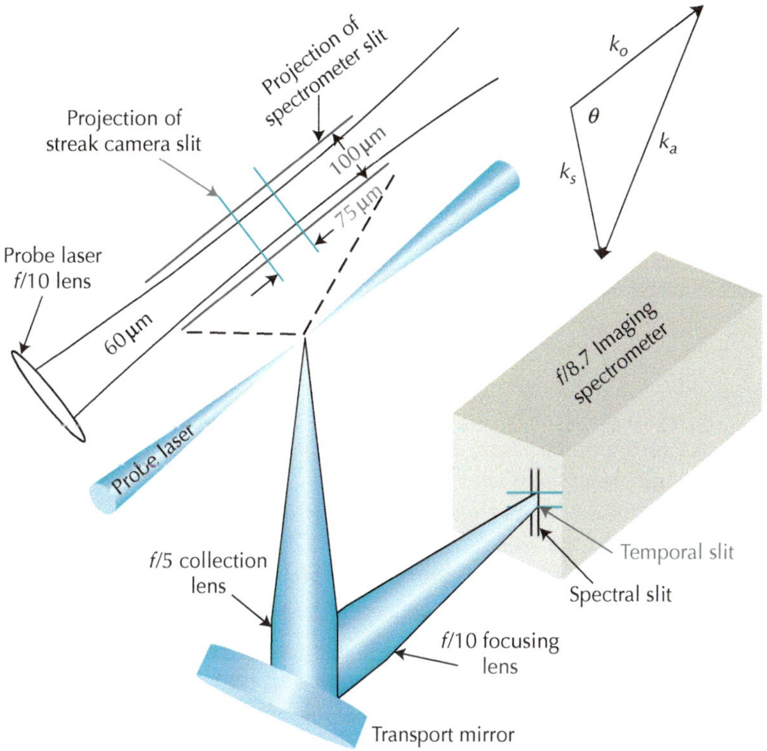

FIGURE 7.21 A typical Thomson scattering setup is shown. The spectrometer and streak camera slits are imaged into the plasma plane defining the Thomson scattering volume. The scattering vector diagram is shown. (Courtesy of Froula *et al.* (2006a).)

a scattering angle of $\theta = 90°$, Eqn (7.1.3) gives the separation between the ion-acoustic features of $\Delta\lambda/\lambda_i = 2.5 \times 10^{-3}$. To resolve this separation, a modern system typically uses a 1 m $f/8.7$ imaging spectrometer with a 3600 line/mm grating (110×110 mm^2). The practical resolving power [see Eqn (7.4.4)] for an optical scattering setup $\lambda_i = 0.26$ μm is typically $R_p = \lambda_i/(\Delta q D_G) \approx 10^5$, where $\Delta q \simeq 15$ μm is the pixel size ("output slit width") and $D_G \approx 2 \times 10^{-7}$ is the linear dispersion. In order to collect enough photons, a spectral slit on the order of $\Delta r = 100\,\mu$m is typical. This reduces the effective resolving power $R_p = \lambda_i/(\Delta r D_G) \approx 10^4$. It is the resolving power given by the spectrometer slit width, which typically dominates in these types of Thomson scattering configurations.

Two typical detector configurations are discussed; imaging Thomson scattering uses a charge-coupled device (CCD) to measure the time integrated profile of the Thomson scattering spectrum along the probe beam, while streaked Thomson scattering employs a streak camera to record the temporal evolution of the scattered signal from a small volume in the plasma (Fig. 7.21). Typical imaging Thomson scattering (ITS) data are recorded with a gated intensified 16 bit CCD camera, while time resolved Thomson scattering (STS) data use a high dynamic range streak camera typically with a $\Delta s \approx 150\,\mu\text{m}$ temporal slit width. The temporal resolution is typically limited by the spectrometer (see Section 7.3.3).

7.8.2 Choice of Optics/Defining a Scattering Volume

The focal lengths of the collection and focusing lenses are chosen to maximize the coupling through the system and determine the size of the Thomson scattering volume (V_{TS}). To maximize the coupling through the spectrometer, the f number of the focusing lens ($f^{\#}_{\text{focus}}$) is chosen to be comparable with the f number of the spectrometer. The Thomson scattering volume is determined by the beam waist of the probe laser (ϕ) and the projection of the streak camera slit into the plasma plane; the widths of the spectral and temporal slits in the plasma plane are determined by the magnification of the system given by the ratio of the f numbers of the lenses $M \simeq f^{\#}_{\text{col}}/f^{\#}_{\text{focus}}$. Typically, high power nanosecond long optical probe lasers use an $f_{\text{probe}} \sim 10$, which produces a $\phi \simeq 60\,\mu\text{m}$ diameter beam waist. An $f^{\#}_{\text{col}} = 5$ collection lens used in connection with an $f^{\#}_{\text{focus}} = 10$ focusing lens define a cylindrical volume $V_{TS} = \pi(30\,\mu\text{m})^2 \times 75\,\mu\text{m}$ (see Fig. 7.21). Minimizing the Thomson scattering volume reduces the spectral broadening by gradient effects and defines a precise location in the plasma, but comes at the cost of reducing the amount of scatter light that is collected.

The measured Thomson scattering spectrum is broadened by the range of ion-acoustic waves probed ($\delta\lambda_{\text{optics}}$). When measuring a thermal scattering spectrum from ion-acoustic waves, this broadening can be calculated from Eqn (5.4.3) to be

$$\frac{\delta\lambda_{\text{optics}}}{\lambda_i} = \frac{1}{2\tan(\theta/2)}\left(\frac{1}{f^{\#}_{\text{col}}} + \frac{1}{f^{\#}_{\text{probe}}}\right)\left(\frac{\Delta\lambda}{\lambda_i}\right). \tag{7.8.1}$$

For scattering with a 4ω ($\lambda_i = 265$ nm) probe laser from a typical fully ionized nitrogen plasma ($T_e = 1\,\text{keV}$, $T_e/T_i = 3$, $n_e = 10^{20}\,\text{cm}^{-3}$, and $Z = 7$) at a scattering angle of $\theta = 90^\circ$ and the optical configuration discussed above, the spectral smearing resulting from the optics is $\delta\lambda_{\text{optics}}/\lambda_i = 4 \times 10^{-4}$; this is convolved with the instrument function of the spectrometer ($1/R_p \approx 10^{-4}$). The result, $\delta\lambda/\lambda_i \approx 5 \times 10^{-4}$, is compared to the separation of the ion-acoustic features $\Delta\lambda/\lambda_i = 2.5 \times 10^{-3}$, when attempting to measure the electron temperature via the sound speed [see Eqn (7.1.3)], or to the width of the ion-acoustic feature, when trying to measure the ion temperature. To improve the resolution of this system, the f number of the collection lens can be increased, but a factor of 2 improvement requires a factor of 4 increase in the f number, which results in a factor of 16 reduction in the scattering signal. Alternately, the f number of the probe can be increased at the cost of increasing the beam diameter, which increases the scattering volume.

Optical Alignment

One of the most critical aspects of a successful Thomson scattering measurement is optical alignment. The projection of the instrument slits into the plasma must overlap the incident probe beam typically requiring alignment precision better than 50 μm. Standard alignment practices use 100 μm spheres accurately positioned to the location of the desired Thomson scattering volume. A low power alignment laser, operating at the same frequency as the probe beam, is used to backlight the alignment ball. With the diagnostic slits open, an image of the ball is then projected through the spectrometer and onto the Thomson scattering detector. Using the transport mirrors, the ball is centered between the slits, and they are closed around the ball. The probe laser is then aligned to the ball. It is often necessary to perform this final alignment under vacuum.

Probe Laser Requirement for Ion-Acoustic Feature

For a given experimental arrangement, the required probe laser energy necessary to achieve a suitable signal-to-noise (S/N) ratio can be estimated. There are three criteria that are considered when selecting the necessary laser power: the power of the heater beams in the experiment that produce background light, bremsstrahlung radiation, and detector sensitivity. For experiments at large laser facilities (greater than two beams), the background is typically dominated by stray heater beam light, which is not completely rejected by the bandpass filters and the spectrometer grating. At small facilities, the bremsstrahlung radiation from the plasma and the detector sensitivity are the most important factors in determining a detectable Thomson scattering signal. Integrating Eqn (5.4.2) over the ion-acoustic features in the collective regime relates the probe laser energy E_0 to the scattered energy E_s,

$$E_s = \frac{\pi r_0^2 n_e L}{8 f_{col}^{\#2}} E_0, \tag{7.8.2}$$

where L is the length of the Thomson scattering volume along the axis of the probe laser. For a beam diameter ø and optical magnification $M = f^{\#}_{col}/f^{\#}_{focus}$, the scattered fluence at the spectrometer entrance slit is

$$\Phi_{sT} = \frac{E_T}{M^2 L ø}, \tag{7.8.3}$$

where T is the total transmission of the optical system.

In this geometry (see Fig. 7.21), the optimal spectral slit width is $\Delta r \sim Mø$ (a small slit width reduces the scattering signal while a larger slit width increases the noise). Assuming the separation of the ion-acoustic features are well resolved, and their widths are instrument limited by the spectral slit width, the fluence at the detector is half of the fluence a the slit and the total number of photoelectrons per pixel is

$$N_{pec} = \frac{\Phi_s}{2}\frac{\sigma\eta}{h\nu_i} = \frac{E_s T}{2LM\Delta r}\frac{\sigma\eta}{h\nu_i} \tag{7.8.4}$$

where σ is the area of a single pixel and η is the quantum efficiency (photoelectrons/photon) of the detector.

Equations (7.8.2) and (7.8.4) yield the necessary probe laser energy required for the (time integrated) imaging spectrometer system (ITS),

$$E_0^{\mathrm{ITS}} = \frac{16 f_{\mathrm{col}}^2}{r_0^2 n_e}\frac{hcN_{pec}}{\lambda_i \sigma\eta}\frac{M\Delta r}{T}. \tag{7.8.5}$$

The laser energy required for the streaked spectrometer system (STS) is

$$E_0^{\mathrm{STS}} = E_0^{\mathrm{ITS}}\frac{\Delta t}{\Delta T}, \tag{7.8.6}$$

where ΔT is the temporal slit width and Δt is the pulse width of the laser probe. Assuming a system limited only by quantum statistical noise, Eqn (6.4.3) states that for $N_{pec} = 100$ we expect to have a singal-to-noise of 10. Therefore a typical minimum probe energy required is $E_0^{\mathrm{ITS}} = 0.5\,\mathrm{mJ}$ for $f_{\mathrm{col}} = 5$, $\lambda_i = 265$ nm, $n_e = 10^{20}$ cm^{-3}, $\eta = 20\%$, $T = 30\%$, $M = 2$, $\Delta r = 100$ µm, $\sigma = 225$ µm^2. This of course neglects noise sources other than the detector (see Section 6.4).

PROBLEMS

7.1. You wish to configure a heterodyne receiver system for a microwave scattering system and wish to determine if the expense of purchasing a high-frequency preamplifier is warranted. To determine this, consider the following two configurations:

(a) Preamplifier (15 dB gain, 2 dB NF) followed by a mixer (5 dB CL, 5 dB NF)

(b) Mixer (5 dB CL, 5 dB NF) followed by an IF amplifier (15 dB gain, 2 dB NF)

What is the overall gain, NF, and noise temperature for the two cases?

7.2. You have a 10- to 11-µm heterodyne receiver and wish to perform a hot–cold noise temperature calibration. It views a 290 K background. The HgCdTe diode photomixer has the following detector properties:

$\lambda_{\mathrm{cutoff}} = 15$ µm, $\eta = 40\%$ for $\lambda < 13$ µm, size $\sim 100 \times 100$ µm, depletion region width ~ 1 µm at operating bias reverse bias impedance $\sim 150\,\Omega$, dielectric constant $= 10$.

The receiver properties are LO: CO_2 laser reflected off 10% reflectivity diplexer. Diplexer transmits 90% of incoming signal. Photomixer output to 150 Ω input impedance amplifier.

Amplifier gain $=100$, BW$=3\times10^9$ Hz, and output noise current: $<I_N^2>^{1/2}=33.2A/\text{Hz}^{1/2}$.

The receiver is tested by looking at hot and cold sources:

500 K source: $V_H=1.040$ V
300 K source: $Vc=1.000$ V

(a) IF bandwidth: Is it limited by amplifier or the RC time constant of the mixer?
(b) Does the receiver operate in the thermal or quantum noise limit?
(c) What is the noise temperature of the amplifier?
(d) What is the laser power required to overcome the amplifier noise?
(e) If $P_L=20$ times 0.68 mW (the minimum), what is the conversion gain?
(f) What is the expected system noise temperature?
(g) What is the measured system noise temperature?

7.3. Consider an experiment using a 6 J, 25-ns ruby laser to scatter light at 90° from a plasma with $n_e=5.10^{13}\,\text{cm}^{-3}$ and T_e around 1 keV. Let the scattering volume have $L=1$ cm and $A=0.3\times0.7\,\text{cm}^2$. Let the collection solid angle be 2×10^{-2} sr, and for dispersion, use a 1200 line/mm grating of area $10\times10\,\text{cm}^2$ in first order with $\varphi_2=20°$ and a 1-cm wide entrance slit.

(a) Calculate the consistent width (a_s(cm)) value for the fiber optic channels and final lens (f_2(cm)) focal length to be able to resolve $1/10^{\text{th}}$ of the scattered light's spread at $T_e=1$ keV, when the grating etendue is twice the collection etendue.
(b) Using photomultiplier characteristics from Fig. 6.7 for the short wavelength side of the scattered light, and assuming an overall transmission factor of $T=0.1$, calculate the average number of photons per channel and the statistical noise.
(c) Comment on how this setup will fare as the temperature varies from 0.1 to 5 keV. Allow for the downward shift of the spectrum as the temperature is raised.

7.4. Consider an experiment to detect magnetic modulation in a plasma with $n_e=2\times10^{15}\,\text{cm}^{-3}$ and $T_e=20$ eV and a magnetic field of 1.6 T in a setup similar to that discussed in Section 4.7 with a 5 J ruby laser. The collection solid angle is 0.9×10^{-3} sr, and the collection etendue is $2.5\times10^{-4}\,\text{cm}^2$ sr.

(a) Calculate the Fabry–Perot Section spectrometer characteristics that would give an instrument etendue twice the collection etendue and also the required values for f_2 and Δr_1 to yield a minimum resolution ($\Delta\lambda_{\text{mm}}$) of a tenth the separation of the cyclotron peaks (see Section 7.6).
(b) Calculate the number of photoelectrons measured assuming $L=1$ cm, $T=2\times10^{-2}$, $\eta\approx0.25$, and $C=0.04$.

(c) Show that this setup satisfies Eqns (4.7.4) and (4.7.5) when $\Delta\theta_s \cong 3 \times 10^{-2}$ rad, $\theta = 30°$, and $\theta_s = 15°$.

7.5. Consider now the use of a CO_2-pulsed laser from Table 6.1. ($\lambda_i = 10.59\,\mu$m, 17 J in 1 μs) for measuring magnetic modulation in a plasma with $B = 3$ T, $n_e = 5.10^{15}\,\text{cm}^{-3}$, and $T_e = 1\,\text{keV}$. The detector will have $D^* \cong 10^{10}\,\text{W}^{-1}\,\text{cm s}^{-1/2}$, response time $< 1\,\mu$s, bandwidth $\Delta f \cong 10^6$ Hz, and area $A = 10\,\text{cm}^2$ (see Section 6.7.3), and $L = 0.5\,\text{cm}$.

(a) For 90° scattering, calculate the values of $\Delta\theta_s$ and θ_s that would satisfy the constraints set by Eqns (4.7.4) and (4.7.5).

(b) Calculate the minimum laser power to exceed the noise equivalent power $P_N = (A\Delta f)^{1/2}/D(\text{W})$ for the case $T = 0.25$ and $C = 10^{-2}$.

(c) Calculate the number of photoelectrons for $\eta \approx 0.5$.

(d) Compare the scattered and bremssrtrahlung signals, Eqn (6.5.3) for $L = 0.5\,\text{cm}$ and the plasma volume viewed, $V_p = 50\,\text{cm}^3$.

SOLUTIONS TO ODD PROBLEMS

7.1. (a) 10 dB, NF = 2.18 dB, and noise temperature = 189.45 K.

(b) 10 dB, NF = 7.00 dB, and noise temperature = 1163.44 K.

7.3. (a) The half width of the scattered light is given by Eqn (7.1.1) as

$$\Delta\lambda_{1/e} = 19.3\,(T_e(\text{eV}))^{1/2}\,\text{Å}.$$

At 1 keV, $\Delta\lambda_{1/e} = 610\,\text{Å}$. The linear dispersion, Eqn (7.4.3), is $D_G = (d\cos\varphi_2)/f_2\,\text{Å/cm}$, where the focal length is in centimeter, and $d = 8300\,\text{Å}$.

Equating $\Delta\lambda_{1/e}/10$ to $a_s D_G$ yields $a_s/f_2 = 61/(d\cos\varphi_2) = 7.8 \times 10^{-3}$.

The collection etendue $E_c = 0.42 \times 10^{-2}\,\text{cm}^2 \cdot \text{sr}$, and the grating etendue (Section 7.4)

$$E_G = 100 a_s/f_2^2\,\text{cm}^2 \cdot \text{sr}.$$

Equating $2 \times E_c$ and E_G yields $a_s/f_2^2 = 0.84 \times 10^{-4}$.
Consequently, $f_2 = 93$ cm and $a_s = 0.72$ cm.

(b) The number of photoelectrons detected in a typical channel is given by Eqn (6.5.5).

$N_{pec} \approx (W_i/h\nu_i) r_0^2 d\Omega n_e L T \eta C$ photoelectrons, where C is the number of channels.

For this example, assume $\eta \approx 0.25$ as indicated in Fig. 6.7.

$N_{pec} \approx 4200$ pe and the statistical error Eqn (6.4.3) is ±65 pes.

(c) At the lowest temperature $\Delta\lambda_{1/e} = 193$ Å, there would only be around 3 or 4 channels; however, the signal levels would be higher and enough for an estimate of the temperature and density. At the highest temperature, $\Delta\lambda_{1/e} = 1365$ Å, and from Eqn (4.7.4), $\Delta\lambda_m \approx 486\,\text{Å}$, while the collection optics only spans 610 Å. These numbers suggest that it would be better to have a number of spectrometers tailored to fit the expected temperature ranges, as a function of both time and position in the plasma.

7.5. **(a)** $\Omega_e = 5.28 \times 10^{11}$ rad s^{-1} and $\omega_i = 1.78 \times 10^{14}$ rad s^{-1}, $a/c = 6.3 \times 10^{-2}$, and from Eqn (4.7.4) we need $\cos\theta_s \leq 74.7°$, $\Delta\theta_s \leq 0.049$ rad, and $\Delta\Omega \cong (\Delta\theta_s)^2 \leq 2.4 \times 10^{-3}$.

(b) For the example given, $P_N = 3.16 \times 10^{-7}$ W, and $P_s = P_i r_0^2 d\Omega L n_e TC$ W, which requires $P_i \geq 2.6 \times 10^8$ W.

(c) $N_{pec} \approx (W_i/h\nu_i) r_0^2 d\Omega n_e LT\eta$ photoelectrons $= 1.1 \times 10^8$ pes.

(d) The spectral half width from Eqn (7.1.1) is $\Delta\lambda_{1/e} = 4 \times 10^{-3}$ $\lambda_i[T_e(\text{eV})]^{1/2} \sin(\theta/2)$.

$N_{pe}/N_{plB} = (P_i r_0^2 \lambda_i 4\pi L)/(2.09 \times 10^{-36} n_e 8 \times 10^{-3} \sin(\theta/2) V_p) = 3$, which suggest that a higher power laser would be desirable.

Chapter 8

Techniques

Chapter Outline Head

8.1 INTRODUCTION

Key developments in Thomson scattering techniques, since the first edition of this book came out, are discussed in Bretz (1997), Luhmann *et al.* (2008), Donné *et al.* (2008), Muraoka (2006), and Glenzer & Redmer (2009).

The earliest Thomson scattering measurements in the laboratory were made one point at a time in the plasma. The television Thomson scattering system (TVTS), developed at PPPL, and applied on PLT and TFTR, Johnson *et al.* (1985), allowed the collection of scattered signals from many points along the input beam, for example, 74 points with 1 cm resolution. This approach made it easier to obtain two-dimensional plots of n_e and T_e. In general, compared to when the first edition of this book came out in 1975, a far wider range of efficient detectors exist today. In the wavelength range of 115 to 1040 nm, the quantum efficiency of various photocathodes runs from 15 to 30%. In the infrared region, the quantum efficiency varies from 0.15 to 1.0 and is typically 0.5. The use of micromachining and thin-film lithography fabrication with new materials and antenna coupling schemes resulted in sensitive high-speed

DOI: 10.1016/B978-0-12-374877-5.00008-7

microbolometers. These have found widespread application in the millimeter wave and THz region. Detectors are discussed in Section 6.7.

A second important advance was the use of a high-repetition rate YAG laser at Garching (Röhr *et al.*, 1982) which permitted data to be obtained at 100 Hz or more. This laser plays an important role in measurements on industrial plasmas. Sources are discussed in Section 6.2.

A further advance was the light detection and ranging (LIDAR) system, which was jointly developed by the University of Stuttgart and JET (Salzmann *et al.*, 1987). It used a very short laser pulse 0.3 ns duration (length 90 mm) and collected backscattered light, permitting measurements to be made along a chord through the plasma.

A triple grating spectrograph with a mask used to filter out the central wavelength radiation was used by Kono & Nakatani (2000). van der Mullen *et al.* (2004) applied the technique in measurements on a magnetic multipole plasma with electron temperature in the range 1.5–5 eV and density 10^{18}–$5 \times 10^{18}\,\mathrm{m}^{-3}$. The application to industrial plasmas is discussed in Section 9.1.

Over the laser 35 years, inertial confinement fusion has matured and Thomson scattering has become a routine method to measure plasma conditions at densities between $10^{19}\mathrm{cm}^{-3} - 10^{21}\mathrm{cm}^{-3}$ and study laser-plasma interactions through the direct probing the instabilities (Froula *et al.*, 2010).

Measurements in warm dense matter have burgeoned; Glenzer & Redmer (2009) discuss the use of x-ray lasers. This area is discussed in Chapter 11.

8.1.1 Some of the Interesting Techniques

Combined LIDAR, YAG, 2-D

An example of combining a number of these developments – LIDAR, YAG laser, 2D measurements – is shown in Fig. 8.1. A short laser pulse was reflected back and forth through the plasma with a long enough time delay that the spectrum could be measured over two dimensions with fewer polychromators (Sumikawa *et al.*, 2007).

Pulsed Polarimetry

Pulsed polarimetry combines the scattering of radiation with the Faraday effect to measure the magnetic field in a plasma (Smith, 2008). Specifically, using existing lasers with a LIDAR-like approach, it offers the opportunity to measure the local poloidal field in a tokamak, and it could achieve spatial resolution of millimeters in high-energy density plasma such as obtained in a field reversed configuration. Important features include the capability for measuring simultaneously the local T_e, n_e, and $B_{||}$ along the line of sight, with a resiliency to refractive effects. Mirnov *et al.* (2007) show that relativistic effects are important in high-temperature plasmas and change the Faraday rotation angle.

Interference Polychromators

Polychromators using high performance interference filters (transmission $\lesssim 80\%$, rejection of laser wavelength $> 10^5$) are commonly used since they were

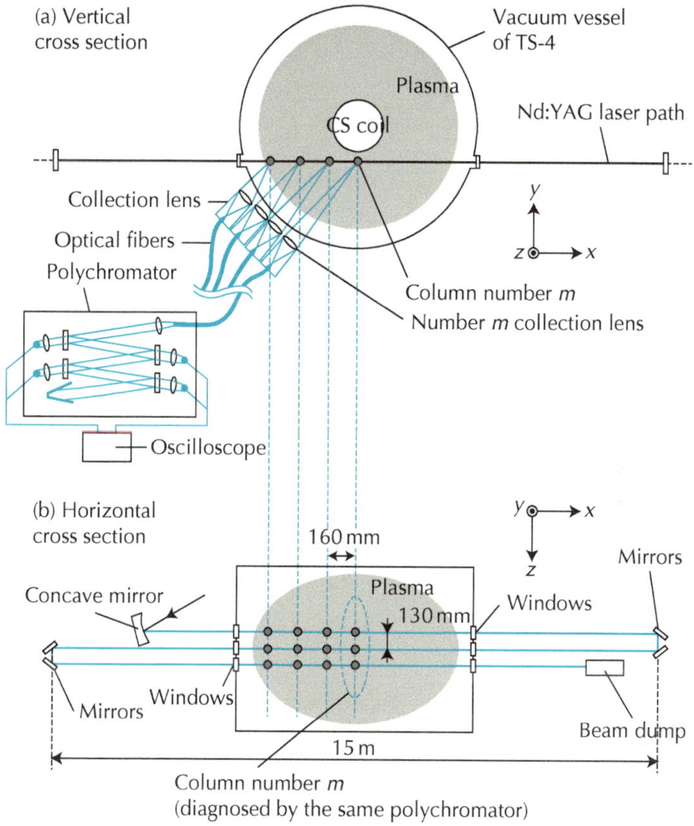

FIGURE 8.1 Vertical and horizontal cross section of the 2D Thomson scattering system on TS-4. (Courtesy of Sumikawa *et al.* (2007) and Japan Society of Plasma Science and Nuclear Fusion Research.)

applied on the ASDEX tokamak (Röhr *et al.*, 1987). Coupled with repetitively pulsed YAG lasers, they have become a standard approach in magnetic confinement systems, for example, on DIII-D, Carlstrom *et al.* (1990, 1992), see Figs. 8.2 and 7.13 which show respectively a schematic of the polychromator and the transmission curves for the DIII-D system; and Hughes et al. (2001), Fig. 8.3. It is proposed for W 7-X, Cantarini *et al.* (2009), and for the ITER edge, Kajita *et al.* (2009), see Fig. 8.4. A proposed layout for edge measurements in ITER is shown in Fig. 8.4.

Phase Conjugate Mirror

Multipass Thomson scattering systems are used to increase the measurement area, the length of time of the measurement, and to increase the measured signal. High reflectivity mirrors are important to realize these gains. A phase conjugate mirror based on stimulated Brillouin scattering (SBS-PCM) has been used to

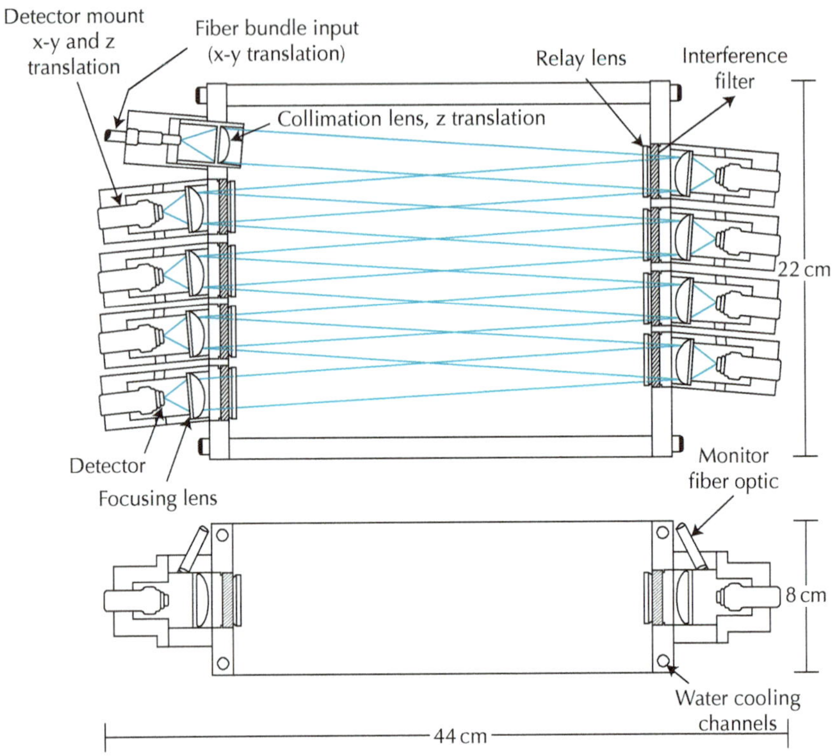

FIGURE 8.2 Schematic of an $f/1.75$ interference filter polychromator, using 3 cm optics. (Courtesy of Carlstrom *et al.* (1992) and The American Institute of Physics.)

achieve 95% reflectivity on the JT-60U scattering system (Hatae *et al.*, 2006). The average power was increased by a factor of > 8 from 45 W (1.5 J, 30 Hz) to 343 W (7.46 J, 50 Hz). A liquid fluorocarbon (trade name 3M Fluorinert, FC-75) was used as the working liquid. A schematic of the system is shown in Fig. 8.5. The estimated performance is shown in Fig. 8.6.

Polarization Interferometer

In a polarization interferometer, Thomson scattered light that traverses a polarizer is incident on a birefringent plate whose fast axis is oriented at 45° to the polarization direction. The plate splits the incident scattered scalar wave component, relatively delaying nominally equal amplitude components by the time τ before they recombine at a final polarizer and are focused on a detector (Howard, 2006; Hatae *et al.*, 2007; Howard & Hatae, 2008). The orthogonally polarized outputs at the final polarizing splitter form complementary or antiphase interferometric images of the input radiation. By suitably choosing the optical delay, these independent outputs provide sufficient information

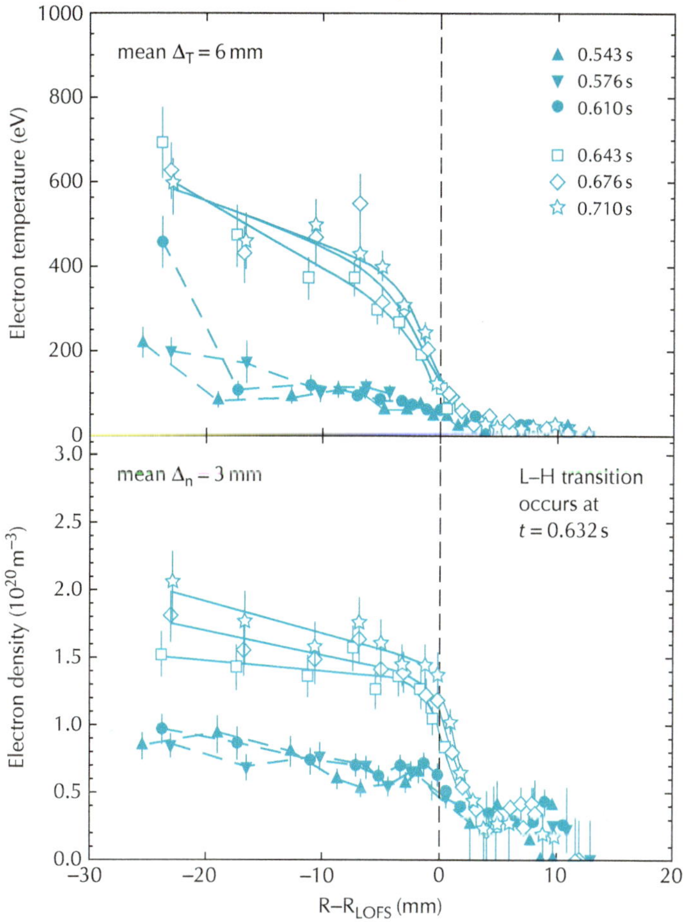

FIGURE 8.3 Edge profiles of T_e and n_e before and after the L–H transition. Error bars on T_e and n_e reflect estimated statistical uncertainties; also, n_e error bars account for uncertainties in the absolute calibration. (Courtesy of Hughes *et al.* (2001) and The American Institute of Physics.)

to determine both the electron temperature and density. A schematic of the diagnostic is shown in Fig. 8.7.

The main advantages of this Fourier transform spectrometer are high throughput, high S/N, and a simple and compact system. Calculated scattered spectra and their interferograms are shown in Fig. 8.8. The estimated resolution dT_e/T_e is typically 0.04 or better with a simple wave plate.

Number of Channels and Accuracy

Yoshida *et al.* (1999b) describe the JT-60U high spatially resolved multipoint Thomson scattering system: 60 points, 8 mm resolution, repetition rate 0.5 Hz,

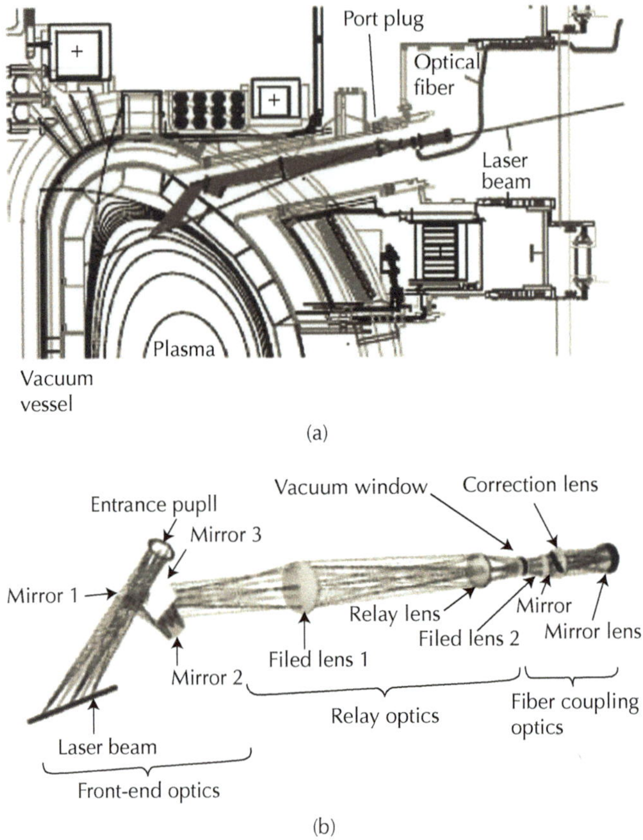

FIGURE 8.4 (a) An illustration of the present design of the ITER edge Thomson scattering system. (b) Schematic diagram of the collection optical system consists of front-end optics, relay optics, and fiber coupling optics. (Courtesy of S. Kajita (2009) and Elsevier.)

and 2 ms interval in burst operation, using 2 ruby lasers. The layout is illustrated in Fig. 8.9. The wavelength coverage of the various channels is illustrated in Fig. 8.10. The relative errors for core and edge measurements are shown in Fig. 8.11. The photodiode array used is discussed in Yoshida *et al.* (1999a). An analysis of how many channels are needed for a given accuracy is given in Naito *et al.* (1999).

8.2 FAR FORWARD SCATTERING, PHASE SCINTILLATION IMAGING, AND PHASE CONTRAST IMAGING

We have thus far been concerned with cases where the scattering angle exceeded the divergence angle of the probe beam. As we shall see in the following, by relaxing this constraint, the maximum measurable wavelength λ_{max} of the

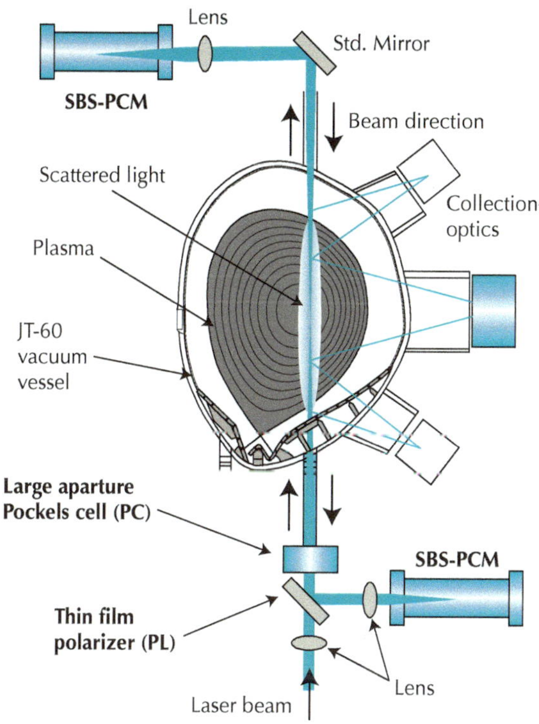

FIGURE 8.5 Schematic of the multipass Thomson scattering method using the SBS-PCM. (Courtesy of Hatae *et al.* (2006) and the American Institute of Physics.)

fluctuations can be extended beyond the familiar $\lambda_{max} = \pi \omega_i c/2$ dictated by the criterion for minimum scattering angle $\theta_{s\ min} = 2\theta_d$, where $\Delta\theta_s = \theta_d = \lambda_i/(\pi w)$ is the divergence angle of a Gaussian beam due to diffraction effects. The lite rature contains references to three primary methods employed for detection of long wavelength fluctuations which are closely related: far forward scattering, phase scintillation imaging, and phase contrast imaging. The basic physical principle upon which these are based is that in traversing the refractive plasma medium, the probe electromagnetic wave undergoes phase and amplitude modifications, which are related to the wavelength, frequency, amplitude, and position of the phase fluctuations of interest. It is the manner in which the desired information contained in the diffracted wave fronts is extracted which distinguishes the techniques. Detailed reviews of the techniques may be found in Donné *et al.* (2008) and Luhmann *et al.* (2008).

Figure 8.12 schematically illustrates the experimental arrangement for far-forward scattering where the plasma density fluctuations result in modifications of the probe beam which are detected in the far field (Evans *et al.*, 1982). Here, a plasma wave is assumed to propagate in the x-direction, which is transverse

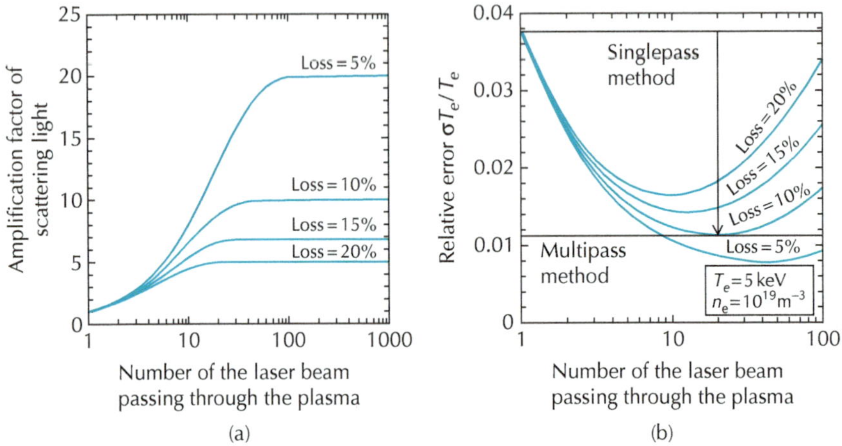

FIGURE 8.6 Estimated performance of the multipass Thomson scattering method. (a) Amplification factor of the scattered light versus the number of laser beams passing through the plasma. (b) Relative error for the electron temperature versus number of laser beams passing through the plasma. As for the relative error estimation, it takes into account the noise characteristics of the detectors. (Courtesy of Hatae *et al.* (2006) and the American Institute of Physics.)

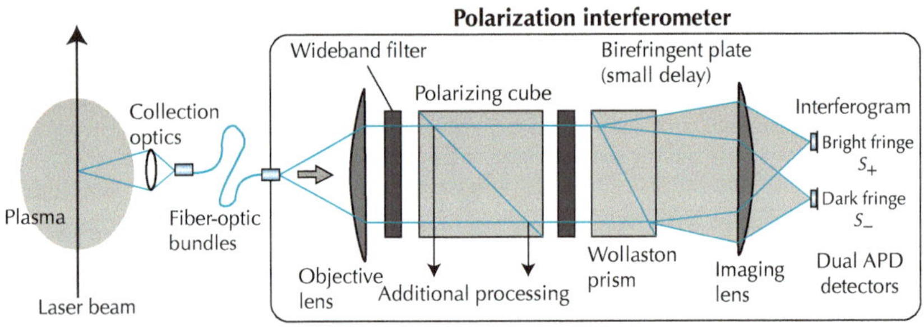

FIGURE 8.7 Schematic of a polarization interferometer for Thomson scattering diagnostic. (Courtesy of Hatae *et al.* (2007) and Japan Society of Plasma Science and Nuclear Fusion Research.)

to the Gaussian probe beam propagating in the z-direction. The plasma wave acts as a moving sinusoidal phase grating $\phi(x)=\phi_0\sin(kx-\omega t)$ resulting in the diffraction of the probe beam. The amplitude and phase of the first-order diffracted beam are given by (James & Yu, 1985)

$$I_1=\frac{P_0\Delta\phi}{\pi w_2^2}e^{-(u_s^2+u_v^2)}\left[e^{-v^2/2}\left(e^{2u_s v}+e^{-2u_s v}-2\cos\rho v^2\right)^{1/2}\right] \quad (8.2.1)$$

$$I_1\sin\phi_1=\frac{P_0\Delta\phi}{\pi w_2^2}e^{-(u_s^2+u_v^2)}\left[e^{-v^2/2}\left(e^{u_s v}\cos\left[\rho v\left(u_x-\frac{v}{2}\right)\right]\right.\right.$$
$$\left.\left.+e^{-u_s v}\cos\rho v\right)\left(u_x+\frac{v}{2}\right)\right], \quad (8.2.2)$$

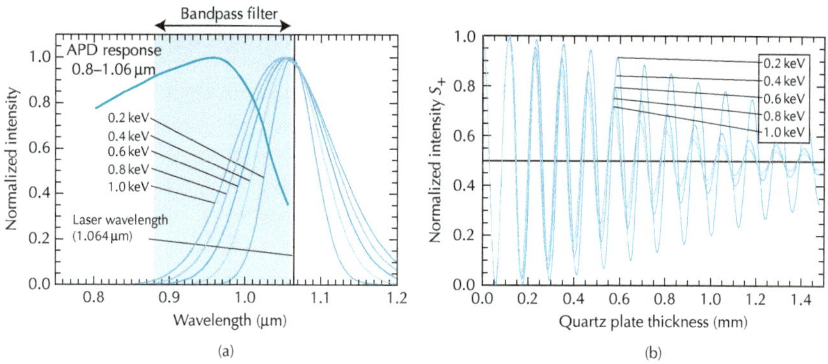

FIGURE 8.8 (a) Thomson scattered spectra for temperatures in the range of 0.2–1.0 keV. (b) The calculated interferograms S_+. (Courtesy of T. Hatae and the JSPF.)

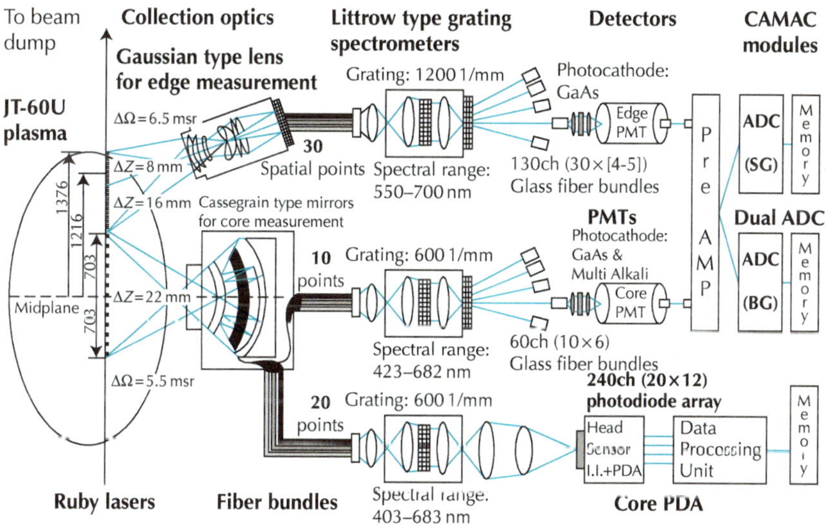

FIGURE 8.9 Schematic of collection and fiber optics, spectrometers, detector systems and data acquisition systems on JT-60U. (Courtesy of Yoshida *et al.* (1999b) and The American Institute of Physics.)

with P_0 the incident power, $u_x = x_2/w_2$, $u_y = y_2/w_2$, $v = w_1 k/(1+\rho^2)$, and $\rho = z_1/z_{r1} + k_0 w_1^2 (L_1 - L)/L_1^2$. Here, w_1 and w_2 are the $1/e$ radii of the probe beam intensity profiles at the wave location and detector location, respectively, and are related to the beam waist radius w_0 by $w_1 = w_0\sqrt{1+(z_1/z_n)^2}$, $w_2 = (L_1 L_2/L)\sqrt{1+\rho^2}/w_1$, where $z_n = k_0 w_0^2$ is the Rayleigh length, $L = (1/L_1 + 1/L_2 - 1/f)^{-1}$, and L_1 and L_2 are the distances from the lens (focal length f) to the waves and detector, respectively. This intensity distribution can be

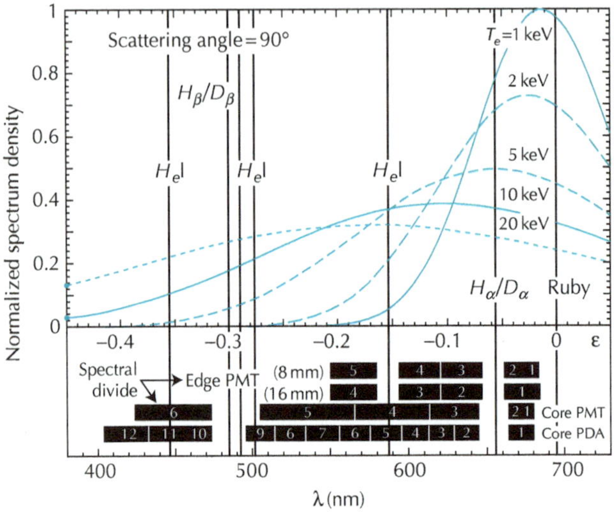

FIGURE 8.10 Thomson scattering spectrum density and spectral divide of edge and core photomultipliers (PMT) and core photodiode detector (PDA) systems. A figure in each band means the divided spectral channel number and means the normalized wavelength shift defined as $(\lambda_s - \lambda_i)/\lambda_i$. (Courtesy of Yoshida *et al.* (1999b) and The American Institute of Physics.)

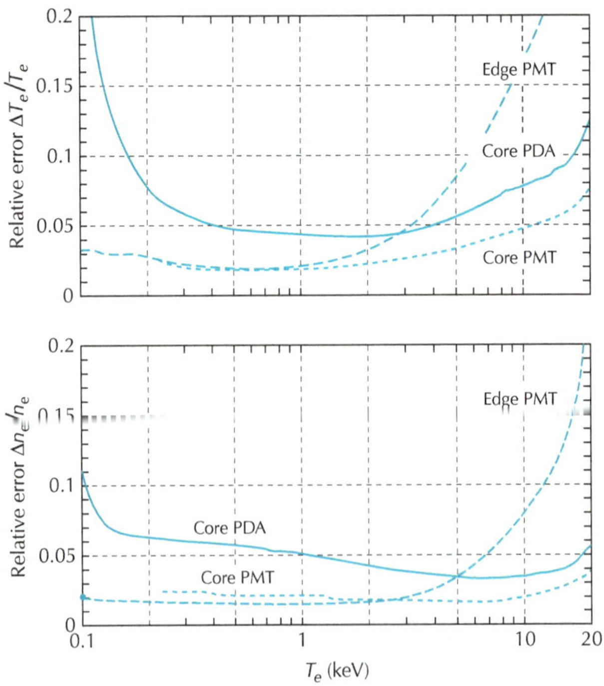

FIGURE 8.11 Overall relative error of T_e and n_e calculated with actual calibration data and detector characteristics. (Courtesy of Yoshida *et al.* (1999b) and The American Institute of Physics.)

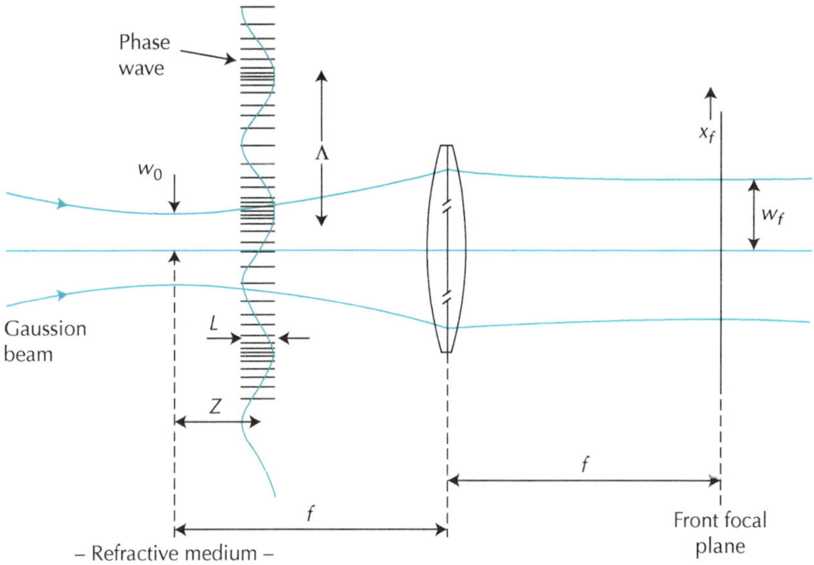

FIGURE 8.12 Schematic of arrangement for far forward scattering (Evans *et al.*, 1982).

interpreted as the result of the two scattering terms beating with the unperturbed term which plays the role of a local oscillator where the component at the wave frequency corresponds to the IF signal from the heterodyne detector. The wave number of the fluctuations can be deduced from the envelope profile of the intensity envelope I_1. Here, it is noted that the scattering occurs in the Raman-Nath regime, i.e., the inequality $Lk^2/k_i = 1$ is satisfied (Klein & Cook, 1967; Chen & Chatterjee, 1996).

The far-forward scattering technique was used to measure long wavelength turbulence on the TOSCA (Evans *et al.*, 1983) and TEXTOR (Vanandel *et al.*, 1987) tokamaks, compressional Alfven waves in a laboratory device (Vonhellermann & Holzhauer, 1984), and wave damping in laboratory experiments (Yu *et al.*, 1988). This method has several attractive features: it is insensitive to vibrations and has experimental simplicity. However, it is quite sensitive to any minute departure from the assumed Gaussian beam profile. The technique is particularly useful for measuring long wavelength coherent waves that are difficult to resolve with conventional scattering systems and is also suitable for long wavelength microturbulence measurements.

Phase scintillation imaging is quite similar to techniques employed in radio astronomy to measure interplanetary and interstellar fluctuations (Thompson *et al.*, 1986; Sharp, 1983; Nazikian & Sharp, 1987). Although related to far forward scattering, phase scintillation imaging differs in that the measurements of phase fluctuations are made in the near field where the measured phase fluctuation spectrum bears a direct relation to that of the plasma density fluctuation. These two techniques correspond, respectively, to the near-field and far-field

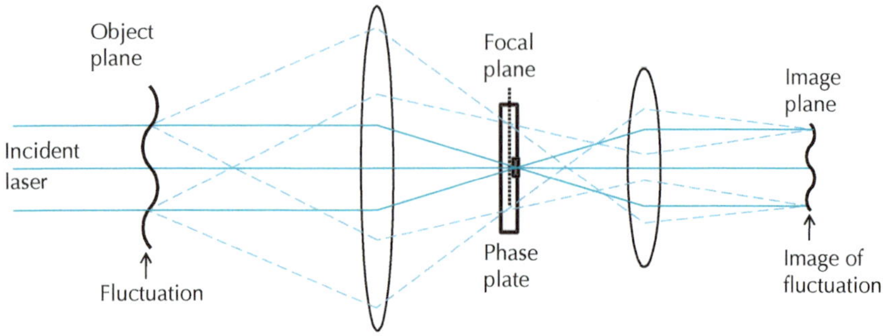

FIGURE 8.13 Optical arrangement for phase contrast imaging. Solid lines indicate nonscattered radiation. Dashed lines indicate scattered radiation.

detection of radiation diffracted by density fluctuations and both can be treated by a generalized theory of diffraction (James & Yu, 1985, 1987; Howard & Sharp, 1992). In most phase scintillation systems, a two-path interferometer is employed to measure the phase modulation of the probe beam. Mechanical vibrations can cause problems in this approach.

By incorporating a phase contrast plate, one may realize an internal phase reference. This near-field detection technique is referred to as phase contrast imaging (Weisen, 1986, 1988). Figure 8.13 illustrates the experimental arrangement (Weisen, 1988). Here, it should be noted that this technique is relatively immune to mechanical vibrations. Another point to note is that the use of a phase plate means the maximum resolvable wavelength is less than the half waist radius of the probe beam.

This technique has been employed in numerous areas ranging from microturbulence (Weisen, 1988; Tanaka *et al.*, 1993, 2003; Coda *et al.*, 2000; Vyacheslavov *et al.*, 2005) to externally driven rf waves (Weisen *et al.*, 1989a,b; Lin *et al.*, 2005). A detailed discussion of this technique may be found in Donné *et al.* (2008); here, we restrict the discussion to some representative systems.

We begin with a description of the CO_2 laser-based systems developed by Massachusetts Institute of Technology (MIT) researchers and implemented on the DIII-D tokamak at General Atomics, San Diego, CA (Coda *et al.*, 1992; Porkolab *et al.*, 2006), and the Alcator C-Mod tokamak at MIT (Lin *et al.*, 2006; Porkolab *et al.*, 2006). Figure 8.14 contains a simplified schematic of the optical layout, which was employed on both tokamaks. Both systems utilized a phase plate with a 16-element photovoltaic HgMnTe detector array used on DIII-D, while a 32-element HgCdTe photoconductive detector array was used on C-Mod. The required $\pi/2$ phase-shifted LO beam is produced when the expanded CO_2 laser beam (10–20 cm wide) is reflected from the $\lambda/8$ step groove thereby experiencing a total path difference of $\lambda/4$ (90° phase shift) as shown in Fig. 8.14.

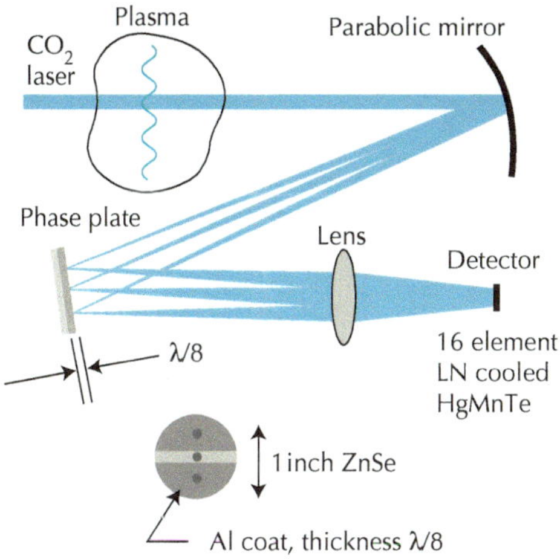

FIGURE 8.14 Conceptual layout of the phase contrast imaging (PCI) optics, including the phase plate and detector array. Actual optical design is much more complex.

The scattered components are subsequently mixed in the detector array with the phase-shifted central component thereby producing an amplitude variation. Porkolab *et al.* (2006) employ the following simplified set of equations for illustrative purposes:

Wave fields before the phase plate:

$$E_0(1+i\Delta\phi). \tag{8.2.3}$$

Wave fields after the phase plate:

$$E_0(i+i\Delta\phi). \tag{8.2.4}$$

Field intensity on the detector array:

$$I \propto |E|^2 - E_0^2(1+2\Delta\phi), \tag{8.2.5}$$

where $\Delta\phi(R) \simeq \int(\epsilon^{1/2}-1)dz \propto \int n(R,z)dz$.

Consequently, the intensities incident upon the detector array have an intensity distribution proportional to the chord integrated line density with position on the detector array directly related to spatial position perpendicular to the propagation direction of the CO_2 laser beam. Therefore, this diagnostic provides the component of wavelength and correlation length perpendicular to the laser beam propagation. There exists a practical long wavelength detection limit since for sufficiently long-wavelength plasma waves, the scattered light cannot be distinguished from the unscattered light. For both systems, this is $\sim$0.7 cm. The beam paths for the two systems are shown in Fig. 8.15 together with calculated traces of signals from loudspeaker launched sound waves. Table 8.1 summarizes the specifications of both phase contrast imaging (PCI) systems. By

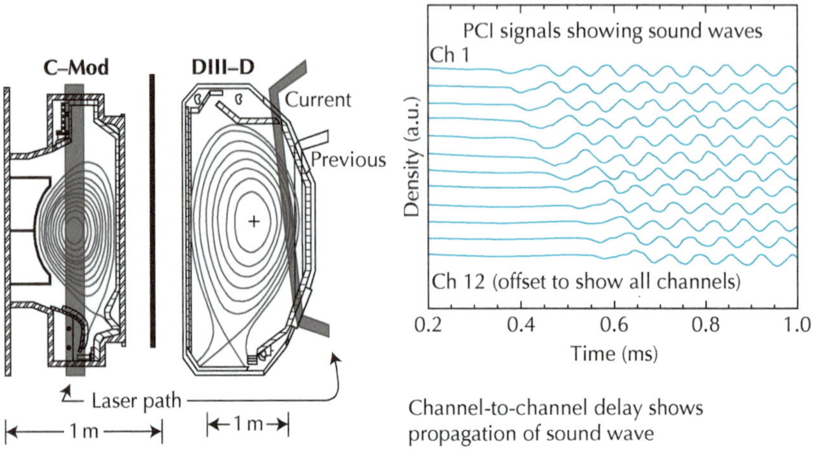

FIGURE 8.15 Laser beam paths in the Alcator C-Mod tokamak (left picture) and in the DIII-tokamak (middle picture). Right side shows sound wave propagation ($k \sim 2\,\text{cm}^{-1}$) as detected by the PCI diagnostic.

TABLE 8.1 DIII-D and C-Mod PCI System Specifications

	DIII-D	C-Mod
Detector Array	16 element HgMnTe	32 element HgCdTe
Detector Element Separation	650 μm	850 μm
Laser Diameter in Plasma	5 cm	6–12 cm
Phase Plate Groove Width	450 μm	400 μm
Optical Magnification	0.125–0.62	0.21–0.81
Nyquist k-Limit	6.5–30 cm^{-1}	8–32 cm^{-1}
Low k-Cutoff	0.7 cm^{-1}	0.7 cm^{-1}
Laser Power	20 W	60 W

partially masking the phase plate, Lin *et al.* (2006) showed on the C-Mod system that they can realize a degree of vertical localization by means of the magnetic pitch angle variation. Figure 8.16 shows scattering data from the so-called quasi-coherent (QC) mode showing the capability of differentiating between the top and bottom of the plasma cross section.

The final example of a PCI system is the two-dimensional (2D) instrument (see Fig. 8.17) employed on the large helical device (LHD) at the National Institute for Fusion Science, Japan (Tanaka *et al.*, 2008; Michael *et al.*, 2006; Vyacheslavov *et al.*, 2005; Sanin *et al.*, 2004). By employing a two-dimensional detector array and making use of magnetic shear, they can simultaneously measure fluctuations from different positions within LHD. The detector array comprised an 8×6 element matrix of HgCdTe LN-cooled photoconductors with an element size of $0.5 \times 0.5\,\text{mm}^2$ and spacing of 1.4 and 0.62 mm. The

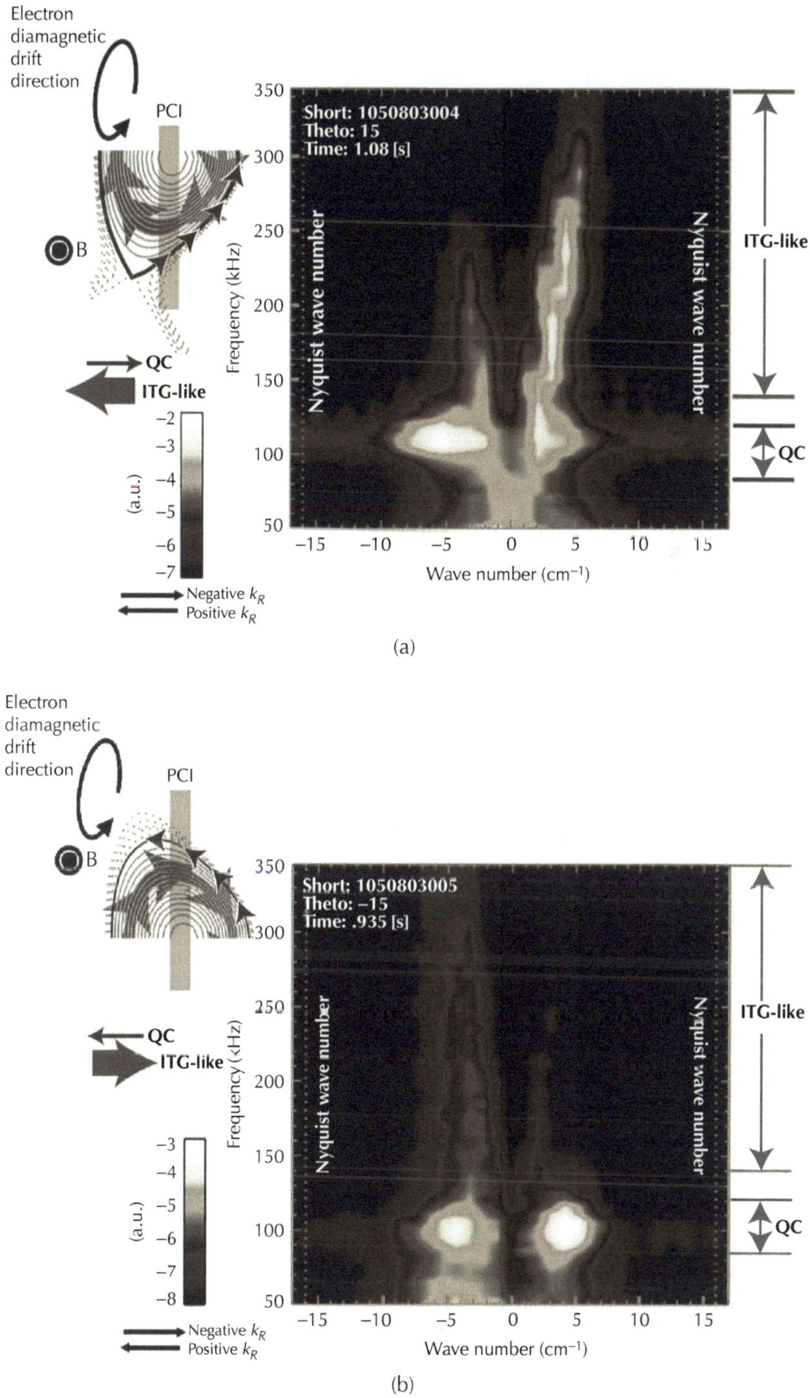

FIGURE 8.16 Frequency/wave number spectra of plasma fluctuations measured by the C-Mod PCI diagnostic. On the left corner of each plot, the diagnostic line of sight is shown according to the masking phase plate setup: (a) the bottom plasma view and (b) the top plasma view.

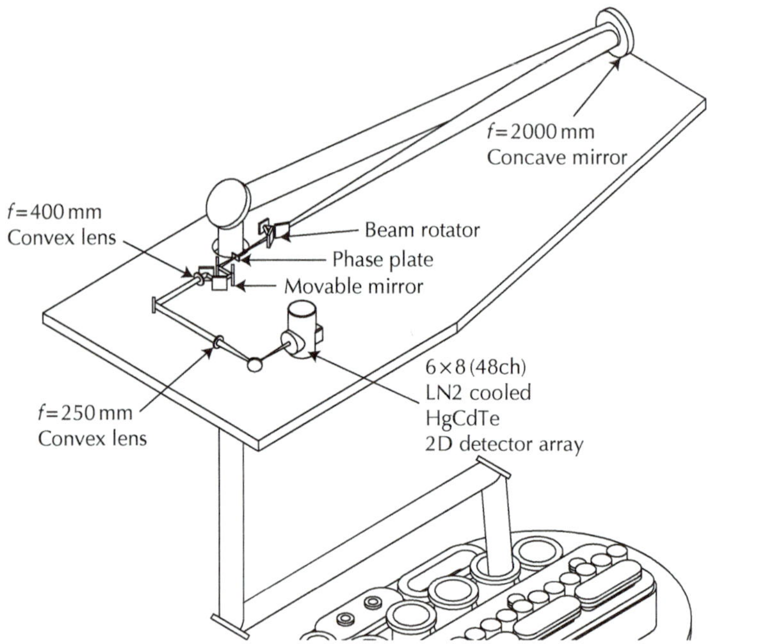

FIGURE 8.17 Schematic of the detection system for the LHD 2D PICI system.

optical system provides a magnification of 5.68 resulting in x- and y-direction (radial and toroidal) sample sizes in the plasma of 3.52 and 7.95 mm, respectively, which correspond to Nyquist wave numbers of 0.89 and 0.40 mm^{-1}. The probe beam utilizes approximately 20% of the 8 W output of a CO_2 laser with the remainder employed for an interferometer and has a full width in the plasma center of 62 mm (toroidal direction) and 72 mm (major radial direction). In their 2006 upgrade, they included the capability to change the system magnification, by sampling only a portion of the probing CO_2 beam, dependent on whether they were interested in high k or low k fluctuations. Figure 8.18 displays an example of the real and imaginary parts of the integrated 2D autocorrelation function $\Gamma(\delta x, \delta y, \delta\omega)$ taken at 78 kHz ± 15 kHz. Employing a 2D maximum entropy method, they convert $\Gamma(\delta x, \delta y, \delta\omega)$ to obtain the line integrated wave number spectrum $\partial\langle N^2\rangle/\partial k_x \partial k_y \partial\omega$.

Figure 8.19 contains an example of the local wave number spectrum $\partial\langle n^2\rangle/\partial k$ obtained from the 2D MEM shown in Fig. 8.20. The authors also obtain the local fluctuation phase velocity in the laboratory frame by converting $\partial\langle n^2\rangle/\partial k\partial\omega$ to $\partial\langle n^2\rangle/\partial k\partial \mathrm{v}$ using $\mathrm{v}=\omega/k$ and $d\mathrm{v}/d\omega=1/k$. Figure 8.20 displays the phase velocity profile $\partial\langle n^2\rangle/\partial \mathrm{v}$ where the spectrum is integrated over k. The white lines are the phase velocity of the fluctuations at maximum fluctuation power at each location. From these data, it is seen that the fluctuations

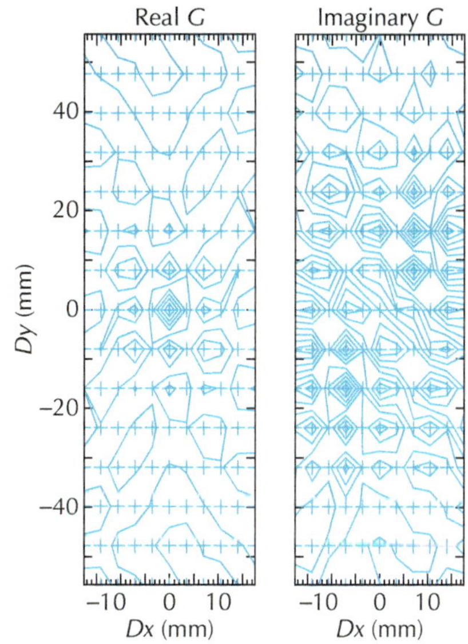

FIGURE 8.18 Integrated 2D autocorrelation function $G(Dx, Dy, w)$. Real and imaginary parts are shown. 78 kHz ± 15 kHz components are shown.

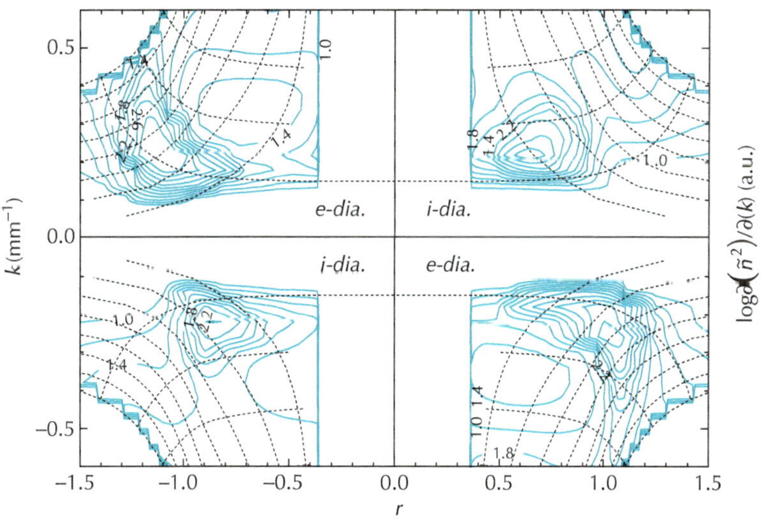

FIGURE 8.19 Local wave number spectrum in flux surface coordinates. The spectrum is frequency integrated for 20–500 kHz.

propagate in the ion diamagnetic direction in the core region ($|\rho| \leq 0.7$) and in the electron diamagnetic direction in the edge region ($|\rho| \geq 0.7$) resulting in strong velocity shear.

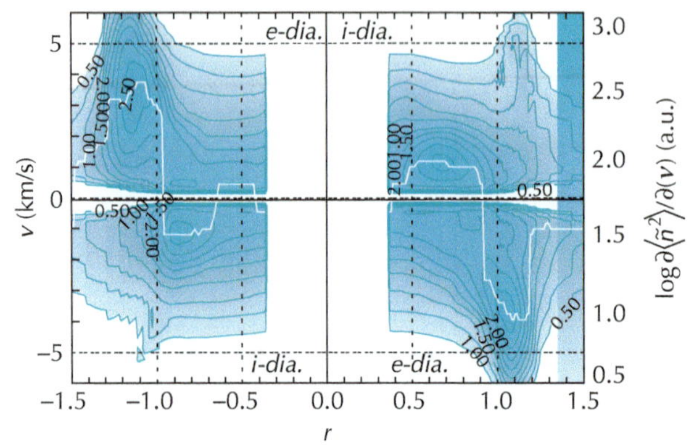

FIGURE 8.20 Phase velocity profile $\partial\langle n^2\rangle/\partial \mathrm{v}$. The spectrum is integrated over k. The white lines are phase velocity of fluctuation at maximum fluctuation power at each location.

8.3 THOMSON SCATTERING FROM DRIVEN (NONTHERMAL) WAVES

The unique ability for Thomson scattering to probe a defined wave vector allows the investigation of driven or nonthermal instabilities (see e.g., Walsh & Baldis, 1982; Giles & Offenberger, 1983; Clayton *et al.*, 1985; Baker *et al.*, 1996; Baldis & Labaune, 1996; Depierreux *et al.*, 2000; Glenzer *et al.*, 2001b; Rousseaux *et al.*, 2006). For example, stimulated Brillouin scattering (SBS), is a three-wave instability that results from the resonant coupling of an intense laser pulse (k_{ib}), a scattered light wave, and an ion-acoustic plasma wave (k_{driven}). The process is resonant for a given ion-acoustic wave copropagating to the incident interaction beam (A further discussion of SBS and other laser plasma instabilities is given in Section 12.7). By selecting a Thomson scattering geometry such that the probed wave vector is matched to the wave vector of the driven wave, the driven ion-acoustic wave can be probed. For SBS-driven ion-acoustic waves, the driven wave vector is well defined [see Eqn (5.4.3)], $k_{\mathrm{driven}} \simeq 2k_{ib}$, and the Thomson scattering geometry is given by

$$\theta \simeq 2\sin^{-1}\left(\frac{k_{ib}}{k_{\mathrm{TS}}}\right). \tag{8.3.1}$$

where k_{TS} is the wave vector of the Thomson probe beam. Using Thomson scattering to simultaneously probe the driven ion-acoustic waves and the counter-propagating thermal fluctuations, as is the case for stimulated Brillouin scattering, the absolute amplitude of the driven ion-acoustic wave can be measured (Froula, 2002b; Froula *et al.*, 2003a). The absolute amplitude of the driven ion-acoustic ($\delta n/n_{\mathrm{e}}$) is measured by comparing the scattered power from thermal fluctuations, with the power scattered by the driven ion-acoustic wave. The power scattered into the thermal ion-acoustic waves is given by substituting

Eqn (5.5.7) into Eqn (5.1.1) and integrating over the ion-acoustic features,

$$P_{\text{thermal}} = \frac{1}{2} r_0^2 n_e L P_i d\Omega, \tag{8.3.2}$$

where L is the length of the scattering volume along the direction of the probe beam.

The power scattered from a driven coherent fluctuation ("wave") leads to a considerable enhancement of the level of scattered power in the direction that satisfies conservation of momentum, and the power is given by

$$P_{\text{driven}} = \frac{1}{4} r_0^2 n_e^2 \lambda_i^2 L L_c P_i \left(\frac{\delta n}{n_e}\right)^2, \tag{8.3.3}$$

where $\delta n/n_e$ is the amplitude of the driven wave and L_c is the correlation length along the direction of the probe beam. The ratio of the power scattered into the driven wave to the power scattered into the thermal wave is then simply

$$\frac{P_{\text{driven}}}{P_{\text{thermal}}} = \sqrt{\pi} n_e \lambda_i^2 L_c \sqrt{d\Omega_{\max}} \left(\frac{\delta n}{n_e}\right)^2, \tag{8.3.4}$$

where $d\Omega_{\max}$ is the minimum solid angle between the light scattered from the driven wave and the collection optics. For scattering from SBS-driven waves, the solid angle for the driven wave is simply given by the f-number of the interaction beam, $f_{ib}^2 = \pi/(4d\Omega_{ib})$, and the correlation length can be estimated by the transverse size of the speckle generated by the interaction beam $L_c = f_{ib}\lambda_{ib}$.

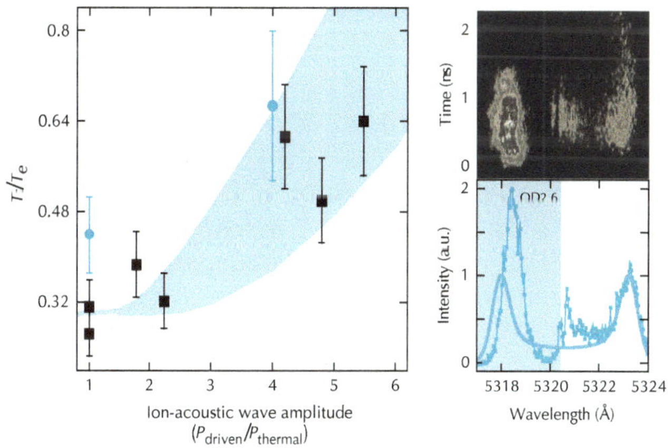

(a) (b)

FIGURE 8.21 (a) The ratio of the electron and ion temperature as a function of the normalized ion-wave amplitude [Eqn (8.3.4)] shows evidence of the generation of hot ions by ion trapping in Be/Au plasmas. (b) A frequency shift is measured on the ion-acoustic wave driven by SBS; the solid line is the measured spectrum from a second Thomson scattering diagnostic that is not affected by the SBS process. Half of the spectrum is filtered at the slit of the streak camera using an OD2.6. (Courtesy of Froula *et al.* (2002, 2004).)

Froula *et al.* (2002, 2003b) have used a multiple Thomson scattering configuration, combined with a multiple-ion species plasma to measure the ion temperature (see Section 5.4.3), to provide direct quantitative evidence of hot ions created by trapping in low-Z laser produced plasmas (Fig. 8.21a). Furthermore, by measuring the absolute frequency of the driven ion-acoustic waves, this team used a multiple Thomson scattering configuration to measure the frequency shift that results from the trapped ions (Fig. 8.21b).

8.4 DIRECT FORWARD THOMSON SCATTERING

In conventional large-angle Thomson scattering setups, the scattering angle is limited by the collection and probe beam apertures. For example, with $f/20$ focusing and a reasonable f-number collection lens, a significant fraction of the probe beam will enter the collection optics at a scattering angle of 25 mrad. This limitation is highlighted in experiments aimed at studying small k plasma waves. For example, to probe relativistic CO_2-laser-driven electron plasma waves with optical light, near exact forward scattering is required ($k_{epw} \sim 1/100\,\mu m^{-1} \ll k_{probe}$). Therefore, the scattered light propagates along the same path as the probe light, making separation between the two very difficult. Furthermore, at low plasma densities of $\sim 10^{16}\,cm^{-3}$, the scattering efficiencies are extremely small (typically $10^{-11} - 10^{-9}$), and the wavelength shift is typically less than 1 nm.

Filip *et al.* (2003) present a novel Thomson scattering system capable of detecting direct forward scattered Thomson scattered light over a wide range of densities by separating the weak scattered light from the intense probe light by a spatial-spectral filter. This system attenuates the unshifted probe light by a factor of 10^9 allowing the Thomson scattered light shifted by 3–24 Å to be spectrally resolved. Figure 8.22a shows the experimental setup where a beat wave produced by a two-frequency CO_2 laser pulse drives a relativistic plasma wave in a He gas fill (Filip *et al.*, 2004). The plasma wave is probed using the second harmonic of a Nd:YAG laser with 2-ns long, linearly polarized, 50 mJ, $\lambda_i = 532\,nm$ pulses. The probe beam is injected into the axis of the CO_2 beam through a hole in the off-axis parabollic (OAP) mirror and is focused at the interaction point (IP) to a spot size of 90 μm. The IR and visible beams copropagate within an angle of less than 0.4°. Approximately 10^{-9} photons acquire a shift in frequency of $\pm\omega_{pe}$, corresponding to a shift in wavelength of ±8 Å. The residual probe photons, which now represent unwanted stray light, propagate out of the interaction chamber and into the Thomson scattering diagnostic consisting of two primary parts. The first is a spatial-spectral filter that is used to reject the stray light at λ_i, and the second is a spectrometer-streak camera combination to frequency and time resolve the scattering light (Filip *et al.*, 2003).

Initially, the unshifted probe light is reduced by a factor of 1000 by triple passing a Fabry–Perot etalon (see Section 7.5). The etalon reflects both the

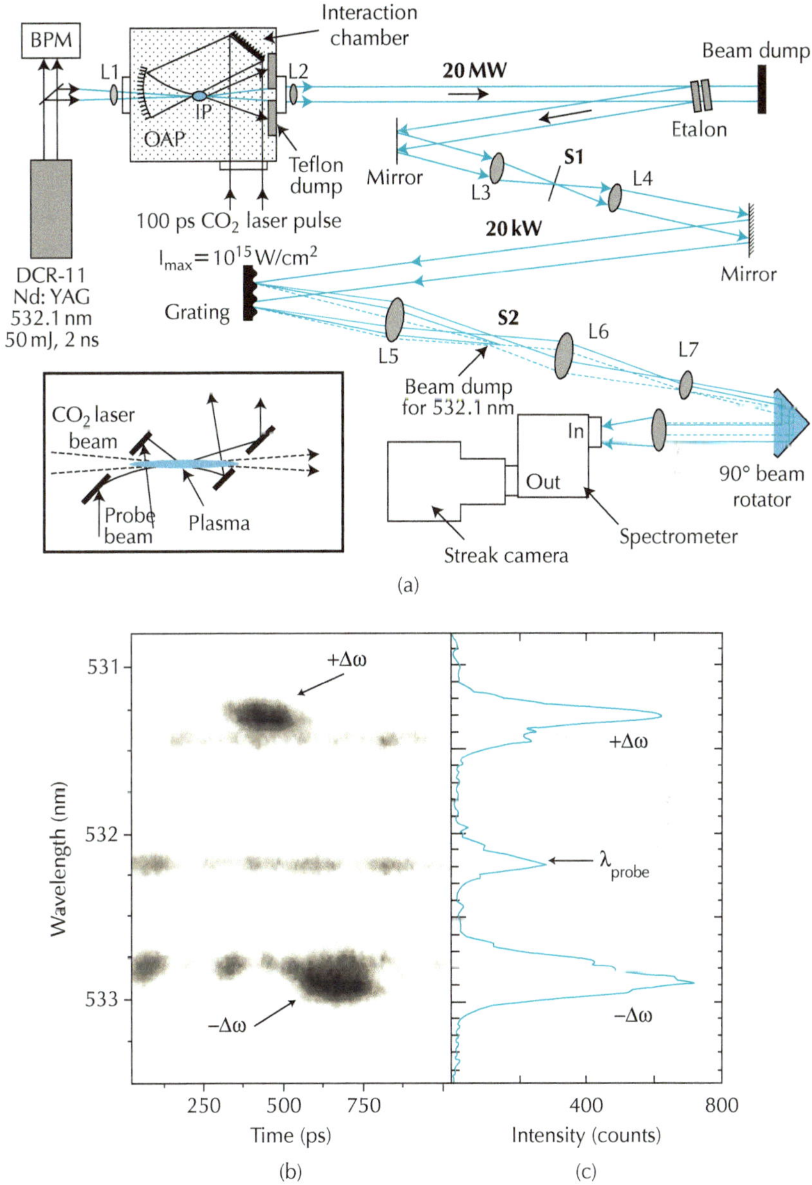

FIGURE 8.22 (a) The schematic of a direct forward Thomson scattering diagnostic. L1-L7-Lenses; S1, S2-spatial filters; OAP-off-axis parabolic mirror; IP-interaction point; BPM-beam position monitor. (b) Time-resolved spectrum of Thomson scattered light recorded in He at $10^{16}\,cm^{-3}$. (b) Lineout from (a) at the peak of the Thomson scattered sidebands. (Courtesy of C. Joshi in Filip *et al.* (2003) and Review of Scientific Instruments.)

red- and blue-shifted scattered light and transmits the probe light. The etalon was designed with a transmission width (FWHM) of 2.6 Å and a free spectral range of ∼27 Å. Due to the parasitic interference produced by the antireflection-coated external surfaces of the etalon, 10% of the incoming probe light is still reflected as residual light.

Following the etalon, a spatial filter is used to block the halo generated by the high-power probe beam on all 23 optical surfaces in the beam path. Note that the 1000 times reduction of the power in the stray light by the etalon is crucial for operation of the spatial filter as the power must be sufficiently low to avoid plasma break down in the spatial filter that would flood the entire spectrum with stray light. To further reduce the stray probe light, a holographic grating along with a knife edge is used reducing the overall probe light from 20 MW to less than 2 mW. Finally, the light is imaged to the slit of a spectrometer and recorded by a streak camera.

Figure 8.22b shows a Thomson scattered spectra obtained with a scattering angle of less than 0.4 mrad. The short, intense sidebands at $\lambda_i \pm 8.1$ Å are the Thomson scattered signal. It is noted that these two sidebands are actually coincident in time but appear shifted because of the different path lengths for the red and blue components through the spatial-spectral filter. The lineout in Fig. 8.22b shows that, even at these levels of power, the signal-to-noise ratio is on the order of 100. When the plasma density was reduced to $2 \times 10^{15}\,\mathrm{cm}^{-3}$, the signal level approached the noise level.

8.5 (ω, k)-RESOLVED THOMSON SCATTERING

Thomson scattering has been used to resolve the frequency-wave number spectrum for a variety of laser-plasma instabilities: the harmonics of driven ion-acoustic waves (Walsh & Baldis, 1982), mode coupling between electron plasma waves and ion-acoustic waves (Clayton *et al.*, 1985; Everett *et al.*, 1995), the Langmuir decay instability (Depierreux *et al.*, 2002; Kline *et al.*, 2005), two-ion decay (Niemann *et al.*, 2004). This technique recognizes that each radial position on the collection lens corresponds to a different scattered wave vector and that each of these positions corresponds to a plasma wave. The magnitude of the wave is given in general by the Law-of-Cosines, and for ion acoustic waves ($k_s \simeq k_i$), this can be simplified to illustrate the dependence of the wave vector on the scattering angle (see Section 5.4),

$$k_{ia} \simeq 2k_i \sin\left(\frac{\theta}{2}\right). \qquad (8.5.1)$$

By imaging this range of angles (i.e., the collection lens) onto the slit of a spectrometer, the frequency for a range of wave vectors can be resolved.

A team at Los Alamos National Laboratory lead by J. Kline and D. S. Montgomery used this technique at the Trident Laser Facility to observe both fluid (wave–wave) and kinetic (wave–particle) nonlinearties associated with

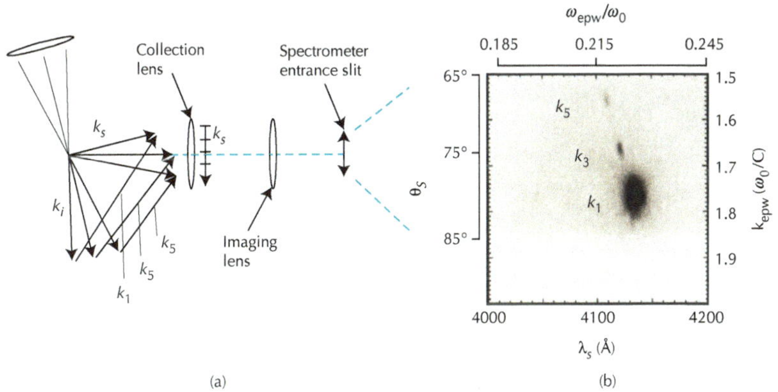

FIGURE 8.23 A schematic of a Thomson scattering system designed to resolve the ω, k spectrum. The scattered light wavelength and the scattering angle are directly related to (ω, k). (b) Angle-resolved spectrum $[(\omega, k)$ spectrum]. The stimulated Raman scattered daughter electron plasma wave (k_1) and two copropagating electron plasma waves driven by the Langmuir decay instability (k_3, k_5) are indicated. (Courtesy of Montgomery *et al.* (2004).)

stimulated Raman scattering (Kline *et al.*, 2005). Figure 8.23 shows a schematic of their Thomson scattering setup, where a 351 nm, 0.5 J Thomson probe beam with a 200 ps pulse duration was used. The Thomson probe beam was focused using an $f/4$ lens, and the Thomson-scattered light was collected by an $f/2.4$ lens. The collection lens was then imaged onto the slit of a 1/4-m imaging spectrometer that was coupled to a charge-coupled device (CCD). The system provided a 200 ps frame of the angle-resolved spectra, and the data were directly correlated to the (ω, k)-spectrum of driven electron plasma waves (Fig. 8.23b).

8.6 SUBPICOSECOND TIME RESOLVED THOMSON SCATTERING OF STIMULATED RAMAN SCATTERING

In stimulated Brillouin (SBS) and Raman (SRS) scattering, laser pump radiation scatters off ion-acoustic or electron plasma waves. The nonlinear damping of these waves leads to non-Maxwellian velocity distributions and alters the wave damping coefficients (Afeyan *et al.*, 1998). Kline *et al.* (2005) found a transition from fluid to kinetic effects at $k\lambda_{De} = 0.29$, where k is the wave number for the most unstable SRS-driven electron plasma wave. A difficulty in making measurements on this phenomenon is the very short growth time of the plasma instabilities – a few picoseconds. Rousseaux *et al.* (2006) have reported Thomson scattering measurements with sub-picosecond resolution of ion-acoustic and electron plasma waves driven by SBS and SRS.

A 1.059 μm laser, FWHM of 550 ps and delivering 10^{14} W-cm^{-2}, was used to ionize helium gas. The main interaction beam, 1.5 TW, 1.059 μm, and 1.5 ps FWHM, interacts with the plasma at a peak electron density 4×10^{19} cm^{-3}

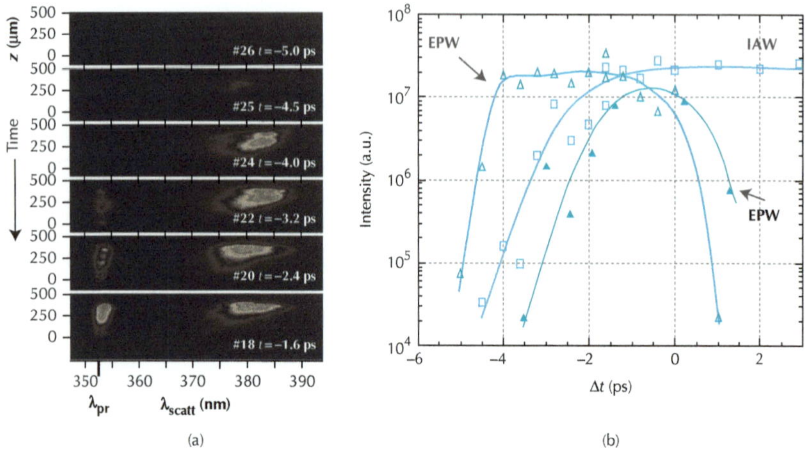

FIGURE 8.24 The figures show the temporal evolution of ion-acoustic (IAW) and electron plasma (EPW) waves driven by the 1.5 ps laser pulse ($I = 9 \times 10^{17}$ W-cm^{-2}). The EPW curve on the far right was obtained at a lower intensity ($I = 7 \times 10^{15}$ W-cm^{-2}), for which the IAW signal remained below the detection level. (Courtesy of Rousseaux *et al.* (2006) and Physical Review Letters.)

(0.04 n_{cr}) and T_e (T_i) close to 300 eV (50 eV). A small fraction of the main pulse is time-compressed and frequency tripled to yield a 2 mJ Thomson probe beam with FWHM of 0.3 ps and a spectral width of, typically, 1.2 nm.

The evolution of the ion-acoustic and electron plasma waves is shown in Fig. 8.24. It shows that the electron plasma waves are quickly damped near the peak of the pulse, regardless of the mean laser intensity. This indicates that over the picosecond timescale of this experiment, nonlinear kinetic effects such as trapping and the associated deformation of the electron distribution are the main cause for SRS disruption.

In subsequent work, Rousseaux *et al.* (2009) used two-dimensional Thomson scattering to show that electron plasma waves driven by SRS are increasingly oblique as a function of time. Particle-in-cell (2D) simulations quantitatively reproduce the measured spectra and reveal that oblique wave numbers mainly result from wave front bowing, while large amplitude electron plasma waves are destroyed by secondary instabilities. A further discussion of these instabilities is given in Section 12.7.

8.7 MULTIPLE ION-ACOUSTIC THOMSON SCATTERING DIAGNOSTICS TO MEASURE (T_e, n_e)

In collective scattering, resolving the ion-acoustic features to measure the electron temperature has been a standard Thomson scattering measurement for more than 30 years (see Section 5.4.3), while the weaker collective electron features (see Section 5.4.2), which provide a measure of the electron density, have been more elusive. Here, we discuss a technique that was first presented by

Froula *et al.* (2005), which exploits the dispersion of ion-acoustic fluctuations and allows the density and temperature to be measured using multiple Thomson scattering diagnostics probing two significantly different ion-acoustic wave vectors.

The two ion-acoustic frequencies can be measured by either using two probe wavelengths or using a single probe laser with two different scattering angles. A small angle diagnostic k_1 can be chosen to provide a good measure of the electron temperature with a small dependence on the density ($k_1\lambda_{De} < 1$), while a large angle diagnostic k_2 would provide a good measure of the electron density ($k_2\lambda_{De} > 1$). There is a limitation for large angles (for a given probe wavelength) given by the constraint of remaining in the collective Thomson scattering regime ($ZT_e/3T_i \gg k^2\lambda_{De}^2$, see Section 5.3.5), while there is a practical limit for small angles given by the instruments ability to resolve the spectral peaks [i.e., the wavelength separation scales with the angle, see Eqn (5.4.10)].

This technique was demonstrated (Froula *et al.*, 2005) using two 0.5 J probe lasers at different wavelengths, $\lambda_i^{2\omega} = 532\,\text{nm}$ and $\lambda_i^{4\omega} = 266\,\text{nm}$. The setup is shown in Fig. 8.25a. The 2ω and 4ω probe beams were focused using a $f/5$ and $f/10$ lens, respectively, to a minimum diameter of 75 μm. The scattering geometry is shown in Fig. 8.25b, where both systems have the same scattering angle ($\theta = 120°$) but probe significantly different wave numbers ($k_{4\omega}/k_{2\omega} = 2$).

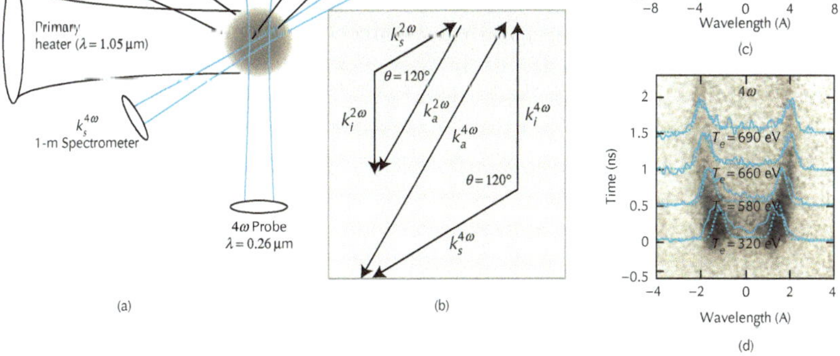

FIGURE 8.25 (a) An experimental setup that employees two Thomson scattered probe beams that scattering light from significantly different wave vectors. (b) Thomson scattering spectra from the same scattering volume show the dispersive nature of the plasma. As the plasma temperature increases with time, the 2ω spectral features increase faster than the 4ω features. The theoretical form factor [Eqn (5.1.2)] is fit for each time (dashed curves). (Courtesy of Froula *et al.* (2005) and Physical Review Letters.)

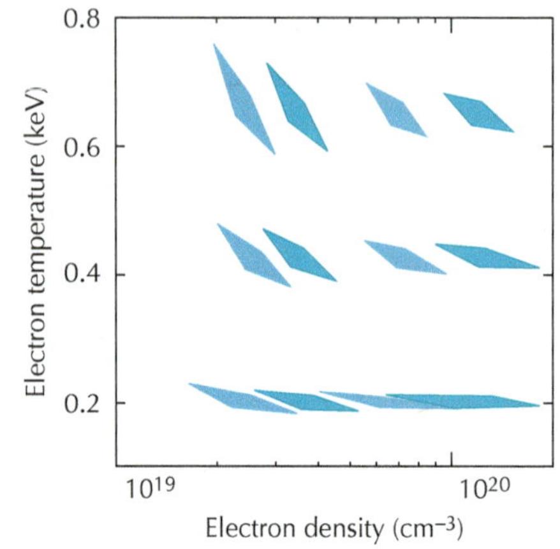

FIGURE 8.26 Each parameter space was calculated for an electron temperature and density using typical measurement errors for each Thomson scattering diagnostic. (Courtesy of Froula *et al.* (2005) and Physical Review Letters.)

Two $f/5$ collection lenses collimated light scattered from a single Thomson scattering volume in a nearly fully ionized nitrogen plasma. The scattered light was focused onto the slit of a 3/4-m (for 2ω) and a 1-m (for 4ω) imaging spectrometer using two $f/10$ focusing lenses. The spectrometer was coupled to streak cameras. The spectrometer and streak camera slits overlapping the probe beams in the plasma defined a $50\,\mu m \times 100\,\mu m \times {\sim}75\,\mu m$ volume. Figures 8.25c–d show Thomson scattered spectra from this setup where it is evident that the difference in scattered wavelength between the two systems is more than just a factor of two $[\Delta\lambda_{2\omega} > 2\Delta\lambda_{4\omega}]$ – the result of a dispersive plasma.

Figure 8.26 shows a series of calculations where the parameter space around the measured electron temperature and density was determined using the actual measurement uncertainties for each Thomson scattering diagnostic ($\delta\lambda_{4\omega} = 0.014\,Å$ and $\delta\lambda_{2\omega} = 0.05\,Å$). When the electron temperature is low, the wavelength separation is small. If $k\lambda_{De}$ is also small, the diagnostic is insensitive to density, and the error in the density results is dominated by the error in the measurement of the ion-acoustic peaks (bottom right area in Fig. 8.26).

For a typical inertial confinement fusion plasma ($T_e = 5\,keV$, $n_e = 5 \times 10^{20}\,cm^{-3}$), the optimal scattering angles for the two collection optics are $40° < \theta_{k2} < 80°$ ($0.4 < k_2\lambda_{De} < 0.7$) and $\theta_{k1} > 140°$ ($k_1\lambda_{De} > 0.9$); using these scattering angles a single 4ω probe laser, and typical instrument resolutions, the local density could be measured to better than 25% with an electron temperature measurement to within 10%.

Chapter 9

Industrial Plasmas and Scattering from Energetic Ions

Chapter Outline Head

9.1 INDUSTRIAL AND OTHER LOW-TEMPERATURE PLASMAS

Plasmas have a wide range of applications outside fusion energy research, for example, in fluorescent lighting, plasma displays, semiconductor processing, making coatings, cleaning surfaces, destroying noxious chemicals, and plasma cloaking. Thomson scattering is an attractive diagnostic option because it can make local measurements without perturbing the plasma.

Key characteristics of such plasmas are a low temperature in the ~0.1–30 eV range and electron densities in the range of 10^{10} to 10^{18} cm^{-3}, often with $n_n \gg n_e$. As discussed in the excellent reviews of Muraoka *et al.* (1998), Warner & Hieftje (2002), and van de Sande (2002), such parameters present a challenge for using Thomson scattering as a diagnostic, because of the low photon count rate at the lower densities, the narrow scattering line width, large Rayleigh scattering, and stray light and background plasma light (see Fig. 9.1; Bowden *et al.*, 1999). In addition, for etching, chemically-reactive gases are commonly used.

DOI: 10.1016/B978-0-12-374877-5.00009-9

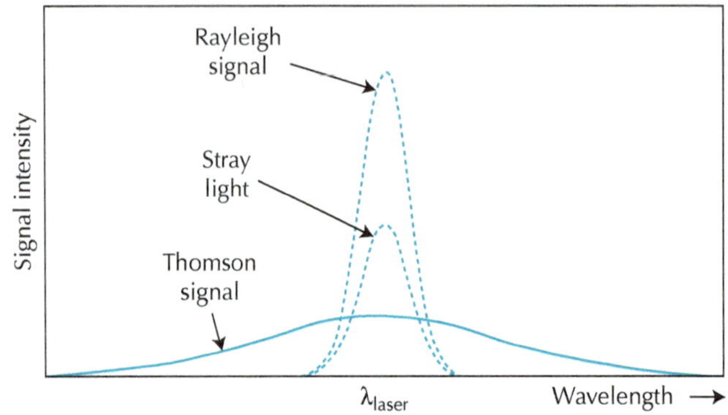

FIGURE 9.1 Different components of the scattered spectrum. (Courtesy of K. Uchino in Bowden *et al.* (1999) and the Institute of Physics Publishing.)

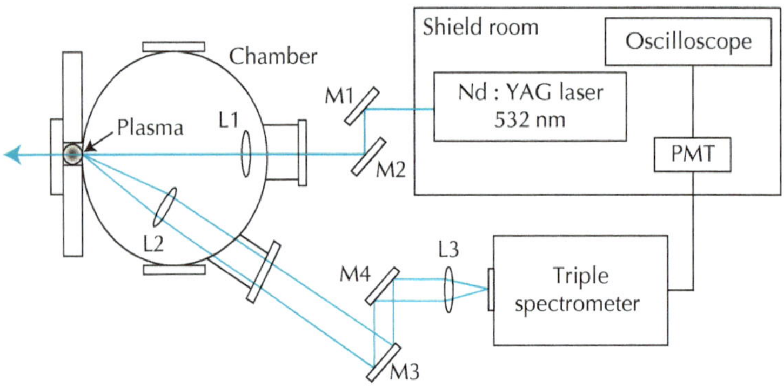

FIGURE 9.2 Experimental arrangement. (Courtesy of K. Tomita (2009) and the JSPF.)

The industrial plasma are usually operated in steady-state; consequently, the approach that has been used with pulsed sources is to

- integrate the signals over as many as 10^3 pulses,
- measure the Rayleigh component separately and subtract it, and
- have two or three monochromators in series to reduce the stray light.

An example setup for a Z-pinch plasma used in extreme ultraviolet lithography is shown in Figs. 9.2 and 9.3. A mask may be used to block out the central part of the spectrum (Fig. 9.4).

Measurements have been made on

- impulse breakdown plasma in atmospheric air (Uchino *et al.*, 1982),
- electron cyclotron resonance (ECR) sources (Bowden *et al.*, 1993, 1999),
- radio frequency inductively coupled sources (Hori *et al.*, 1996a,b, 1998); an example of electron temperature measurements is shown in Fig. 9.5,
- magnetic neutral loop discharges (Sakoda *et al.*, 1997),
- capacitively coupled radio frequency sources (Wesseling & Kronast, 1996),

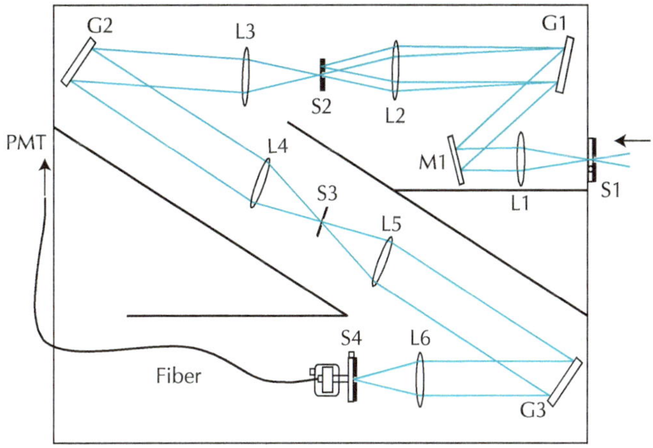

FIGURE 9.3 Schematic diagram of the triple grating spectrometer. S1–S4: slits, M1: flat mirror, G1–G3: gratings, L1–L6 lenses. (Courtesy of K. Tomita (2009) and the JSPF.)

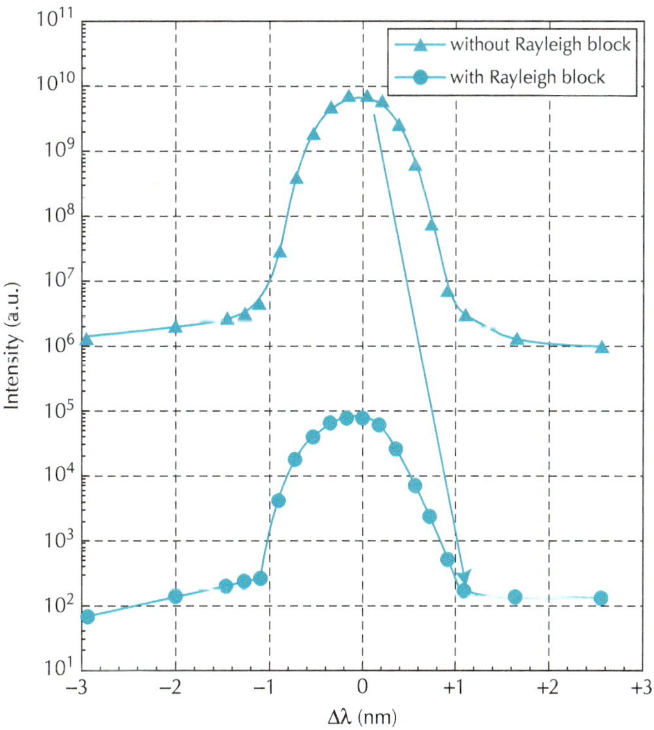

FIGURE 9.4 A mask may be used to block out the central part of the spectrum. (Courtesy of K. Uchino in Hassaballa *et al.* (2004) and the IEEE (2004).)

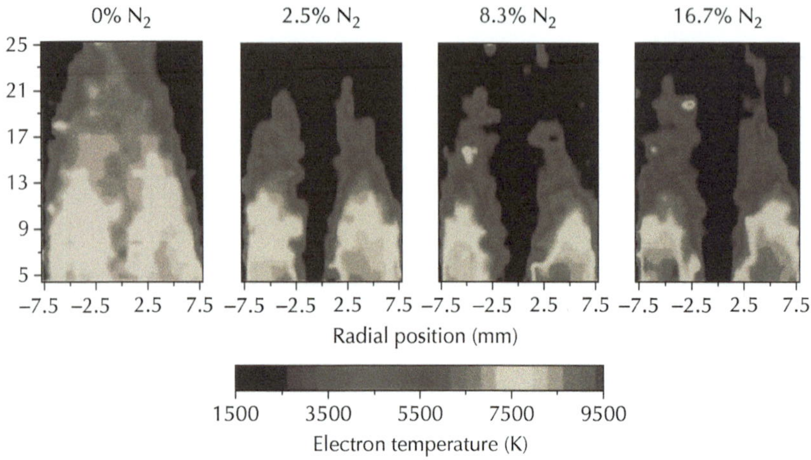

FIGURE 9.5 The electron temperature for various percentages of nitrogen added to the central-channel flow of an argon ion cyclotron plasma is shown. (Courtesy of Warner & Hieftje (2002) and Elsevier.)

- reactive plasmas in a capacitively coupled RF-discharge and a GEC reference cell (Kazcor & Soltwisch, 1999, 2001).
- microdischarge plasmas, as used in a plasma display panel (Noguchi *et al.*, 2001, 2002; Hassaballa *et al.*, 2004, 2005).
- neon–mercury positive column (Bakker & Kroesen, 2001),
- atmospheric argon plasma (Zaidi, 2001),
- tin vapor discharge (Kieft *et al.*, 2004, 2005),
- microwave plasma torch (van der Mullen, 2004),
- magnetic multipole (Maurmann *et al.*, 2004),
- EUV light sources (Tomita *et al.*, 2009).

Hori *et al.* (1996a) demonstrated the ability to measure non-Maxwelliian electron energy distributions in an inductively coupled plasma. Similar results were found in measurements on a glow discharge (Gamez *et al.*, 2003). The deviation from a Maxwellian shape as the gas pressure is lowered is shown in Fig. 9.6. The system used for studying an ECR plasma by Bowden *et al.* (1999) is shown in Fig. 9.7.

The ECR plasma was operated in argon at 1 mTorr (3.5×10^{13} particles/cm^3). The electron density and temperature determined from the fitted Gaussian spectrum ranged from $2.1-6.2 \times 10^{11}$ cm^{-3} and 2.5–3.1 eV, respectively. In one experiment, they operated a YAG laser at 532 nm and at 1 kHz repetition rate with 1 mJ pulses of duration 7 ns and beam divergence of 0.5 mrad. To improve the signal, they used spherical mirrors to reflect the laser light up to 26 times through the scattering volume. A difficulty with this approach was the relative increase in background light because of the longer duration of the measurement. The lower limit on density measurement capability was estimated as 5×10^{10} cm^{-3}.

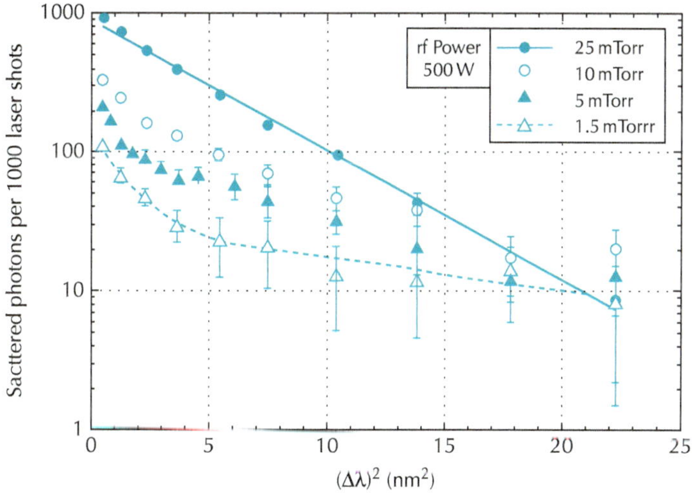

FIGURE 9.6 Thomson scattering from the plasma at various pressures. (Courtesy of K. Uchino in Hori *et al.* (1998) and The American Institute of Physics.)

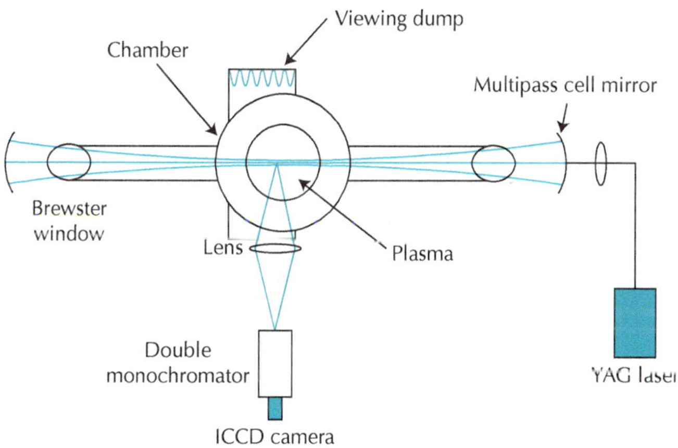

FIGURE 9.7 Schematic diagram of the apparatus. (Courtesy of K. Uchino in Bowden *et al.* (1999) and the Institute of Physics Publishing.)

A more satisfactory approach was to operate the laser at 10 Hz repetition rate with 0.5 J pulses of 10 ns duration and a beam divergence of 0.5 mrad. The estimated lower limit on density capability was 1×10^{10} cm^{-3}. Measured spectra, and without the plasma, are shown in Fig. 9.8.

Scotti & Kado (2009) measured electron temperatures as low as 0.05 eV at densities in the 10^{12} cm^{-3} range, showing agreement with theoretical models of a diverter plasma. Electron density ($n_e \sim 0.7$–5.5×10^{11} cm^{-3}) and temperature (0.9–2 eV) profiles were measured in a neutral loop discharge by Sakoda *et al.* (1997).

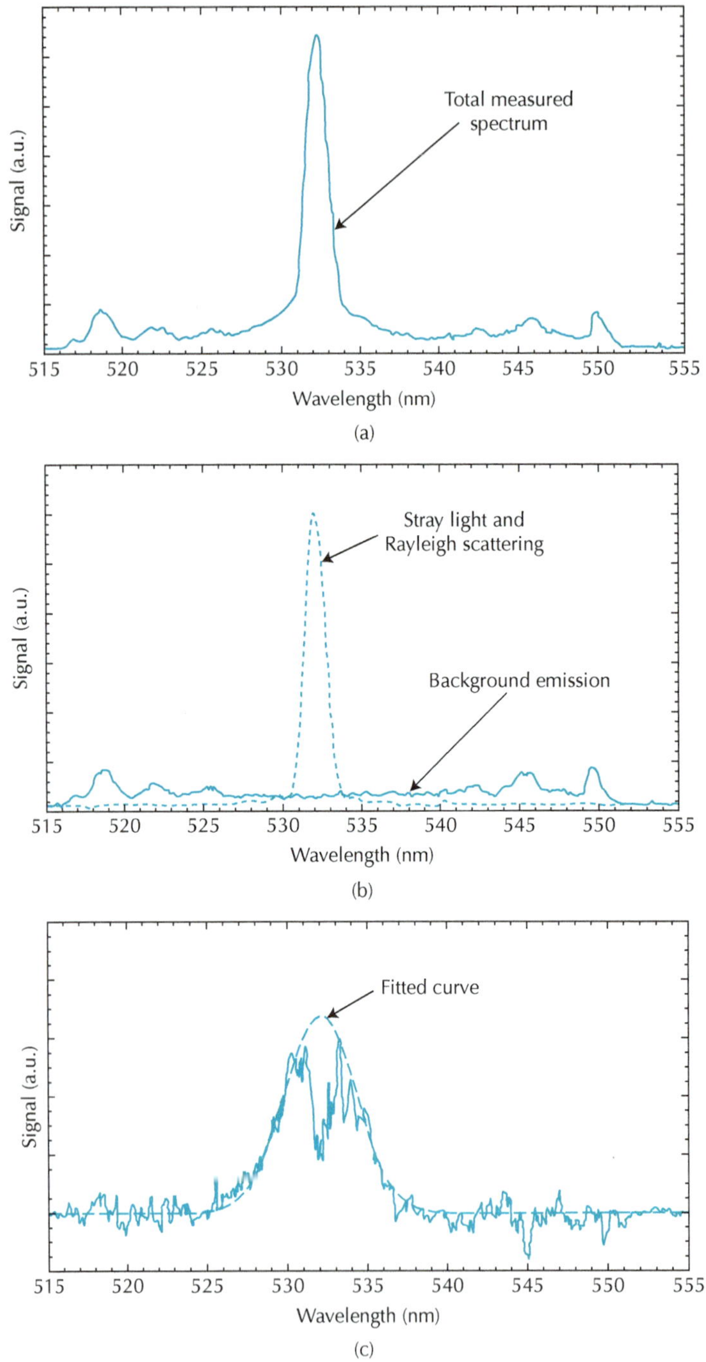

FIGURE 9.8 Example of spectra measured with the ICCD camera. (a) is the full spectrum measured when the laser light is being scattered by the plasma. (b) shows the two additional spectra, which must be measured in order to determine the Thomson scattering component from the spectrum in (a). (c) Thomson scattering signal component of the spectrum shown in (a). The dashed line is a Gaussian fitted curve. (Courtesy of K. Uchino in Bowden *et al.* (1999) and the Institute of Physics Publishing.)

A tin vapor discharge is one of the sources under development for extreme ultraviolet lithography. During the prepinch phase, the electron temperature and density are, respectively, in the range of 5–30 eV and 10^{17}–$10^{18}\,\text{cm}^{-3}$. Collective scattering measurements on such plasmas have been made by Kieft *et al.* (2004, 2005); see Fig. 9.9.

Collective Thomson scattering measurements of n_e, T_e, and Z in the argon plasma of an EUV light source have been made by Tomita *et al.* (2009), as shown in Fig. 9.10.

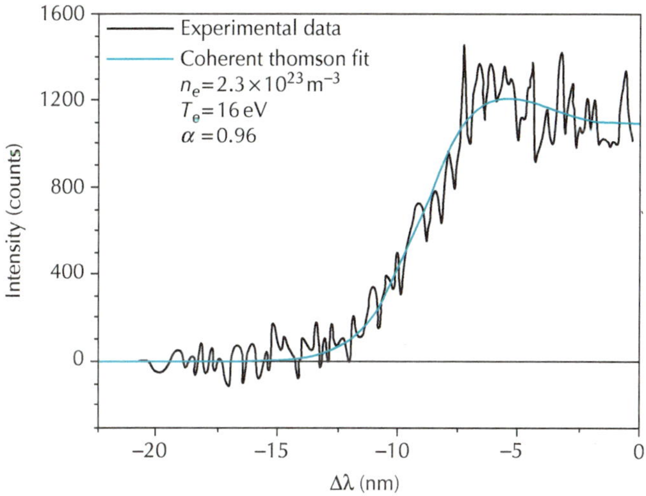

FIGURE 9.9 Example of a measured collective Thomson scattering spectrum. The negative counts at the lower wavelength end are due to the background subtraction. (Courtesy of E. Kieft (2004) and The American Physical Society.)

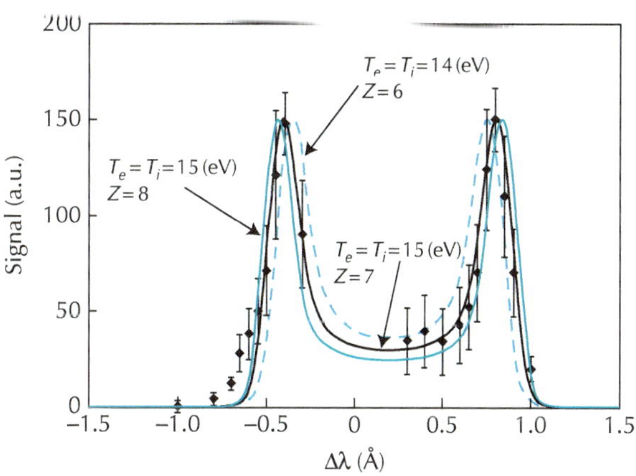

FIGURE 9.10 Observed ion spectrum from an Ar gas discharge and fitted theoretical curve. (Courtesy of K. Tomita (2009) and the JSPF.)

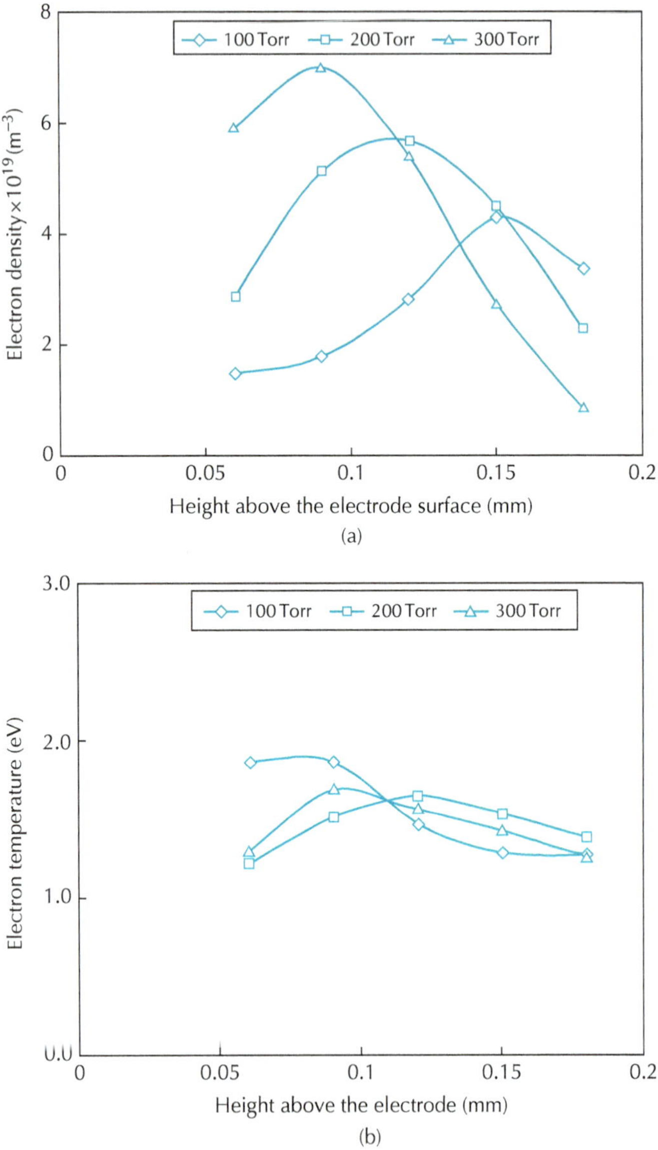

FIGURE 9.11 (a) Spatial dependence of the electron density in the Z direction above the electrode surface at 220 ns after starting the discharge pulse. (b) Spatial dependence of the electron temperature in the Z direction above the electrode surface at 220 ns after starting the discharge pulse. (Courtesy of K. Uchino in Hassaballa *et al.* (2004) and the IEEE (2004).)

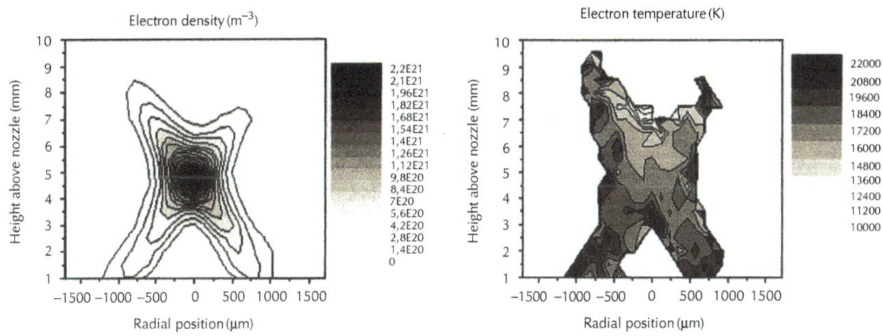

FIGURE 9.12 Electron density and temperature maps for the argon microwave plasma torch. (Courtesy of J. van der Mullen (2004) and Elsevier.)

Noguchi *et al.* (2001, 2002) and Hassaballa *et al.* (2004) made measurements in small scale (0.1–1 mm) plasmas of a microdischarge, as used in a plasma display (PD). Again a key to their success was the suppression of strong laser stray light with the aid of a triple-grating spectrometer. The plasma parameters were $n_e \sim 8 \times 10^{12}\,\mathrm{cm}^{-3}$ and $T_e \sim 0.5{-}1.4\,\mathrm{eV}$. Such plasma have a very small depth of about 2 mm, and measurements require a resolution ~ 0.1 mm. Detailed measurements were made of n_e and T_e as shown in Fig. 9.11. In a separate experiment, Hassaballa *et al.* (2005) measured n_e and T_e in a striated PD plasma.

High-resolution maps of n_e and T_e were made in a microwave plasma torch by van der Mullen *et al.* (2004) and van der Mullen (2004), using a 2D Thomson scattering system (see Fig. 9.12).

9.2 SCATTERING FROM ENERGETIC IONS

9.2.1 Fast Ion Distribution Function

Energetic ions, with energy $E_f > T_i$, derive from two main sources, heating by neutral beams and radio-frequency waves (RF), and from fusion. In the case of RF heating, the ions are accelerated from a background distribution to give a high-energy tail. In the case of beams and fusion, the ions start at an energy E_{f0} and then slow down by collisions, initially with electrons and then with other ions. In general, $\mathrm{v}_i < \mathrm{v}_f < \mathrm{v}_e$ for thermal ions, fast ions, and electrons, respectively. The simplest form of their slowing down velocity distribution function is

$$f_f(\mathbf{v}) = \frac{F_0}{\mathrm{v}_f^3 + \mathrm{v}_c^3}, \tag{9.2.1}$$

where $F_0 = 3/(4\pi \ln(1 + \mathrm{v}_f^3/\mathrm{v}_c^3))$ for $\mathrm{v}_f \le \mathrm{v}_{f0}$. In the above, $\mathrm{v}_c = (3\pi^{1/2}\hat{z}/4)^{1/3}$ is the speed at which the drag from the electrons equals the drag from the plasma ions, and $\hat{z} = \sum_j z_j^2 n_j / n_e m_i / m_e$.

However, there are also ions heated above their injection energy by the faster moving electrons, thereby giving rise to a high-energy tail of the form,

$$f_f(\mathbf{v}) \propto \exp(-(E_f - E_{f0})/T_e) \tag{9.2.2}$$

for $v_f > v_{f0}$.

In practice, the situation can be considerably more complicated. In the case of toroidal geometry, ions may be trapped in mirror field regions and also diffuse radially (see Cordey, 1976), where an analytic solution is given. A general discussion of the behavior of fast ions in plasmas may be found in Chapter 14 by Goldston & Rutherford (1995).

9.2.2 Scattering from Fast Ions

Suggestions for using scattering of electromagnetic radiation to measure the alpha particle distribution function were made by Hutchinson *et al.* (1985) for CO_2 laser light and by Woskov (1987) for millimeter waves. The total scattered spectrum is

$$S(\mathbf{k},\omega) = S_e(\mathbf{k},\omega) + S_i(\mathbf{k},\omega) + S_\alpha(\mathbf{k}\omega), \tag{9.2.3}$$

where, for a Maxwellian background plasma in a magnetic field, the electron and ion terms are given by Eqn (10.3.9), and the dielectric function $\epsilon = 1 + H_e + H_i + H_\alpha$. For unmagnetized alphas,

$$H_\alpha = \frac{\omega_{p\alpha}^2}{k^2}\int d^3\mathrm{v}\frac{\mathbf{k}\cdot\partial f_\alpha/\partial\mathbf{v}}{(\omega - \mathbf{k}\cdot\mathbf{k} + i\delta)}. \tag{9.2.4}$$

Vahala *et al.* (1986, 1988) calculated $S_\alpha(\mathbf{k},\omega)$ for the velocity distribution function in Eqn (9.2.1)

$$S_\alpha(\mathbf{k},\omega) = \left|\frac{\mathbf{H_e}}{\epsilon_L}\right|^2 \frac{4n_{\alpha 0}}{n_{e0}}\frac{2\pi}{|\mathbf{k}|}f_\alpha^{(1)}(\omega/k), \tag{9.2.5}$$

where $f_\alpha^{(1)}$ is the one-dimensional distribution, with velocity components perpendicular to $\mathbf{k}$ being integrated out.

Vahala *et al.* pointed out that there is an enhancement of scattered power in the region of the lower hybrid resonance; $\omega_{LH} = \left[\omega_{pi}^2/(1+\omega_{pe}^2/\Omega_e^2\right]^{1/2}$. As discussed in the literature, the bulk ions usually dominate this part of the spectrum near the lower hybrid slow wave frequency: however, in the region of the lower hybrid fast wave, the fast ion feature is not obscured by the bulk ions. The spectrum in this region is not adequately described in the electrostatic approximation, and it is necessary to allow for electromagnetic effects. Subsequent papers include those of Aamodt & Russell (1990, 1992), Chiu (1991), Bindslev *et al.* (2003), and Bindslev (1996). In the latter paper, Bindslev shows that it is necessary to account not only for scattering from magnetic fluctuations but also for the first-order moment of the velocity distribution function. This is because the scattering off density and magnetic fluctuations at the fast magnetosonic wave cancel each other almost perfectly, lowering the spectral power density significantly. A discussion of relativistic effects is given by Bindslev (1991).

The dispersion relations derived by Chiu for the lower hybrid waves are

Fast Waves:

$$\left(\frac{ck_{\perp}}{\omega}\right)^2_{\text{FW}} = \frac{\left(S-n_{||}^2\right)^2 - D^2}{S-n_{||}^2} + \frac{D^2}{|\epsilon_{zz}|^2}\text{Re}(\epsilon_{zz})\frac{n_{||}^2}{S-n_{||}^2} \tag{9.2.6}$$

Slow Waves:

$$\left(\frac{ck_{\perp}}{\omega}\right)^2_{\text{SW}} = \text{Re}(\epsilon_{zz})\frac{S-n_{||}^2}{S}, \tag{9.2.7}$$

where $n_{||} = ck_{||}/\omega$,

$$S \approx 1 - \sum_i \frac{\omega_{pi}^2}{\omega^2} + \frac{\omega_{pe}^2}{\Omega_e^2},$$

$$D \approx \frac{\omega_{pe}^2}{\omega\Omega_e}, \quad \text{and}$$

$$\epsilon_{zz} \approx \left(2\omega_{pe}^2/\omega^2\right)\xi_e^2\left[1+\xi_e Z(\xi_e)\right], \tag{9.2.8}$$

where $\xi_e = \omega/k_{||}v_e$.

It is informative at this point to put in some numbers for the example given by Chiu: a D-T plasma with $n_D = n_T$, $n_e = 1\times10^{14}\,\text{cm}^{-3}$, $T_e = T_D = T_T = 30\,\text{keV}$, $B = 4.5\,\text{T}$, $\omega = 0.8\omega_{LH}$, and $n_\alpha/n_e = 2.27\times10^{-2}$. $\omega_{pi} = 8.34\times10^9\,\text{rad s}^{-1}$, $\omega_{pe} = 5.64\times10^{11}\,\text{rad s}^{-1}$, $\Omega_i = 1.72\times10^8\,\text{rad s}^{-1}$, $\omega_{LH} = 6.8\times10^9\,\text{rad s}^{-1}$, and $\omega = 5.4\times10^9\,\text{rad s}^{-1}$. Then, $S = -0.85$, $D = 75.6$. Table 9.1 shows the relative values of scattered power from the electrons, plasma ions, and alphas as a function of the value of $n_{||}$ from the Chiu paper, normalized to the alpha power.

For the temperatures in fusion plasmas, relativistic effects must be included, as discussed in Chapters 2, 3, 4 and 5. In fact, finite velocity effects should also be taken into account for very energetic fusion ion products and energetic beam ions, such as the 1 MeV deuteron beams proposed for ITER (see Table 9.2).

TABLE 9.1 Relative Scattered Power Normalized to the Value for Alphas [from Chiu (1981)]

$n_{\|}$	0	1	1.5	2.5	3
Alphas	1	1	1	1	0
Electrons	0	0	1.4	5.2	–
Plasma Ions	9.4	0.14	0	0	–

TABLE 9.2 Fast Ion Speed Relative to the Velocity of Light

Fast Ion	v_{fk0}/c (%)
1 MeV Deuteron	3.3
3.5 MeV Alpha	4.3
3.02 MeV Proton	8.0
14.7 MeV Proton	17.7

As shown in Eqn (2.3.17), there is a correction factor for $S(\mathbf{k}, \omega)$ to first order in ω/ω_i of $(1+2\omega/\omega_i)$. Here, the correction factor is given approximately by

$$\left(1+2\frac{\omega}{\omega_i}\right) \simeq \left[1+4\sin\left(\frac{\theta}{2}\right)\right]\frac{v_{fk0}}{c}. \tag{9.2.9}$$

Therefore, the correction at the birth velocity will be important for larger scattering angles, such as the backscattering used on TEXTOR and ASDEX-UG, and proposed for ITER (see Bindslev *et al.*, 2004; Meo *et al.*, 2004).

9.2.3 Experimental Data and Plans for ITER

To date, the experimental tests have involved measurements of ions resulting from neutral beam injection. The first attempts used CO_2 lasers, as proposed by Hutchinson *et al.* (1985). However, a major problem with using 10.6 μm light is the requirement for the very small scattering angle needed to obtain a sufficiently high value of $\alpha = 1/k\lambda_{De}$ so that the scattered signal from the alphas will be equal to that from the electrons. For the phase velocities $\omega/k = v_\alpha$,

$$\frac{P_{s\alpha}}{P_{se}} = \pi^{1/2}\frac{a}{v_\alpha}\frac{Zn_\alpha}{n_e}\alpha^4\frac{1+I_\omega^2(x_e)}{\exp(-x_e^2)}, \tag{9.2.10}$$

where $I_\omega(x_e) = \pi^{1/2}\exp(-x_e^2)$ and $x_e = v_\alpha/a$.

For example, when $Z = 2$, $n_\alpha = 7.5\times10^{11}\ \mathrm{cm^{-3}}$, and $n_e = 1.2\times10^{14}\ \mathrm{cm^{-3}}$, $P_{s\alpha}/P_{se} = 0.193\alpha^4$, requiring $\alpha = 1.67$ for $P_{s\alpha}/P_{se} = 1$. With 10.6 μm light, this means $\theta = 0.70°$ (see Section 8.4 for an example of a small angle scattering experiment).

As discussed in Donné & Barth (2006), a proof of principle experiment was undertaken on Advanced Toroidal Facility (ATF) with beams (Richards *et al.*, 1993). Subsequently, a system was installed on JT-60U (Kondoh *et al.*, 2003; Richards *et al.*, 2003). The results were hampered by large stray light levels, electrical noise from the laser, and higher order laser modes. An improved CO_2 laser with high-energy (~ 17 J), high repetition rate (15 Hz), single mode, and feed back control has improved the signal-to-noise ratio (Kondoh *et al.*, 2006, 2007).

As Luhmann *et al.* (2008) mention, a more successful approach has been to use longer wavelength radiation in the millimeter range, readily available following the development of the gyrotron (Woskoboinikow, 1986). At the lower end of the millimeter-wave range, there are essentially no limitations on scattering angle, and the ion feature always dominates, thereby allowing the scattering geometry to be selected to meet other objectives. In this frequency range, the challenge is the relatively strong electron cyclotron emission (ECE) raising the background noise. Consequently, certain millimeter-wave frequencies are not usable because of excessive ECE or absorption of the probe and scattered radiation. For electron temperatures up to ~ 10 keV in toroidal fusion plasmas with aspect ratios of 3 or more, CTS systems with the probe frequency between the fundamental- and second-harmonic features of the ECE spectrum may be used.

The requirements for measuring alphas in ITER are discussed in the ITER Final Design Report (2001): time resolution 100 ms, spatial resolution $a/10 \approx 20$ cm, energy range 0.1 to 3.5 MeV, alpha density range 10^{11} to 2×10^{13} cm^{-3}, and accuracy 20%. The energy resolution was not defined. The reference H-mode plasma has $B = 5.3$ T, $n_e = 1 \times 10^{14}$ cm^{-3}, $T_e = T_i = 25$ keV, and the alpha density is 5×10^{11} cm^{-3}. Bindslev *et al.* (2004) have analyzed four sources as shown in Table 9.3.

TABLE 9.3 Alternative Sources for Alpha Measurement In ITER

Probe Frequency	Relation to ECE Spec	Likely Source	Main Concern	Maximum θ
60 GHz	X-Mode, $\omega_i < \omega_{ce}$	Gyrotron	Refraction, ECE	No Limit
170 GHz	O-Mode, $\omega_{ce} < \omega_i < 2\omega_{ce}$	Gyrotron	ECE	50°
3 THz	In Upper Tail of ECE Spec.	Optically Pumped FIR	Source, ECE	4°
28 THz	Far Above ECE	CO_2 Laser	Source, Small Scattering Angle	0.4°

Courtesy of Bindslev et al. (2004).

For the higher temperature expected in ITER, they conclude that the best option is to work at 55 to 60 GHz, which is below the fundamental ECE resonance. This option could meet the ITER requirements and be able to look both parallel (co and counter) to the magnetic field and perpendicular to it. They also discount the 170 GHz option, which would require an unreasonably high power level. The problem for the 3 THz approach is the current lack of the required source (6 J and 600 ns pulse length, according to their estimates). However, it should be noted that there are no technical obstacles preventing the development of a megawatt level FEL achieving such power levels. The 28 THz (CO_2 laser) approach has been attempted on JT-60 with scattering at small angles, but appears to be impractical for ITER.

The system proposed for ITER by the Risϕ group (Bindslev *et al.*, 2003, 2004; Meo *et al.*, 2004) would use a 60 GHz gyrotron with $\phi = 10°$ and $\theta = 20°$. It is assumed that the measurement would be made in the presence of 1 MeV deuterons from neutral beam heating. A calculated spectrum in Fig. 9.13 shows that the alpha particles could be distinguished from the beam heating ions by the asymmetry of the scattered spectrum.

The first gyrotron-based fast ion measurements, coming from neutral beams, were made on JET (Bindslev *et al.*, 1999a). Subsequently, a similar system

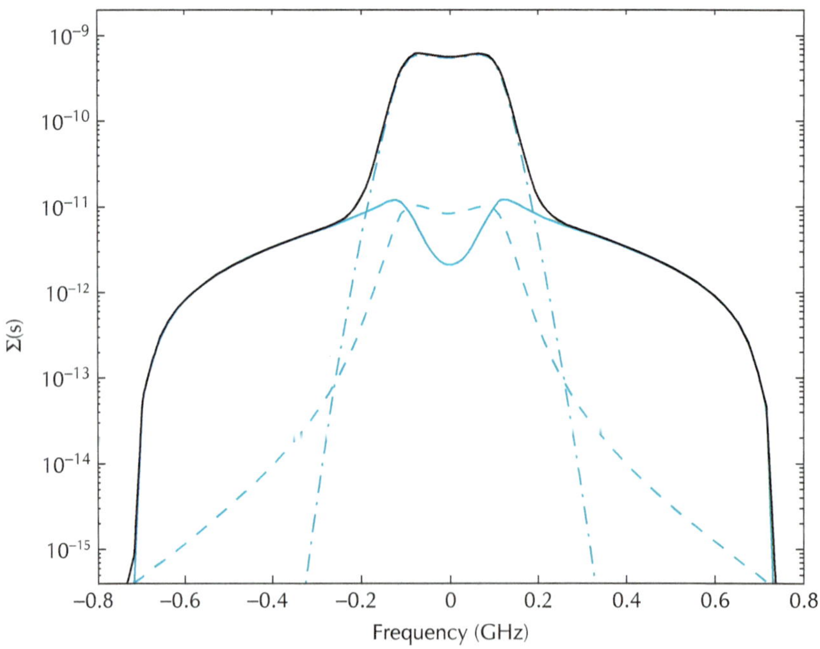

FIGURE 9.13 Computed spectrum for ITER. Descending from the top center of the figure, we have the total, ion, electron, and alpha (prominent dip in the center) contributions to the scattering spectrum. The alphas contribute most of the pronounced shoulders. (Courtesy of Bindslev *et al.* (2003).)

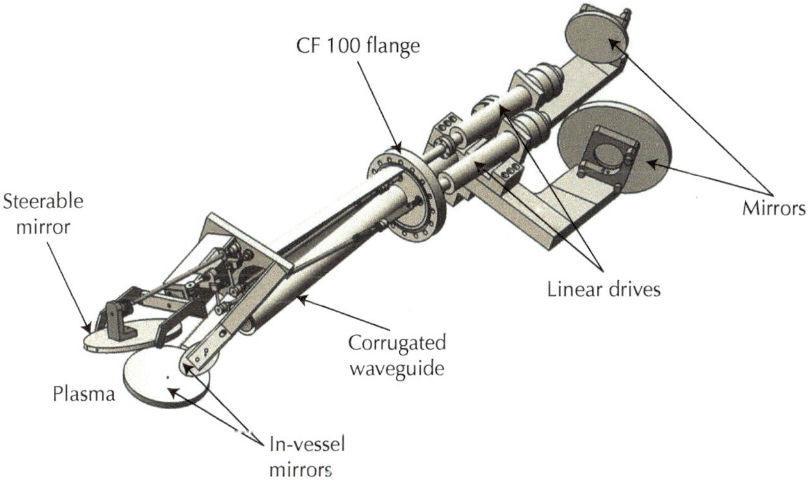

FIGURE 9.14 Scattering set up on TEXTOR. (Courtesy of Bindslev *et al.* (2007) and the Japanese Society of Plasma Science and Nuclear Fusion research.)

was deployed to measure beam ions on TEXTOR (Bindslev *et al.*, 2006, 2007; Nielsen *et al.*, 2008) and also on ASDEX-U (Michelsen *et al.*, 2004; Korsholm *et al.*, 2006). The experimental setup on TEXTOR is shown in Fig. 9.14.

The system used a 200 kW, 110 GHz gyrotron source, of 0.2-s duration. The scattering angle was 150° to 170°, and the receiver had 42 channels with a bandwidth of 106.3–113.4 GHz. A pair of 60 dB each, wavelength notch filters were used in the receiver front end to reject stray gyrotron radiation. Also, a rapidly switched attenuator was used to block the receiver signal during the gyrotron transient on and off times, when spurious gyrotron modes were present. The gyrotron was operated with a train of up to 100, 2-ms duration pulses with an adjustable time delay between pulses to temporally track fast ion evolution during the plasma shot.

The injected ions are born at a particular velocity and spread out in space as discussed above. The scattered radiation comes from the velocity component in the direction of **k**. Time traces at various frequencies for TEXTOR are shown in Fig. 9.15. The rotation velocity of thermal ions is shown in Fig. 9.15b. Figure 9.16 shows the logarithm of the ion velocity distribution.

A 77-GHz system to measure both the bulk and tail ion distribution has been designed for the LHD stellarator by Nishiura *et al.* (2008).

9.3 BURNING PLASMAS

Diagnostics of D-T burning plasmas presents numerous challenges for optical systems. In addition to the problem of material ablated by the plasma coating optical surfaces – usually a mirror or window – the fusion neutrons require much more complicated final optics to protect sensitive equipment. Neutrons

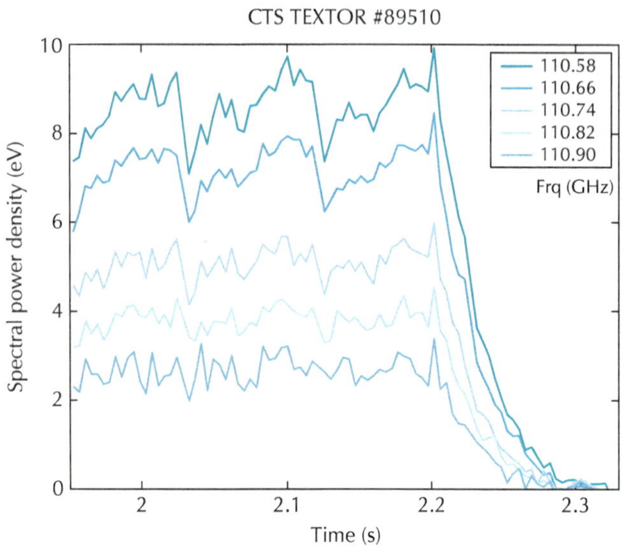

(a)

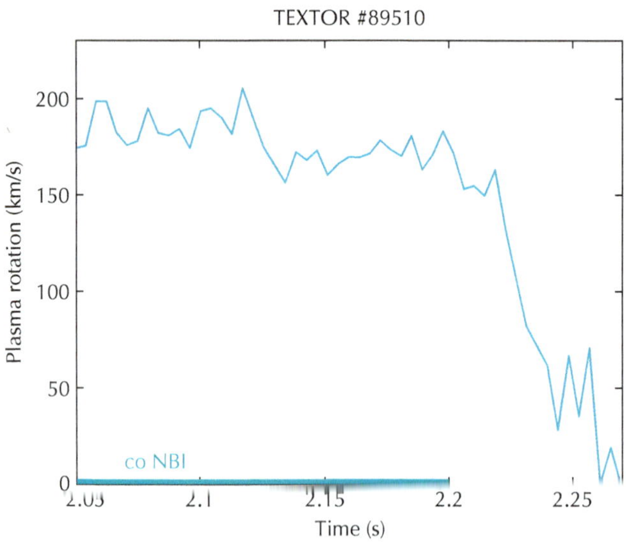

(b)

FIGURE 9.15 (a) Time traces of the collective Thomson scattering (CTS) spectral power density in five channels. The signals respond, respectively, to the frequencies shown in the legend, top right. (b) The toroidal rotation velocity of the thermal ion population versus time inferred from the wings of the thermal ion CTS spectral feature. Auxiliary heating is on for 2.2 s. (Courtesy of Bindslev *et al.* (2007) and the JSPF.)

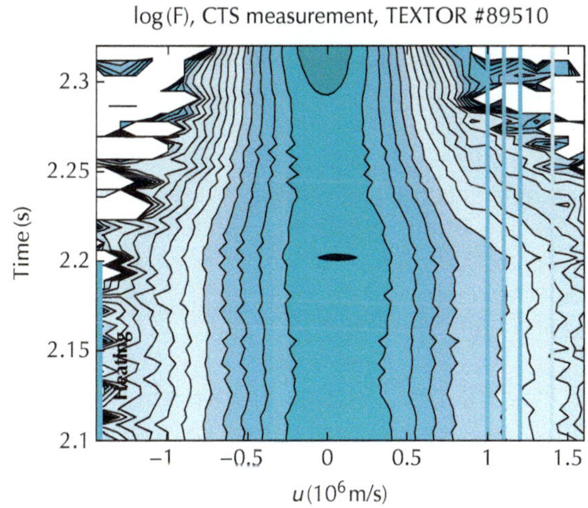

FIGURE 9.16 Contour plot of the logarithm of the measured ion velocity distribution. The auxiliary heating was turned off at $t = 2.2$ s. (Courtesy of Bindslev *et al.* (2007) and the JSPF.)

can scatter around corners, and a rule of thumb is that the neutrons should go through three or four right angle scatterings to reduce their flux to safe levels before they reach an area with hands-on maintenance. This problem has already been faced at low 14 MeV neutron flux and fluence in magnetic fusion in TFTR and JET and in the National Ignition Facility (NIF), NOVA Facility, and Omega Facility in inertial fusion. Gammas and, in inertial fusion, X-rays must also be accounted for. The system must also be designed to prevent tritium leakage. All components close to the plasma will become activated and require remote handling. Calibration will become more difficult, because even when there is no plasma, the interior of the vacuum vessel will have radiation from the induced radioactivity.

9.3.1 Magnetic Fusion

Major deuterium-tritium (D-T) experiments are proposed for JET in the 2014-2015 time-frame and in the early 2020s in the International Thermonuclear Experiment ITER. A performance evaluation of ITER Thomson scattering systems has been made by Walsh & al. (2008). The main systems considered are for the divertor region, mid-plane core, and the edge region. Measurements are required of electron temperatures from ~ 1 eV in the divertor, in the presence of strong background light, to as high as 40 keV in the plasma center, where the blue shift will be very large (Chapter 4). Of concern is that in the core the spectrum may deviate from Maxwellian, possibly owing to the presence of a high-energy tail coming from auxiliary heating. Non-Maxwellian distributions have been seen on TFTR and JET, accompanying a difference between electron cyclotron emission measurements of T_e and those from Thomson scattering (see de la Luna *et al.*, 2003; Taylor & Harvey, 2008, 2009). To assist in the analysis of this situation, scattered spectra for relativistic and noncollective spectra for generalized Lorentzian distributions have been calculated by Naito *et al.* (1996) (see also Naito *et al.*, 1997).

A Nd-YAG 1064 nm laser is favored for the core LIDAR measurement in ITER. With the extreme blue shifts expected, measurements are desirable down to 300 nm. Components of this system are expected to include rhodium-coated mirrors, and multiple 5 J, 150 ps pulse length lasers with a 6–8 channel polychromator. A Nd-YAG laser delivering 5 J at 100 Hz is under development for edge measurements (Hatae *et al.*, 2007).

9.3.2 Other Examples of Systems Proposed for ITER

Bindslev *et al.* (2003) discuss a system to measure the alpha particle distribution in ITER (see Section 9.2.2).

Smith (2008) has suggested the use of pulsed polarimetry to measure the magnetic field in a plasma. Important features include the capability for measuring simultaneously the local T_e, n_e, and $B_{||}$ along the line of sight, with a resiliency to refractive effects (see Section 8.2). Kajita *et al.* (2009) have proposed an edge plasma scattering system (see Fig. 8.5).

9.3.3 Inertial Fusion

Inertial confinement fusion (ICF) uses photons to ablate the shell of a capsule filled with cryogenically cooled deuterium and tritium. The radially outward ablation compresses the capsule creating a central "hot spot" that initiates thermal nuclear burn. The dense shell that is assembled during the compression confines the alpha particles allowing fusion to be sustained until enough of the fuel has been depleted to produce a substantial gain over the input energy (Nuckolls *et al.*, 1972). There are two main approaches to ICF; direct drive directly illuminates the capsule with laser light (McCrory *et al.*, 2008), while indirect drive uses a radiation cavity "hohlraum" to convert the laser light into soft x-rays which in turn drive the capsule (Lindl *et al.*, 2004).

With the completion of the National Ignition Facility (Moses & Wuest, 2004) at the Lawrence Livermore National Laboratory in 2009, experiments have started (Glenzer *et al.*, 2010) to compress matter to densities and temperatures higher than the interior of the sun which will initiate nuclear fusion and burn. The NIF consists of 192 laser beams with 1.8 MJ of 0.35 μm laser light which have been arranged to irradiate a hohlraum from the top and bottom hemispheres. This "indirect-drive" laser geometry has been chosen for the first experiments. The hohlraums use a low-Z cryogenic gas fill to prevent fast wall plasma blow off which may impose asymmetric capsule implosion conditions. Consequently, the laser beams have to initially propagate through up to 1 cm of dense low-Z plasma before they deposit their energy in the hohlraum wall. To produce the desired soft x-ray profiles, indirect drive requires efficient and reproducible laser beam propagation through this under-dense plasma, and predictive modeling of the laser-plasma interaction processes requires detailed understanding of instabilities including laser backscattering by stimulated Brillouin scattering (SBS) and stimulated Raman scattering (SRS), laser beam deflection, beam filamentation, and self-focusing (see Sections 8.3, 8.6, and 12.7). The ability to mitigate these instabilities requires a solid understanding of the plasma conditions (e.g., T_e, T_i, n_e) encountered by the incident laser beams (Froula *et al.*, 2010), and therefore, Thomson scattering has become a critical diagnostic at large inertial fusion faculties.

Inertial confinement fusion research typically requires high-density plasmas with electron densities of $n_e \sim 10^{21}\,\text{cm}^{-3}$. In this regime, absorption of optical laser light with wavelengths of $0.35\,\mu\text{m} < \lambda_i < 0.8\,\mu\text{m}$ becomes important. Furthermore, dispersion of optical light is nonnegligible at high densities so that the phase and the group velocity of the light wave are affected. In order to reduce both effects, absorption of the probe laser and dispersion of the light wave, a short wavelength laser is desirable. For these reasons, several large scale laser facilities have implemented a high-energy, high-power 4ω probe laser.

NOVA Thomson Scattering

The 4ω probe laser (operating wavelength: $\lambda_i = 263.3\,\text{nm}$) was implemented at the 30 kJ Nova laser facility (Glenzer *et al.*, 1999a). The Nova laser was a Nd:glass laser operating at $1.055\,\mu\text{m}$ (1ω), which was frequency converted to 2ω or 3ω with potassium dihydrogen phosphate (KDP) crystals mounted at the target chamber. For the 4ω beam line, the central unused part of one Nova beam was separated out between the main amplifier chain and the frequency converting crystals and redirected to a separate dedicated port of the Nova target chamber.

This setup resulted in a laser beam with a maximum measured energy of 220 J at 1ω before the frequency converting crystals. A type II doubler and a type I quadrupler KDP crystals (area: $15\,\text{cm}^2$, thickness: 18 mm for the doubler, 8 mm for the quadrupler) were used to convert the beam to 4ω. After the KDP crystals, a 20-cm diameter fused silica lens focuses the beam through a 1.9-cm thick debris shield, which is also used as the vacuum barrier, onto the target ($f/16.6$ with a beam diameter of 15 cm). These optics were made of Excimer grade fused silica (Corning 7980, grade A, "Ø") to survive the high fluence level at 4ω. Under routine operating conditions, a frequency conversion efficiency of 20% and a maximum energy of 50 J at 4ω was achieved.

Glenzer *et al.* (2001a) used the NOVA 4ω beam to scatter light from a hohlraum plasma and measure the electron temperature. The 4ω beam was focused inside the hohlraum, and Thomson scattered light was collected through a $400\,\mu\text{m} \times 400\,\mu\text{m}$ diagnostic hole cut in the side of the hohlraum (Fig. 9.17a). Figure 9.17b shows a least-squares fit to the spectrum by applying the theoretical form factor $S(\mathbf{k}, \omega)$ [Eqn (5.1.2)]. As can be seen, the spectrum was fit rather well by assuming scattering on a pure gold plasma where the width of the ion-acoustic peaks is determined by the instrument function and modest velocity gradients in the Au plasma. For these conditions, the fit is not sensitive to the

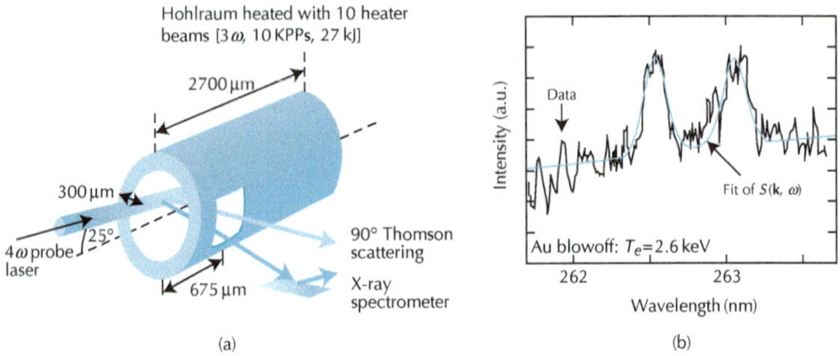

FIGURE 9.17 (a) A schematic of the experiment where the 4ω beam was used to measure the plasma conditions in the Au blow-off plasma inside of a hohlraum. (b) The Thomson scattering spectrum is fit with the theoretical form factor with an electron temperature of $T_e = 2.6\,\text{keV}$ and $Z = 51$ measured by x-ray spectroscopy. (Courtesy of Glenzer *et al.* (2001a).)

choice of the electron density and ion temperature, and therefore, parameters from the radiation-hydrodynamic modeling were used, $n_e = 1.4 \times 10^{21}\,\text{cm}^{-3}$ and $T_i = 0.8\,\text{keV}$, respectively. These results indicated the importance of developing steady-state kinetic codes for high-Z plasmas to better understand the x-ray production, and specifically, the Thomson scattering results indicated that the ionization distribution was shifted to a higher charge state compared to detailed atomic physics simulation.

This 4ω Thomson scattering system provided many important measurements over its years of operation (Glenzer *et al.*, 1997a,b, 1999b, 2001a,b; Wharton *et al.*, 1998, 1999; Glenzer *et al.*, 2000a; Ford *et al.*, 2000a,b; Geddes *et al.*, 2003; Fournier *et al.*, 2006).

OMEGA Thomson Scattering

MacKinnon *et al.* (2004) implemented a 4ω Thomson scattering probe beam at the Omega Laser Facility by diverting one of the 60 existing 2ω beams (diameter = 280 mm, maximum energy on target = 400 J in 1 ns) to a dedicated port. A 30-cm diameter aspheric lens with a 150-cm focal length focused the beam on target. To convert the probe to the fourth harmonic, a 30-cm diameter type I KDP crystal, thickness 10 mm, was inserted after the final 2ω turning mirror in the system. The maximum conversion from 2ω to 4ω was measured to be nearly 70%. At peak conversion, the system produced an energy on target of 260 J of 264 nm light in a 1 ns pulse. When a 200 ps pulse is used, the maximum energy on target is 60 J.

The Thomson scattered light was collected with an achromatic fused silica $f/10$ lens with a focal length of 50 cm. The optic was mounted in a Ten Inch Manipulator (TIM) at a distance of 50 cm from the plasma. The angle between the TIM and the probe beam was 101°. A fused silica blast shield coated to reject the light from the heater beams 3ω was mounted before the collection optic. The collected light was then transported out the back of the TIM to a series of turning mirrors that direct the scattered light a diagnostic table (see Fig. 9.18). The

FIGURE 9.18 A schematic of the Thomson scattering diagnostic at the OMEGA Laser Facility. (Courtesy of J. S. Ross.)

light was split by a 50:50 poke-a-dot splitter to a 1-m and 0.3-m spectrometer. The light directed to the 0.3-m spectrometer was focused with an $f/5$ focusing mirror. A 600 lines/mm grating with a 200 μm entrance slit was used to resolve the electron plasma features. The light directed to the slit of the 1-m spectrometer is focused by a mirror with a 75-cm focal length with a magnification of 1.5. The spectrometer uses a 3600 lines/mm grating and a 100 μm entrance slit. An intensified, gated 16-bit charge-coupled device (ICCD) or a streak camera is coupled to the output of the spectrometer. When the ICCD is employed, the slit of the spectrometer is rotated to be parallel to the probe beam.

Ross *et al.* (2006) used this system to measure the electron temperature profile in an asymmetrically heated gas-filled hohlraum. Figure 9.19 shows the spatially resolved Thomson scattering spectrum used to measure the electron temperature gradient produced by only heating the target with the heater beams on one side of the hohlraum. Over a 1-mm region inside the hohlraum, the electron temperature varies from 1.5 to 3 keV. These temperatures were measured by fitting a multiple ion species form factor to the experimental data.

This system continues to be operational and to date has provided many important measurements (Froula *et al.*, 2006, 2007b; Heater *et al.*, 2007; Niemann *et al.*, 2008; Petrasso *et al.*, 2010).

NIF Thomson Scattering

A 4ω laser capability is planned on the NIF to be used as a Thomson scattering probe. A quad of beams will be converted to 4ω and used to measure the plasma conditions in an ignition hohlraum (see Fig. 9.20). The most stringent

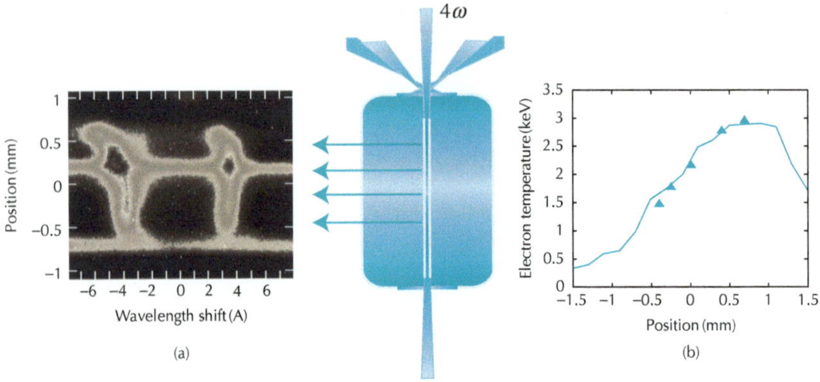

FIGURE 9.19 (a) The Thomson scattering spectrum shows a large temperature gradient along the axis of the hohlraum. (b) The electron temperature profile (triangles) shows close agreement with the HYDRA simulation (solid line) using a flux-limited diffusion model with a flux limiter of 0.05. (Courtesy of Ross *et al.* (2006) and Review of Scientific Instruments.)

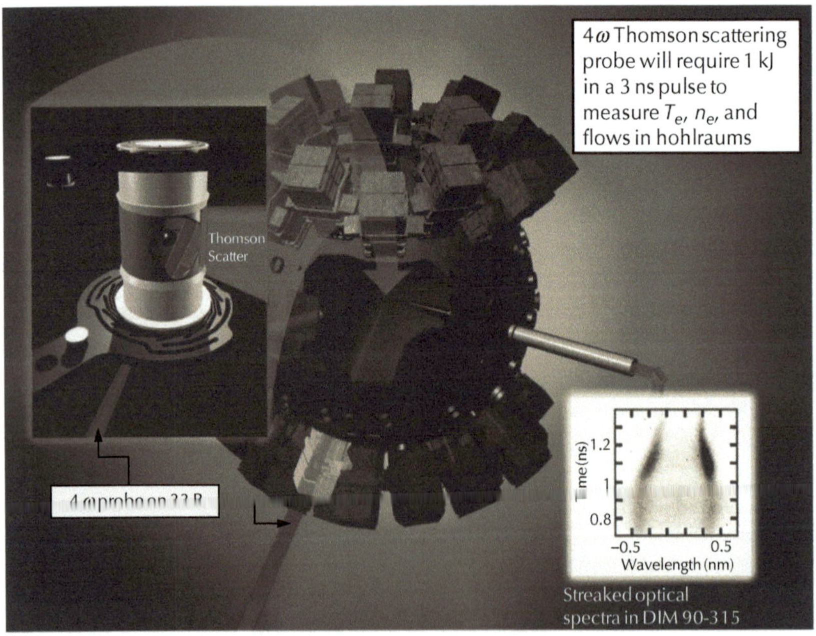

FIGURE 9.20 A schematic of the NIF target chamber is shown. An ignition hohlraum (~ 1 cm long cylinder) is shown with a diagnostic hole cut to allow scattered light from a 4ω probe beam that enters through the bottom to be collected by an telescope located in the 90-315 Diagnostic Instrument Manipulator (DIM). The light is transported out of the back of the DIM where it is spectrally dispersed and recorded. (Courtesy of C. Dahlen.)

demands on the 4ω probe beam in terms of energy and alignment are driven by Thomson scattering applications in hohlraums. Scaling from previous hohlraum experiments where ~ 100 J of 4ω are focussed at $< 100\,\mu$m in an OMEGA hohlraum heated by 20 kJ of 3ω light, the 4ω beam on the NIF will require ~ 2 kJ when scattering from hohlraum experiments using 400−1800 kJ. To resolved plasma gradients, a spot size of 0.2–0.5 mm is required. Pointing to within 150 μm will allow the Thomson scattering volume to be well defined. The bandwidth is to remain as small as possible, no high-frequency modulator smoothing by spectral dispersion (SSD). The pulse length is typically 2–10 ns square. Clearly, open-geometry experiments will require less energy, typically of order 100–500 J to measure the temperatures and density with Thomson scattering or magnetic fields with Faraday rotation.

PROBLEMS

9.1. Rayleigh scattering will play a significant role in low-temperature plasmas, in which there is incomplete ionization of the plasma. Compare Rayleigh and Thomson scattering power at $\lambda_i = 532$ nm, for an industrial argon plasma with $T_e \sim 0.5$ eV and a percentage ionization of 0.08%.

9.2. Using the cold-fluid theory, show that the dispersion relation for the cold lower hybrid wave is approximately given by

$$\omega^2 \cong \omega_{LH}^2 \left[1 + \frac{\omega_p^2}{\Omega_p^2}\frac{n_{||}^2}{n_\perp^2}\right] = \omega_{LH}^2 \left[1 + \frac{m_i}{m_e}\frac{n_{||}^2}{n_\perp^2}\right]$$

where $\omega_{LH}^2 = \dfrac{\Omega^2}{\left(1+\frac{\omega_{pe}^2}{\omega^2}\right)}$ and $n_{||}^2 \ll n_\perp$.

Hint: These are electrostatic waves so the dispersion relation can be written:

$$k_\perp^2 \varepsilon_{xx} + k_{||}\varepsilon_{zz} = 0$$

9.3. Consider an unmagnetized, uniform laboratory argon plasma with density $n_e = 10^{11}\,\text{cm}^{-3}$ and $T_e = 5$ eV and $T_i = 0.5$ eV. You wish to calculate the effect of adding a helium impurity. Calculate the total ion-acoustic wave damping and real part of the frequency for helium concentrations of 0, 0.1, 1, 5, 10, 50, and 100%. Plot your results.

The effect of adding a light ion impurity can be understood from the figure below. This contains a schematic illustration of the ion wave phase velocities superimposed on an electron and two ion velocity distribution functions. The functions have been normalized at v = 0. The increased damping produced by a small population of light ions is obvious.

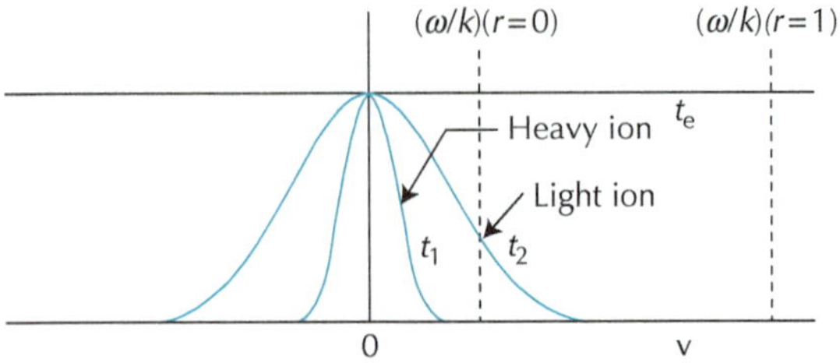

9.4. It can be shown that the X0-mode dispersion relation can be cast in the form

$$\frac{c^2k^2}{\omega^2} = 1 - \frac{\omega_{pe}^2}{\omega^2-\omega_c^2} - \frac{\Omega_p^2}{\omega^2-\Omega_c^2} - \frac{\left(\frac{\omega_c}{\omega}\frac{\omega_{pe}^2}{\omega^2-\omega_c^2} - \frac{\Omega_c}{\omega}\frac{\Omega_p^2}{\omega^2-\Omega_c^2}\right)^2}{1-\frac{\omega_{pe}^2}{\omega^2-\omega_c^2}-\frac{\Omega_p^2}{\omega^2-\Omega_c^2}},$$

Using this dispersion relation:

(a) Show that the lower hybrid resonance ω_{LH} always lies above the ion cyclotron frequency Ω_c.
(b) Show that ω_{LH} always lies below the left-hand cutoff ω_L.
(c) Draw a qualitative $v_{ph}/c^2 = \omega^2/c^2k^2$ plot for the X0-mode as we have done in class but now including the ion motions.

9.5. Compare the speed of fusion alpha particles to the phase velocity of fast and slow LHR for the example given in Section 8.2. Let $n_\parallel = 1.5$.

9.6. For the case $P_{s\alpha}/P_{se} = 1$, calculate the maximum scattering angle for alpha (3.5 MeV) scattering measurements in ITER (the reference H-mode plasma has $B = 5.3$ T, $n_e = 1\times10^{20}\,\text{m}^{-3}$, $T_e = T_i = 25$ keV, and $Z = 1.5$, and the alpha density is $5\times10^{17}\,\text{m}^{-3}$). Use the following source wavelengths: 10.6 μm, 385 μm, and 5 mm.

SOLUTIONS TO ODD PROBLEMS

9.1. For Thomson scattering, $d\sigma_T/d\Omega = r_0^2(1-\sin^2\theta\cos^2\phi)$. For Rayleigh scattering, $d\sigma_R/d\Omega = (\pi^2\alpha_p^2/\varepsilon_0^2\lambda_i^4)(1-\sin^2\theta\cos^2\phi)$, where the polarizability $\alpha_p = (3\varepsilon_0/n_\mu)(\mu^2-1)/(\mu^2+2)$ and μ is the refractive index of the gas measured at the density n_μ.

For argon, $\alpha_p = 1.85\times10^{-40}$, and $r_0^2 = 7.95\times10^{-30}\,\text{m}^2$.

Consequently, $d\sigma_R/d\Omega/d\sigma_T/d\Omega = 6.8\times10^{-3}$. But for the scattered power, we must multiply by $n_n/n_e \sim 1.25\times10^3$ leading to scattered power from Rayleigh scattering 8.5 times higher than from Thomson scattering.

9.3. We recall that our dispersion relation is give by:

$$1 = \sum \frac{\omega_{pj}^2}{k^2 v_{tj}^2} Z'\left(\frac{\omega}{k v_{tj}}\right).$$

We use $Z'(\xi_e) \simeq -2$ and $\omega_{pe}^2/v_{te}^2 = k^2\lambda_{De}^2$ to yield

$$\frac{2k^2}{k^2\lambda_{De}^2} = -2 + \sum \frac{\Omega_{pj}^2}{\omega_{pe}^2}\frac{v_{te}^2}{v_{tj}^2} Z'(\xi_j)$$

or

$$\frac{k^2}{k^2\lambda_{De}^2} = -1 + \sum_j \frac{\alpha_j}{2}\frac{T_e}{T_j} Z'(\xi_j) \simeq 0 \quad \text{for} \quad k\lambda_{De} \ll 1$$

$$1 = \sum_j \frac{\alpha_j}{2}\frac{T_e}{T_j} Z'(\xi_j)$$

and assuming equal ion temperatures, T_j, this becomes

$$\frac{2T_i}{T_e} = \sum_j Z'(\xi_j)\alpha_j = 2\theta.$$

This problem has been treated by Fried *et al.* (1971). They considered ion-acoustic wave propagation in a multi-ion species plasma. Again, the dispersion relation is

$$\frac{2k^2\lambda_{De}^2}{k_1^2} = \frac{1}{\theta(1-f)} Z'\left[s\left(\frac{m_e}{m_1}\right)^{1/2}\theta^{-1/2}\right] + Z'(s) + \frac{f}{1-\rho} Z'\left[s\left(\frac{m_2}{m_1}\right)^{\frac{1}{2}}\right],$$

where $k_1^2 = \frac{4\pi n_1 e^2}{\kappa T_1}$, $s = \frac{\omega}{ka_1}$, $T_1 = T_2$, $\theta = \frac{T_e}{T_1}$, $f = \frac{n_2}{n_e} = 1 - (n_1/n_e)$, and $a_i = (2\kappa T_i/m_i)^{1/2}$

$$f_1^{(0)} = \left(\pi^{\frac{1}{2}} a_1\right)^{-1} \exp[-(v/a_1)^2]$$

$$f_2^{(0)} = \left(\pi^{\frac{1}{2}} a_2\right)^{-1} \exp[-(v/a_2)^2]$$

$$f_e^{(0)} = \left(\pi^{\frac{1}{2}} a_e\right)^{-1} \exp[-(v/a_e)^2].$$

Now, for the case of interest, the temperature ratio is 10 resulting in phase velocities large compared to both ion thermal velocities. We have both $s \gg 1$ and $s \gg (m_1/m_2)^{1/2}$. Consequently, you may use the asymptomic expressions for the ions Z' functions. We then have

$$-\frac{\operatorname{Im}(s)}{\operatorname{Re}(s)} = \frac{\pi^{\frac{1}{2}} s^3}{1-f+(m_1/m_2)f}\left[\theta^{-\frac{3}{2}}\left(\frac{m_e}{m_1}\right)^{\frac{1}{2}} + 2f\left(\frac{m_2}{m_1}\right)^{\frac{1}{2}} \exp\left(-\frac{m_2}{m_1}s^2\right) + 2(1-f)\exp(-s^2)\right]$$

and $\quad s^2 = \left(1 + \frac{m_1}{m_2}\frac{\theta(1-f)}{2}\right).$

The figure below from Samec *et al.* slows the damping as a function of ion species composition for two difference temperature ratios where argon and helium are the two ion species.

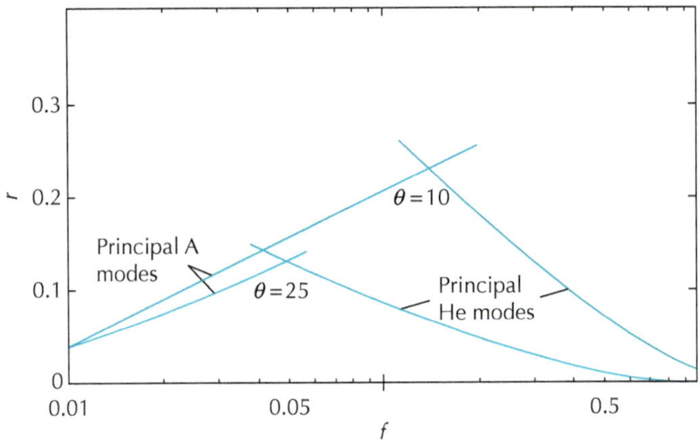

9.5. Start with Eqns (9.2.6)–(9.2.9).
A D-T plasma with $n_D = n_T$, $n_e = 1 \times 10^{20}\ \text{m}^{-3}$, $T_e = T_D = T_T = 30\,\text{keV}$, $B = 4.5\,\text{T}$, $\omega = 0.8\omega_{\text{LH}}$, and $n_\alpha/n_e = 2.27 \times 10^{-2}$.

$$\omega_{pi} = 8.33 \times 10^9 \,\text{rad s}^{-1}, \omega_{pe} = 5.65 \times 10^{11} \,\text{rad s}^{-1},$$

$$\Omega_e = 7.92 \times 10^{11} \,\text{rad s}^{-1}, \Omega_i = 1.73 \times 10^8 \,\text{rad s}^{-1},$$

$$\omega_{\text{LH}} = 6.8 \times 10^9 \,\text{rad s}^{-1}, \text{ and } \omega = 5.11 \times 10^9 \,\text{rad s}^{-1}.$$

Then, $S = -0.85$, $D = 74.1$.
For $n_\| = 1.5$, $\xi_e = 1.92$, $\varepsilon_{zz} \sim \operatorname{Re}(\epsilon_{zz}) \sim 8 \times 10^4$. Since $\varepsilon_{zz} \sim \operatorname{Re}(\epsilon_{zz}) \gg D \gg |S - n_\|^2|$, we obtain

$$\frac{\left(\frac{ck_\perp}{\omega}\right)^2_{\text{SW}}}{\left(\frac{ck_\perp}{\omega}\right)^2_{\text{FW}}} = \frac{\left[\frac{\omega}{k_\perp}\right]_{\text{FW}}}{\left[\frac{\omega}{k_\perp}\right]_{\text{SW}}} \approx -\frac{\operatorname{Re}(\epsilon_{zz})\left(S-n_\|^2\right)^2}{D^2 S} \approx 165$$

$[\omega/k_\perp]_{\text{SW}} \approx 5.6 \times 10^5\ \text{m s}^{-1}$, $[\omega/k_\perp]_{\text{FW}} \approx 9.2 \times 10^7\ \text{m s}^{-1}$, and the alphas slow down from $v_\alpha = 1.29 \times 10^7 \text{m s}^{-1}$.

Chapter 10

Scattering from a Magnetized Plasma

Chapter Outline Head

DOI: 10.1016/B978-0-12-374877-5.00010-5

10.1 INTRODUCTION[1]

It is fair to say that most plasmas contain a magnetic field. In magnetic fusion research, plasmas are contained and stabilized by magnetic fields. In addition, there are fields generated by currents in the plasma. In nature, we have, for example, the magnetic field of the earth in the ionosphere and magnetosphere and the field of the sun immersed in the solar wind. Clearly, it is necessary to determine the effect of the magnetic field on the scattered spectrum. From a negative point of view, it enables us to establish the conditions for the neglect of magnetic field effects. More important of course is the possibility it gives us of obtaining information about the field.

The radiation is scattered by the plasma density fluctuations, and these are associated with the longitudinal modes in the plasma, through Poisson's equation (3.3.3). In the previous calculations where $\mathbf{B}_0 = 0$, we have in fact only considered these modes. The transverse modes enter through the Maxwell curl equations which we have neglected. This was a reasonable restriction for the unmagnetized, nonrelativistic plasma for which the longitudinal and transverse modes may be decoupled. In the presence of an external magnetic field, coupling occurs, and consequently, the dispersion relation for the longitudinal modes is modified. An example of when the full solution is required is given in Section 9.2.2. It is important to understand where the coupling of the modes enters the theory and to establish when the effects are negligible. This is discussed in Section 10.5.

10.1.1 Kinetic Equations

We will consider a plasma in equilibrium containing a magnetic field $B_0\hat{z}$. The coordinate system is shown in Fig. 10.1. The basic orbit of a given charge is

$$\mathbf{r}(t) = \mathbf{r}(0) - \rho_q \cos\varphi\hat{x} + \rho_q \sin\varphi\hat{y} + \mathrm{v}_\parallel t, \qquad (10.1.1)$$

where $\rho_q = \mathrm{v}_\perp / \Omega_q$ is the cyclotron radius and $\Omega_q = qB_0/m_q c$, $\varphi = \Omega_q t + \varphi(0)$. If we look in the direction of the magnetic field (the positive z axis), the electron rotation is clockwise, and the ion rotation is anticlockwise. Each charge suffers small perturbations to this orbit owing to interaction with other charges. The full picture of the behavior of the magnetized plasma comes from studying the kinetic equations. In the collisionless case, as in Eqns (2.2.3), (2.2.4), and Appendix B,

$$\frac{\partial F_{0q}}{\partial t} + \mathbf{v}\cdot\frac{\partial F_{0q}}{\partial \mathbf{r}} + \frac{q}{m}\left(\frac{\mathbf{v}}{c} \times \mathbf{B}_0\right)\cdot\frac{\partial F_{0q}}{\partial \mathbf{v}} = 0, \qquad (10.1.2)$$

[1] The calculations presented in this chapter are based on the papers by Bernstein (1958), Salpeter (1961a,b), Farley *et al.* (1961), and Dougherty & Farley (1963b).

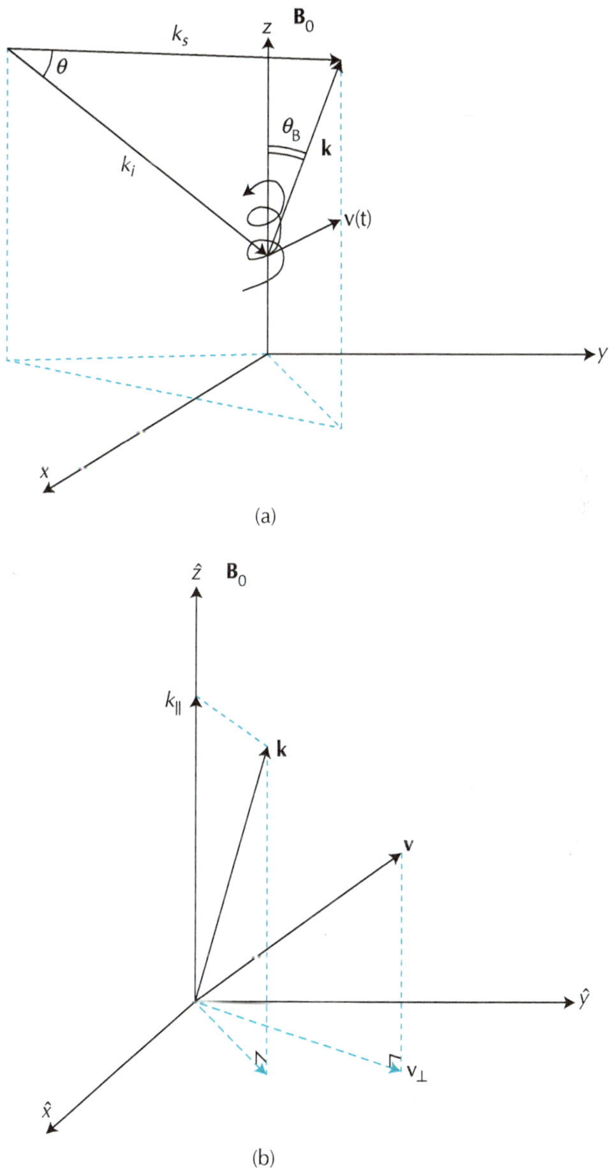

FIGURE 10.1 (a) Scattering coordinate system; (b) wave vector components.

$$\frac{\partial F_{1q}}{\partial t} + \mathbf{v} \cdot \frac{\partial F_{1q}}{\partial \mathbf{r}} + \frac{q}{m}\left(\frac{\mathbf{v}}{c} \times \mathbf{B}_0\right) \cdot \frac{\partial F_{1q}}{\partial \mathbf{v}} + \frac{q}{m}\mathbf{E}_1 \cdot \frac{\partial F_{0q}}{\partial \mathbf{v}} = 0, \tag{10.1.3}$$

where for a homogeneous plasma, $\partial F_{0q}/\partial \mathbf{r} = 0$, and for a stationary system, Eqn (10.1.2) then describes the basic helical orbits of the charges. We note that[2]

$$\frac{\mathbf{v}}{c} \times \mathbf{B}_0 \cdot \frac{\partial F_{1q}}{\partial \mathbf{v}} = -\frac{B_0}{c}\frac{\partial F_{1q}}{\partial \varphi}. \tag{10.1.4}$$

10.1.2 Useful Identities

There are a number of mathematical identities that may be usefully applied in calculations involving magnetized plasmas. To help the reader, they are assembled below.

A and S[3] $$e^{iz\sin\varphi} = \sum_{l=-\infty}^{+\infty} J_l(z) e^{il\varphi} \tag{10.1.5}$$

A and S $$J_{l-1}(z) + J_{l+1}(z) = (2l/z) \cdot J_l(z) \tag{10.1.6}$$

A and S $$\sum_{m=-\infty}^{+\infty} J_m^2(z) = 1 \tag{10.1.7}$$

A and S $$J_{-l}(z) = (-)^l J_l(z), \qquad I_{-l}(z) = I_l(z) \tag{10.1.8}$$

W[4] (p. 395) $$\int_0^\infty J_l^2(bt)\exp(-p^2t^2)\,t\,dt = \frac{1}{2p^2}\exp\left[-\left(\frac{b^2}{2p^2}\right)\right] I_l\left(\frac{b^2}{2p^2}\right) \tag{10.1.9}$$

---

[2]Equation (10.1.4) follows because

$$\left(\frac{\mathbf{v}}{c} \times \mathbf{B}_0\right) \cdot \frac{\partial F_{1q}}{\partial \mathbf{v}} = \left(\frac{v_y B_0}{c}\right)\frac{\partial F_{1q}}{\partial v_x} - \left(\frac{v_x B_0}{c}\right)\frac{\partial F_{1q}}{\partial v_y} + (0)\frac{\partial F_{1q}}{\partial \mathbf{v}_z}$$

$$\frac{\partial F_{1q}}{\partial V_x} = \frac{\partial v_\perp}{\partial v_x}\frac{\partial F_{1q}}{\partial v_\perp} + \frac{\partial \varphi}{\partial v_x}\frac{\partial F_{1q}}{\partial \varphi}$$

and

$$\frac{\partial v_\perp}{\partial v_x} = \frac{v_x}{v_\perp}, \quad \frac{\partial \varphi}{\partial v_x} = -\frac{v_y}{v_x^2}\cos^2\varphi, \quad \frac{\partial v_\perp}{\partial v_y} = \frac{v_y}{v_\perp}, \quad \frac{\partial v_\perp}{\partial v_y} = \frac{v_y}{v_\perp}, \quad \frac{\partial \varphi}{\partial v_y} = \frac{\cos^2\varphi}{v_x}$$

$$\frac{\partial F_{1q}}{\partial v_y} = \frac{\partial v_\perp}{\partial v_y}\frac{\partial F_{1q}}{\partial v_\perp} + \frac{\partial \varphi}{\partial v_y}\frac{\partial F_{1q}}{\partial \varphi}.$$

[3]See Abramowitz & Stegun (1965)

[4]See Watson (1958).

A and S For small argument, $I_l(z) \cong (z/2)/|l|!$ (10.1.10)

A and S For large argument, where $\mu = 4l^2$

$$I_l(z) \cong \frac{e^z}{(2\pi z)^{1/2}} \left\{1 - \left(\frac{\mu-1}{8z}\right) + \frac{(\mu-1)(\mu-9)}{2!(8z)^2} - \cdots\right\} \tag{10.1.11}$$

W (p. 358)
$$\sum_l J_l^2(z) e^{-il\Omega_e t} = J_0[2^{1/2} z(1-\cos\Omega_e t)^{1/2}] \tag{10.1.12}$$

A and S
$$\sum_l J_l(z) e^{il\varphi} \cos\varphi = \sum_l \frac{l}{z} J_l(z) e^{il\varphi}$$
$$\sum_l J_l(z) e^{il\varphi} \sin\varphi = \sum_l \frac{\partial J_l(z)}{dz} e^{il\varphi} \tag{10.1.13}$$

A and S
$$\int_0^\infty \mathrm{v}_\perp d\mathrm{v}_\perp \exp[-(\mathrm{v}_\perp^2/a^2)] J_0(k_\perp \mathrm{v}_\perp t) = (a^2/2)\exp[-(a^2 k_\perp^2 t^2/4)] \tag{10.1.14}$$

A and S
$$\sum_{l=-\infty}^{+\infty} I_l(z) = e^z \tag{10.1.15}$$

A and S
$$\sum_{m=1}^{\infty} m^2 I_m(z) e^{-z} = z/2 \tag{10.1.16}$$

obtained using the recurrence relation

$$I_{m-1}(z) - I_{m+1}(z) = \frac{2m}{z} I_m(z).$$

10.2 CALCULATION OF THE SPECTRAL DENSITY FUNCTION $S(\mathbf{k},\omega)$

10.2.1 Calculation of Fluctuating Density

The spectral density function for a magnetized plasma may be evaluated in the same way as that used in Chapter 3, but with one proviso, namely that we must specify carefully the ordering of the characteristic lengths and times of the system. Put simply the charges must be seen to be gyrating if we are to detect the effects of their gyrations.

The calculation immediately below is for the case where both the electrons and ions are magnetized. This requires that $\rho_e, \rho_i \ll L$, the dimension of the scattering volume, and that $2\pi/\Omega_e, 2\pi/\Omega_i \ll \tau_c, \tau_I$, where $\tau_c = \lambda_i^2/2c\Delta\lambda_i$ is the coherence time of the source and τ_I is the integration time of the detector. The conditions are of course hardest to meet for the ions. Consider the situation for protons. The mean proton perpendicular speed is $\mathrm{v}_\perp \cong (2\kappa T_i/m_i)^{1/2}$, and from

Eqns (1.4.6) and (1.4.7), the first condition requires

$$\frac{1.4\times 10^2\,[T_\mathrm{i}(\mathrm{eV})]^{1/2}}{B_0(G)\ll L}\mathrm{cm}. \tag{10.2.1}$$

This is attainable in some laboratory experiments, e.g., $T_\mathrm{i}\lesssim 100\,\mathrm{eV}, B_0\lesssim 10\,\mathrm{kG}, L=1\,\mathrm{cm}$.

The time condition may be written as

$$\frac{\Delta\lambda_\mathrm{i}/\lambda_\mathrm{i}\ll\lambda_\mathrm{i}\,(\mathrm{cm})B_0(\mathrm{G})}{2\times 10^6}. \tag{10.2.2}$$

This is harder to meet, e.g., with a ruby laser source $\lambda_\mathrm{i}=6.943\times 10^{-5}\,\mathrm{cm}$ and $B_0=10\,\mathrm{kG}$, and we require $\Delta\lambda_\mathrm{i}\ll 2.4\times 10^{-3}$ Å. As we have seen in Chapter 5, the ion-acoustic feature is in many cases only observable at longer wavelengths or higher densities where we can satisfy the condition $\alpha=1/k\lambda_\mathrm{De}>[ZT_\mathrm{e}/3T_\mathrm{i}-1]^{-1/2}$, and at these longer wavelengths, it is easier to satisfy Eqn (10.2.2). The complete magnetized calculation is therefore worthwhile. We treat the case of unmagnetized ions but magnetized electrons at the end of this section.

The fluctuating density is given by Eqn (3.3.1):

$$n_{1q}(\mathbf{k},\omega)=\sum_\mathrm{v} d\mathbf{v}F_{1q}(\mathbf{k},\mathbf{v},\omega).$$

The first step in our calculation is therefore to obtain $F_{1q}(\mathbf{k},\mathbf{v},\omega)$ by taking the Fourier–Laplace transform of Eqn (10.1.3). We find

$$-F_{1q}(\mathbf{k},\mathbf{v},0)+[i\omega+\lambda-i\mathbf{k}\cdot\mathbf{v}]F_{1q}(\mathbf{k},\mathbf{v},\omega)$$
$$-\frac{qB_0}{m_qc}\frac{\partial F_{1q}(\mathbf{k},\mathbf{v},\omega)}{\partial\varphi}+\frac{q}{m_q}\mathbf{E}_1(\mathbf{k},\omega)\cdot\frac{\partial F_{0q}}{\partial\mathbf{v}}=0. \tag{10.2.3}$$

Now,

$$\mathbf{k}\cdot\mathbf{v}=k_\perp\mathrm{v}_\perp\cos(\phi-\delta)+k_\parallel\mathrm{v}_\parallel,$$

where

$$k_\perp=\left[(\mathbf{k}\cdot\hat{x})^2+(\mathbf{k}\cdot\hat{y})^2\right]\quad\text{and}\quad\tan\delta=\frac{\mathbf{k}\cdot\hat{x}}{\mathbf{k}\cdot\hat{y}}. \tag{10.2.4}$$

With the aid of Eqn (10.1.5), we may rearrange Eqn (10.2.3) to obtain

$$F_{1q}(\mathbf{k},\mathbf{v},\omega)=-\int^{\varphi}d\varphi'\exp\left[-(i\omega+\gamma-ik_\parallel\mathrm{v}_\parallel)\varphi'/\Omega_q\right]\sum_{l=-\infty}^{+\infty}J_l(k_\perp\rho_q)e^{il\varphi'}$$
$$\times\left[F_{1q}(\mathbf{k},\mathbf{v},0)-(q/m_q)\mathbf{E}_1(\mathbf{k},\omega)\cdot\partial F_{0q}/\partial\mathbf{v}\right]$$
$$\div\,\Omega_q\exp\left\{\left[-(i\omega+\gamma-ik_\parallel\mathrm{v}_\parallel)\varphi+ik_\perp\rho_q\sin\varphi\right]/\Omega_q\right\}. \tag{10.2.5}$$

We will only consider situations in which F_{0q} is independent of φ, and then,

$$\mathbf{k}\cdot\frac{\partial F_{0q}}{\partial\mathbf{v}}=k_\parallel\frac{\partial F_{0q}}{\partial\mathrm{v}_\parallel}+k_\perp\cos\varphi\frac{\partial F_{0q}}{\partial\mathrm{v}_\perp}.$$

We use the identity (10.1.6) and define

$$\mathbf{k}\cdot\frac{\partial F_{0q}}{\partial \mathbf{v}^*}\equiv k_{\parallel}\frac{\partial F_{0q}}{\partial \mathrm{v}_{\parallel}}+\frac{l}{\rho_q}\frac{\partial F_{0q}}{\partial \mathrm{v}_{\perp}}. \qquad (10.2.6)$$

The integral over φ' is completed, and we obtain

$$F_{1q}(\mathbf{k},\mathbf{v},\omega)=-\sum_{l=-\infty}^{+\infty}\sum_{m=-\infty}^{+\infty}\left\{-F_{1q}(\mathbf{k},\mathbf{v},0)-(iq/m_q)\mathbf{E}_1(\mathbf{k},\omega)\cdot\partial F_{0q}/\partial\mathbf{v}^*\right\}$$
$$\times J_l(k_{\perp}\rho_q)J_m(k_{\perp}\rho_q)e^{i(l-m)\varphi}/(\omega-k_{\parallel}\mathrm{v}_{\parallel}-l\Omega_q-i\gamma). \qquad (10.2.7)$$

(The phase factor is not explicitly shown for this low-temperature case.)

10.2.2 Longitudinal Approximation

It is at this point that we make the longitudinal approximation and use only Poisson's equation to determine the electric field. From Eqns (3.3.3) and (3.3.6), we find

$$\mathbf{E}_1(\mathbf{k},\omega)=\frac{4\pi i}{k^2}\mathbf{k}\rho_1(\mathbf{k},\omega), \qquad (10.2.8)$$

where $\rho_1(\mathbf{k},\omega)=Zen_{1\mathrm{i}}(\mathbf{k},\omega)-en_{1\mathrm{e}}(\mathbf{k},\omega)$ is the Fourier–Laplace transform of the fluctuating charge density. Substituting in Eqn (10.2.7) leads to

$$F_{1q}(\mathbf{k},\mathbf{v},\omega)=\sum_l\sum_m\left\{-iF_{1q}(\mathbf{k},\mathbf{v},0)-(4\pi q/m_qk^2)\rho_1(\mathbf{k},\omega)\mathbf{k}\cdot\partial F_{0q}/\partial\mathbf{v}^*\right\}$$
$$\times J_l(k_{\perp}\rho_q)J_m(k_{\perp}\rho_q)e^{i(l-m)\phi}/(\omega-k_{\parallel}\mathrm{v}_{\parallel}-l\Omega_q-i\gamma), \qquad (10.2.9)$$

where for electrons,

$$q/m_q=-e/m_\mathrm{e},\quad \Omega_\mathrm{e}=-eB_0/m_\mathrm{e}c,\quad \rho_q=\rho_\mathrm{e},\quad F_{0\mathrm{e}}=n_{\mathrm{e}0}f_{0\mathrm{e}};$$

and for ions,

$$q/m_q=Ze/m_\mathrm{i},\quad \Omega_\mathrm{i}=ZeB_0/m_\mathrm{i}c,\quad \rho_q=\rho_\mathrm{i},\quad F_{0\mathrm{i}}=(n_{\mathrm{e}0}/Z)f_{0\mathrm{i}}.$$

We make these substitutions in Eqn (10.2.9) and sum the two resultant equations over all velocities to obtain expressions for $n_{1\mathrm{e}}$ and $n_{1\mathrm{i}}$. We eliminate $n_{1\mathrm{i}}$ in a manner analogous to that used in Section 3.3 and obtain

$$n_{\mathrm{ie}}(\mathbf{k},\omega)$$
$$=-i\left[\left(1-\frac{H_\mathrm{e}}{\varepsilon_\mathrm{L}}\right)\sum_{j=1}^{\mathrm{N}}e^{i\mathbf{k}_0\cdot\mathbf{r}_j(0)}\sum_{l,m}\frac{J_l(k_{\perp}\rho_e(0))J_m(k_{\perp}\rho_e(0))}{\omega-k_{\parallel}\mathrm{v}_{\parallel j}(0)-l\Omega_\mathrm{e}-i\gamma}e^{i(l-m)\phi(0)}\right.$$
$$\left.+\frac{ZH_\mathrm{e}}{\varepsilon_\mathrm{L}}\sum_{h=1}^{\mathrm{N/Z}}e^{i\mathbf{k}\cdot\mathbf{r}_h(0)}\sum_{l,m}\frac{J_l(k_{\perp}\rho_i(0))J_m(k_{\perp}\rho_i(0))}{\omega-k_{\parallel}\mathrm{v}_{\parallel h}(0)-l\Omega_i-i\gamma}e^{i(l-m)\phi(0)}\right], \qquad (10.2.10)$$

where the longitudinal dielectric function is

$$\varepsilon_{\mathrm{L}}(\mathbf{k},\omega)=1+H_{\mathrm{e}}(\mathbf{k},\omega)+H_{\mathrm{i}}(\mathbf{k},\omega). \tag{10.2.11}$$

Now, $d\mathbf{v}=\mathrm{v}_\perp d\mathrm{v}_\perp d\mathrm{v}_\parallel d\phi$ and integration over ϕ in $H_{\substack{\mathrm{e}\\ \mathrm{i}}}$ gives

$$\sum_{l,m}\int_0^{2\pi} d\varphi e^{i(l-m)\phi}=\underset{l=m}{2\pi}+\underset{l\neq m}{0}. \tag{10.2.12}$$

Therefore,

$$H_{\mathrm{e}}(\mathbf{k},\omega)=\int_{-\infty}^{+\infty} d\mathbf{v}\frac{4\pi e^2 n_0}{m_{\mathrm{e}}k^2}\sum_l \frac{J_l^2(k_\perp\rho_{\mathrm{e}})\mathbf{k}\cdot\partial f_{0\mathrm{e}}/\partial\mathbf{v}^*}{\omega-k_\parallel \mathrm{v}_\parallel-l\Omega_{\mathrm{e}}-i\gamma}$$

$$H_{\mathrm{i}}(\mathbf{k},\omega)=\int_{-\infty}^{+\infty} d\mathbf{v}\frac{4\pi Ze^2 n_0}{m_{\mathrm{i}}k^2}\sum_m \frac{J_m^2(k_\perp\rho_{\mathrm{i}})\mathbf{k}\cdot\partial f_{0\mathrm{i}}/\partial\mathbf{v}^*}{\omega-k_\parallel \mathrm{v}_\parallel-m\Omega_{\mathrm{i}}-i\gamma}. \tag{10.2.13}$$

The procedure of Section 3.4 is followed, and we obtain, with the aid of Eqn (10.1.7),

$$S(\mathbf{k},\omega)=2\lim_{\gamma\to 0}\gamma\left|1-\frac{H_{\mathrm{e}}}{\varepsilon_{\mathrm{L}}}\right|^2\int_{-\infty}^{+\infty}\frac{d\mathbf{v}\sum_l J_l^2(k_\perp\rho_{\mathrm{e}})f_{0\mathrm{e}}(\mathbf{v})}{(\omega-k_\parallel \mathrm{v}_\parallel-l\Omega_{\mathrm{e}})^2+\gamma^2}$$
$$+Z\left|\frac{H_{\mathrm{e}}}{\varepsilon_{\mathrm{L}}}\right|^2\int_{-\infty}^{+\infty}\frac{d\mathbf{v}\sum_m J_m^2(k_\perp\rho_{\mathrm{i}})f_{0\mathrm{i}}(\mathbf{v})}{(\omega-k_\parallel \mathrm{v}_\parallel-m\Omega_{\mathrm{i}})^2+\gamma^2}. \tag{10.2.14}$$

10.2.3 Unmagnetized Ions

The form in which $\varepsilon(k,\omega)$ [Eqn (10.2.11)] and $S(\mathbf{k},\omega)$ [Eqn (10.2.14)] are written brings out the independent nature of the way in which the electrons collectively shield, respectively, other electrons and ions. For the case of unmagnetized ions, we simply replace $H_{\mathrm{i}}(\mathbf{k},\omega)$ by $\chi_{\mathrm{i}}(\mathbf{k},\omega)$ from Eqn (3.3.10)

$$\chi_{\mathrm{i}}(\mathbf{k},\omega)=\int_{-\infty}^{+\infty} d\mathbf{v}\frac{4\pi Ze^2 n_0}{m_{\mathrm{i}}k^2}\frac{\mathbf{k}\cdot\partial f_{0\mathrm{i}}/\partial\mathbf{v}}{(\omega-\mathbf{k}\cdot\mathbf{v}-i\gamma)} \tag{10.2.15}$$

and replace the latter part of Eqn (10.2.14) with the corresponding part of Eqn (3.4.3)

$$S(\mathbf{k},\omega)$$
$$=2\lim_{\gamma\to 0}\gamma\left|1-\frac{H_{\mathrm{e}}}{\varepsilon_{\mathrm{L0}}}\right|^2\int_{-\infty}^{+\infty}\frac{d\mathbf{v}\sum_l J_l^2(k_\perp\rho_{\mathrm{e}})f_{0\mathrm{e}}(\mathbf{v})}{(\omega-k_\parallel \mathrm{v}_\parallel-l\Omega_{\mathrm{e}})^2+\gamma^2}+\frac{2\pi Z}{k}\left|\frac{H_{\mathrm{e}}}{\varepsilon_{\mathrm{L0}}}\right|^2 f_{\mathrm{i0}}\left(\frac{\omega}{k}\right), \tag{10.2.16}$$

where

$$\varepsilon_{L0}=1+H_e+\chi_i. \tag{10.2.17}$$

In fact, in the second term, the dependence on $f_{i0}(\omega/k)$ means that the term is important primarily at low frequencies for which H_e is relatively independent of the magnetic field.

10.3 $S(\mathbf{k},\omega)$, MAXWELLIAN DISTRIBUTION FUNCTIONS

The Maxwellian distributions are

$$f_{e0}=\exp(-\mathrm{v}^2/a^2)/(\pi a^2)^{3/2}, \quad f_{i0}=\exp(-\mathrm{v}^2/b^2)/(\pi b^2)^{3/2}, \tag{10.3.1}$$

where the mean thermal speeds are

$$a=(2\kappa T_e/m_e)^{1/2}, \qquad b=(2\kappa T_e/m_i)^{1/2}. \tag{10.3.2}$$

These are substituted in Eqn (10.2.14) with $\mathrm{v}^2=\mathrm{v}_\perp^2+\mathrm{v}_{||}^2$, and the integral in the numerator is evaluated with the aid of Eqn (10.1.9)

$$\lim_{\gamma\to 0}\gamma\int_0^{2\pi}\frac{d\varphi}{(\pi a^2)^{3/2}}\int_0^{\infty}\mathrm{v}_\perp d\mathrm{v}_\perp\sum_l J_l^2(k_\perp\rho_e)\exp\left(\frac{-\mathrm{v}^2}{a^2}\right)$$

$$\int_{-\infty}^{+\infty}\frac{d\mathrm{v}_{||}\exp(-\mathrm{v}_{||}^2/a^2)}{(\omega-k_{||}\mathrm{v}_{||}-l\Omega_e)^2+\gamma^2}$$

$$=\pi^{1/2}\sum_{l-\infty}^{+\infty}\exp(-k_\perp^2\bar{\rho}_e^2)I_l(k_\perp^2\bar{\rho}_e^2)\cdot\frac{\exp\{-[(\omega-l\Omega_e)/k_{||}a]^2\}}{k_{||}a}, \tag{10.3.3}$$

where

$$\rho_e=\mathrm{v}_\perp/\Omega_e, \quad \bar{\rho}_e=a/2^{1/2}\Omega_e, \quad \bar{\rho}_i=b/2^{1/2}\Omega_i. \tag{10.3.4}$$

Similarly, with the aid of Eqns (10.1.7) and (10.1.9),

$H_e(\mathbf{k},\omega)$

$$=\lim_{\gamma\to 0}\int_0^{2\pi}d\varphi\int_0^{\infty}\mathrm{v}_\perp d\mathrm{v}_\perp\int_{-\infty}^{+\infty}d\mathrm{v}_{||}\frac{4\pi e^2 n}{m_e k^2}\sum_l\frac{J_l^2(k_\perp\rho_e)\left(k_{||}\frac{\partial f_{0e}}{\partial \mathrm{v}_{||}}+\frac{l\Omega_e}{\mathrm{v}_\perp}\frac{\partial f_{0e}}{\partial \mathrm{v}_\perp}\right)}{\omega-k_{||}\mathrm{v}_{||}-l\Omega_e-i\gamma}$$

$$=\lim_{\gamma\to 0}\alpha^2\left[1-\sum_{l-\infty}^{+\infty}\exp(-k_\perp^2\bar{\rho}_e^2)I_l(k_\perp^2\bar{\rho}_e^2)\right.$$

$$\left.\times\int_{-\infty}^{+\infty}\frac{d\mathrm{v}_{||}\omega\exp(-\mathrm{v}_{||}^2/a^2)}{(\pi a^2)^{1/2}(\omega-k_{||}\mathrm{v}_{||}-l\Omega_e+i\gamma)}\right]$$

$$(k_{||}=0)=\alpha^2\left[1-\sum_l\exp(-k_\perp^2\bar{\rho}_e^2)I_l(k_\perp^2\bar{\rho}_e^2)\left(\frac{\omega}{\omega-l\Omega_e}\right)\right] \tag{10.3.5}$$

$$(k_{\|} \neq 0) = \alpha^2 \left[1 - \sum_l \exp(-k_\perp^2 \bar{\rho}_e^2) I_l(k_\perp^2 \bar{\rho}_e^2) \left(\frac{\omega}{\omega - l\Omega_e} \right) \times \left\{ 2x_{el} \exp\left(-x_{el}^2\right) \int_0^{x_{el}} \exp(p^2)\, dp + i\pi^{1/2} x_{el} \exp\left(-x_{el}^2\right) \right\} \right], \tag{10.3.6}$$

where

$$x_{el} = (\omega - l\Omega_e)/k_{\|}a, \qquad \alpha = 1/k\lambda_{De}.$$

The function in braces is essentially the plasma dispersion function discussed in Section 5.2 [see Eqn (5.2.7) and (5.2.8)], and the imaginary part is the Landau damping term.

$$H_i(\mathbf{k}, \omega) = \frac{ZT_e}{T_i} \alpha^2 \left[1 - \sum_m \exp(-k_\perp^2 \bar{\rho}_i^2) I_m(k_\perp^2 \bar{\rho}_i^2) \left(\frac{\omega}{\omega - m\Omega_i} \right) \times \left\{ \begin{array}{ll} 1 & \text{for} \quad k_{\|} = 0 \\ \left\{ 2x_{im} \exp\left(-x_{im}^2\right) \int_0^{x_{im}} \exp(p^2) dp + i\pi^{1/2} x_{im} \exp\left(-x_{im}^2\right) \right\} & \text{for} \quad k_{\|} \neq 0 \end{array} \right\} \right], \tag{10.3.7}$$

where

$$x_{im} = (\omega - m\Omega_i)/k_{\|}b, \qquad \lambda_{De} = (\kappa T_e / 4\pi e^2 n_e)^{1/2} \text{ (Gaussian units)}. \tag{10.3.8}$$

The spectral density function is

$$S(\mathbf{k}, \omega) = 2\pi^{1/2} \left| 1 - \frac{H_e}{\varepsilon_L} \right|^2 \sum_{l=-\infty}^{+\infty} \exp(-k_\perp^2 \bar{\rho}_e^2) I_l(k_\perp^2 \bar{\rho}_e^2) \frac{\exp\{-[(\omega - l\Omega_e)/k_{\|}a]^2\}}{k_{\|}a} + 2\pi^{1/2} Z \left| \frac{H_e}{\varepsilon_L} \right|^2 \sum_{m=-\infty}^{+\infty} \exp(-k_\perp^2 \bar{\rho}_i^2) I_m(k_\perp^2 \bar{\rho}_i^2) \times \frac{\{-[(\omega - m\Omega_i)/k_{\|}b]^2\}}{k_{\|}b} \tag{10.3.9}$$

with $\varepsilon_L = 1 + H_e + H_i$.

10.3.1 Unmagnetized Ions

$$\chi_i = Z \frac{T_e}{T_i} \alpha^2 \left[1 - 2x_i \exp\left(-x_i^2\right) \int_0^{x_i} \exp(p^2) dp - i\pi^{1/2} x_i \exp\left(-x_i^2\right) \right], \tag{10.3.10}$$

where $x_{\rm i}=\omega/kb$. Substituting in Eqn (10.2.16) yields

$$S(\mathbf{k},\omega)=2\pi^{1/2}\left|1-\frac{H_{\rm e}}{\varepsilon_{\rm L0}}\right|^2\sum_l \exp\left(-k_\perp^2\bar{\rho}_{\rm e}^2\right)I_l\left(k_\perp^2\bar{\rho}_{\rm e}^2\right)\exp\frac{\{-[(\omega-l\Omega_{\rm e})/k_{||}a]^2\}}{k_{||}a}$$
$$+2\pi^{1/2}Z\left|\frac{H_{\rm e}}{\varepsilon_{\rm L0}}\right|^2\frac{\exp(-{\rm v}^2/b^2)}{kb}, \tag{10.3.11}$$

where $\varepsilon_{\rm L0}=1+H_{\rm e}+\chi_{\rm i}$.

10.4 COLLISIONAL MAGNETIZED PLASMA

10.4.1 Arbitrary Distribution Functions

The calculation presented below follows the pattern of the calculation for $B=0$ in Section 3.7; therefore, only the main steps are shown. The collision term $-\nu_q\left(F_{1q}-n_{1q}f_{0q}(\mathbf{v})\right)$ is added to the right-hand side of the kinetic equation (10.1.2). The longitudinal approximation is used, and the Fourier–Laplace transform of the equation is then

$$-F_{1q}(\mathbf{k},\mathbf{v},0)+[i\omega+\nu_q-i\mathbf{k}\cdot\mathbf{v}]F_{1q}(\mathbf{k},\mathbf{v},\omega)-\frac{qB_0}{m_q c}\frac{\partial F_{1q}(\mathbf{k},\mathbf{v},\omega)}{\partial\varphi}$$
$$+\frac{i4\pi q}{k^2 m_q}\rho_1(\mathbf{k},\omega)\mathbf{k}\cdot\frac{\partial F_{0q}}{\partial\mathbf{v}}=\nu_q n_{1q}(\mathbf{k},\omega)f_{0q}-n_{1q}(\mathbf{k},0)f_{0q}. \tag{10.4.1}$$

The initial value of the collision term has been added [see Eqn (3.7.5)]. The equation is rearranged and integrated over φ.

$$F_{1q}(\mathbf{k},\mathbf{v},\omega)=-\sum_l\sum_m\left[\frac{\begin{array}{r}\left\{iF_{1q}(\mathbf{k},\mathbf{v},0)+\frac{4\pi}{k^2}\frac{q}{m}\rho_1(\mathbf{k},\omega)\,\mathbf{k}\cdot\frac{\partial F_{0q}}{\partial\mathbf{v}^*}\right.\\ \left.+i\nu_q n_{1q}(\mathbf{k},\omega)F_{0q}-in_{1q}(\mathbf{k},0)F_{0q}\right\}\end{array}}{(\omega-k_{||}{\rm v}_{||}-l\Omega_q-i\nu_q)}\right]$$
$$\times J_l(k_\perp\rho_q)J_m(k_\perp\rho_q)e^{i(l-m)\varphi}. \tag{10.4.2}$$

The electron and ion versions of this equation are summed over velocity and

$$n_{1\rm e}(\mathbf{k},\omega)(1+U_{\rm e})$$
$$=-i\sum_{j=1}^N e^{i(\mathbf{k}\cdot\mathbf{r}_j(0))}\left\{\sum_{l,m}\frac{J_l(k_\perp\rho_{\rm e}(0))J_m(k_\perp\rho_{\rm e}(0))}{\omega-k_{||}{\rm v}_{||j}(0)-l\Omega_{\rm e}-i\nu_{\rm e}}e^{i(l-m)\varphi(0)}-\frac{U_{\rm e}}{i\nu_{\rm e}}\right\}$$
$$+\frac{H_{\rm e}}{e}\rho_1(\mathbf{k},\omega), \tag{10.4.3}$$

$$n_{1\mathrm{i}}(\mathbf{k},\omega)(1+U_{\mathrm{i}}) = -i\sum_{h=1}^{N/Z} e^{i(\mathbf{k}\cdot\mathbf{r}_h(0))}\left\{\sum_{l,m}\frac{J_l(k_\perp\rho_{\mathrm{i}}(0))J_m(k_\perp\rho_{\mathrm{i}}(0))}{\omega-k_{||}v_{||h}(0)-l\Omega_{\mathrm{i}}-i\nu_{\mathrm{i}}}e^{i(l-m)\varphi(0)}-\frac{U_{\mathrm{i}}}{i\nu_{\mathrm{i}}}\right\} - \frac{H_{\mathrm{i}}}{Ze}\rho_{\mathrm{i}}(\mathbf{k},\omega), \tag{10.4.4}$$

where

$$U_q = i\nu_q\sum_l\int\frac{d\mathbf{v}\,J_l^2(k_\perp\rho_q)f_{0q}}{\omega-k_{||}v_{||}-l\Omega_q-i\nu_q}. \tag{10.4.5}$$

Between these two equations $n_{1\mathrm{i}}$ is eliminated, and we obtain $n_{1\mathrm{e}}(\mathbf{k},\omega)$ and then

$$S(\mathbf{k},\omega)=\frac{2\nu_q}{V}\frac{\left\langle|n_{1\mathrm{e}}(\mathbf{k},\omega)|^2\right\rangle}{n_0},$$

$$S(\mathbf{k},\omega)=2\left|1-\frac{L_{\mathrm{e}}}{\varepsilon_{\mathrm{L}}}\right|^2 M_{\mathrm{e}}+2Z\left|\frac{L_{\mathrm{e}}}{\varepsilon_{\mathrm{L}}}\right|^2 M_{\mathrm{i}}, \tag{10.4.6}$$

where

$$M_q=\frac{\nu_q}{\left|1+U_q\right|^2}\sum_l\int\frac{d\mathbf{v}f_{0q}(\mathbf{v})J_l^2(k_\perp\rho_q)}{\left(\omega-k_{||}v_{||}-l\Omega_q\right)^2+\nu_q^2}-\frac{\left|U_q\right|^2}{\nu_q^2} \tag{10.4.7}$$

and

$$\varepsilon_{\mathrm{L}}=1+\frac{H_{\mathrm{e}(\nu_{\mathrm{e}})}}{1+U_{\mathrm{e}}}+\frac{H_{\mathrm{i}}(\nu_{\mathrm{i}})}{1+U_{\mathrm{i}}}=1+L_{\mathrm{e}}+L_{\mathrm{i}}. \tag{10.4.8}$$

10.4.2 Maxwellian Distribution Functions

We may follow the procedures used in Sections 3.7.2 and 10.3 to obtain

$$U_{\mathrm{e}}=i\sum_l\exp\left(-k_\perp^2\bar\rho_{\mathrm{e}}^2\right)I_l\left(k_\perp^2\bar\rho_{\mathrm{e}}^2\right)\frac{\nu_{\mathrm{e}}}{(\omega-l\Omega_{\mathrm{e}}-i\nu_{\mathrm{e}})} \times\left\{2y_{el}\exp\left(-y_{el}^2\right)\int_0^{y_{el}}\exp(p^2)\,dp+i\pi^{1/2}y_{el}\exp\left(-y_{el}^2\right)\right\}. \tag{10.4.9}$$

$$L_{\mathrm{e}}=\frac{\alpha^2}{(1+U_{\mathrm{e}})}\left[1-\sum_l\exp\left(-k_\perp^2\bar\rho_{\mathrm{e}}^2\right)I_l\left(k_\perp^2\bar\rho_{\mathrm{e}}^2\right)\frac{(\omega-i\nu_{\mathrm{e}})}{(\omega-l\Omega_{\mathrm{e}}-i\nu_{\mathrm{e}})} \times\left\{2y_{el}\exp\left(-y_{el}^2\right)\int_0^{y_{el}}\exp(p^2)\,dp+i\pi^{1/2}y_{el}\exp\left(-y_{el}^2\right)\right\}\right] \tag{10.4.10}$$

$$M_e = \sum_l \frac{\exp(-k_\perp^2 \bar{\rho}_e^2) I_l(-k_\perp^2 \bar{\rho}_e^2)}{|1+U_e|^2}$$
$$\times \operatorname{Im}\left\{ \frac{2y_{el}\exp(-y_{el}^2)\int_0^{y_{el}} \exp(p^2)\,dp + i\pi^{1/2} y_{el}\exp(-y_{el}^2)}{\omega - l\Omega_e - i\nu_e} \right\}$$
$$- \frac{|U_e|^2}{\nu_e |1+U_e|^2}, \tag{10.4.11}$$

where

$$y_{el} = (\omega - l\Omega_e - i\nu_e)/k_{||}a.$$

There are similar expressions for U_i, L_i, and M_i.

We do not need to differentiate between the cases $k_{||} = 0$ and $k_{||} \neq 0$ because the collisions provide damping in all situations. For small ν_e and ν_i, these expressions may be expanded using the expansions of the unmagnetized case (Section 3.7.2).

10.5 TRANSVERSE MODES

The complete calculation of the spectrum of waves in a magnetized plasma is too lengthy to present here; therefore, we will only look at the dispersion relation for the waves. This helps us to find the situations in which coupling between transverse and longitudinal waves will be important. Sitenko (1967) gives an extensive review of the topic. The case of the lower hybrid resonance is discussed in Section 9.2.2. We first combine Maxwell's equations (1.3.4) to eliminate **B** and obtain

$$\nabla \times (\nabla \times \mathbf{E}) + \frac{1}{c^2}\frac{\partial^2 \mathbf{E}}{\partial t^2} = -\frac{4\pi}{c^2}\frac{\partial \mathbf{J}}{\partial t}, \tag{10.5.1}$$

where for our case, the current density is

$$\mathbf{J} = \sum_q q \int F_{1q} \mathbf{v}\, d\mathbf{v}. \tag{10.5.2}$$

We look for solutions of the form $\exp[i(\mathbf{k}\cdot\mathbf{r} - \omega t)]$. Substituting and rearranging leads to

$$\mathbf{E} - \frac{k^2 c^2}{\omega^2}[\mathbf{E} - (\hat{k}\cdot\mathbf{E})\hat{k}] = -i\frac{4\pi}{\omega}\sum_q q \int F_{1q}(\mathbf{k}, \mathbf{v}, \omega)\mathbf{v}\, d\mathbf{v}. \tag{10.5.3}$$

Now, $F_{1q}(\mathbf{k}, \mathbf{v}, \omega)$ is given by Eqn (10.2.5). We substitute again neglecting the initial conditions (these are not needed for this calculation). The component equations in the $\hat{x}$, $\hat{y}$, and $\hat{z}$ directions are then obtained, and finally, we arrive at the dispersion relation by equating the determinant of the three equations, with the three unknowns E_x, E_y, E_z, to zero.

For convenience, we take $\hat{k}$ in the xz plane, $\mathbf{B}_0 = B_0\hat{z}$. We assume that F_{0q} is a function only of the magnitude of $|\mathbf{v}|$ in which case

$$\frac{\partial F_{0q}}{\partial \mathrm{v_i}} = \frac{\partial \mathrm{v}}{\partial \mathrm{v_i}}\frac{\partial F_{0q}}{\partial \mathrm{v}} = \frac{\mathrm{v}_\perp}{\mathrm{v}}\frac{\partial F_{0q}}{\partial \mathrm{v}},$$

and then,

$$\mathbf{E}_1 \cdot \frac{\partial F_{0q}}{\partial \mathbf{v}} = \frac{E_{1x}\mathrm{v}_x + E_{1y}\mathrm{v}_y + E_{1z}\mathrm{v}_z}{\mathrm{v}}\left(\frac{\partial F_{0q}}{\partial \mathrm{v}}\right).$$

In addition, we may use the identities (10.1.13).

The dispersion relation is

$$\left[1 - \left(\frac{\omega}{ck}\right)^2 \varepsilon_{yy}\right]\left\{-\left(\frac{\omega}{ck}\right)^2 \underline{\underline{\left[\eta_\perp^2\,\varepsilon_{xx} + 2\eta_\parallel\,\eta_\perp\,\varepsilon_{xz} + \eta_\parallel^2\,\varepsilon_{zz}\right]}}\right.$$
$$\left. + \left(\frac{\omega}{ck}\right)^4 \varepsilon_{xx}\,\varepsilon_{zz} - \left(\frac{\omega}{ck}\right)^4 \varepsilon_{xz}^2\right\} + \left(\frac{\omega}{ck}\right)^4 \left[\eta_\perp \varepsilon_{xy} - \eta_{||}\varepsilon_{yz}\right]^2$$
$$- \left(\frac{\omega}{ck}\right)^6 \left[\varepsilon_{zz}\varepsilon_{xy}^2 + \varepsilon_{xx}\varepsilon_{yz}^2 + 2\varepsilon_{xy}\varepsilon_{yz}\varepsilon_{xz}\right] = 0, \qquad (10.5.4)$$

where

$$\eta_{||} = \hat{k}\cdot\hat{z}, \quad \eta_\perp = \hat{k}\cdot\hat{x}.$$

The electrostatic dispersion relation is underlined. The results are written in the form used by Callen & Guest (1973), who have discussed this coupling in relation to instabilities in mirror-confined plasmas. The dielectric function is given by

$$\boldsymbol{\varepsilon} = \delta_{ij} - \sum_q \frac{\omega_{pq}^2}{\omega^2}\delta_{ij} - \sum_l \int d\mathbf{v} \frac{\left(k_{||}\frac{\partial f_{0q}}{\partial \mathrm{v}_{||}} + l\frac{\Omega_q}{\mathrm{v}_\perp}\frac{\partial f_{0q}}{\partial \mathrm{v}_\perp}\right)}{\omega - k_{||}\mathrm{v}_{||} - l\Omega_q}\boldsymbol{\pi}, \qquad (10.5.5)$$

where

$$\boldsymbol{\pi} = \begin{vmatrix} \left(\frac{l\Omega}{k_\perp}\right)^2 J_n^2 & i\mathrm{v}_\perp\left(\frac{l\Omega}{k_\perp}\right)J_lJ_l' & \mathrm{v}_{||}\left(\frac{l\Omega}{k_\perp}\right)J_l^2 \\ -i\mathrm{v}_\perp\left(\frac{l\Omega}{k_\perp}\right)J_lJ_l' & \mathrm{v}_\perp^2(J_l')^2 & -i\mathrm{v}_{||}\mathrm{v}_\perp J_lJ_l' \\ \mathrm{v}_{||}\left(\frac{l\Omega}{k_\perp}\right)J_l^2 & i\mathrm{v}_{||}\mathrm{v}_\perp J_lJ_l' & \mathrm{v}_{||}^2J_l^2 \end{vmatrix} \qquad (10.5.6)$$

It can be seen that the criterion for the neglect of coupling of transverse to longitudinal modes is that for all ε_{ij} we have

$$(\omega^2/k^2)\varepsilon_{ij} \ll c^2.$$

For application to scattering, it is convenient to rewrite this in terms of α; thus,

$$\frac{\alpha^2}{2}\frac{\omega^2}{\omega_{pe}^2}\cdot\varepsilon_{ij}a^2 \ll c^2, \qquad \text{where} \quad \frac{a^2}{c^2} \cong 4\times 10^{-6}\,T_e\,(\mathrm{eV}).$$

The important point to notice is that while coupling may occur, it is not necessarily in a realizable scattering region because of practical limitations on α. For stable plasmas, we may expect to find $\varepsilon_{ij} \lesssim \omega_{pe}^2/\omega^2$ [see Eqn (10.5.5)]. Therefore, as a rough rule for the neglect of coupling, we need

$$\alpha \ll 700/[T_e(\text{eV})]^{1/2}.$$

(It is not possible to make general statements for unstable situations.) An example, discussed by Weinstock (1965a,b), is that of the electron plasma frequency satellites. The complete dispersion relation gives

$$\omega^2 = \omega_{pe}^2 + \Omega_e^2 \sin^2\theta + (\omega_{pe}^2 \Omega_e^2/c^2k^2)\sin^2\theta,$$

while the longitudinal approximation is $\omega_{app}^2 = \omega_{pe}^2 + \Omega_e^2 \sin^2\theta$ [see Eqn (10.8.15)] for $\omega_{pe}^2/\Omega_e^2 \gg 1$. The latter is valid if

$$(\Omega_e^2/c^2k^2)\sin^2\theta \qquad \text{or} \qquad \omega_{pe}^2/c^2k^2 \ll 1;$$

clearly, this is the criterion discussed above.

10.6 GENERAL FEATURES OF THE MAGNETIZED SPECTRUM

10.6.1 Parameters

The magnetized spectrum is relatively complicated, and we will therefore only look in detail at the quasi-equilibrium case (Maxwellian velocity distributions, but arbitrary T_e/T_i), and then only at a limited number of situations:

(a) Inspection of Eqns (10.3.5)–(10.3.8), (10.3.9), and (10.4.9)–(10.4.11) shows that the important parameters are

$$\alpha = 1/k\lambda_{De} \cong \lambda/\lambda_{De}, \qquad k_{||} = k\cos\theta_B, \qquad k_\perp = k\sin\theta_B,$$
$$k_\perp \bar{\rho}_e \cong \bar{\rho}_e/\lambda, \qquad k_\perp \bar{\rho}_i \cong \bar{\rho}_i/\lambda$$

and in the collisional case ν_e/Ω_e, ν_i/Ω_i.

(b) $\alpha \ll 1$: In this situation, $H_e, H_i \Rightarrow 0$, and we recover the noncollective spectrum, Eqn (4.6.12).

(c) $\alpha \gtrsim 1$: In this case, plasma effects are important, and the spectrum consists of two terms:

  (i) the electron feature [the first term in Eqn (10.3.4)], which the inspection of the numerator of Eqn (10.3.9) shows will be important when $\omega \lesssim k_{||}a$;

  (ii) the ion feature [the second term in Eqn (10.3.9)], which is important in the lower frequency range $\omega \lesssim k_{||}b$. [We assume that $T_i/T_e \ll (m_i/m_e)^{1/2}$.] This situation is similar to the unmagnetized case (see Section 5.2), but now, both the numerator and denominator consist of a series of terms at multiples of the cyclotron frequencies. Sometimes, the terms blend to produce spectra similar to the unmagnetized case, but under other conditions, they give rise to a distinctive modulation.

(d) $\theta_B \to 0$: In this limit, $k_{||} \to k$ and $k_{\perp} \to 0$. With the aid of identity (10.1.10), we see that

$$\exp\left(-k_{\perp}^2 \bar{\rho}^2\right) I_l\left(k_{\perp}^2 \bar{\rho}^2\right) = \begin{cases} 1, & l=0 \\ 0, & l \neq 0 \end{cases} \tag{10.6.1}$$

and consequently, the spectrum Eqn (10.3.9) reduces to the unmagnetized spectrum Eqn (5.2.2).

(e) $\theta_B \to \pi/2$: In this limit, $k_{||} \to 0, k_{\perp} \to k$, and the difference between the magnetized and unmagnetized spectra becomes very marked. The most striking effects are observed when we look at waves propagating exactly perpendicular to the field, for then, there is little blurring of the modulation. For $k^2 \bar{\rho}^2 \gg 1$, we find with the aid of Eqn (10.1.11) that

$$\exp\left(-k_{\perp}^2 \bar{\rho}^2\right) I_l\left(k_{\perp}^2 \bar{\rho}^2\right) \cong \left(\frac{2}{\pi}\right)^{1/2} \frac{1}{k^2 \bar{\rho}} \left\{1 - \left(\frac{4l^2 - 1}{8k_{\perp}^2 \bar{\rho}^2}\right) + \cdots\right\}. \tag{10.6.2}$$

This quantity has a significant value for a wide range of l; therefore, there are a large number of peaks.

For $k_{\perp}^2 \bar{\rho}^2 \ll 1$, we only obtain a significant contribution from the $l=0$ term; thus, the spectrum is concentrated around zero frequency. The physical explanation is simply that the electrons are tied to the field lines, we are looking on a scale λ much greater than the cyclotron radius, the electrons do not appear to move much, and the Doppler shift in frequency is negligible.

(f) $k_{\perp}\bar{\rho}$: It is useful to rewrite this factor in different ways:

$$k_{\perp}\bar{\rho}_e = \frac{k_{\perp} a}{\sqrt{2}\Omega_e} = \left(\frac{\omega_{pe}}{\Omega_e}\right) \frac{\sin\theta_B}{\alpha} = \frac{c}{a} \beta_e^{1/2} \frac{\sin\theta_B}{\alpha}, \tag{10.6.3}$$

$$k_{\perp}\bar{\rho}_i = \frac{k_{\perp} b}{\sqrt{2}\Omega_i} = \left(\frac{\omega_{pe}}{\Omega_e}\right) \left(\frac{m_i T_i}{m_e T_e}\right)^{1/2} \frac{\sin\theta_B}{\alpha}, \tag{10.6.4}$$

where $\beta_e = 4\pi n\kappa T_e/B^2$ is the ratio of the electron gas pressure to the magnetic pressure. We note that $(c/a\alpha) = ck/\sqrt{2}\omega_{pe}$ and that

$$\frac{\omega_{pe}}{\Omega_e} = \frac{3.2 \times 10^{-3}[n_e(\text{cm}^{-3})]^{1/2}}{B(G)}.$$

Typically,

$$m_i T_i / m_e T_e \gg 1.$$

Therefore, $k_{\perp}\bar{\rho}_i \gg k_{\perp}\bar{\rho}_e$. Examples of laboratory plasmas are shown in Fig. 1.1. For fusion reactor conditions, we might expect $c/a \cong 10, \beta_e^{1/2} \cong 0.3$, so $k_{\perp}\bar{\rho}_e \cong (3\sin\theta_B)/\alpha$.

(g) *Collisions*: For $\theta_B = \pi/2$, the spectrum reduces to a series of delta function spikes, and in addition, the Landau damping term disappears [see Eqns (10.3.5) and (10.3.7)]. This is discussed in Section 10.6.2. In real life because of collisions, we do not obtain such a spiky distribution. Each peak reduces to the Lorentzian form

$$\nu_q/[(\omega - l\Omega_q)^2 + \nu_q^2]. \tag{10.6.5}$$

For $k_\| \neq 0$, these cyclotron peaks become broader as θ_B decreases from $\pi/2$. The breadth of each peak is approximately $k_\| a, k_\| b$, in the collisionless case, and as discussed in Section 4.7.2, these peaks are resolvable only if $k_\| a = ka\cos\theta_B < \Omega_e$, or for the ions $k_\| b < \Omega_i$. For a collisional plasma, there is of course the additional collisional broadening, and from Eqn (10.6.5), we see that we require $\nu_q \ll \Omega_q$ in order for the magnetic field to exercise a significant effect on the spectrum. When this condition is not satisfied, the charges can no longer complete cyclotron orbits, and they are effectively not magnetized. At these high collision frequencies, we recover the spectra shown Chapter 5 (see Figs 10.5a,b).

10.6.2 Limit $B_0 \rightarrow 0$ ($k_\| = 0$)

This case is interesting because for finite B_0, there is no Landau damping [see (Eqns 10.3.5) and (10.3.7)], yet in the equivalent unmagnetized case, there is damping. The simple physical picture is that in the magnetized case, each charge loses energy as it moves against a given wave in one-half of the cyclotron orbit, but because the process is reversible, it gains it back again in the other half of the orbit. Therefore, as long as we look on a time much longer than a cyclotron period (or treat the system as stationary), we will see no damping. The damping only occurs on a timescale less than $1/\Omega$ or in the stationary case when $B_0 = 0$. This topic is discussed in detail by Baldwin & Rowlands (1966).

10.7 TOTAL CROSS SECTION, $S_T(\mathbf{k})$

It is instructive to look at the total cross section before studying the frequency spectrum because some important features of the scattered radiation emerge. The evaluation of

$$S_T(\mathbf{k}) = \int_{-\infty}^{+\infty} d\omega S(\mathbf{k}, \omega) \tag{10.7.1}$$

has been discussed by Dougherty & Farley (1963b) for the case of an equilibrium magnetized plasma.

$$T_e - T_i$$

We may rewrite $S(\mathbf{k}, \omega)$ of Eqns (10.3.9) and (10.4.6)–(10.4.11) for $T_e = T_i$ in the form

$$S(\mathbf{k}, \omega) = -\frac{2}{\alpha^2 \omega}\left|\frac{1+H_i}{\varepsilon_L}\right|^2 \mathrm{Im}(H_e) - \frac{2}{\alpha^2 \omega}\left|\frac{H_e}{\varepsilon_L}\right|^2 \mathrm{Im}(H_i). \tag{10.7.2}$$

For the collisional case, we substitute L_e for H_e and L_i for H_i. It is convenient to retain collisions in these expressions to avoid problems when $k_{\parallel}=0$, for in the collisionless case, H_e and H_i are then real. Equation (10.7.2) may be rewritten as

$$S(\mathbf{k},\omega)=-\frac{2}{\alpha^2\omega}\mathrm{Im}\left[\frac{(1+L_i)L_e}{1+L_e+L_i}\right]. \tag{10.7.3}$$

The integral over frequency is easily evaluated by contour integration (see Section 5.6 and Appendix A). For a stable plasma, ε_L has no poles in the upper half-plane, and we pick up only the pole at $\omega=0$

$$S_T(\mathbf{k})=\frac{2\pi}{\alpha^2}\left[\frac{(1+L_i(0))L_e(0)}{1+L_e(0)+L_i(0)}\right]=\frac{2\pi(1+Z\alpha^2)}{(1+\alpha^2+Z\alpha^2)}. \tag{10.7.4}$$

This is the same total cross section as in the unmagnetized case Eqns (5.6.3) and (5.6.6).

In Sections 5.3 and 5.6, we found that it was possible to separate out the properties of the electron and ion features (the "Salpeter approximation"). A similar approximation may be used here (Salpeter, 1961b). The essential difference in this magnetized case lies in the form of the numerator in the terms of Eqn (10.3.9). In the electron feature, we have the factor

$$\sum_{l=-\infty}^{+\infty}\exp\left(-k_{\perp}^2\bar{\rho}_e^2\right)I_l\left(k_{\perp}^2\bar{\rho}_e^2\right)\exp\frac{[-((\omega-l\Omega_e)/k_{||}a)^2]}{k_{||}a}.$$

We may separate the term $l=0$ from the terms $l\neq 0$. The $l\neq 0$ terms are important when $\omega\cong l\Omega_e$, and in general, these are high frequencies for the ions, and they cannot respond; therefore, $L_i, H_i=0$.

The $l=0$ term is important when $\omega\lesssim k_{||}a$, and although this includes $\omega=0$, if $k_{||}a\gg\Omega_i$, i.e., weak magnetic fields, then since $(\omega-m\Omega_i)/k_{||}b\gg 1$ we may also set $L_i, H_i=0$ for the bulk of the frequency range covered by this term. On the other hand, if $\theta_B\to\pi/2$ and $k_{||}a<\Omega_i$, i.e., $(1/k_{\perp}\bar{\rho}_i)(m_e/m_i)\gg\cos\theta_B$, we must retain some part of $L_i(0), H_i(0)$ in the electron feature.

$k_{||}a\gg\Omega_i$

(For example, $\theta_B\neq\pi/2$, small magnetic field, $k_{\perp}\bar{\rho}_e\gg m_e/m_i$.)

$$S_e(\mathbf{k},\omega)\cong\frac{2\pi^{1/2}}{|1+H_e|^2}\sum_l\exp\left(-k_{\perp}^2\bar{\rho}_e^2\right)I_l\left(k_{\perp}^2\bar{\rho}_e^2\right)\exp\frac{[-((\omega-l\Omega_e)/k_{||}a)]}{k_{||}a}.$$

From Eqn (10.3.6),

$$S_e(\mathbf{k},\omega)=\frac{2}{\alpha^2\omega}\mathrm{Im}\left(\frac{1}{1+H_e}\right). \tag{10.7.5}$$

By contour integration,

$$S_e(\mathbf{k})=2\pi/(1+\alpha^2). \tag{10.7.6}$$

Thus, for small magnetic fields and $\theta_B \neq \pi/2$, the electron feature cross section is the same as in the unmagnetized case. The ion feature is obtained by subtracting Eqn (10.7.5) from Eqn (10.7.4) [see Eqn (5.6.3)]

$$S_i(\mathbf{k}) = 2\pi Z\alpha^4/(1+\alpha^2)(1+\alpha^2+Z\alpha^2). \tag{10.7.7}$$

$k_{||}a < \Omega_i$

(For example, $\theta_B \cong \pi/2$, large magnetic field, $k_\perp \bar{\rho}_e < m_e/m_i$.) Consider the $l=0$ term. Then, $\omega \lesssim k_{||}a < \Omega_i, \Omega_e, k_\perp \bar{\rho}_e \ll 1$. For small ω, we may not drop H_i in the *denominator*, but since $x_{el}, x_{im} \gg 1$, we may drop all the terms with $l, m \neq 0$. In addition, $\omega \cong k_{||}a \gg k_{||}b$; therefore,

$$2x_{i0}\exp\left(-x_{i0}^2\right)\int_0^{x_{i0}}\exp(p^2)dp + i\pi^{1/2}x_{i0}\exp\left(-x_{i0}^2\right) = 1, \quad x_{i0} = \omega/k_{||}b$$

and

$$\begin{aligned} S_{e0}(\mathbf{k},\omega) \cong \frac{2\pi^{1/2}}{k_{||}a}&\left\{\left[\left|1+Z\alpha^2\left[1-\exp\left(-k_\perp^2\bar{\rho}_i^2\right)I_0\left(k_\perp^2\bar{\rho}_i^2\right)\right]\right|^2\right.\right.\\ &\times\exp\left(-k_\perp^2\bar{\rho}_e^2\right)I_0\left(k_\perp^2\bar{\rho}_e^2\right)\exp[-(\omega/k_{||}a)^2]\Big]\\ &\div\left|1+Z\alpha^2\left[1-\exp\left(-k_\perp^2\bar{\rho}_i^2\right)I_0\left(k_\perp^2\bar{\rho}_i^2\right)\right]\right.\\ &+\alpha^2\left[1-\exp\left(-k_\perp^2\bar{\rho}_e^2\right)I_0\left(k_\perp^2\bar{\rho}_e^2\right)\right]\\ &\times\left.\left.\left.\left[2x_{e0}\exp\left(-x_{e0}^2\right)\int_0^{x_{e0}}\exp(p^2)dp + i\pi^{1/2}x_{e0}\exp\left(-x_{e0}^2\right)\right]\right]\right|^2\right]\right\}\\ = -(2/\alpha^2\omega)&\left|1+Z\alpha^2\left[1-\exp\left(-k_\perp^2\bar{\rho}_i^2\right)I_0\left(k_\perp^2\bar{\rho}_i^2\right)\right]\right|^2\mathrm{Im}(1/\varepsilon_L). \end{aligned} \tag{10.7.8}$$

$$\begin{aligned} S_{e0}(\mathbf{k}) &= \mathrm{Im}\frac{2\pi i}{\alpha^2}\left\{\frac{1}{\varepsilon_L(\infty)} - \frac{1}{\varepsilon_L(0)}\right\}[|1+Z\alpha^2(1-X_{i0})|^2] \\ &= \frac{2\pi X_{e0}|1+Z\alpha^2(1-X_{i0})|^2}{[1+Z\alpha^2(1-X_{i0})+\alpha^2][1+Z\alpha^2(1-X_{i0})+\alpha^2(1-X_{e0})]}, \end{aligned} \tag{10.7.9}$$

where

$$X_{q0} = \exp\left(-k_\perp^2\bar{\rho}_q^2\right)I_0\left(k_\perp^2\bar{\rho}_q^2\right).$$

For the ion feature, $\omega \lesssim k_{||}b \ll k_{||}a$ so $H_e = \alpha^2$, and

$$S_{i0}(\mathbf{k},\omega) \cong -\frac{2\alpha^2}{\omega}\mathrm{Im}\left[1+\alpha^2+Z\alpha^2\left\{1-X_{i0}\left(x_{i0}\exp\left(-x_{i0}^2\right)\int^{x_{i0}}\exp(p^2)\,dp\right.\right.\right.$$
$$\left.\left.\left.+i\pi^{1/2}x_{i0}\exp\left(-x_{i0}\right)\right)\right\}\right]^{-1} \tag{10.7.10}$$

$$S_{i0}(\mathbf{k}) = 2\pi Z\alpha^4 X_{i0}/[1+\alpha^2+Z\alpha^2(1-X_{i0})][1+\alpha^2+Z\alpha^2]. \tag{10.7.11}$$

Now for $k_\perp^2\bar{\rho}_e^2 \ll 1, X_{e0} \Rightarrow 1$ and then,

$$S_{e0}(\mathbf{k})+S_{i0}(\mathbf{k}) \cong 2\pi\left(\frac{1+Z\alpha^2}{1+\alpha^2+Z\alpha^2}\right) = S_T(\mathbf{k}). \tag{10.7.12}$$

Thus for **k** close to perpendicular to B_0, and large enough magnetic fields, we find all the intensity is invested in the central lines set about zero frequency, and this is essentially independent of the value of $k_\perp\bar{\rho}_i$, Salpeter (1961b) gives the criterion as

$$\left(\frac{m_e}{m_i}\right)^{1/2}\frac{1}{k_\perp\bar{\rho}_i} \cong \frac{m_e}{m_i}\frac{1}{k_\perp\bar{\rho}_e} \gg \cos\theta_B \gg \left(\frac{m_e}{m_i}\right)^{1/2} \tag{10.7.13}$$

and $k_\perp^2\bar{\rho}_e^2 \gg 1$. We find all the intensity is invested in the central lines set about zero frequency, and this is essentially independent of the value of $k_\perp\bar{\rho}_i$. The reason is simply that the electrons are tied to the field lines, and if we look close to the perpendicular direction on a scale $\lambda \gg \bar{\rho}_e$, we do not see a significant change in the average position of the electrons; therefore, the Doppler shift is small. Further, since the electron motion is coupled to the ions only over a Debye length, then for $\lambda_{De} \ll \lambda \ll \bar{\rho}_e$ even this electron motion is restricted, and the Doppler shifts are small.

To summarize:

1. The total cross section is the same as in the unmagnetized case.
2. For $k_{||}a \gg \Omega_i$, i.e., a weakly magnetized plasma and $\theta_B \neq \pi/2$, all the terms in the numerator contribute, and the electron and ion features have the same individual cross sections as in the unmagnetized case. (In fact, barring the modulation, the envelope of the spectrum is similar to the unmagnetized case.
3. For $k_{||}a \ll \Omega_i$, i.e., strongly magnetized and $\theta_B \cong \pi/2$, the motion of the electrons is restricted, the Doppler shifts are small, and most of the spectral power resides in the region of zero frequency.

10.8 HIGH-FREQUENCY SPECTRUM

10.8.1 $k_{||} = 0, \omega \cong \omega_{pe}, \Omega_e$

The numerator in the spectrum function Eqn (10.3.9) determines the frequency range over which each term will be important. For high frequencies $\omega \cong \omega_{pe}, \Omega_e$, the main contribution comes from the electron feature. The characteristic form of the collective spectrum $\alpha \gtrsim 1$ is governed by the behavior of the dielectric function $S(\mathbf{k}, \omega) \propto 1/|\varepsilon_L|^2$, and peaks in the spectrum occur at the resonances given by $|\varepsilon_L|^2 = 0$. These spectra have been discussed by numerous authors, notably, Bernstein (1958), and Salpeter (1961a,b). We will treat the case $k_{||} = 0$ first as this has the most unusual features of the magnetized spectra. The dispersion relation for longitudinal waves from Eqns (10.3.5), (10.3.7), and (10.3.9) is

$$\varepsilon_L = 1 + \alpha^2 \left[1 - \sum_l \exp(-k_\perp^2 \bar{\rho}_e^2) I_l(k_\perp^2 \bar{\rho}_e^2) \left(\frac{\omega}{\omega - l\Omega_e} \right) \right] + Z \frac{T_e}{T_i} \alpha^2 \left[1 - \sum_m \exp(-k_\perp^2 \bar{\rho}_i^2) I_m(k_\perp^2 \bar{\rho}_i^2) \left(\frac{\omega}{\omega - m\Omega_i} \right) \right] - 0. \tag{10.8.1}$$

From the form of Eqn (10.8.1), we see that there will be a multiplicity of roots, known as "Bernstein modes." The waves exhibit no damping for this collisionless case as discussed in Section 10.6.2, but in real life, there are collisions. We see from Eqn (10.4.10) that the factor $\omega/(\omega - l\Omega_q)$ then changes to $(\omega - i\nu_q)/(\omega - l\Omega_q - i\nu_q)$.

$k_\perp^2 \bar{\rho}_e^2 \ll 1$

We see that $(\omega/\omega - m\Omega_i) \to 1$, and with the aid of Eqn (10.1.15), we realize that the ion component of ε_L is negligible. Rearranging Eqn (10.8.1) then leads to

$$\varepsilon_L = 1 - 2\alpha^2 \sum_{l=1}^{\infty} \exp\left(-k_\perp^2 \bar{\rho}_e^2\right) I_l(k_\perp^2 \bar{\rho}_e^2) \frac{l^2}{(\omega/\Omega_e)^2 - l^2}. \tag{10.8.2}$$

Using Eqn (10.1.10), we obtain

$$\varepsilon_L = 1 - 2\alpha^2 (1 - k_\perp^2 \bar{\rho}_e^2) \left\{ \frac{k_\perp^2 \bar{\rho}_e^2}{2[(\omega/\Omega_e)^2 - 1]} + \frac{k_\perp^2 \rho_e^2}{2[(\omega/\Omega_e)^2 \quad 4]} + \cdots \right\} = 0. \tag{10.8.3}$$

We now want to solve this for ω/Ω_e, and there are two cases:

1. If ω/Ω_e is not close to an integer, then because $k_\perp^2 \bar{\rho}_e^2 \ll 1$, we may reject all but the first term in the series and obtain the dominant root of Eqn (10.7.4), which we will call ω_0/Ω_e

$$(\omega_0/\Omega_e)^2 = 1 + \alpha^2 k_\perp^2 \bar{\rho}_e^2 \quad \text{or} \quad \omega_0^2 = \omega_{pe}^2 + \Omega_e^2. \tag{10.8.4}$$

This is usually called the "upper hybrid frequency."

It is clear that for the magnetized plasma, the frequencies $\pm\omega_0$ correspond to the plasma frequency features of the unmagnetized case (Section 5.4.2). For $\lambda_{De}^2 \ll \bar{\rho}_e^2$, Eqn (10.8.4) reduces to the plasma frequency.

By analogy with the derivation of Eqn (10.7.5), we see that the integrated intensity of each satellite is

$$S(\mathbf{k})\Big|_{\omega_0} \cong \int d\omega \frac{2}{\alpha^2\omega}\mathrm{Im}\left(\frac{1}{\varepsilon_{\mathrm{L}}}\right) = \frac{2\pi k_\perp^2 \bar{\rho}_e^2}{1+\alpha^2 k_\perp^2 \bar{\rho}_e^2}, \tag{10.8.5}$$

where we have acted as if there is a low level of collisions so that ε_{L} has an imaginary component. For $\lambda_{\mathrm{De}}^2 \ll \bar{\rho}_e^2$, this reduces to Eqn (10.7.6). Finally, we must comment on the other roots of (10.8.3); these correspond to frequencies for which $(\omega/\Omega_e) - l^2$ is small enough to make the lth term comparable to the first term. However, the intensity of these terms is

$$S(\mathbf{k})\Big|_{\omega_l} \cong \frac{2\pi (k_\perp^2 \bar{\rho}_e^2)^l}{2^{(l-1)} l! (1+\alpha^2 k_\perp^2 \bar{\rho}_e^2)},$$

and this is small compared to Eqn (10.8.5).

2. If ω_0/Ω_e lies close to an integer l, then the lth term contributes in the evaluation of the dominant frequency. We set $\omega_0/\Omega_e = l \pm \Delta\omega/\Omega_e$, $\Delta\omega/\Omega_e \ll l$ and find

$$\frac{\Delta\omega}{\Omega_e} \cong \frac{\alpha^2}{(l-1)!}\left(\frac{k_\perp^2 \bar{\rho}_e^2}{2}\right)^l. \tag{10.8.6}$$

There are consequently two lines on each side of $\omega_0/\Omega_e = \pm l$, and these share the intensity of Eqn (10.8.5). The coupling of the $l=1$ and $l=2$ modes is discussed in some detail by Platzman *et al.* (1968) (see Fig. 10.2).

$k_\perp^2 \bar{\rho}_e^2 \gg 1$

In this case, we can use expansion Eqn (10.6.2) and

$$\varepsilon_{\mathrm{L}} \cong 1 - 2\alpha^2 \left(\frac{2}{\pi}\right)^{1/2} \frac{1}{k_\perp \bar{\rho}_e}\left(\frac{1}{(\omega/\Omega_e)^2 - 1} + \frac{4}{(\omega/\Omega_e)^2 - 4} + \cdots\right). \tag{10.8.7}$$

It can be seen that for $k_\perp \bar{\rho}_e \gg 2l^2\alpha^2$, the resonances occur with $\omega/\Omega_e \cong l$. Surko *et al.* (1972) have driven Bernstein modes in a plasma with $k_\perp \bar{\rho}_e \cong 0.3-5$ and have observed the scattering of CO_2 radiation from them. Their apparatus is shown in Fig. 10.3. The waves were excited by the application of a 268-MHz voltage to a probe in the plasma. They studied waves propagating perpendicular to the magnetic field and used a heterodyne detection system (see Section 6.7.5), to improve the signal-to-noise ratio ($n_e \cong 4 \times 10^{10}\,\mathrm{cm}^{-3}$). The detector was fixed to receive a particular wave number; a typical signal received as the magnetic field was swept is shown in Fig. 10.4a. The modulation in the signal level is caused by the variation with magnetic field of the phase angle between the scattered and local oscillator fields. The maximum in the envelope of the spectrum corresponds to coincidence of resonant plasma wave number and detected wave number. A comparison of the observed and

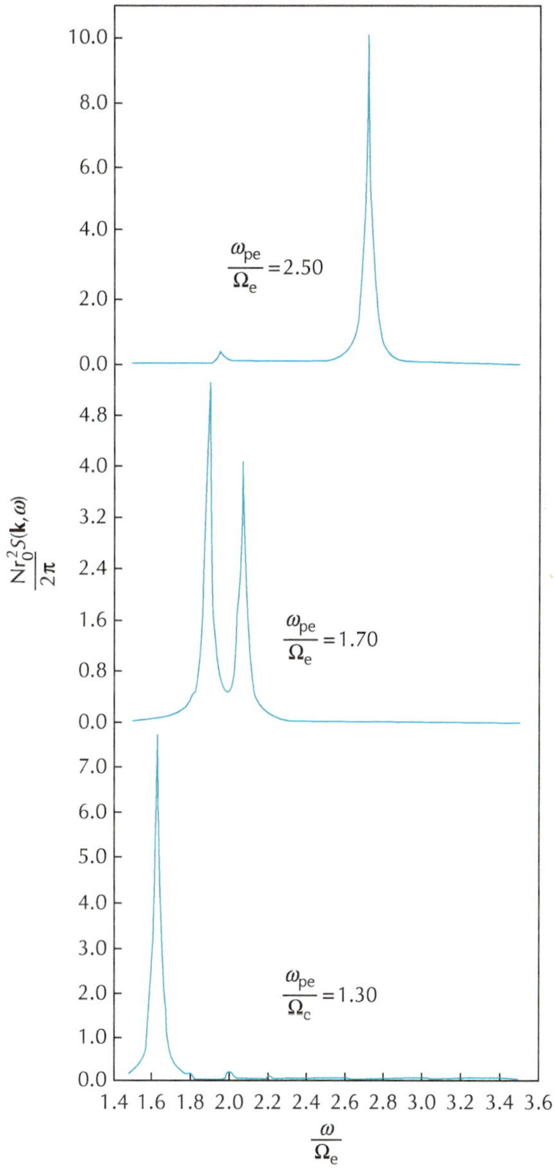

FIGURE 10.2 The scattering cross section versus ω/Ω_e for $\alpha = 14$ and $\Omega_e/\nu_e = 50$. For $\omega_{pe}/\Omega_e = 1.3$, the intensity is mainly in the upper hybrid mode at $\omega = 1.65\,\Omega_e$. For $\omega_{pe}/\Omega_e = 1.7$, the upper hybrid and cyclotron modes have similar intensity. For $\omega_{pe}/\Omega_e = 2.5$, the intensity is in the upper hybrid mode at $\omega = 2.7\,\Omega_e$. (Courtesy of Platzman *et al.* (1968).)

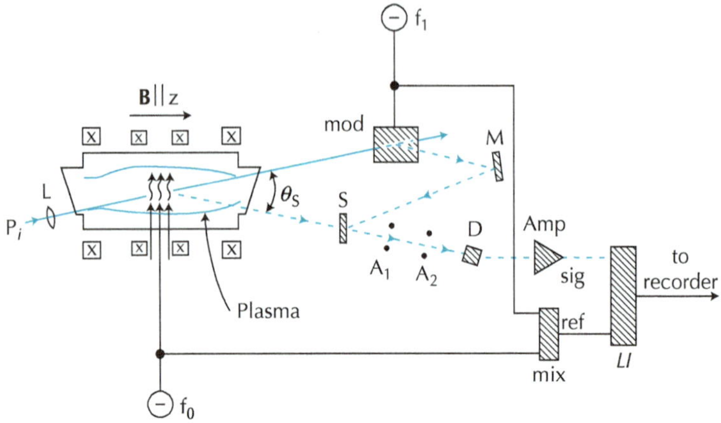

FIGURE 10.3 Schematic diagram of apparatus for light mixing spectroscopy of 10.6-μm radiation scattered from cyclotron harmonic waves. (Courtesy of Surko *et al.* (1972).)

theoretical dispersion relations $(\omega/\Omega_e)^2$ versus $k_\perp\bar{\rho}_e/\sqrt{2}$ [Eqns (10.8.3) and (10.8.7) at the two limits] is shown in Fig. 10.4b.

10.8.2 $k_{||}\neq 0, \omega\cong\omega_{pe}, \Omega_e, k_{||}a$

In this case, we obtain for the integral in Eqn (10.7.1) a factor similar to the plasma dispersion function (Section 5.2), and we regain the Landau damping term; this comes from the motion of the charges along the field lines.

$$\varepsilon_L\Big|_{k_{||}\neq 0} = 1+\alpha^2\left\{1-\sum_{l=-\infty}^{+\infty}\exp\left(-k_\perp^2\bar{\rho}_e^2\right)I_l\left(k_\perp^2\bar{\rho}_e^2\right)\frac{\omega}{(\omega-l\Omega_e)}\right.$$
$$\left.\times\left[2x_{el}\exp\left(-x_{el}^2\right)\int_0^{x_{el}}\exp(p^2)\,dp+i\pi^{1/2}x_{el}\exp\left(-x_{el}^2\right)\right]\right\}$$
$$+Z\frac{T_e}{T_i}\alpha^2\left\{1-\sum_{m=-\infty}^{+\infty}\exp\left(-k_\perp^2\bar{\rho}_i^2\right)I_m\left(k_\perp^2\bar{\rho}_i^2\right)\frac{\omega}{(\omega-m\Omega_i)}\right.$$
$$\left.\times\left[2x_{im}\exp\left(-x_{im}^2\right)\int_0^{x_{im}}\exp(p^2)\,dp+i\pi^{1/2}x_{im}\exp\left(-x_{im}^2\right)\right]\right\}, \tag{10.8.8}$$

where

$$x_{el}=(\omega-l\Omega_e)/k_{||}a, \quad x_{im}=(\omega-m\Omega_i)/k_{||}b.$$

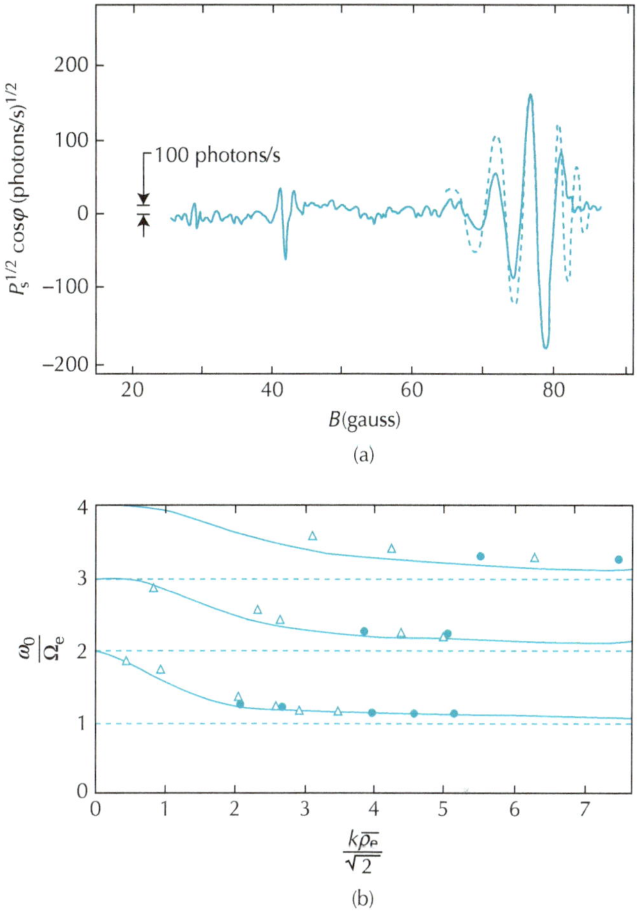

FIGURE 10.4 (a) Solid curve is for light mixing signal $P_s^{1/2} \cos\varphi$ at the output of *LI* as a function of *B* (φ is the phase angle between the scattered and local oscillator electric fields). The detector was positioned at the Bragg angle for 1.5-mm waves. Dashed curve is for calculated signal. (b) Light scattering data (closed circles) and probe data using an rf interferometer (open triangles). Solid lines show theoretical dispersion relation of cyclotron harmonic waves for $(\omega_{pe}/\Omega_e)^2 = 120$. (Courtesy of Surko *et al.* (1972).)

The functions $\mathrm{Rw}(x) = 1 - 2x\exp(-x^2)\int_0^x \exp(p^2)dp$ and $\mathrm{Iw}(x) = i\pi^{1/2}x \exp(-x^2)$ are plotted in Fig. 5.1. For $x < 1$ ($v_{ph} < v_{th}$),

$$1 - \mathrm{Rw}(x) \cong 2x^2\left[1 - \frac{2x^2}{3} + \frac{4x^4}{15}\cdots\right], \tag{10.8.9a}$$

for $x \gg 1$ ($v_{ph} \gg v_{th}$),

$$1 - \mathrm{Rw}(x) \cong 1 + \frac{1}{2x^2}\left[1 + \frac{3}{2x^2} + \frac{15}{4x^4}\cdots\right]. \tag{10.8.9b}$$

High-Frequency Components ($\omega \cong \omega_{pe}, \Omega_e, \alpha \gg 1$)

Now, $\omega \gg k_{||}b$, and from Eqns (10.1.10) and (10.1.11), we see that $\exp(-k_\perp^2 \overline{\rho}_i^2) I_m(k_\perp^2 \overline{\rho}_i^2)$ is small unless $m\Omega_i < k_{||}b$; therefore, we have $x_{im} \gg 1$ wherever this factor has a significant value, and we may use the second expansion in Eqn (10.8.9a). In addition, we may drop the ion Landau damping term. With the aid of Eqns (10.1.15) and (10.1.16), we then find

$$H_i \cong -ZT_e\alpha^2 k^2 b^2 / T_i 2\omega^2, \quad k^2 = k_\perp^2 + k_{||}^2. \tag{10.8.10}$$

This result is certainly valid when $k_\perp^2 \overline{\rho}_i^2 \lesssim 1$; therefore, $k_\perp^2 \overline{\rho}_e^2 \ll 1$, and

$$\exp(-k_\perp^2 \overline{\rho}_e^2) I_l(k_\perp^2 \overline{\rho}_e^2) \cong (1 - k_\perp^2 \overline{\rho}_e^2)_{l=0} + (k_\perp^2 \overline{\rho}_e^2/2)_{l=1} + \cdots, \tag{10.8.11}$$

where we have used identity (10.1.10). Now, the resonant frequencies obtained by setting $\varepsilon_L = 0$ are at high frequencies $\sim\omega_{pe}, \Omega_e$, whichever is greatest; therefore $x_{el} \gg 1$ if $\alpha \gg 1$, and in this case, we may again use the second expansion in Eqn (10.8.10) for the dispersion function. The damping is small, and therefore, the resonances are sharp and may be obtained from $\mathrm{Re}(\varepsilon_L) = 0$, which is then given by

$$\mathrm{Re}(\varepsilon_L) \cong 1 - \alpha^2 \left[\frac{k^2 a^2}{2\Omega_e^2} \frac{(\omega/\Omega_e)^2 - \cos^2\theta_B}{[(\omega/\Omega_e)^2 - 1](\omega/\Omega_e)^2} + \frac{ZT_e k^2 b^2}{T_i 2\omega^2} \right] = 0. \tag{10.8.12}$$

We solve for $(\omega/\Omega_e)^2$

$$(\omega/\Omega_e)^2 = \frac{1}{2} \left\{ \left(1 + \alpha^2 k^2 \overline{\rho}_e^2\right) \pm \left[\left(1 + \alpha^2 k^2 \overline{\rho}_e^2\right) - 4\alpha^2 k^2 \overline{\rho}_e^2 \left(Zm_e/m_i + \cos^2\theta_B \right)^{1/2} \right] \right\}, \tag{10.8.13}$$

where we have dropped $m_e/m_i \ll 1$.

$\alpha^2 k^2 \overline{\rho}_e^2 \ll 1$

There are two solutions

$$\omega^2 = \omega_{pe}^2 + \Omega_e^2 - \frac{\omega_{pe}^2 \Omega_e^2}{\omega_{pe}^2 + \Omega_e^2} \left(\frac{Zm_e}{m_i} + \cos^2\theta_B \right). \tag{10.8.14}$$

This is equivalent to (10.8.4) and reduces to it when $\theta_B \to \pi/2\,(m_e/m_i \ll 1)$. The other solution is

$$\omega^2 = \left(\frac{Zm_e}{m_i} + \cos^2\theta_B \right) \frac{\omega_{pe}^2 \Omega_e^2}{\omega_{pe}^2 + \Omega_e^2}. \tag{10.8.15}$$

When $\theta_B \to \pi/2$, because the electron motion is restricted, this root is found at low frequencies

$$\omega^2 \cong \frac{Zm_e}{m_i} \frac{\omega_{pe}^2 \Omega_e^2}{\omega_{pe}^2 + \Omega_e^2}.$$

In fact, if $1/k^2 \overline{\rho}_e^2 \gg \alpha^2 \gg 1$, it reduces to the ion plasma frequency ω_{pi}.

10.9 LOW-FREQUENCY SPECTRUM

10.9.1 $k_{||}=0, \omega \cong \omega_{pi}, \Omega_i$

At low frequencies, the ions can respond readily, and they play a major role in the creation of resonances. For these conditions, the electron component of the dielectric function has the same form as Eqn (10.8.3), and we retain the first term. For $\alpha \ll 1$, we find that the roots of significant intensity have $\omega \gg \Omega_i$; therefore, we rearrange the sum in the ion term of Eqn (10.1.1) and then set

$$(m\Omega_i)^2/(\omega^2 - m^2\Omega_i^2) \cong m^2\Omega_i^2/\omega^2$$

and using Eqn (10.1.15) obtain

$$\varepsilon_L = 1 - \frac{\alpha^2 k_\perp^2 \overline{\rho}_e^2}{(\omega/\Omega_e)^2 - 1} - \alpha^2 \frac{ZT_e}{T_i} k_\perp^2 \overline{\rho}^2 \frac{\Omega_i^2}{\omega^2} = 0. \tag{10.9.1}$$

The solution, provided ω_r/Ω_i is not close to an integer, is

$$\omega_r^2 = \frac{m_i}{m_e} Z\Omega_i^2 \frac{\alpha^2 k_\perp^2 \overline{\rho}_e}{1+\alpha^2 k_\perp^2 \overline{\rho}_e^2} = \frac{m_e}{m_i} \frac{\omega_{pe}^2 \Omega_e^2}{\omega_{pe}^2 + \Omega_e^2}. \tag{10.9.2}$$

This solution is valid for $\cos^2\theta_B \ll Zm_e/m_i$ [see (10.1.15)]. Rearrangement of Eqn (10.9.2) shows that for large $\alpha, \omega_r^2 \gg (kb)^2$; therefore, this resonance contributes mainly through the electron feature. In the limit $1/k_\perp^2 \overline{\rho}_e^2 \gg \alpha^2 \gg 1, \omega_r \to \omega_{pi}$. If ω_r/Ω_i is close to an integer m, we obtain two lines set around $\omega_r/\Omega_i = \pm m$, analogous to Eqn (10.8.6)

$k_\perp \overline{\rho}_i^2, k_\perp^2 \overline{\rho}_e^2 \gg 1$

For this case, $\lambda \ll \rho_e, \rho_i$, that is, weak magnetic fields, and the spectrum consists of a multiplicity of lines, separated by the cyclotron frequencies and many of significant amplitude [see Eqn (10.6.2)]. The envelope of the spectrum is then similar to the field free case (see Fig. 10.5a,b).

10.9.2 $k_{||} \neq 0, \omega \cong \omega_{pi}, \Omega_i, k_{||}b$

For $k_{||} \neq 0$, the main contribution comes from the ion feature, the second term in Eqn (10.3.9). For directions close to perpendicular to $\mathbf{B}_0$, the intensity is invested primarily in the $m=0$ term; the cross section of this term is given by Eqn (10.7.11). For $Z=1, \alpha \gg 1$, it is

$$S_{i0}(\mathbf{k}) = \pi X_{i0}/(2 - X_{i0}). \tag{10.9.3}$$

This is to be compared with the total cross section Eqn (10.7.12) of π. Now, $X_{i0} = \exp(-k_\perp^2 \overline{\rho}_i^2) I_0(k_\perp^2 \overline{\rho}_i^2)$; this quantity decreases from unity, when $k_\perp^2 \overline{\rho}_i^2 = 0$ as $k_\perp^2 \overline{\rho}_i^2$ is increased [see Eqn (10.6.2)]. Therefore, for $k_\perp^2 \overline{\rho}_i^2 \ll 1$, most of the intensity is in the central line. For $k_\perp^2 \overline{\rho}_i^2 \gtrsim 1$, it is spread over a number of terms,

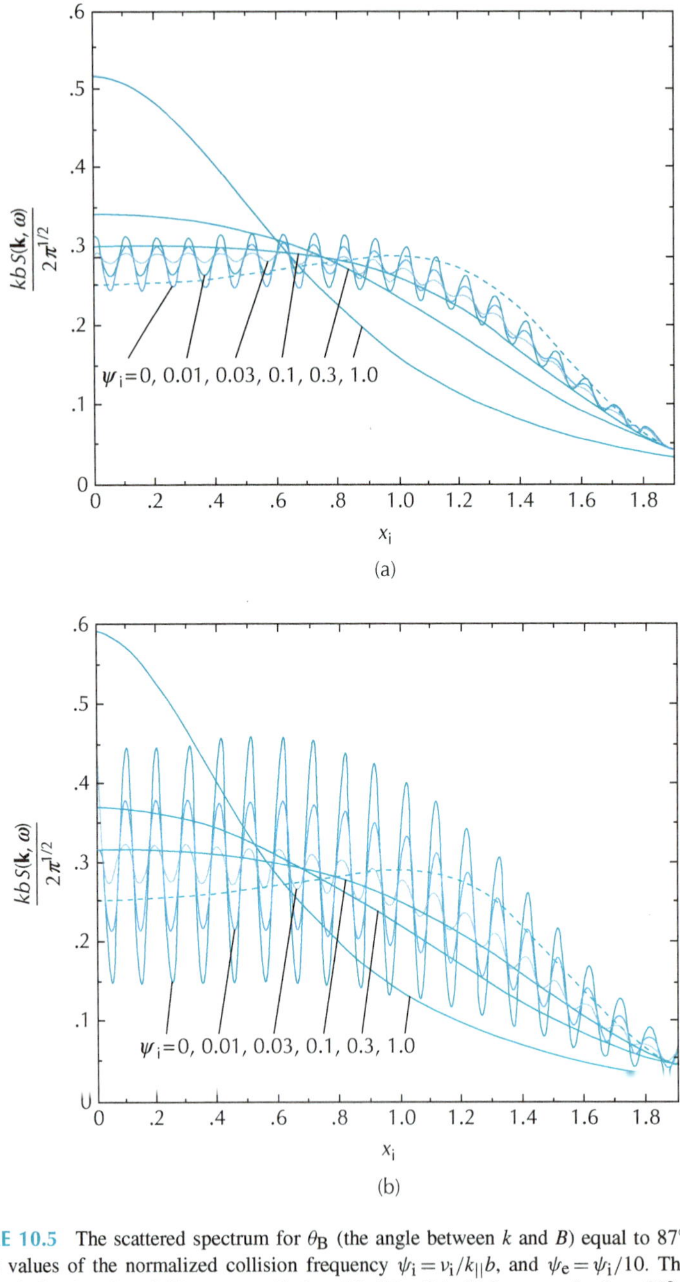

FIGURE 10.5 The scattered spectrum for θ_B (the angle between k and B) equal to 87° and for various values of the normalized collision frequency $\psi_i = \nu_i / k_{||} b$, and $\psi_e = \psi_i / 10$. The dashed spectrum is for $\psi_i = \theta_B = 0$. Here, $x_i = \omega / k_{||} b$ and $\Omega_i / kb = 0.1$. (b) Spectrum for $\theta_B = 88°$, $\Omega_i / kb =$ 0.1. (Courtesy of Dougherty & Farley (1963b).)

and as θ_B decreases from π_2, these terms merge together and approach the field free spectrum.

For $\cos\theta_B \gg (m_e/m_i)^{1/2}$, and at these low frequencies $H_e \Rightarrow \alpha^2$, if also $\cos\theta_B \ll \Omega_i/kb$, the individual peaks are still narrow and resolvable, the damping is small, and the dispersion relation reduces to

$$\varepsilon_L = 1-\alpha^2 - 2Z\alpha^2 \sum_{m=1}^{\infty} \exp(-k_\perp^2 \bar{\rho}_i^2) I_m(k_\perp^2 \bar{\rho}_i^2) \frac{m^2}{(\omega/\Omega_i)^2 - m^2}. \qquad (10.9.4)$$

These points are illustrated in Figs 10.5a,b. In addition, the effects of ion–neutral collisions are shown, and the reduction of the spectra, as the collision frequency is increased, to the collision-dominated unmagnetized one is evident [see Section 10.6.1(g)].

PROBLEMS

10.1. Using a Bessel function identity, show that the magnetized plasma term [H_e in Eqns (10.3.5) and (10.3.6)] reduces to the unmagnetized case Eqns (5.2.5) and (5.2.6) as B goes to zero. [see Eqn (10.1.11)].

10.2. We wish to look at oblique propagation of low-frequency ($\omega \ll \Omega_c$) waves in a magnetized plasma ($\mathbf{B}_0 = \hat{z}B_0$). Let k lie in the x-z plane.

(a) Find the dispersion relation for the fast and slow modes in a cold plasma.

(b) Assume $T_e \neq 0$ and show that there are now *three* modes.

10.3. Consider a cold plasma with a linear density gradient. Derive an expression for the distance Δx between the right-hand cutoff and the X0-mode resonances.

10.4. Derive the dispersion relation for obliquely propagating waves in a cold plasma valid for $\omega \lessapprox \Omega_c$. Show that there is a resonance for

$$\omega^2 = \frac{\Omega_c^2 k_{||}^2 c^2}{k_{||}^2 c^2 + \Omega_p^2}.$$

In addition, show that for $\omega \cong \Omega_c$ that the dispersion relation can be approximately written as

$$n^2 \cong \frac{c^2/v_A^2}{1+\cos^2\theta} \qquad \text{(fast mode)}$$

and $n^2\cos^2\theta \cong \frac{\left(\frac{c^2}{v_A}\right)(1+\cos^2\theta)\Omega_c^2}{\Omega_c^2-\omega^2}$ (slow mode).

10.5. Derive Eqn (10.2.5) from Eqn (10.2.3), noting that the phase factor (δ) is not explicitly shown.

10.6. Find the *electrostatic* dispersion relation for waves propagating obliquely to the magnetic field $\mathbf{B}_0 = \hat{z}B_0$. Assume that $\mathbf{k} = \hat{x}k\sin\theta +$

$\hat{3}k\cos\theta = \hat{x}k_{\perp} + \hat{z}k_{\parallel}, k_{\parallel}v_{thi} \ll \omega \ll k_{\parallel}v_{the}, c_z^2 = \frac{\kappa T_e}{m_i}$, and $\omega_s^2 = k^2c_s^2/(1+k^2\lambda_{De}^2)$. Notice that one root of the dispersion relation lies above Ω_c and the other below Ω_c. Find explicit expression for ω^2 in the limits $\omega_s^2/\Omega_c^2 \ll 1$ and $\omega_s^2/\Omega_c^2 \gg 1$.

10.7. Explain why the individual cyclotron peaks in Eqn (10.4.6) reduce to the Lorentzian form Eqn (10.6.5) when $\theta_B = \pi/2$. Will electron–ion collisions significantly damp the cyclotron peaks for the ionosphere ($n_e = 10^5\,\mathrm{cm}^{-3}$, $T_e = 0.1\,\mathrm{eV}$, and $4\times10^{-5}\,\mathrm{T}$); In an industrial plasma ($n_e = 10^{12}\,\mathrm{cm}^{-3}$, $T_e = 2\,\mathrm{eV}$, and $B = 0.1\,\mathrm{T}$); and a magnetic fusion plasma ($n_e = 10^{14}\,\mathrm{cm}^{-3}$, $T_e = 10\,\mathrm{keV}$, and $B = 6\,\mathrm{T}$)?

10.8. Derive Eqn (10.7.4) from Eqn (10.7.3).

SOLUTIONS TO ODD PROBLEMS

10.1. This problem relates to how the term $\Sigma_l \exp(-k_\perp^2\rho_e^2) I_l(k_\perp^2\rho_e^2)(\omega/(\omega - l\Omega_e))$ behaves as B goes to zero because, while Ω_e also goes to zero, ρ_e goes to infinity.

For large argument, see Eqn (10.1.11), the function $I_l(z) \rightarrow \{\exp(z^2)/(2\pi z)^{1/2}\}\{1-(\mu-1)/8z+\cdots.\}$, where $\mu = 4l^2$, and in this case, $z = k_\perp^2\rho_e^2$.

Substitution for I_l yields $\Sigma_l\{1-(\mu-1)/8z+\cdots.\}/(2\pi z)^{1/2}$, which goes to zero with B.

10.3. We recall that $\omega_{UH} = \sqrt{\omega_{pe}^2 + \omega_c^2}$ and $\omega_R = \frac{1}{2}\left[\omega_c + \sqrt{\omega_c^2 + 4\omega_{pe}^2}\right]$. In terms of density, these yield

$$\frac{\omega_{p2}^2}{\omega^2} = 1 - \frac{\omega_c^2}{\omega^2} \qquad \text{(upper hybrid)}$$

and $\frac{\omega_{p1}^2}{\omega^2} = 1 - \frac{\omega_c}{\omega}$ (R-wave cutoff),
where $\omega_{p1} < \omega_{p2}$ and $B_0 = \text{const}$.

Then, $\frac{\omega_{p2}^2}{\omega_{p1}^2} = \frac{1-\frac{\omega_c^2}{\omega^2}}{1-\frac{\omega_c}{\omega}} = 1 + \frac{\omega_c}{\omega}$.

Letting $n = n_0\left(1+\frac{X}{L}\right)$ with $\omega_{p1}^2 = \frac{4\pi n_e e^2}{m_e}$, we have

$$\frac{\omega_{p2}^2}{\omega_{p1}^2} = \frac{n_e\left(1+\frac{\Delta X}{L}\right)}{n_e} = 1 + \frac{\Delta X}{L} = 1 + \frac{\omega_c}{\omega}$$

or $\boxed{\Delta X = \frac{\omega_c}{\omega}L}$.

10.5. $[i\omega+\gamma-\mathbf{k}\cdot\mathbf{v}]F_{1q}(\mathbf{k},\mathbf{v},\omega)-(qB_0/m_ec)[\delta F_{1q}(\mathbf{k},\mathbf{v},\omega)/\delta\varphi]=$
$F_{1q}(\mathbf{k},\mathbf{v},0)-(q/m_e)E_1(\mathbf{k},\omega).\ F_{0q}(\omega)/\delta\mathbf{v}.\ =\text{RHS}$
where $\mathbf{k}\cdot\mathbf{v}=k_\perp v_\perp\cos(\varphi-\delta)+k_\parallel v_\parallel$, and we will ignore the phase factor for simplicity.

$\delta(\cos(\varphi-\delta))/\delta\varphi=-\sin(\varphi-\delta)$. Also $qB_0/m_ec=\Omega_q$ and $v_\perp/\Omega_q=\rho_q$.

Now, $(i\omega+\gamma-\mathbf{k}\cdot\mathbf{v})\Omega_q=$
$\{\delta\exp[-(i\omega+\gamma-k_\parallel v_\parallel)\varphi/\Omega_q+k_\perp v_\perp\sin\varphi/\Omega_q]/\delta\varphi\}/\exp[-(i\omega+\gamma-k_\parallel v_\parallel)\varphi/\Omega_q+k_\perp v_\perp\sin\varphi/\Omega_q]$.

Using $\delta(ab)/\delta\varphi=a\delta b/\delta\varphi+b\delta a/\delta\varphi$ and $e^{(iz\sin\varphi)}=\sum_{l=-\infty}^{l=+\infty}J_l(z)e^{il\varphi}$, we obtain

$$F_{1q}(\mathbf{k},\mathbf{v},\omega)=-\int^{\varphi}d\varphi'\exp[-(i\omega+\gamma-k_\parallel v_\parallel)\varphi'/\Omega_q]\cdot\sum_{l=-\infty}^{l=+\infty}J_l$$

$(k_\perp\rho_q)e^{il\varphi'}\cdot\text{RHS}/\{\Omega_q.\exp[-(i\omega+\gamma-k_\parallel v_\parallel)\varphi/\Omega_q+k_\perp v_\perp\sin\varphi/\Omega_q]\}$.

10.7. When $\theta_B=\pi/2$ the parallel wave number $k_\parallel=0$. In this case, the term in the denominator of Eqn (10.4.7) may be taken outside the integral over $\mathbf{v}$, leading to the term Eqn (10.6.5) multiplied by a factor that represents the envelope of the scattered spectrum.

Plasma	Ω_e (rad s^{-1})	ν_{ei} (s^{-1})
Ionosphere	7.0×10^6	1.3×10^2
Industrial	1.8×10^{10}	1.1×10^7
Magnetic Fusion	1.1×10^{12}	4.2×10^3

For the conditions chosen electron–ion collisions will not damp the cyclotron peaks significantly.

Chapter 11

X-Ray Thomson Scattering

Chapter Outline Head

11.1 OVERVIEW

With the advent of the field of High Energy Density Physics (Drake, 2006) in this millennium, the need for accurate measurements of the density and temperature conditions, physical properties, and structure of dense plasmas has emerged (Glenzer & Redmer, 2009). Typically, conditions with pressures above $1\,\mathrm{Mbar} = 10^2\,\mathrm{GPa} = 10^5\,\mathrm{J\,cm^{-3}}$ are of interest with densities approaching solid density or higher and with temperatures above several electron volts. Powerful x-ray sources enable these measurements with spectrally resolved x-ray scattering. They require a large photon flux in a short x-ray burst with high spectral purity to resolve the Compton and plasmon features in short-lived states of matter. The first experiments have employed He$-\alpha$ and Ly$-\alpha$ sources to demonstrate Compton and plasmon measurements in isochorically heated matter (Glenzer *et al.*, 2003b, 2007). These proof-of-principle experimental studies have been extended to shock-compressed matter (Regan *et al.*, 2007; Lee *et al.*, 2009a) and the characterization of shocks with picosecond temporal resolution employing K$-\alpha$ x-rays (Kritcher *et al.*, 2008, 2009; LePape, 2010). In parallel to the experimental demonstration, the theory of x-ray scattering has been developed (Gregori *et al.*, 2003b, 2006a; Gregori & Gericke, 2009), including an appropriate description of the inelastic scattering

DOI: 10.1016/B978-0-12-374877-5.00011-7

component beyond the random phase approximation with the Mermin approach and close coupling corrections (Redmer *et al.*, 2005; Höll *et al.*, 2004; Thiele *et al.*, 2008; Gregori *et al.*, 2003a).

Importantly, these x-ray scattering experiments observe quasi-elastic and inelastic scattering features, i.e., the Compton and plasmon scattering features. The latter are signatures of the noncollective and collective scattering regimes that are being accessed in backward and forward scattering geometries, respectively. For plasmas that are ideal or moderately coupled, these features are theoretically understood. In contrast, apart from simple limits, calculations of elastic scattering are still being developed, for example, using hypernetted-chain equations with quantum potentials or density functional theory (Schwarz *et al.*, 2007; Wünsch *et al.*, 2008, 2009). Inelastic scattering has required the application of powerful x-ray sources with energies in the range of $2\,\mathrm{keV} < E < 9\,\mathrm{keV}$ that penetrate through compressed materials having densities of solid and above. Laser-produced He$-\alpha$ and Ly$-\alpha$ radiation from nanosecond laser plasmas or K$-\alpha$ radiation from ultra short pulse laser plasmas fulfills the stringent requirements on photon numbers and bandwidth for spectrally resolved x-ray scattering measurements in single shot experiments. In addition, free-electron laser sources are now being applied for scattering experiments in multishot mode (Höll *et al.*, 2007; Fäustlin *et al.*, 2010b).

In the Compton (backward) scattering regime, the scattering process is noncoherent, and the spectrum shows the Compton downshifted line that is broadened by the thermal motion of the electrons (Glenzer *et al.*, 2003b). In degenerate systems, the width of the Compton scattering line is thus determined by the Fermi energy providing a measurement of the electron density (Lee *et al.*, 2009a). In this case, corrections due to the Pauli exclusion principle and finite collective effects must be included. On the other hand, in nondegenerate heated matter, the electron velocity distribution function transitions to a Maxwell–Boltzmann distribution and yields the electron temperature (Landen *et al.*, 2001b). Thus, for the full interpretation of the elastic and inelastic scattering components from dense plasmas, the theoretical form factor of Chihara (1987, 2000) has been further developed by Gregori *et al.* (2003a,b, 2006a) to account for these effects and to further include free–free and bound–free processes.

In the plasmon (forward) scattering regime, the collective plasmon feature has been observed at small frequency shifts by Glenzer *et al.* (2007), Kritcher *et al.* (2008), Lee *et al.* (2009a), and Neumayer *et al.* (2010). In addition, at higher electron temperature, the upshifted plasmon feature has been measured (Döppner *et al.*, 2009), which allows inferring the temperature via the relation of detailed balance. The plasmon frequency shift is determined by the plasmon dispersion relation with the leading term being determined by the plasma frequency and thus the electron density while the electron temperature being a second-order effect contributing to spatial propagation of plasmon (Langmuir) oscillations (Thiele *et al.*, 2008). In plasmas with accurately known density, the plasmon frequency thus provides an estimate of the electron temperature. In addition, damping of the plasmon feature is an important observable in present experiments. The width of the plasmon is hence sensitive to temperature and collisions. In the latter case, theoretical approximations that use the Mermin approach to account for collisions (Redmer *et al.*, 2005) and local field correction (Fortmann *et al.*, 2010) have provided consistent descriptions of experimental spectra (Glenzer *et al.*, 2007; Neumayer *et al.*, 2010).

11.2 X-RAY SCATTERING RELATIONS

The dense plasma is irradiated by nonpolarized x-rays of energy E and wavelength λ_i from a laser-irradiated foil or by polarized free-electron laser radiation along the direction of the incident wave vector $\mathbf{k}_i$ with $k_i = 2\pi/\lambda_i$. The detector observes the scattered radiation at the scattering angle θ in the direction of the scattered wave vector $\mathbf{k_S}$ at a distance much larger than the plasma size. During the scattering process, the incident x-ray photon transfers, on average, momentum $\hbar\mathbf{k}$ and the Compton energy $E_C = \hbar\omega = \hbar^2 k^2/2m_e = \hbar\omega_i - \hbar\omega_s$ to the electron, where ω_S is the frequency of the scattered radiation. In the nonrelativistic limit ($\hbar\omega \ll \hbar\omega_i$) and for small momentum transfers, we have $\mathbf{k} = 2\mathbf{k}_i \sin(\theta/2)$, and we can express the scattering vector $\mathbf{k}$ in terms of the x-ray probe energy

$$k = |\mathbf{k}| = 4\pi \frac{E}{hc} \sin(\theta/2). \tag{11.2.1}$$

Equation (11.2.1) determines the scale length $\lambda^\star \approx 2\pi/k$ of the electron density fluctuations measured in the scattering experiment. The comparison with the screening length, λ_S, determines the scattering regime, which is characterized by the dimensionless scattering parameter α

$$\alpha = \frac{1}{k\lambda_S}, \tag{11.2.2}$$

The screening length may be calculated numerically using Fermi integrals or analytically using an effective temperature. In the nondegenerate case, we obtain the usual Debye screening length

$$\lambda_S^{-1} \to \lambda_{\text{De}}^{-1} = \left(\frac{n_e e^2}{\varepsilon_0 \kappa T_e}\right)^{1/2}. \tag{11.2.3}$$

For degenerate systems, we obtain the Thomas–Fermi screening length.

$$\lambda_S^{-1} \to \lambda_{TF}^{-1} = \sqrt{\frac{m_e e^2}{\pi \varepsilon_0 \hbar^2} \left(\frac{3n_e}{\pi}\right)^{1/3}}. \tag{11.2.4}$$

This regime is characterized by a degeneracy parameter $\Theta \leq 1$. Θ indicates the role of quantum statistical effects in the system. It is defined by the ratio of the thermal energy and the Fermi energy ϵ_F

$$\Theta = \frac{\kappa T_e}{\epsilon_F}, \quad \epsilon_F = \frac{\hbar^2}{2m_e}\left(3\pi^2 n_e\right)^{2/3}, \qquad (11.2.5)$$

In a degenerate plasma, the Fermi energy is larger than the thermal energy, i.e., $\Theta \leq 1$, and most electrons populate states inside the Fermi sea where quantum effects are of importance.

As seen from Eqns (11.2.3, 11.2.4), λ_S is determined by the plasma conditions, density, and temperature, while $\lambda^\star$ is primarily determined by the x-ray probe energy and the scattering angle [i.e., by the setup of the scattering experiment, Eqn (11.2.1)]. However, corrections to the length of the scattering vector occur due to the finite frequency shift of the scattered radiation and due to the dispersion of the radiation in the dense plasma. The scattering parameter defines the regime of a scattering experiment with $\alpha > 1$ indicating the collective scattering regime where collective plasmon and acoustic oscillations are observed. On the other hand, for $\alpha << 1$, the scattering spectrum reflects the noncollective motion of the electrons.

With the scattering vector **k** defined by Eqn (11.2.1), the scattered power P_s from N electrons into a frequency interval $d\omega$ and solid angle $d\Omega$, i.e., the scattering spectrum, is given by Eqn (2.3.13),

$$P_s(\mathbf{R},\omega)d\Omega d\omega = \frac{P_i r_0^2 d\Omega}{2\pi A} N S(\mathbf{k},\omega)d\omega \times \left|\hat{\mathbf{s}} \times \left(\hat{\mathbf{s}} \times \hat{\mathbf{E}}_i\right)\right|^2. \qquad (11.2.6)$$

In Eqn (11.2.6), P_i denotes the incident x-ray power, A is the plasma area irradiated by the probe x-rays, and $S(\mathbf{k},\omega)$ is the total electron dynamic structure factor. The structure factor is defined as the Fourier transform of the electron–electron density fluctuations [Eqn (2.3.14)], containing the details of the correlated many-particle system, cf. Ichimaru *et al.* (1984) and Ichimaru (1984a). The polarization term reflects the dependence of the scattered power on the polarization of the incident radiation; for nonpolarized laser-produced x-ray sources, Eqn (4.2.3) gives, $\left|\hat{\mathbf{s}} \times \left(\hat{\mathbf{s}} \times \hat{\mathbf{E}}_i\right)\right|^2 = \left(1 - \frac{1}{2}\sin^2\theta\right) = \frac{1}{2}\left(1 + \cos^2\theta\right)$. By multiplying Eqn (11.2.6) with the length of the scattering volume, L, it is seen that the scattered power depends on the setup of the scattering experiment (initial probe power, scattering angle, probe wavelength, probe polarization, detector solid angle), the electron density, the plasma length, and the total dynamic structure factor $S(\mathbf{k},\omega)$.

The full theoretical structure factor accounts for inelastic Compton or plasmon scattering, elastic scattering at the incident x-ray energy, as well as possible contributions of bound electrons to the total inelastic scattering component. The total electron dynamic structure factor $S(\mathbf{k},\omega)$ includes these effects for the full range of temperatures assessed in experiments. Chihara (1987, 2000) derived the form

$$S(k,\omega)=|f_I(k)+q(k)|^2 S_{ii}(k,\omega)+Z_f S_{ee}^0(k,\omega) + Z_b\int d\omega' \widetilde{S}_{ce}(k,\omega-\omega')S_s(k,\omega'). \tag{11.2.7}$$

The first term in Eqn (11.2.7) accounts for the density correlations of electrons that dynamically follow the ion motion. This includes both the bound electrons, represented by the ion form factor $f_I(k)$, and the screening cloud of free (and valence) electrons that surround the ion, represented by $q(k)$. The ion–ion density correlation function, $S_{ii}(k,\omega)$, reflects the thermal motion of the ions and/or the ion plasma frequency. The second term in Eqn (11.2.7) gives the contribution to the scattering from the high-frequency motion of the free electrons that do not follow the ion motion. Here, $S_{ee}^0(k,\omega)$ is the high-frequency part of the electron–electron correlation function, and it reduces to the usual electron feature in the case of an optical probe (see Section 5.2.1). The last term of Eqn (11.2.7) includes inelastic scattering by bound electrons, which arises from Raman transitions to the continuum of core electrons within an ion, $\tilde{S}_{ce}(k,\omega)$, modulated by the self-motion of the ions, represented by $S_s(k,\omega)$.

Simple analytical equations can be derived to illustrate the dependency of the scattering spectra on density and temperature. The frequency shift of a scattered photon by a free electron is determined by the Compton and the Doppler effects

$$\omega = -\hbar k^2/2m_e \pm \mathbf{k}\cdot\mathbf{v}. \tag{11.2.8}$$

The former is due to the transfer of momentum from the incident photon to the electron during the scattering process resulting in the Compton downshift of the scattered radiation to lower x-ray energies. This term defines the inelastic scattering component in fully noncollective x-ray Thomson scattering experiments. Thus, the spectrum of the Compton-downshifted line reflects the velocity component of the electron in the direction of the scattering vector **k**.

To calculate the number of electrons that contribute to the broadening via Eqn (11.2.8), the three-dimensional velocity space is considered. The number of electrons within the velocity interval v_x and v_x+dv_x is proportional to the volume described by an annulus around v_x that must be integrated to include all possible projections of velocity vectors $\mathbf{v}$ onto v_x. With the Fermi distribution function applied for weakly degenerate plasmas, $\kappa T_e < \epsilon_F$, where $v_F=(2\epsilon_F/m_e)^{1/2}$ is the Fermi velocity, and Landen *et al.* (2001b) showed

$$f_0\left(\frac{v_x}{v_F}\right) \propto \int_0^{\pi/2} \frac{(v_x/v_F\cos\beta)^2\tan\beta d\beta}{e^{\left(\left(\frac{v_x}{v_F\cos\beta}\right)^2-\eta\right)/\left(\frac{T_e}{\epsilon_F}\right)}+1}, \tag{11.2.9}$$

with β being the angle between $\mathbf{v}$ and the v_x-axis and η is given by the Sommerfeld expansion

$$\eta = 1 - \left(\frac{\pi^2}{12}\right)\left(\frac{T_e}{\epsilon_F}\right)^2. \tag{11.2.10}$$

Here, the second term accounts for the fact that the chemical potential in the expression for the occupation of states for fermions, $(\exp(E-\mu)/k_B T_e + 1)^{-1}$, has some temperature dependence at finite temperature. For nondegenerate systems, the expression $\eta = (n_e \Lambda_e^3)/2$, where Λ_e is the thermal wavelength, may be used. A general solution is the inversion of Fermi integrals as mentioned after Eqn (11.2.2).

In a degenerate system, applying Eqns (11.2.9) and (11.2.10) will provide the spectrum of the Compton line with the width proportional to $(\epsilon_F)^{1/2}$ and hence to $(n_e)^{1/3}$ providing the density of the plasma. In a nondegenerate plasma, on the other hand, the Compton scattering spectrum will reflect a Maxwell–Boltzmann distribution, and the width will yield a measurement of the electron temperature.

With decreasing scattering angle, the scattering length increases, and the scattering parameter assumes values with $\alpha > 1$. This scattering regime is collective, and Eqn (11.2.7) will describe the collective electron plasmon oscillations and acoustic resonances. The latter are not resolved in present x-ray scattering experiments, and they are restricted to measure relative or absolute intensities of quasi-elastically scattered x-rays (Riley *et al.*, 2000; Saiz *et al.*, 2008a,b; Ravasio *et al.*, 2007; Kritcher *et al.*, 2009; Barbrel *et al.*, 2009; Kugland *et al.*, 2009). On the other hand, the plasmon spectrum has been resolved with laser-produced x-ray and free-electron laser sources. The plasmon frequency shift from E is determined by the plasmon dispersion relation, and the width is determined by Landau damping and collisional damping processes. For small values of k and large values of Θ, using an inversion of Fermi integrals results in a modified analytical Bohm–Gross dispersion relation (Thiele *et al.*, 2008),

$$\omega_{pl}^2 = \omega_{pe}^2 + \frac{3}{2}k^2 a_{th}^2 (1 + 0.088 n_e \Lambda_e^3) + \left(\frac{\hbar k^2}{2m_e}\right)^2, \tag{11.2.11}$$

where $\omega_{pe}\kappa\,\Lambda_e = h/\sqrt{2\pi m_e \kappa T_e}$ is the thermal wavelength.

In Eqn (11.2.11), the first term is a result of electron oscillations in the plasma, and the second term represents the effect on propagation of the oscillation from thermal pressure. The third term includes degeneracy effects from Fermi pressure, and the last term is the quantum diffraction shift.

Independent of the evaluation of the dielectric function, the shape of the structure factor, and of modeling of collisional effects, the dielectric function fulfills the requirement,

$$\frac{S(\mathbf{k},\omega)}{S(-\mathbf{k},-\omega)} = e^{-\hbar\omega/\kappa T_e}. \qquad (11.2.12)$$

This relation is referred to as the detailed balance relation. As a general consequence, the structure factor shows an asymmetry with respect to **k** and ω. In the collective scattering regime and for conditions with sufficiently high temperatures, upshifted plasmons have been observed to provide temperatures (Döppner *et al.*, 2009).

11.3 X-RAY SCATTERING EXPERIMENTS

The first spectrally resolved backward and forward x-ray scattering experiments in warm dense matter was performed on isochorically heated solid-density beryllium. The solid is homogeneously and isochorically heated by L-shell x-rays from a mid-Z foil (1–2 μm thick Mg, Rh, or Ag foils wrapped around the solid), and the dense plasma is probed after $t = 0.5$ ns with the narrow band x-ray He$-\alpha$ or Ly$-\alpha$ probe radiation from titanium or chlorine. About 10 kJ of laser energy was used for producing the narrow band x-ray probe, and the broadband heating radiation produced with killoJoule lasers provides sufficient photons for producing homogeneous High-Energy-Density Plasma (HEDP) states of matter and probing in single shot experiments. The same target platform was subsequently employed to probe isochorically heated carbon. Isochoric heating experiments have the advantage that the mass density is known *a priori* so that the electron density can be directly inferred if the ionization state is known or vice versa.

The back-scattering spectrum from heated beryllium is shown in Fig. 11.1. Elastic scattering from both the titanium He$-\alpha$ radiation at 4.75 keV and Ly$-\alpha$ radiation at 4.96 keV is observed together with the downshifted Compton scattering feature. The Compton scattering feature shows significant broadening due to the thermal motion of the electrons resulting in a full width at half maximum that is a factor of two larger than for scattering spectra from cold beryllium, i.e., no heater laser beams. The spectrum is fit with theoretical spectra employing the measured spectrum of the titanium probe x-rays, spectrometer resolution, and the dynamic structure factor, providing temperatures and densities from the broadening of the Compton downshifted line and from the intensity ratio of elastic and inelastic scattering components. The figure also shows the sensitivity of the shape of the red Compton scattering wing to the electron temperature, which is inferred from these data with an accuracy of about 10%.

The ionization state and therefore the electron density can be inferred from the intensity ratio of elastically to inelastically scattered radiation. In these conditions, with large values of **k** and where the contribution of weakly bound electrons to the intensity of the inelastic scattering component is small, the intensity of the elastic scattering feature approaches f_I^2. For the data in Fig. 11.1,

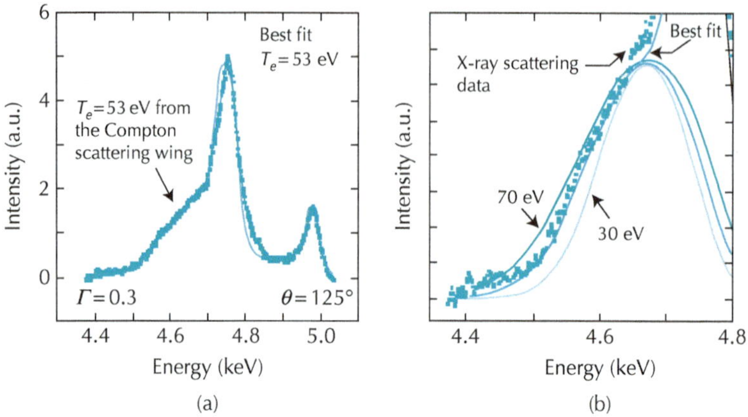

FIGURE 11.1 (a) Scattering spectrum (dots) from isochorically heated beryllium is shown from Glenzer *et al.* (2003b). Elastic and inelastic (Compton) scattering components are observed from both the titanium He$-\alpha$ and Ly$-\alpha$ probe x-rays. (b) The spectrum fits well with theoretical scattering spectra employing the dynamic structure factor (solid line). The fit of the red wing of the Compton scattering spectrum is sensitive to the temperature. These data provide temperatures of $T_e = 53\,\text{eV}$ and densities of $n_e = 3.3 \times 10^{23}\,\text{cm}^{-3}$ characterizing the solid density plasma regime with an error bar of about 10%.

the ratio yields the density with accuracy of about 15%. For lower temperature conditions, however, where bound electrons scatter nonelastically, the contributions from free and weakly bound electrons blend. Nonetheless, the bound–free spectrum is often known and can be distinguished from the Compton scattering spectrum of the free electrons yielding temperature and density information with slightly larger error bars.

Figure 11.2 shows experimental scattering spectra from isochorically heated beryllium measured in forward scattering of the chlorine Ly$-\alpha$ line at 2.96 keV. These experiments access collective scattering in a weakly degenerate solid-density beryllium plasma with electron temperature of the order of the Fermi temperature, $T_e \simeq T_F = 15\,\text{eV}$. They require forward scattering with $\theta = 40°$ and x-ray probe energies of order $E = 3\,\text{keV}$ ($\lambda_i \approx 0.4\,\text{nm}$). In these conditions, the scattering is predominantly probing **k**-vectors with $\text{k} = 1\,\text{Å}^{-1}$. Calculating the screening length at the effective temperature results in $\alpha = 1.6$. In this regime, the scattered light spectrum shows collective effects corresponding to scattering resonances off ion-acoustic waves and off electron plasma waves, i.e., plasmons.

To resolve the plasmon frequency shift and damping, the x-ray bandwidth has to be small compared to that used for Compton scattering measurements. In Fig. 11.2, the plasmon shift of $\Delta E = 28\,\text{eV}$ was resolved by the chlorine Ly$-\alpha$ probe radiation with an effective bandwidth of 7.7 eV and no significant dielectronic satellite radiation on the red wing of the Ly$-\alpha$ doublet. Also shown are theoretical scattering profiles that represent a convolution of the theoretical form

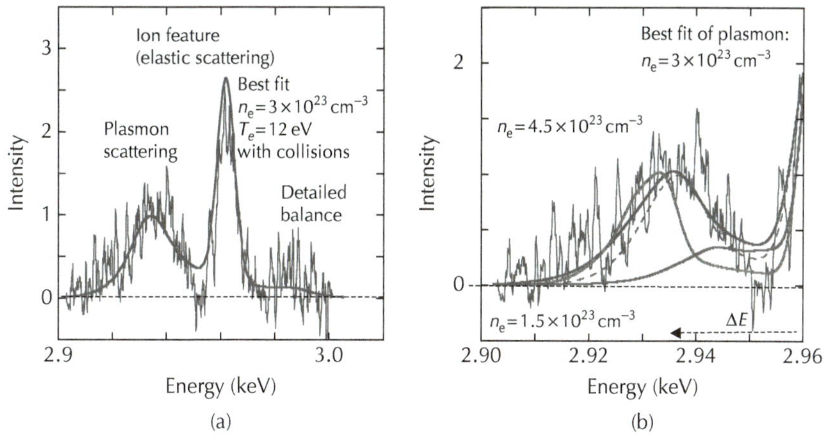

FIGURE 11.2 (a) Scattering spectrum (gray line) from isochorically heated beryllium is shown from Glenzer *et al.* (2007). Plasmon scattering down- and up-shifted in energy is observed, where the intensity of the upshifted plasmon is reduced according to detailed balance, cf. Eqn (11.2.12). The spectrum is best fit with theoretical scattering spectra of Section 11.2 for $T_e = 12$ eV and $n_e = 3 \times 10^{23}$ cm^{-3}. (b) Also shown is the comparison of the plasmon spectrum with calculations for two different densities, $n_e = 4.5 \times 10^{23}$ cm^{-3} and $n_e = 1.5 \times 10^{23}$ cm^{-3}, indicating that the spectrum is best fit for $n_e = 3 \times 10^{23}$ cm^{-3}. The dashed curve represents a calculation for a density of $n_e = 3 \times 10^{23}$ cm^{-3} that is neglecting collision.

factor $S(\mathbf{k}, \omega)$, calculated for the range of **k**-vectors in the experiment, with the spectral resolution of 7.7 eV.

The ion feature is observed as an elastic scattering peak at $E = 2.96$ keV that is not resolved in this experiment. On the lower energy wing of the ion feature, we observe a strong plasmon resonance at 2.93 keV. On the higher energy wing with nearly the same frequency shift, the data show a weak upshifted plasmon signal. Compared to the intensity of downshifted plasmon, the intensity is reduced by the Bose function, cf. Eqn (11.2.12), reflecting the principle of detailed balance. The intensity ratio of these plasmon features is thus sensitive to the temperature.

In $S(\mathbf{k}, \omega)$, collision effects on plasmons are accounted for in the Born Mermin approximation, which also incorporates degeneracy effects. The fit provides a temperature of $T_e = 12$ eV and density of $n_e = 3 \times 10^{23}$ cm^{-3}. Figure 11.2b indicates that the density is accurately determined by the shift of the plasmon. For the small **k**-vectors probed in the present experiment, significant deviations of ω_{pl} from the random phase approximation are not expected yielding an error bar in density that is solely determined by the signal-to-noise ratio and the quality of the fit. The calculations for densities of $n_e = 4.5 \times 10^{23}$ cm^{-3} and $n_e = 1.5 \times 10^{23}$ cm^{-3} indicate that the error bar in density is of order 20%; this value has been subsequently improved with better signal-to-noise ratio, for example, in Döppner *et al.* (2009); Neumayer *et al.* (2010).

Inferring the temperature from the collective scattering spectrum alone requires accurate measurements of the upshifted plasmon and the application of detailed balance, or alternatively, the width of the plasmon may be used. For the conditions in isochorically heated beryllium, collisions account for an additional broadening of approximately 5 eV. The difference can be as large as a factor of 1.5 for scattering spectra when **k**-vector blurring is negligible. Clearly, scattering experiments in a regime where damping by collisions dominates the broadening of plasmons are of interest to test models of the collisionality and conductivity in dense plasmas. Including the corrections due to collisions provides a good fit of the plasmon spectrum for a temperature of 12 eV and with the dynamic collision frequency calculated in Born approximation.

An example of spectrally resolved x-ray scattering spectra from shock-compressed matter is shown in Fig. 11.3. Lithium hydride has been compressed by a nanosecond laser beam, and the density and temperature conditions have been measured with a 10 ps K$-\alpha$ source (Kritcher *et al.*, 2008). The power of the nanosecond laser beam was shaped with a 4 ns-long foot at 10^{13} W cm^{-2} and

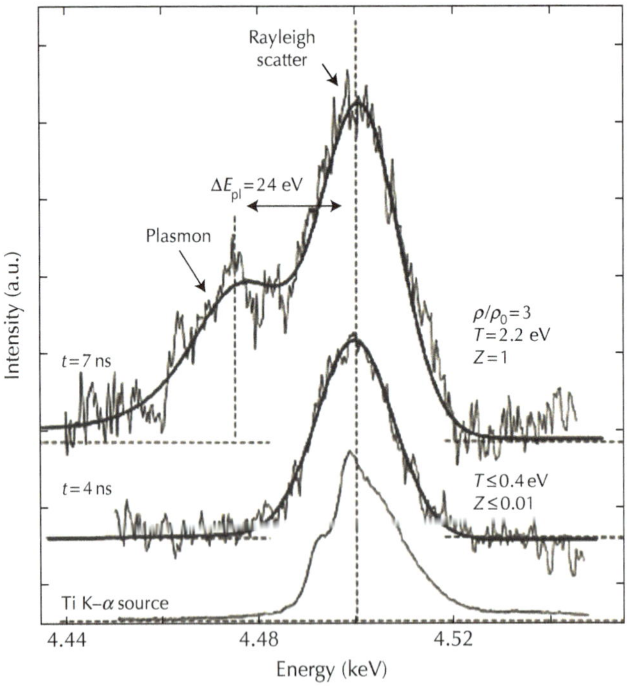

FIGURE 11.3 Experimental and theoretical scattering spectra from shock-compressed LiH are shown from Kritcher *et al.* (2008). At $t = 7$ ns, the scattering parameter is $\alpha \simeq 1$, and the inelastic scattering component shows a plasmon downshifted by 24 eV from the K$-\alpha$ energy. At $t = 4$ ns, the lack of inelastic scattering indicates small temperatures and negligible ionization. Also shown is the K$-\alpha$ x-ray probe spectrum.

a 2 ns-long peak at 3×10^{13} W cm^{-2} launching two shock waves into the target. Radiation-hydrodynamic simulations show that the shock waves coalesce and compress the target by a factor of three over the 600 μm focal spot area at 7 ns after the beginning of the laser drive.

The conditions before and during coalescence are measured with 10 ps resolution by varying the delay between the nanosecond laser beam and the short pulse laser-produced K–α x-rays. The x-rays penetrate through the dense material, and measurements of the elastic and inelastic scattering components at different times during the shock evolution are observed with a HOPG crystal spectrometer at a scattering angle of about 40°.

At $t = 7$ ns, a plasmon is observed that is downshifted in energy by about 24 eV. The shift is dominated by the plasma frequency, i.e., by the electron density, while quantum diffraction results in a correction of order 7 eV. For the low temperatures and high densities in the shock wave, the system is degenerate with $\alpha \simeq 1$, and in this case, the width and shift of the plasmon yields the electron density of $n_e = 1.7 \times 10^{23}$ cm^{-3}. Due to noise, the error bar for the density is about 20%.

The elastic scattering intensity is calibrated by the inelastic scattering component; thus, this fact allows one to infer the temperature history from the dependence of the dynamic structure factor. At early times, $t < 6$ ns, the lack of inelastic scattering is a strong indication of low ionization states, $Z < 0.1$, and simultaneously low temperatures, $T < 0.5$ eV. At shock coalescence, rapid heating to temperatures of $T = 2.2$ eV is observed when the scattering spectra show the collective plasmon oscillations that indicate the transition to the dense metallic plasma state. This 10 ps capability allows for testing of shock coalescence models that vary in this regime by 300 ps.

Free-electron laser scattering spectra are shown in Fig. 11.4. The experiments use the Free Electron Laser in Hamburg (FLASH) (Ackermann *et al.*, 2007) at a photon energy of 91.8 eV. Free-electron laser radiation with 5 Hz pulse repetition rate, average pulse energy on target of 15 μJ, and duration of ~40 fs is focused to a 25 μm spot using an elliptical mirror, yielding intensities of $\sim 8 \times 10^{13}$ W cm^{-2}. The pulses irradiate a liquid hydrogen jet, which has a diameter of 20 μm, an atomic density of 4.2×10^{22} cm^{-3}, and a temperature of 2 meV. The scattering data are measured with a variable line space grating spectrograph (Fäustlin *et al.*, 2010a) at 90° relative to the incident radiation and in the vertical plane since the free-electron laser radiation is horizontally polarized.

The intense free-electron laser pulses deposit their energy into the liquid hydrogen while simultaneously probing the system. Due to the ultrashort pulse duration and the high penetration depth, the liquid hydrogen is heated isochorically. The scattering spectrum for 15-min integration time (4500 pulses) shows asymmetric peaks (plasmons) equally blue and red shifted from the incident photon energy by 0.65 eV. The measured spectrum is mainly broadened by the incident laser bandwidth (1.1 eV full width at half maximum). Further, source broadening effects (0.2 eV corresponding to the 20 μm source diameter) are

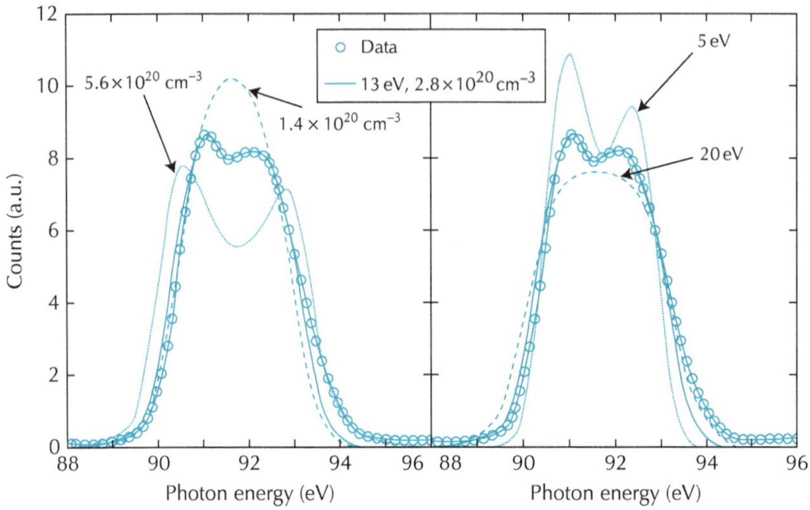

FIGURE 11.4 Experimental spectrum (circles) and the best fit of a calculated spectrum using the Born-Mermin approximation with $n_e = 2.8 \times 10^{20}\,\text{cm}^{-3}$ and $T_e = 13\,\text{eV}$ (solid line) from Fäustlin *et al.* (2010b). Also shown are comparison to theoretical spectra with densities (left graph) of $n_e = 5.6 \times 10^{20}\,\text{cm}^{-3}$ (dashed) and $n_e = 1.4 \times 10^{20}\,\text{cm}^{-3}$ (dash-dotted), with temperatures (right graph) of 20eV (dash-dotted) and 5eV (dashed). (Courtesy of R. R. Fäustlin and T. Tschentscher.)

taken into account. These scattering data indicate a small degree of ionization as a result of the short timescale of the interaction. Although the electronic subsystem is thermalized within the duration of the free-electron laser pulse, no equilibration has been reached between the other components resulting in low degree of ionization at high electron temperature.

11.4 APPLICATIONS

Spectrally resolved x-ray scattering measurements are of central importance for measuring temperatures and densities of dense plasmas produced by isochoric heating of matter or by shock compression of solids. Experimental results from dense plasmas in laser shock-compressed foils are shown in Fig. 11.5. In these studies, laser beams directly illuminate a beryllium foil with intensities in the range of $10^{14}\,\text{W cm}^{-2} < I < 10^{15}\,\text{W cm}^{-2}$ compressing the foil by ablation pressure in the range of 20–60 Mbar.

The Compton scattering data employing a 6.2-keV He$-\alpha$ x-ray probe measured in noncollective scattering with $\alpha \simeq 0.5$ and a scattering vector of $k = 4.4\,\text{Å}^{-1}$ shows a parabolic-like spectrum indicating a nearly Fermi-degenerate plasma, and the width of the Compton spectrum provides the Fermi energy according to Eqn (11.2.5). In addition, the intensity ratio of the elastic to inelastic scattering feature from Fermi-degenerate plasmas is sensitive to the ion

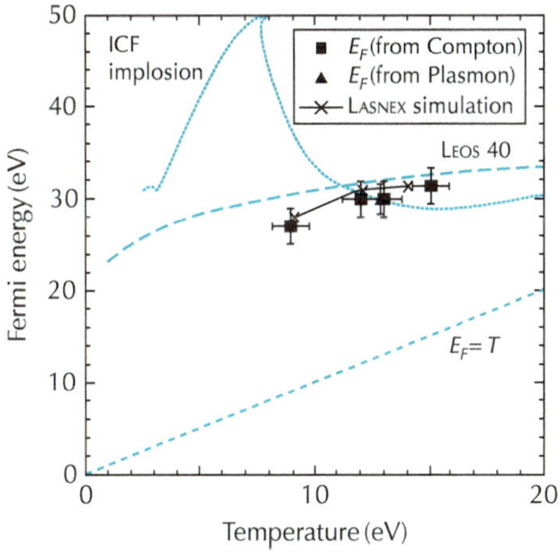

FIGURE 11.5 Fermi energies inferred from the measured densities from the Compton and plasmon scattering data are shown versus temperature from Lee *et al.* (2009a). A high level of degeneracy is found close to predictions from LASNEX (solid curve) and the LEOS data table (long dashed). Also shown are the trajectory of beryllium in inertial confinement fusion implosion modeling (dashed) together with the curve $E_F = T$ (short dashed).

temperature because in this regime elastic scattering is dependent on the ion–ion structure factor.

In addition to the Compton scattering spectrum, plasmons have been measured by accessing the collective scattering regime with $\alpha = 1.56$ and $k = 1.36\,\text{Å}^{-1}$. In contrast to noncollective scattering, the broad inelastic Compton scattering feature is replaced with two small plasmon features which are shifted in energy by $\sim 40\,\text{eV}$ from the two incident x-ray probe features. In this case, the frequency shift due to the Compton effect is small of order 7 eV and the frequency shift of the plasmons is dominated by the frequency of plasma (Langmuir) oscillations.

Experimental scattering spectra have been fit with theoretical spectra and the sensitivity to the plasma conditions has been determined by varying the electron density and the temperature. For the analysis, $T_e = T_i$ and $Z = 2$ has been assumed consistent with calculations and with observations from isochorically heated Be. Density and temperature obtained from the Compton scattering spectra are $n_e = 7.5 \times 10^{23}\,\text{cm}^{-3}$ and $T = 13\,\text{eV}$, respectively. The error bar due to noise for these measurements is $5-10\%$. In addition, the frequency shift of the plasmon provides an accurate measure of the electron density of $n_e = 7.5 \times 10^{23}\,\text{cm}^{-3}$ with an error bar of $\pm 6\%$. Conversely, using the electron density from the 90° scattering data, we obtain an estimate of $T_e = 13 \pm 3\,\text{eV}$ for the electron temperature from the thermal corrections to the plasmon dispersion. This result indicates equal electron and ion temperatures as expected for the dense matter conditions encountered in these shock-compressed foils. In addition, these findings establish the procedure to derive temperatures from the absolute elastic scattering amplitude in the noncollective scattering regime.

The experiments demonstrate a direct measurement of the degeneracy and adiabat in these single-shocked foils. The direct comparison with radiation-hydrodynamic LASNEX simulations (Zimmerman & Kruer, 1975) that use the LEOS equation of state (More *et al.*, 1988) shows agreement with the data. The simulations indicate slightly lower compression than the Hugoniot data predicted by LEOS by taking into account x-ray preheat and averaging over the densities in the shocked region. The agreement with LASNEX indicates that the conditions at low drive show observable effects of gradients. However, the data measured at higher drive agree well with both with LASNEX and LEOS.

Also, shown in Fig. 11.5 are calculations of the conditions predicted for inertial confinement fusion experiments during the implosion phase. These calculations use a beryllium ablator and predict ignition and gain employing a 1.2 MJ laser drive on the National Ignition Facility. The present experiment shows that these conditions are accessible by x-ray Thomson scattering allowing a direct measurement of the degeneracy in implosions.

Ionization balance measurements have been demonstrated with spectrally resolved noncollective x-ray scattering measurements using the ratio of inelastic to quasi-elastic scattering components. Measurements in isochorically heated carbon employing a variety of target platforms that span electron temperatures from cold to fully ionized carbon at 280 eV are shown in Fig. 11.6.

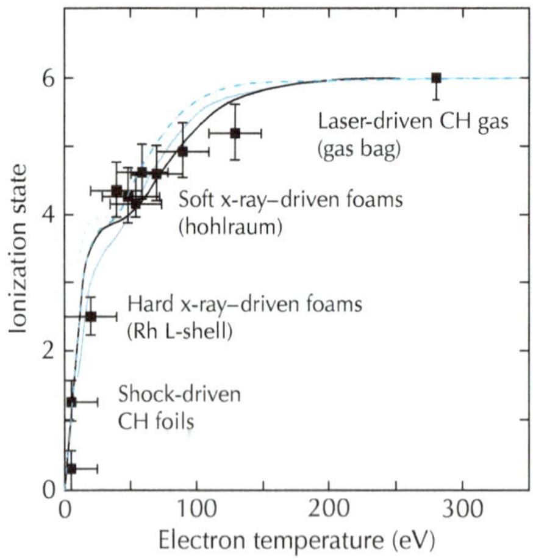

FIGURE 11.6 The measured ionization state of carbon is shown as a function of the measured electron temperature from Gregori *et al.* (2006b). These experiments employ noncollective x-ray scattering on various target platforms. The experimental data points correspond to laser-driven gas bag experiments, soft x-ray–driven carbon foams, hard x-ray–driven carbon foams, and cold plastic foils. Also shown are calculations from kinetic calculations using the code FLYCHK for densities of $10^{21}\,cm^{-3}$ (dotted line), $10^{22}\,cm^{-3}$ (dashed line), and $10^{23}\,cm^{-3}$ (solid blue line) and COMPTRA for carbon at $0.2\,g\,cm^{-3}$ (solid black line).

The mid-to-high T_e data points were obtained on low-density (0.2 g cm^{-3}) carbon foam targets driven by hohlraum soft x-rays (Gregori *et al.*, 2008), while the mid-to-low T_e points correspond to hard x-ray isochorically heated higher density (0.7 g cm^{-3}) carbon foams (Gregori *et al.*, 2004). A low-temperature data point on cold CH plastic foils (1 g cm^{-3}) from Regan *et al.* (2007) and a high-temperature data point from a fully ionized gas bag plasma are also shown in the figure. To extract temperature and ionization state from CH foils, Gregori *et al.* (2006b) extended $S_{ii}(k,\omega)$ for multicomponent plasmas. The complete data set presents the full ionization balance curve for carbon over a large range of electron temperatures from $2\,\text{eV} < T_e < 300\,\text{eV}$ and densities from $10^{21}\,\text{cm}^{-3} < n_e < 10^{23}\,\text{cm}^{-3}$.

The comparison with ionization balance calculations is also given in Fig. 11.6 for COMPTRA (Kuhlbrodt *et al.*, 2005) and FLYCHK (Chung *et al.*, 2003) simulations. The calculation reported in the figure for COMPTRA corresponds to the case of solid density carbon at 0.2 g cm^{-3}. The model shows agreement with the data except in the 120−170 eV range, which is near the transition from helium-like to hydrogen-like carbon. Similar behavior is also shown by FLYCHK for which three calculations are shown. From Saha equilibrium, the calculations for higher densities result in smaller ionization states. With exception of the transition region between helium-like to hydrogen-like carbon, the ionization balance calculations agree well with the experiments. These results may be extended to other species to test the predictive capability of kinetics collisional radiative modeling of plasmas.

Generally, shock-compressed matter experiments have begun delivering data indicating that tests of ion–ion structure factor calculations are possible (Riley *et al.*, 2000; Saiz *et al.*, 2008a,b; Ravasio *et al.*, 2007; Kritcher *et al.*, 2009; Barbrel *et al.*, 2009; Kugland *et al.*, 2009). Experiments by Saiz *et al.* (2008b) on shocked lithium have allowed inferring the ion–ion static structure factor when calibrating the measured elastic scattering amplitude with the measured signal from free electrons; the latter is determined by the f-sum rule,

$$\int_{-\infty}^{\infty} \left(\frac{d^2\sigma}{d\Omega d\omega}\right)_{free} \omega d\omega = \frac{Z\hbar k^2}{2m_e}. \tag{11.4.1}$$

Ion–ion structure factor data obtained in this way have been described by calculations using density functional theory for all electrons coupled with molecular dynamics for the ions. Calculations that essentially use one-component plasma-like behavior underestimate the total structure factor, while calculations that use the potential inferred from density functional theory show agreement with the data.

An extensive study of elastic scattering amplitudes as a function of scattering vector is shown in Fig. 11.7 for shock-compressed lithium hydride. The electron densities and temperatures have been measured from the inelastic x-ray scattering features indicating $1.6\times10^{23}\,\text{cm}^{-3}$ and 1.7 eV, with an error of about

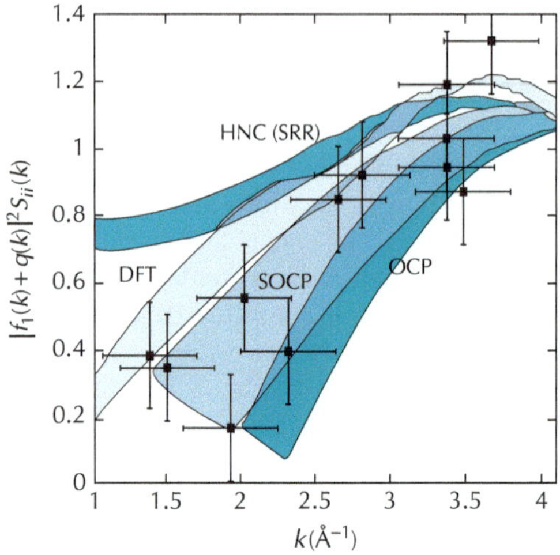

FIGURE 11.7 Measurements of the elastic scattering intensity as a function of the scattering wave vector from shock-compressed LiH targets using Ti K–α x-rays (squares) from Kritcher *et al.* (2009). Also shown are theoretical models for the elastic scattering intensity, and the area labeled SOCP represents the screened one-component plasma (SOCP) model that uses a charged hard spheres representation for the ions with a screening correction to the bare spheres. The orange curve labeled HNC uses the hyper-netted chains equations uses with short range repulsions (SRR), and the area labeled DFT are calculations using density functional theory and molecular dynamics.

±20% due to noise in the experimental data. The absolute scattering intensities take into account the dependency on polarization and are further corrected for variations in the number of source photons in each shot and variations in the source solid angle.

Also shown in Fig. 11.7 are theoretical models for the elastic scattering intensity, dependent on ion–ion structure factor values, use the total two-ion-species Chihara formula to calculate the total ionic scattering component. The band of values for each model are calculated for a range of temperature and density conditions, which account for shot-to-shot variations in the drive beam waveform and intensity. The screened one-component plasma model (Gregori *et al.*, 2006a) calculates the electrons as a neutralizing background, and the ions are modeled as linearly screened, charged hard spheres, to account for short-range strong coupling effects. The hyper-netted chain equation model uses Yukawa potentials to account for screening effects, and short range repulsion was added to the Debye potential as an effective ion interaction potential. Also plotted are density functional theory calculations, coupled with molecular dynamics (MD) simulations (Wünsch *et al.*, 2008).

These calculations use the two-ion species Chihara formula to calculate the ion scattering component and are in agreement with experimental data. Models that use single-ion species total ionic structure factors overestimate the intensity of the scattered ionic feature by a factor of about 1.5 to 3. The data approach the simplistic large k limit, where $q \to 0$ and $Sii(k, \omega) \to 1$, and the total elastic scattering component scales with $Z_b^2 = (1/N)\Sigma N^\alpha (Z_b^\alpha)^2$. Here, N^α and Z_b^α are the atomic number per molecule and number of bound electrons per species α. The parameter N is the total number of atoms. For one-ion-species models, the average number of bound electrons is greater, and the elastic scattering component is overestimated.

Models that ignore screening effects, such as the one-component plasma model (OCP), lie below the data for $k < 3$ Å. Calculations using DFT-MD are in agreement with the data, but are slightly greater than measured values of the ion component. The SOCP model offers better agreement with the measurements, indicating that the screened potentials best approximate the data, which was also observed for compressed Li (Saiz *et al.*, 2008b). The SRR potential overapproximates the elastic scattering feature for small k.

Shock-compressed matter experiments have also been employed to measure optical properties and collisionality of dense plasmas (Neumayer *et al.*, 2010). The plasmon energy shift and width from inelastic x-ray scattering from boron is shown as a function of the scattering wave number k in Fig. 11.8. The experimental results are compared with the RPA and the BMA calculations both with and without local field corrections (Fortmann *et al.*, 2010). The data at the smallest and the largest k vector are in agreement with all three models. At small k, local field corrections, scaling as k^2, are negligible, and the plasmon shift in all three models tends toward the plasma frequency as $k \to 0$. In the strongly noncollective regime, the plasmon dispersion is described by the Compton shift (single-particle ridge) $\Delta E_{\text{Compton}} = \hbar^2 k^2 / 2m_\text{e}$. Within the error margins, the experimental point at $k = 4.3\,\text{Å}^{-1}$ is described by the Compton shift and is in agreement with all three models. In this limit, correlations become less important since distances smaller than the screening length are probed.

Between $k = 1.5\,\text{Å}^{-1}$ and $k = 3\,\text{Å}^{-1}$, the plasmon shift is between $\Delta E_{\text{plasmon}} = 32$ and 34 eV, i.e., it is barely changing as a function of k. This vanishing plasmon dispersion is in good agreement with the BMA+LFC model that takes into account $e{-}e$ correlations. Both RPA and BMA, lacking $e{-}e$ interactions, predict a much stronger increase of ΔE_{pl}.

The horizontal plasma dispersion constitutes an important observation in a shock-compressed plasma and reveals the importance of electron–electron correlations on the dynamical properties of these extreme states of matter. Obviously, the plasmon's group velocity $v_G = d\omega/dk$ vanishes at $\hbar\omega \simeq 32$ eV, i.e., plasmons do not propagate at this energy.

The dispersion data favor an even weaker and possibly negative dispersion as compared to the BMA+LFC dispersion, which predicts larger plasmon shifts (see, e.g., the point at $k = 2.6\,\text{Å}^{-1}$). This complies with plasmon dispersion measurements by inelastic x-ray scattering and electron energy loss spectroscopy in metals, which also report on plasmon dispersions below local field corrected RPA calculations.

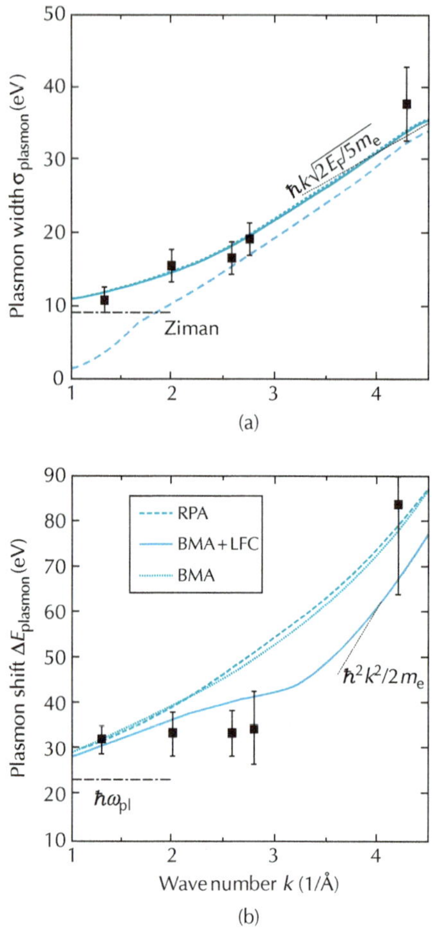

FIGURE 11.8 Plasmon width (a) and dispersion (b) as function of the scattering wave vector k. Experimental data are compared to calculations within the random phase approximation (long dashed), Born-Mermin approximation (short dashed), and Born-Mermin with local field corrections (solid curve) calculations. The straight lines represent the limits of dispersion and damping at small and large k, respectively. (Courtesy of P. Neumayer and C. Fortmann.)

Possible explanations for the remaining discrepancy between data and the BMA+LFC calculation include lowering of the plasmon shift through coupling of the collective mode to inner core electrons as well as remnants of band structure effects in the still near-ordered correlated ion system. It is also possible that the additional plasmon lowering is a signature of higher order terms in the electron-ion interaction, e.g., multiple scattering. Within the dipole approximation applied so far, $e-i$ collisions do not influence the plasmon dispersion, since only terms independent of k are considered.

Finally, the plasmon width as a function of k is determined. A measure of the intrinsic plasmon width (i.e., without instrumental broadening) is the root-mean-square (rms) width σ_{plasmon}. Since the plasmon scattering signal is the convolution of the dynamical structure factor with the instrumental function, its mean-square deviation is the sum of both, $\sigma^2_{\text{total}} = \sigma^2_{\text{inst}} + \sigma^2_{\text{plasmon}}$, which yields σ_{plasmon} knowing $\sigma_{\text{inst}} = 8.5\,\text{eV}$ and extracting σ^2_{total} from the data after subtraction of the quasi-elastic ion feature. Results are shown in Fig. 11.8a and compared to RPA and BMA with and without local field corrections. The two points at $k = 1.3\,\text{Å}^{-1}$ and $2.0\,\text{Å}^{-1}$ corresponding to collective scattering show that plasmon damping under these shock-compressed conditions (strongly degenerate + strongly collective) is mainly due to electron–ion collisions. Conversely, the RPA severely underestimates the plasmon width. This complies with results from collective scattering experiments on isochorically heated Be (see Fig. 11.2).

At the present parameters, the critical wave number up to which plasmons are a well-defined mode is $k_c = 1.7\,\text{Å}^{-1}$. Above k_c, the plasmon merges with the single-pair continuum giving rise to Landau damping in the RPA theory. At wave vector $k = 1.3\,\text{Å}^{-1}$, we find the plasmon resonance at $E_{\text{pl}} = 32\,\text{eV}$, which lies beyond the cutoff energy of the single-pair continuum $\hbar\omega_{\text{pair}} \leq \hbar^2(k^2 + 2kk_F)/2m_e = 29.1\,\text{eV}$. The sharp edge of the pair continuum reflects the steplike Fermi distribution of degenerate electrons. Hence, both modes cannot couple (Landau damping), and the plasmon in the RPA is undamped. These studies show promise for accurately measuring collisionality and conductivity in dense plasmas as needed for understanding transport and material properties.

PROBLEMS

11.1. Consider a photon of energy $\hbar\omega_1$ and momentum $\hbar c\vec{k}_1$ scattering on an electron (rest mass $m_e c^2$) at rest. After the scattering process, the photon is detected under an angle θ with respect to k_1.

(a) Proof the following expression for the energy $\hbar\omega_2$ of the scattered photon (Compton shift):

$$\hbar\omega_2 = \frac{\hbar\omega_i}{1 + \frac{\hbar\omega_i}{m_e c^2}(1 - \cos\theta)}.$$

Hints: Make use of the conservation laws for both total energy and momentum and take into account the rest mass of the electron.

(b) Show that for the nonrelativistic limit ($\hbar\omega_i/m_e c^2 \ll 1$), the momentum transferred to the electron is given by

$$p_e = \hbar k = \left|\vec{k}_i - \vec{s}_s\right| = 2\frac{\hbar\omega_1}{\hbar c}\sin\theta/2.$$

(c) Show that in this limit, the Compton shift is given by $\Delta\omega_C = \omega_i - \omega_s = \hbar k^2/2m_e$.

(d) Calculate the Compton shift for a photon of wavelength $\lambda_i = 0.24$ nm scattering on an electron under $\theta = 140^{\circ}$ angle. Without repeating the calculation, estimate the Compton shift for the same wavelength and angle but with the electron replaced by a proton.

11.2. The RPA dielectric function that was calculated in exercise 1 gives the dielectric response of a collisionless plasma. A simple way to include collisions is the Drude model. Within this approach, the following ansatz for the "optical" dielectric function $\varepsilon(0,\omega)$ is given:

$$\varepsilon^{\mathrm{Drude}}(0,\omega) = 1 - \frac{\omega_{\mathrm{pe}}^2}{\omega(\omega + i\nu)}.$$

The parameter ν is called the collision frequency.

(a) Using Maxwell's equations and Ohm's law $\vec{j} = \sigma\vec{E}$, show that the collision frequency ν is related to the dynamical conductivity $\sigma(\omega)$ by

$$\sigma(\omega) = \frac{\omega_{\mathrm{pe}}^2 \varepsilon_0}{\nu} \frac{1}{1 - i\omega/\nu}.$$

Hints: Write down Maxwell's equations for the fields $\vec{E}(\vec{r},t)$ and $\vec{B}(\vec{r},t)$. Make use of the relation $\nabla\times\nabla\times\vec{E}(\vec{r},t) = -\nabla^2\vec{E}(\vec{r},t)$ to eliminate the $\vec{B}(\vec{r},t)$ fields. Eliminate the current density $\vec{j}(\vec{r},t)$ using Ohm's law. Furthermore, consider an E field of the form $E(\vec{r},t) = \vec{E}_0 e^{i(\vec{k}\cdot\vec{r}-\omega t)}$ and make use of $\nabla^2\vec{E}(\vec{r},t) = -k^2\vec{E}(\vec{r},t)$, and $\frac{\partial}{\partial t}\vec{E}(\vec{,}t) = -i\omega\vec{E}(\vec{r},t)$. You will then find a relation between k^2, σ, and ω^2. Compare to the dispersion relation for electromagnetic waves in a plasma, $k^2c^2 = \omega^2\varepsilon(k,\omega)$ to establish the wanted expression.

(b) Calculate the electrical conductivity at the plasma frequency $\sigma(\omega_{\mathrm{pl}})$ for $n_{\mathrm{e}} = 10^{23}\,\mathrm{cm}^{-3}$ and $\nu = 10^{15}\mathrm{s}^{-1}$.

(c) Calculate the full width at half maximum (FWHM) and the resonance frequency for the spectrum $S(\omega) \propto \mathrm{Im}\,\frac{1}{\omega\varepsilon^{\mathrm{Drude}}(0,\omega)}$.

11.3. The dielectric function governs the response of charged particles to electromagnetic fields. It is defined as

$$\varepsilon_c^{\mathrm{RPA}}(k,\omega) = 1 - V_{cc}(k)\sum_{\vec{p}} \frac{n_{\mathrm{F}}^{(c)}(\vec{p}+\hbar\vec{k}/2) - n_{\mathrm{F}}^{(c)}(\vec{p}-\hbar\vec{k}/2)}{\hbar\omega - \hbar\frac{\vec{k}\cdot\vec{p}}{m_e} + i\eta}.$$

Here, c denotes the particle species, e.g., electrons, protons, and ions, and $n_{\mathrm{F}}^{(c)}(p)$ is the Fermi momentum distribution function. $\eta > 0$ is a small but finite frequency to ensure convergence of the sum.

In the following, we consider the classical limit of the RPA dielectric function for electrons.

(a) Give the definition of the Fermi distribution function for electrons. Show that for high temperatures, the Fermi distribution can be approximated by the Maxwell–Boltzmann distribution

$$f(\vec{p}) = n_e \lambda_e^3 \exp(-p^2/2m_e \kappa T),$$

and determine the normalization constant λ_e (thermal de Broglie wavelength).

(b) For soft photons, i.e., in the limit $\hbar\vec{k} \to 0$, the RPA dielectric function for the classical electron gas takes the form

$$\varepsilon^{\mathrm{RPA}}(k,\omega) = 1 - \frac{e^2}{\varepsilon_0 k^2} \int \frac{d^3p}{(2\pi\hbar)^3} \frac{\vec{k}\cdot\vec{\nabla} f(\vec{p})}{\vec{k}\cdot\vec{p}/m_e - \omega - i\eta}.$$

Derive this result.

(c) Evaluate the above integral using Cartesian coordinates. Show that after integration of the p_x and p_y directions, one is left with the expression

$$\varepsilon^{\mathrm{RPA}}(k,\omega) = 1 + \frac{\kappa_D^2}{k^2}(1 + ix\sqrt{\pi}\,w(x)), \qquad x = \frac{\omega}{k}\sqrt{\frac{m_e}{2\kappa T}},$$

where $x = \frac{\omega}{k}\sqrt{\frac{m_e}{2\kappa T}} \kappa_D^2 = n_e e^2/\kappa T \varepsilon_0$ is the Debye screening parameter, and

$$w(z) = \frac{i}{\pi} \int_{-\infty}^{\infty} \frac{e^{-t^2}\,dt}{z - t} = e^{-z^2}\left(1 + \frac{2i}{\sqrt{\pi}} \int_0^z e^{t^2}\,dt\right), \qquad \mathrm{Im}\, z > 0,$$

is Dawson's integral.

In the following, we will make use of the series expansions of Dawson's integral:

- Taylor series:

$$w(x) = \sum_{n=0}^{\infty} \frac{(ix)^n}{\Gamma(n+1/2)}.$$

- Asymptotic expansion at $x \to \infty$:

$$-ix\sqrt{\pi}\,w(x) \simeq \left(1 + \sum_{m=1}^{\infty} \frac{1\cdot 3\cdot 5\cdots(2m-1)}{(2x^2)^m}\right).$$

(d) Using these expansions, show that in the static limit $\omega \to 0$ for finite k, the familiar screening function

$$\varepsilon^{\mathrm{RPA}}(k,0) = 1 + \frac{\kappa_D^2}{k^2}$$

is obtained. Furthermore, show that in the long-wavelength limit $k \to 0$ at finite ω, the "optical" dielectric function

$$\varepsilon^{\mathrm{RPA}}(0,\omega) = 1 - \frac{\omega_{\mathrm{pe}}^2}{\omega^2}$$

(e) Show that at $k=0$, the collective longitudinal plasma oscillation (plasmon) occurs at $\omega = \omega_{\mathrm{pe}}$. Hint: Collective resonances appear at those k and ω, where $\mathrm{Re}\,\varepsilon^{\mathrm{RPA}}(k,\omega) = 0$.

(f) Show that at small wave numbers (i.e., $k \ll \kappa_D$), the plasmon resonance lies approximately at

$$\omega(k) = \omega_{\mathrm{pe}}\left(1 + \frac{3k^2}{2\kappa_D^2}\right).$$

This relation is often referred to as the Bohm–Gross plasma dispersion relation.

(g) Why does this formula not contain the Compton shift $\Delta\omega_C = \frac{\hbar^2 k^2}{2m_e}$?

(h) Show that collective oscillations occur only for those wave numbers that fulfill $k < k_c$ with the critical wave number $k_c \simeq 0.9\,\kappa_D$.

(i) Determine ω_{pe}, κ_D, k_c, k, and $\omega(k)$ for the following experiment: photons of wavelength 0.24 nm scattering under 30° scattering angle on a plasma with $n_e = 3 \times 10^{23}\,\mathrm{cm}^{-3}$ and $\kappa T = 10\,\mathrm{eV}$. Is the scattering collective, i.e., do plasmons show up in the spectrum? Are the assumptions made in deriving the dielectric function justified?

(j) Demonstrate the symmetrie relations of the dielectric function:

$$\mathrm{Re}\,\varepsilon^{\mathrm{RPA}}(k,-\omega) = \mathrm{Re}\,\varepsilon^{\mathrm{RPA}}(k,\omega)$$
$$\mathrm{Im}\,\varepsilon^{\mathrm{RPA}}(k,-\omega) = -\mathrm{Im}\,\varepsilon^{\mathrm{RPA}}(k,\omega).$$

The dynamical structure factor is related to the dielectric function via the Fluctuation-Dissipation theorem

$$S(k,\omega) = -\frac{\hbar\varepsilon_0 k^2}{\pi n_e e^2}\frac{\mathrm{Im}\,\varepsilon^{\mathrm{RPA}}(k,\omega)}{\exp(\hbar\omega/\kappa T) - 1}.$$

(k) State the symmetrie relations for $S(k,\omega)$.

(l) Calculate $S(k,0)$.

(m) Show the "detailed balance condition" $S(k,-\omega)/S(k,\omega) = \exp(\hbar\omega/\kappa T)$.

11.4. **(a)** How is the Fermi momentum distribution function $n_F(p)$ defined?

(b) In practice, one is interested in the thermodynamical properties of plasmas at given density n and temperature T. Give expressions for the internal energy and the average momentum for an ideal Fermi gas at finite T. Explain how to calculate the chemical potential μ.

State your answers in terms of the Fermi integrals $F_\ell(\eta)$,

$$F_\ell(\eta)=\frac{1}{\Gamma(\ell+1)}\int_0^\infty dx\frac{x^\ell}{e^{x-\eta}+1}.$$

(c) Demonstrate the following limiting cases for Fermi integrals:

$$\lim_{\eta\to-\infty}F_\ell(\eta)=\exp(\eta)$$

$$\lim_{\eta\to+\infty}F_\ell(\eta)=\frac{\eta^{\ell+1}}{(\ell+1)\Gamma(\ell+1)}$$

(d) Calculate the chemical potential, internal energy, and average momentum for a system of electrons at density n_e and temperature T in both limits.

(e) For both cases, state two examples of astrophysical plasmas that correspond well to these limits.

(f) How much energy is needed to compress 1 mg of cryogenic hydrogen from 0.1 g/cc to 100 g/cc?

11.5. Consider a plasma of electrons ($q_e=-e$) and ions ($q_i=+Ze$). For an ion placed at $\vec{r}=0$, we want to calculate the effective potential, that a given charged particle "feels", accounting for the screening of the central ion due all other charges.

(a) Justify that the effective Poisson equation for the potential $\Phi(\vec{r})$ for this problem reads

$$\nabla^2\Phi(\vec{r})=-\frac{Ze}{\varepsilon_0}\delta(\vec{r})-\frac{1}{\varepsilon_0}\sum_c q_c n_c(T,\mu_c-q_c\Phi(\vec{r})).$$

(b) Expand the density $n_c(T,\mu_c-q_c\Phi(\vec{r}))$ into a Taylor series around $q_c\Phi(\vec{r})=0$. Neglect all terms of order q_c^2 and higher, and show that the linearized Poisson equation reads

$$\nabla^2\Phi(\vec{r})=-\frac{Ze}{\varepsilon_0}\delta(\vec{r})-\kappa^2\Phi(\vec{r}),$$

with the screening parameter

$$\kappa^2=\sum_c\frac{q_c^2}{\varepsilon_0}\frac{\partial}{\partial\mu_c}n_c(T,\mu_c).$$

Solve the linearized Poisson equation (11.5.2). To do this:

(c) Perform a Fourier transformation on both sides of the equation. The Fourier transform of the delta function is $\int d^3\vec{r}\delta(\vec{r})=(2\pi)^3$.

(d) Show that the Fourier transformed potential reads

$$\Phi(\vec{k})=\frac{Ze}{\varepsilon_0(k^2+\kappa^2)}.$$

(e) Calculate $\Phi(\vec{r})$ by inverse Fourier transform of $\Phi(\vec{k})$. For the inverse Fourier transform, make use of the integral

$$\int_0^\infty dx \frac{x^{2m+1}\sin(ax)}{(x^2+z)^{n+1}} = \frac{\pi}{2}\frac{(-1)^{n+m}}{n!}\frac{d^n}{dz^n}\left(z^m e^{-a\sqrt{z}}\right)$$

and show that

$$\Phi(\vec{r}) = \frac{Ze}{4\pi\varepsilon_0 r}\exp(-\kappa r).$$

(f) What is the physical meaning of κ_D?

(g) Calculate κ_D for the case of a classical, fully ionized hydrogen plasma (Debye screening) at $T = 10^6$K and $n_e = n_i = 10^{20}\,\mathrm{cm}^{-3}$ and for an electron gas at zero temperature (Thomas–Fermi screening) with $n_e = 10^{23}\,\mathrm{cm}^{-3}$.
Hint: Use the relation $\frac{\partial}{\partial\eta}F_\ell(\eta) = F_{\ell-1}(\eta)$ for Fermi integrals.

SOLUTIONS TO ODD PROBLEMS

11.1. (a) From conservation of energy and momentum, we have

$$mc^2 + \hbar c k_i = \sqrt{m_e^2c^4 + p_e^2c^2} + \hbar c k_s$$

$$\hbar\vec{k}_i = \hbar\vec{k}_s + \vec{p}_e.$$

Square both sides of these equations and eliminate p_e^2 in the energy conservation. The solution follows after elementary operations.

(b) Neglecting all terms $\hbar\omega_1/m_ec^2$ in the expression for p_e^2, we find

$$p_e^2 = \frac{\hbar^2\omega_i^2}{c^2}(2 - 2\cos\theta).$$

Making use of $1 - \cos\theta = 2\sin\theta/2$ we find the solution.

(c) Expanding the result for $\omega_i - \omega_s$ in first order in $\hbar\omega_i/m_ec^2$, we get

$$\Delta\omega_c = \omega_i - \omega_s = \frac{\hbar\omega_i^2}{m_ec^2}(1 - \cos\theta).$$

Using again the relation for $1 - \cos\theta$ and the solution given in 11.1(b), we find the solution.

11.3. (a) Fermi distribution is defined as $n_F^{(e)}(\vec{p}) = [\exp(p^2/2m_e\kappa T - \mu_e/\kappa T) + 1]^{-1} = \exp(\mu_e/\kappa T)\ [\exp(p^2/2m_e\kappa T) + \exp(\mu_e/\kappa T)]^{-1}$. For high temperatures, the chemical potential tends towards $-\infty$, and the fugacity $\exp(\mu_e/\kappa T)$ can be neglected in the denominator, yielding $\lim_{T\to\infty} n_F^{(e)}(\vec{p}) = \exp(\mu_e/\kappa T)\exp(-p^2/2m_e\kappa T)$. Identifying

$\exp(\mu_e/\kappa T) = n_e\lambda_e^3/2$, and summing over spin degrees of freedom (factor 2), we find the equation given in problem 11.3(a).

Employing the normalization condition

$$\int \frac{d^3p}{(2\pi\hbar)^3} f(\vec{p}) = n_e,$$

and using the integral $\int_{-\infty}^{\infty} dx \exp(-x^2/2\alpha^2) = \sqrt{2\pi}\alpha$, we find $\lambda_e = \sqrt{2\pi\hbar^2/m_e\kappa T}$.

(b) For soft photons, we have

$$f(\vec{p}+\hbar\vec{k}/2) - f(\vec{p}-\hbar\vec{k}/2) \simeq \hbar\vec{k}\cdot\vec{\nabla}f(\vec{p})$$

and hence the solution (note that the sum over momentum and spin states has to be replaced by $2\Omega_0 \int d^3p/(2\pi\hbar)^3$, and Ω_0 is the normalization volume).

(c) We set $\vec{p} = \vec{e}_z p$. Integration over x and y components yields a factor λ^{-2}. Intermediate result is

$$\varepsilon(k,\omega) = 1 - \frac{e^2}{\varepsilon_0 k^2} \frac{n_e}{\sqrt{2\pi m_e \kappa T}} \int_{-\infty}^{\infty} dp \frac{kp}{m_e \kappa T} \frac{\exp(-p^2/2m_e\kappa T)}{\omega - kp/m_e + i\eta}.$$

After rearrangement of terms, substituting the integration variable $t = p/\sqrt{2m_e\kappa T}$, and identifying Dawson's integral, we find the solution.

(d) Neglecting all frequency dependent terms immediately yields the static limit.

(e) In the limit $k \to 0$, the asymptotic expansion of Dawson's integral has to be employed. Neglecting terms of order x^{-4} and higher gives the solution.

(f) Employing the longitudinal dispersion relation $\mathrm{Re}\,\varepsilon(0,\omega) = 0$ gives the relation $\omega = \pm\omega_{\mathrm{pl}}$.

(g) Expanding Dawson's integral up to the term x^{-4} and $\mathrm{Re}\,\varepsilon(k \to 0), \omega(k)) = 0$ gives a quadratic equation for $\omega^2(k)$:

$$\omega^4(k) - \omega^2(k)\omega_{\mathrm{pe}}^2 - 3k^2\omega_{\mathrm{pe}}^2 \frac{\kappa T}{m_e} = 0.$$

Elementary solving for ω^2 (k) yields

$$\omega^2(k) = \frac{\omega_{\mathrm{pe}}^2}{2}\left(1 + \sqrt{1 + 12\frac{k^2}{\kappa_D^2}}\right).$$

Expanding $\sqrt{1+x} \simeq 1 + x/2$ yields the solution.

(h) In order for $\mathrm{Re}\,\varepsilon(k,\omega)$ to have a root, the function $1 - 2x\sqrt{\pi}$ $\exp(-x^2)\int_0^{\infty} \exp(t^2)dt$ has to become negative. At $x = 0.92$, the function $\exp(-x^2)\int_0^\infty \exp(t^2)dt$ takes on its maximum of 0.54. Solving for $\min \mathrm{Re}\,\varepsilon(k,\omega) < 0$ yields $k \le 0.8724\kappa_D$.

(i) All terms in $\hbar$ have been omitted; hence, the Compton shift $\propto \hbar$ is not contained in the classical RPA.

(j) $\omega_{\mathrm{pe}} = 3.09 \times 10^{16}\,\mathrm{s}^{-1} = 20.34\,\mathrm{eV}$. $\kappa_D = 2.32 \times 10^{10}\,\mathrm{m}^{-1} = 1.23 a_{\mathrm{B}}^{-1}$; hence, $k_c \simeq 2.1 \times 10^{10}\,\mathrm{m}^{-1}$. $k = 1.36 \times 10^{10}\,\mathrm{m}^{-1}$; hence, scattering is collective, $\alpha \simeq 0.6$, Bohm–Gross dispersion $\omega(k) = 27.33\,\mathrm{eV}$. Degeneracy parameter $\vartheta = T/T_F \simeq 0.61$; hence, quantum effects are important, terms $\propto \hbar$ have to be considered, e.g., Compton shift.

(k) $S(k,-\omega) = \exp(\hbar\omega/\kappa T)S(k,\omega)$.

(l) $S(k,0) = \frac{1}{\pi}\sqrt{\frac{m_e}{2\kappa T}}\frac{1}{k}\frac{1}{(1+\kappa_D^2/k^2)^2}$.

Chapter 12

Scattering from Unstable Plasmas

Chapter Outline Head

DOI: 10.1016/B978-0-12-374877-5.00012-9

12.1 INTRODUCTION

12.1.1 Unstable Plasmas

It is easy to find cases of plasmas that are not in thermodynamic equilibrium. In fact, the high temperature that characterizes the majority of plasmas really precludes the equilibrium state in the strict sense. We usually mean by "equilibrium" a transient state in which the plasma does not change significantly on the timescale of interest.

In the preceding chapters, we have in fact considered small departures from equilibrium, such as unequal electron and ion temperatures, non-Maxwellian distributions, and small relative drifts of electrons and ions. Nevertheless there has always been the understanding that while these conditions might modify the fluctuations $n_1, F_1, S(\mathbf{k}, \omega)$, they would not lead to rapid or significant changes in the average conditions n_0, F_0, T_e, T_i. Put another way we required that the "free energy" was small enough that the system would relax quietly to equilibrium.

Now, we turn our attention to cases where the free energy is large so that the system may return to equilibrium in an "explosive" or "unstable fashion."

Instabilities in plasmas are important because they can modify the microscopic and global behavior of the plasma. In magnetized plasmas, they enhance the transport of charged particles across the field (see Section 12.6). In the warm dense plasmas of inertial fusion, they can limit the compression of the target through the Rayleigh–Taylor mode and enhance the reflection of incident laser light, e.g., through ion-acoustic modes (see Section 12.7). Today's wide range of sources and detectors allow a number of these instabilities to be studied by scattering; see reviews by Chen (1987), Ross (1989), Horton (1999), Glenzer & Redmer (2009), and Hammett (2009).

12.1.2 Enhanced Scattering

The main feature from the point of view of scattering is that the unstable modes in a plasma grow to a large amplitude, and for these modes, the spectral density function is enhanced to well above the thermal level at discrete angles. As a consequence, measurements can often be made in parameter regions, which are inaccessible because parasitic radiation would swamp the thermal spectrum, i.e., at very low densities and for very small scattering angles. Equally the unstable modes can obscure the thermal spectrum; this is well illustrated by the work of Daehler & Ribe (1967) and Daehler *et al.* (1969). They measured the ion feature on a theta pinch plasma by the forward scattering of ruby laser light ($\theta = 5.5° - 17°$); the apparatus is shown in Fig. 12.1. A more detailed drawing of the novel multichannel Fabry–Perot etalon spectrometer is shown in Section 7.6. The electron density $n_{\text{peak}} = 2.8 \pm 0.4 \times 10^{-16}\,\text{cm}^{-3}$, temperature $T_e = 345 \pm 40\,\text{eV}$, and ion temperature $T_i = 2\,\text{keV}$ were determined by subsidiary measurements. The spectrum, which is in Fig. 12.2, was enhanced

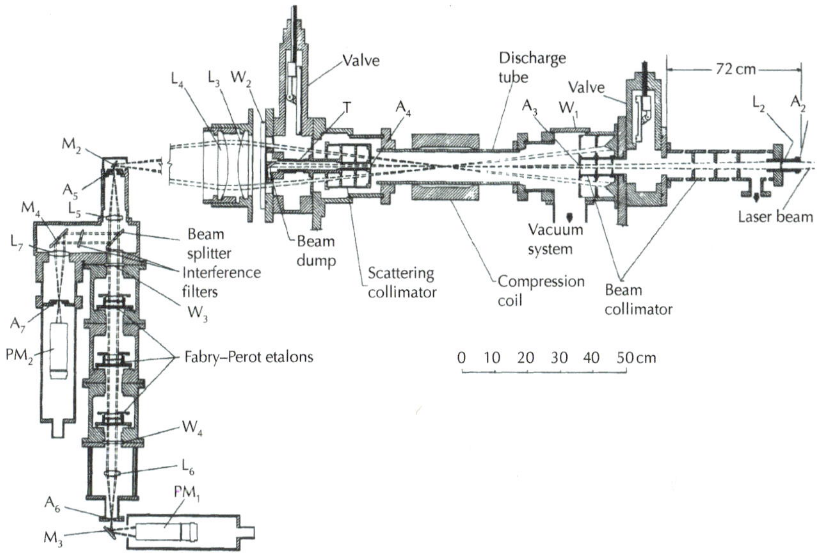

FIGURE 12.1 Side view of theta pinch apparatus and scattering equipment. (Courtesy of Los Alamos Sci. Lab. and Daehler & Ribe (1967).)

15 times above the thermal level. This enhancement was attributed to microinstabilities, which are generated during the production of the hot plasma.

12.1.3 Instability Scale Length

Plasmas can be both macroscopically (scale length $\gg \lambda_{De}, \rho_i$) and microscopically (scale length $\cong \lambda_{De}, \rho_e, \rho_i$) unstable. There is no general theory of the evolution of an instability or of the turbulent state to which most instabilities lead. We will concentrate here on microscopically unstable plasmas, as this topic has been the subject of a number of interesting experimental studies involving scattering as a diagnostic, and it is possible to make a limited comparison of theory and experiment. A very brief review of the relevant instability theory is given in Section 12.2, and experimental work on the various regions of instability is discussed in the subsequent sections.

Section 12.3: *Onset of Instability*: Experimental work on scattering from a plasma in a state of marginal stability with respect to ion-acoustic modes is discussed.

Section 12.4: *Quasi-linear Theory of a Weakly Unstable Plasma*: Work on a "gentle bump" instability is discussed.

Section 12.5: *Weak Turbulence and Strong Turbulence Theories*: A discussion is given of the results of scattering from turbulence in shock waves.

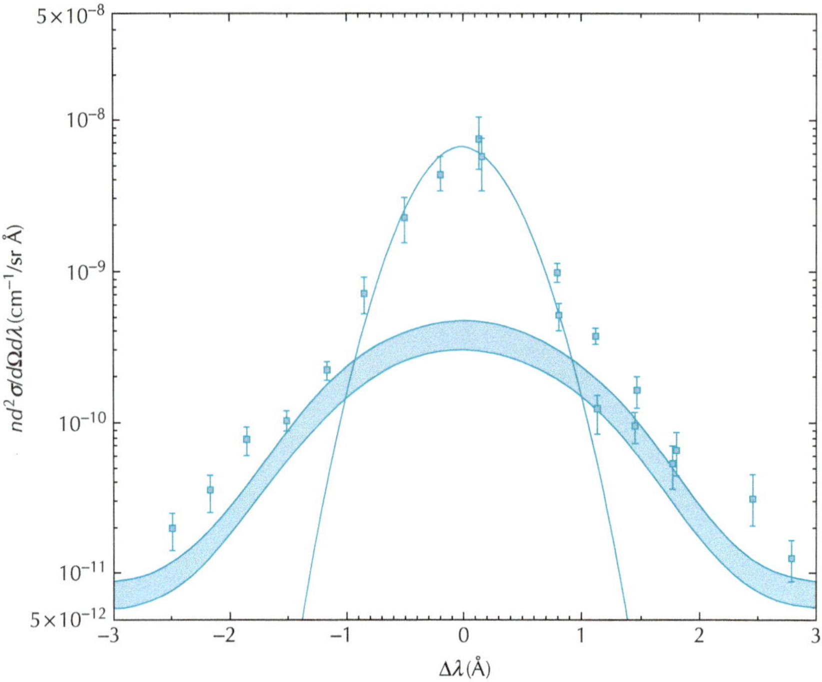

FIGURE 12.2 Logarithmic plot of scattered spectrum. The shaded band is the theoretical curve corresponding to $T_i = 2\,\text{keV}$, allowing for expected density variations. The thin curve is the Gaussian fit. (Courtesy of Los Alamos Sci. Lab. and Daehler & Ribe (1967).)

12.2 MICROSCOPIC INSTABILITY THEORY

12.2.1 Basic Equations

In any plasma, we can see fluctuations at some level at all frequencies and wavelengths, consistent of course with observation times and scale lengths. The level of each mode of fluctuation is limited by the damping mechanisms that operate; in a plasma, this means collisional and Landau damping.

In an unstable situation, the system has "free energy." For example, in a plasma, this might be due to an anisotropy in the velocity distributions, to a relative drift of electrons and ions, or to the presence of strong spatial gradients. This free energy is available to promote the growth of particular modes. We may determine whether a system is stable or unstable by studying the temporal behavior of each mode in the plasma.

For a homogeneous, nonrelativistic plasma, the distribution functions satisfy Eqns (B.7.9) and (B.7.10). For simplicity, we consider an unmagnetized plasma

($\mathbf{B}_0 = 0$), and the Fourier-transformed equations are

$$\frac{\partial F_{0q}(\mathbf{v}, t)}{\partial t} = \frac{q}{m_e}\int \frac{d\mathbf{k}}{(2\pi)^3} i\mathbf{k}\cdot\frac{\partial}{\partial \mathbf{v}}\langle F_{1q}(\mathbf{k}, \mathbf{v}, t), \varphi_1(-\mathbf{k}, t)\rangle \tag{12.2.1}$$

$$\times\left(\frac{\partial}{\partial t} - i\mathbf{k}\cdot\mathbf{v}\right) F_{1q}(\mathbf{k}, \mathbf{v}, t) + \frac{iq}{m_e}\varphi_1(\mathbf{k}, t)\mathbf{k}\cdot\frac{\partial F_{0q}}{\partial \mathbf{v}}$$

$$= \frac{iq}{m_e}\int \frac{d\mathbf{k}'}{(2\pi)^3}(\mathbf{k}-\mathbf{k}')\cdot\frac{\partial}{\partial \mathbf{v}}(\varphi_1(\mathbf{k}-\mathbf{k}', t)F_{1q}(\mathbf{k}', \mathbf{v}, t)) \tag{12.2.2}$$

$$\mathbf{E}_1(\mathbf{k}, t) = i\mathbf{k}\varphi_1(\mathbf{k}, t), \quad \varphi_1(\mathbf{k}, t) = \frac{4\pi}{k^2}\sum_q q\int d\mathbf{v}\, F_{1q}(\mathbf{k}, \mathbf{v}, t). \tag{12.2.3}$$

The Fourier transforms are defined as

$$F(\mathbf{k}, t) \equiv \int d\mathbf{r}\, e^{i\mathbf{k}\cdot r} F(\mathbf{r}, t). \tag{12.2.4}$$

Integrating (12.2.2) with respect to time yields

$$F_{1q}(\mathbf{k}, \mathbf{v}, t) = -F_{1q}(\mathbf{k}, \mathbf{v}, 0)\, e^{i\mathbf{k}\cdot\mathbf{v}t} - \frac{iq}{m_e}\int_0^t dt'\, \varphi_1\, e^{i\mathbf{k}\cdot\mathbf{v}(t-t')}\,\mathbf{k}\cdot\frac{\partial F_{0q}}{\partial \mathbf{v}}$$
$$+ \frac{iq}{m_e}\int \frac{d\mathbf{k}'}{(2\pi)^3}\int_0^t dt'\, e^{i\mathbf{k}\cdot\mathbf{v}(t-t')}\,\varphi_1(\mathbf{k}-\mathbf{k}', t')\cdot[\mathbf{k}-\mathbf{k}']\cdot\frac{\partial}{\partial \mathbf{v}}F_{1q}. \tag{12.2.5}$$

Now, $n_{1e}(\mathbf{k}, t) = \int d\mathbf{v} F_{1e}(\mathbf{k}, \mathbf{v}, t)$, and the spectral density function

$$S(\mathbf{k}, t) \propto \langle |n_{1e}(\mathbf{k}, t)|^2\rangle.$$

The terms on the right-hand side of (12.2.5) that contribute to the fluctuations at **k** are, respectively, (a) the initial fluctuations, (b) the fluctuations generated by the charges as they follow their basic orbits, and (c) a nonlinear term that involves mode coupling by which mechanism energy is transferred from one mode to another. In the framework of strong turbulence, this term is treated from the point of view of its effect on perturbing the basic particle orbits.

12.2.2 Onset of an Instability

The onset conditions may be obtained by solving the linearized version of Eqns (12.2.1) and (12.2.2) because until the instability is established, the fluctuations are close to the thermal level. Some techniques for assessing microscopic stability are discussed by Jackson (1960), Mikhailovskii & Laing (1992), and Mikhailovskii (1998).

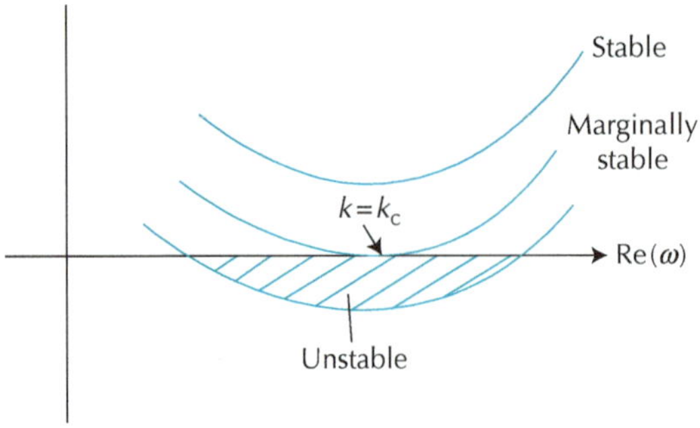

FIGURE 12.3 Curves for $\varepsilon(\omega_c, k_c)=0$.

A brief discussion of the stability of longitudinal plasma oscillations, when $\mathbf{B}_0=E_0=0$, is given in Section A.4. The solution of Eqns (12.2.1) and (12.2.2) for the fluctuating electric field is

$$E_1(k,t)=i\frac{4\pi e}{k^2}\sum_c e^{i\omega_c t}\,\text{Residue}\left[\frac{\int d\text{v}\,F_0(k,\text{v},0)/[(\omega/k)-\text{v}]}{\varepsilon(k,\omega)}\right]_{\omega=\omega_c}, \tag{12.2.6}$$

where $\omega=\omega_c$ is the solution of the dispersion relation $\varepsilon(k,\omega)=0$. If $\text{Im}(\omega_c)<0$, then the mode at (k_c, ω_c) grows exponentially with time, that is, this mode is unstable. If for all modes $\text{Im}(\omega_c)>0$, the system is stable.

As the system changes from stable to unstable, the solution of the dispersion relation will move in the ω plane across the real axis; this is sketched in Fig. 12.3. Note that instability will occur first at some quite definite value of $\mathbf{k}_c$ given by $\varepsilon(k_c, \omega_c)=0$, and if we are to investigate the onset by scattering, it is this $\mathbf{k}_c$ we must study.

Particles moving with a velocity near the phase velocity of the wave take energy from it or give energy to it according to the sign of the slope of the distribution function f. Thus, for a single species, a distribution function that is monotonically decreasing from the mean velocity is stable. Conversely, a distribution function with a bump on it has the opposite characteristic; it has a region of Landau growth. Equally, if the electrons and ions have a relative drift, even though each distribution decreases monotonically from the mean species velocity, when $\mathbf{k}\cdot\partial f/\partial\mathbf{v}$ at the phase velocity of a wave is positive, this wave is unstable (see Fig. 12.4).

In the case of the drift instability, waves propagating at a phase velocity between the peak of the ion and electron distributions are Landau damped on the ion distribution but gain energy from the electrons, "Landau growth." One effect of increasing the relative drift velocity even prior to instability is an increase in the fluctuation level at the most unstable natural mode $[\text{Re}(\varepsilon(k,\omega))\cong 0]$ because the damping term, the imaginary part of $\varepsilon(k,\omega)$ [Eqn (12.2.6)], also approaches zero (see Section 5.3.6).

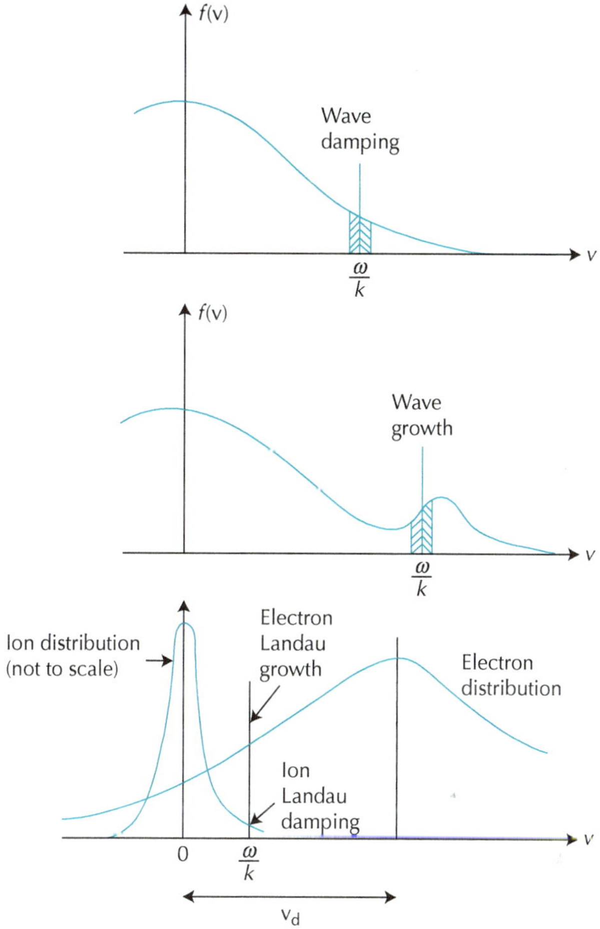

FIGURE 12.4 Illustration of some stable and unstable distribution functions

12.2.3 Initial Evolution of an Instability

Once the instability is established, the nonlinear terms, which vary as $\exp[2\,\mathrm{Im}\,(\omega_c)t]$, become important. They affect the system in a number of ways.

a. The collision term in Eqn (12.2.1) is modified by the contribution from the growing waves, and this leads to an increase in the diffusion in velocity space, which changes F_{0q}. Rogister & Oberman (1968) have derived a modified form of the Lenard–Balescu equation (Section B.8.2), which includes the contribution of the previously neglected, weakly damped modes in the stable case and growing modes in the marginally unstable case. This, coupled with an equation describing the evolution of the field energy

$I(k,t) \cong |\varphi_1(k,t)|^2$, may be used to follow the evolution of the distribution functions from the stable to the marginally unstable region. The application to scattering has been discussed by Joyce & Salat (1971).

b. The change in F_{0q} in turn leads to a change in F_{1q} through the second term on the right-hand side of Eqn (12.2.5).
c. Finally, the nonlinear term in Eqn (12.2.5) becomes important, and this also affects F_{1q}.

Eventually, if the system is driven, e.g., $\mathbf{E}_0 \neq 0$, we may reach a quasi-stationary state in which there is a balance between the promotion of growth by the driver and some "nonlinear" damping mechanism, and there is a continuous dissipation of the input energy of the driver. If the system is not driven, the enhanced fluctuations will eventually die down, and it will revert to a stable stationary state. The "free energy" will have been dissipated by heating the plasma or by increases in the losses to the boundaries of the system. During the evolution of the instability, there may well be quasi-stationary states, in which nonthermal fluctuations exist in the plasma.

In most cases, it is not possible to follow the details of this evolution. The nonlinear equations are unmanageable except with drastic approximation, and we are forced to guess the asymptotic behavior. We therefore look for quasi-stationary states in which some plausible mechanism balances the growth.

12.2.4 Quasi-Linear Theory[1]

The most drastic approximation we can make is to drop the nonlinear term in Eqn (12.2.2) and look for solutions in which the growth of the instability is finally limited by the modification of F_{0q} caused by the enhancement of the velocity diffusion term in Eqn (12.2.1).

The quasi-linear equations are

$$\frac{\partial F_{0q}}{\partial t} = \frac{\partial}{\partial \mathrm{v}_\alpha}\left(D_{\alpha\beta}\frac{\partial F_{0q}}{\partial \mathrm{v}_\beta}\right) \qquad (10.2.7\text{a})$$

with the diffusion coefficient

$$D_{\alpha\beta} = \frac{e^2}{m_e^2}\sum_k \frac{k_\alpha k_\beta \langle\varphi(k,t)\cdot\varphi(-k,t)\rangle}{i(\mathbf{k}\cdot\mathbf{v}-\omega)} \qquad (10.2.7\text{b})$$

and

$$\frac{\partial}{\partial t}\langle\varphi(k,t)\varphi(-k,t)\rangle = 2\,\mathrm{Im}(\omega)\langle\varphi(k,t),\varphi(-k,t)\rangle \quad \mathrm{Im}(\omega) < 0.$$

This approximation is valid if the wave amplitude remains very small so that F_{0q} changes on a slow timescale compared to F_{1q} and also so that the mode coupling terms are weak and act on a long timescale compared to the time for the

[1] Numerous essentially noninteracting modes.

instability to saturate. This theory was applied by Drummond & Pines (1962b) and Vedenov *et al.* (1961) to the case of an electron distribution with a "bump" on the tail (see Fig. 12.4 and Section 12.4).

12.2.5 Weak Turbulence[2]

In the next level of approximation, the mode-coupling terms are included. These terms are expanded in powers of the ratio of the fluctuating electric field energy density to the thermal energy density, and terms to order $|E_k|^4$ are retained. The conditions for the validity of this approach are

a. $\mathrm{Im}(\omega)/\omega \ll 1$;
b. $\sum |E_k|^2/n_e\kappa T \ll 1$;
c. there must be a continuous wave spectrum at a low level with initially random phases. The mode coupling terms include (see Aamodt & Drummond, 1965)
   (1) Resonant three-wave interactions in which two waves combine to give a third wave which is a natural mode of the plasma, i.e., satisfies the dispersion relation. Thus, $\mathbf{k}''+\mathbf{k}'=\mathbf{k}$ and $\omega''+\omega'=\omega$. This includes the merging of two waves to give one wave and decay of one wave to give two waves.
   (2) Nonresonant three-wave interactions in which the beat wave is not a natural mode, but this virtual wave has a phase velocity near the thermal speed of one charge species and can be Landau damped. This is sometimes called "nonlinear Landau damping." Note that the wave can in turn combine with waves in the plasma and regenerate the original waves so that the net damping is reduced.

These terms both transfer energy from the unstable part of the spectrum to a region where damping occurs and quasi-stationary turbulent states can be found in which the growth is balanced by these mechanisms. This theory is discussed by Kadomstev (1965), Vedenov (1968), and Tsytovich (1972). Possibly the most interesting application has been to the interpretation of the anomalous (nonclassical) dissipation in shock fronts (see Section 12.5).

12.2.6 Strong Turbulence[3]

The theory of situations in which the nonlinear terms are strongly active even during the development of the instability and F_{0q} and F_{1q} evolve on a similar timescale is not a well-developed topic. The extension to wave amplitudes exceeding those appropriate to weak turbulence but with F_{0q} still evolving on a slower timescale than F_{1q} can be achieved by treating the nonlinear term in

---

[2]Numerous weakly interacting modes.

[3]Numerous strongly interacting modes.

Eqn (12.2.2) as a source which perturbs the basic particle trajectories. The inclusion of a diffusion in the particle trajectories modifies the fluctuations so that a quasi-stationary state may be achieved in the presence of a nonthermal spectrum. This theory was developed by Dupree (1966) (and see Einaudi & Sudan, 1969).

12.3 SCATTERING FROM A MARGINALLY STABLE PLASMA

12.3.1 Onset of the Drift Instability

This topic has been discussed by Ichimaru (1962) and Ichimaru *et al.* (1962a) for the case of the ion-acoustic drift instability. The dispersion equation $\varepsilon(\mathbf{k},\omega)=0$ has the form [see Eqn (5.3.8)]

$$k^2\lambda_{\mathrm{De}}^2 = -\mathrm{Rw}(x_{\mathrm{e}}-x_{\mathrm{d}}) - \frac{ZT_{\mathrm{e}}}{T_{\mathrm{i}}}\mathrm{Rw}(x_{\mathrm{i}}) + i\pi^{1/2}\left\{(x_{\mathrm{e}}-x_{\mathrm{d}})\exp[-(x_{\mathrm{e}}-x_{\mathrm{d}})^2] + \frac{ZT_{\mathrm{e}}}{T_{\mathrm{i}}}x_{\mathrm{i}}\exp(-x_{\mathrm{i}}^2)\right\}, \tag{12.3.1}$$

where $x_{\mathrm{e}}=\omega/ka, x_{\mathrm{i}}=\omega/kb$, the relative ion-electron drift velocity is $\mathrm{v_d}$, and $x_{\mathrm{d}}=\mathrm{v_d}/a$. The case $k||\mathrm{v_d}$ has been chosen as this is the most unstable case.

$$\mathrm{Rw}(x) = 1-2x\exp(-x^2)\int_0^x \exp(p^2)dp \rightarrow \begin{cases} 1-2x^2, & x<1 \\ -1/2x^2, & x\gg 1. \end{cases}$$

The state of marginal stability occurs when the growth terms in ε balance the damping terms $\mathrm{Im}(\omega)=0$, x is real and

$$(x_{\mathrm{e}}-x_{\mathrm{d}})\exp[-(x_{\mathrm{e}}-x_{\mathrm{d}})^2] + \frac{ZT_{\mathrm{e}}}{T_{\mathrm{i}}}x_{\mathrm{i}}\exp(-x_{\mathrm{i}}^2)=0.$$

A detailed analysis of this situation has been made by Jukes & Shaffer (1960). For $ZT_{\mathrm{e}}/T_{\mathrm{i}}\ll 1, \mathrm{v_d}/a \gtrsim 1$ and for $ZT_{\mathrm{e}}/T_{\mathrm{i}}\gg 1, \mathrm{v_d}/b\ll 1$, they obtained analytic expressions for the onset conditions. In the latter case $x_{\mathrm{e}}-x_{\mathrm{d}}\ll 1$ (the ion-acoustic instability), the instability boundary occurs when

$$x_{\mathrm{d}} = x_{\mathrm{e}} + \frac{ZT_{\mathrm{e}}}{T_{\mathrm{i}}}x_{\mathrm{i}}\exp(-x_{\mathrm{i}}^2).$$

For $x_{\mathrm{i}}\gg 1$,

$$k^2\lambda_{\mathrm{De}}^2 \cong -1 + \frac{ZT_{\mathrm{e}}}{2T_{\mathrm{i}}x_{\mathrm{i}}^2} \quad \text{or} \quad \omega_{\mathrm{ac}} \cong \left(\frac{ZT_{\mathrm{e}}}{m_{\mathrm{i}}}\right)^{1/2}.$$

For $\mathrm{v_d}/a > (m_{\mathrm{e}}Z/2m_{\mathrm{i}})^{1/2}$, the boundary occurs when $k_c^2=0$, i.e., $x_{\mathrm{i}}^2 = ZT_{\mathrm{e}}/2T_{\mathrm{i}}$, and the initial drift velocity for onset is

$$\frac{\mathrm{v_c}}{a} \cong \left(\frac{m_{\mathrm{e}}Z}{2m_{\mathrm{i}}}\right)^{1/2} + 2\left(\frac{ZT_{\mathrm{e}}}{2T_{\mathrm{i}}}\right)^{3/2} e^{-(ZT_{\mathrm{e}}/2T_{\mathrm{i}})}. \tag{12.3.2}$$

For $\mathrm{v_d}/a<(m_eZ/2m_i)^{1/2}$, the boundary occurs at finite k, and

$$\mathrm{v_c}/a \cong x_i(m_eT_i/m_iT_e)^{1/2}, \qquad (12.3.3)$$

where

$$x_i^2 \cong \tfrac{3}{2}\ln(ZT_e/T_i)+\tfrac{1}{2}\ln(m_i/m_eZ)+0(2\ln x_i).$$

The critical k is given by

$$k_c^2\lambda_{De}^2 \cong (Zm_e/2m_i)/(a^2/\mathrm{v_d^2})-1.$$

For the light elements, the condition $\mathrm{v_d}/a\cong(m_eZ/2m_i)^{1/2}$ occurs at $T_e/T_i\cong 10$.

Effect of Collisions

In practice, there are of course collisions in a plasma so that the case $k^2=0$, which means $\lambda\to\infty$, cannot occur because waves are damped for $\lambda \gtrsim \lambda_{coll}$. Further, finite geometry effects enter the problem. Therefore, onset occurs for $\mathrm{Im}(k^2)=0$ and $0<\lambda<\lambda_{coll}, L$. The effect of collisions was discussed by Ichimaru (1962). For the case $T_e/T_i\gg 1$, the imaginary part of ε is given by

$$\mathrm{Im}(\varepsilon)\cong -\alpha^2\left[\pi^{1/2}\left(x_e-\frac{\mathrm{v_d}}{a}\right)+\frac{\psi_i}{2x_i^3}\frac{ZT_e}{T_i}+\frac{ZT_e}{T_i}\pi^{1/2}x_i\exp(-x_i^2)\right], \qquad (12.3.4)$$

where $\psi_i=\mathrm{v_i}/kb$ and electron collisions are neglected (see Section 5.5.2) [remember that $(\psi_i/2x_i^3)(ZT_e/T_i)-2x_e\psi_e>0$].

Spectrum near Onset, $T_e/T_i\gg 1, \alpha^2\gg 1$, and $x_i^2\gg 1$

We may use Eqn (5.3.7), and under the above conditions, the first term dominates and reduces to

$$S(k_c,\omega)_{res}=\frac{2\pi^{1/2}}{k_c a}\frac{\omega_c^2}{4(\omega-\omega_{ac})^2+\pi\omega_{ac}^2\left[\underbrace{\left(\frac{\mathrm{v_c}-\mathrm{v_d}}{a}\right)}_{\text{electron Landau resonance}}+\underbrace{\frac{ZT_e}{T_i}\frac{\omega_{ac}}{kb}\exp\left(-\frac{\omega_{ac}}{k_c b}\right)^2}_{\text{ion Landau damping}}\right]^2}, \qquad (12.3.5)$$

where $\mathrm{v_c}=\omega_{ac}/k_c$ and k_c is parallel to the drift velocity $\mathrm{v_d}$.

Instability occurs when the drift velocity exceeds the velocity $\mathrm{v_c}$ by a sufficient margin to overcome the ion Landau damping.

12.3.2 Observation of Ion-Acoustic Drift Instability near Onset

This instability was investigated, for a marginally stable positive column, by Arunasalam & Brown (1965), for the case $\mathrm{v_d}\cong\mathrm{v_c}$ and $k_c\cong 0$ and consequently $\omega_c\cong 0$. Subsequently, Mase & Tsukishima (1975) measured ion

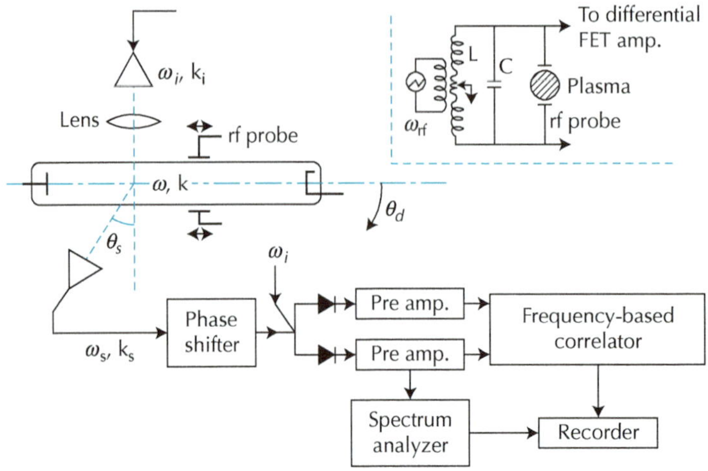

FIGURE 12.5 Schematic diagram of the experimental setup. (Courtesy of Mase & Tsukishima (1975) and The American Institute of Physics.)

wave turbulence in a cylindrical mercury discharge using microwave scattering (20 mW at 35 GHz). The experimental arrangement is shown in Fig. 12.5, and the wave number spectra is shown in Fig. 12.6. In the caption, θ_s is the angle between $\mathbf{k}_i$ and $\mathbf{k}_s$, and L is the distance from the grid whose potential was kept floating. At $L = 28$ cm, the total wave energy is saturated. The spectrum at $L = 15$ cm is shown for reference. It can be seen that the wave energy is transferred toward the lower k region as the turbulence develops. The results are consistent with the $|E(\mathbf{k})|^2 \propto k^{-5}$ prediction of Mannheimer (1971).

12.4 SCATTERING FROM A WEAKLY UNSTABLE PLASMA

12.4.1 Beam–Plasma Instability

The nonlinear behavior of a plasma containing a small high-energy electron beam has been studied theoretically by Drummond & Pines (1962b) and Vedenov *et al.* (1961) and for the case of a monoenergetic beam by Shapiro (1963). This "gentle bump instability" may be treated using the quasi-linear equations (10.2.1) and (10.2.7). It is found that a spectrum of waves develops in the narrow band of positive gradient of $f(\mathrm{v})$ (Fig. 12.4) with frequencies $\omega \cong \omega_{pe}$. The electrons in this region react strongly with the waves, and the resultant diffusion in velocity space acts to flatten the bump thereby stabilizing the instability. Provided the bump is small, which requires $(n_{beam}/n_{plasma})^{1/3} < 1$, the mode coupling terms exert no influence on the evolution of the instability up to and including the final quasi-linear stationary state.

A secondary effect that occurs is the trapping of some electrons, with velocity close to the resonant wave velocity, in the potential well of these waves.

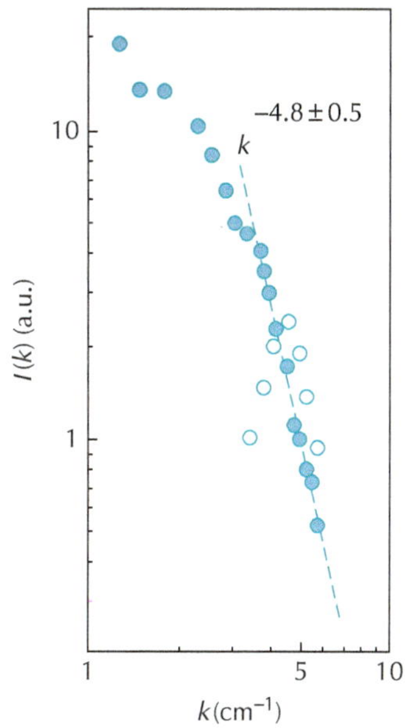

FIGURE 12.6 Wave number spectrum $I(k)$ for $\theta_s = 0$. The dots are for $L = 28\,\text{cm}$, and the curves are for $L = 15\,\text{cm}$. (Courtesy of Mase & Tsukishima (1975) and The American Institute of Physics.)

These electrons oscillate (bounce) in the well and generate a satellite wave at the bounce frequency $\Delta\omega \cong \omega_{pe}(n_B/3n_e)^{1/4}$.

This theory predicts, for given initial conditions n_e, T_0, n_B, v_d:

a. the energy density of unstable waves near the plasma frequency in the final state;
b. the range of perpendicular wave numbers;
c. the number of spatial e-folds for the waves to reach the final quasi-stationary state;
d. the width of the enhanced spectrum

$$P(\omega) \propto \frac{\gamma}{\pi\left[(\omega-\omega_{pe})^2+\gamma^2\right]}, \quad \text{where} \quad \gamma \cong \frac{\sqrt{3}}{2}\left(\frac{n_B}{2n_e}\right)^{1/3}\omega_{pe}$$

is the linear growth rate of the plasma wave;
e. the heating of the plasma electrons owing to the transfer of energy from the bump as it flattens.

12.4.2 Scattering from an Unstable Beam–Plasma System

The scattering of microwaves has been employed as a diagnostic in investigations of beam–plasma instabilities by Bohmer & Raether (1966), Malmberg &

Wharton (1969), Arunasalam *et al.* (1971), and Bollinger & Bohmer (1972). A brief review of the detailed study of the gentle bump instability conducted by Arunasalam *et al.* is given below. The argon and neon plasmas were produced in a linear mirror machine by electron cyclotron resonance heating; this leads to a plasma with an electron distribution corresponding in the main to $T_e \cong 15$ eV but having a small high-energy tail corresponding to a temperature $T_1 \cong 680$ eV. A gentle bump was produced by injecting an electron beam axially. The characteristic frequencies were $\omega_{ce} \cong 1.8$ GHz, $\omega_{pe} \cong 0.5 - 1.2$ GHz.

The scattering was made with an incident beam of 6 mW of 8 mm μ-waves; the scattering volume was $\sim 2\,\text{cm}^3$, and the scattering vector **k** is arranged to be parallel to v_d. Data are shown in Fig. 12.7.

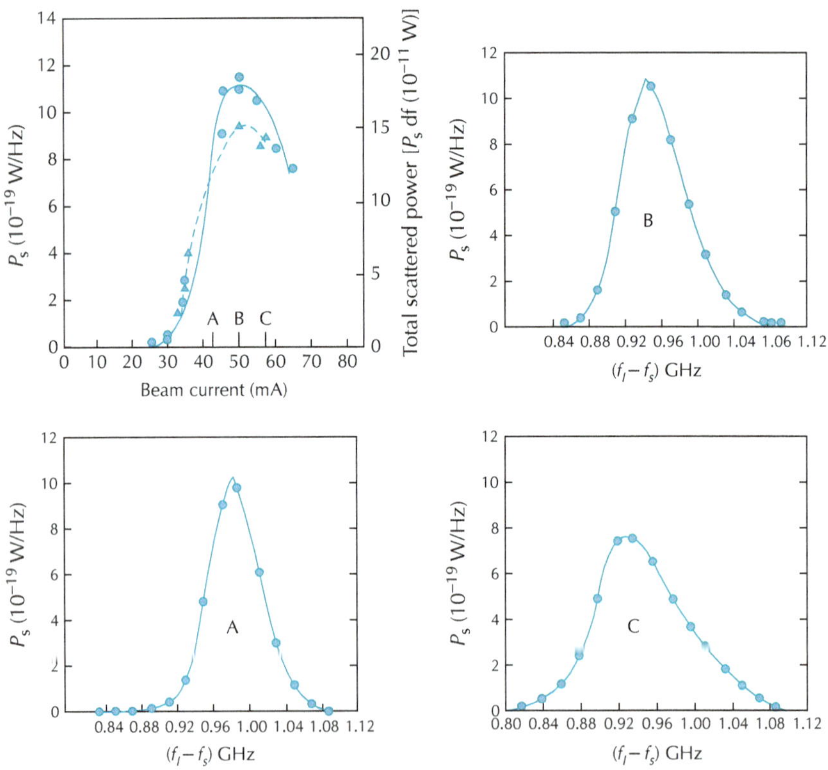

FIGURE 12.7 Typical behavior of the scattered power P_s (filled circle) and the shape of the scattered plasma line (filled triangle) as a function of the beam current, keeping all other parameters fixed. Curve A illustrates the line shape prior to the quasi-linear, steady-state conditions; B at the quasi-linear steady-state; and C beyond the quasi-linear steady-state, showing nonlinear mode coupling effects. (Courtesy of Arunasalam *et al.* (1971).)

12.5 SCATTERING FROM MICROTURBULENCE IN SHOCK FRONTS

12.5.1 Introduction

In an unstable plasma, the free energy is often converted into random energy in the plasma, that is, the instability acts to heat the plasma. An interesting example of this phenomena occurs in shock fronts, which propagate across the magnetic field in a magnetized plasma.

We may, by applying the laws of conservation of mass, momentum, and energy, calculate the change across a shock of basic plasma parameters, e.g., density, magnetic field, without discussing in detail how these changes were achieved. From such calculations, we can determine the amount of energy dissipated in the shock front. The way in which this random energy is shared between electrons and ions is a separate issue, but for some low Mach number shocks $M_A < 3$, $\mathbf{v}_{\text{shock}} \cdot \mathbf{B}_0 = 0$, most of the energy goes to the electrons ($M_A \equiv$ shock speed/magnetoacoustic speed). In this situation, the dissipation mechanism is resistivity. There is a jump in magnetic field B_z across the shock, and consequently, a current flows parallel to the front $j_y \cong \partial B_z/\partial x$. The current is opposed by electron–ion collisions, and some of the directed energy is dissipated, with the lighter electrons being heated.

It was discovered by Paul *et al.* (1967), for the perpendicular shock in the Tarantula experiment, that dissipation at the classical rate with the observed current density profile would only raise the electron temperature by ~5 eV. The jump to satisfy the conservation laws was, however, ~40 eV, and electron temperatures at this level were measured by the noncollective scattering of ruby laser light.

The increase of the dissipation rate over the classical level is attributed to turbulence generated by microinstabilities in the shock front. The instabilities occur because the relative electron–ion drift velocity exceeds the critical level for the onset of a number of instabilities. There are gradients n and T; consequently, a large number of instabilities are possible sources of the turbulence. In addition, there is the question of whether the situation fits weak or strong turbulence. Some of the experimental work on this phenomenon is discussed below.

12.5.2 Experiment

a. A diagram of the Tarantula apparatus and scattering equipment is shown in Fig. 12.8. Included is a vector diagram indicating the scattering geometry. The initial magnetic field is in the z direction with $B_{z0} = 0.12\,\text{Wb/m}^2$, $n_e = 6.4 \times 10^{14}\,\text{cm}^{-3}$, $T_{e0} \cong T_{i0} \cong 1\,\text{eV}$. The shock wave propagates radially, $M_A = 2.5$, the final electron temperature is 44 eV, the shock width is ~1 mm, and the shock velocity $v_s = 2.4 \times 10^7\,\text{cm}\,\text{s}^{-1}$. The current associated with the jump in magnetic field across the shock is in the azimuthal θ direction.

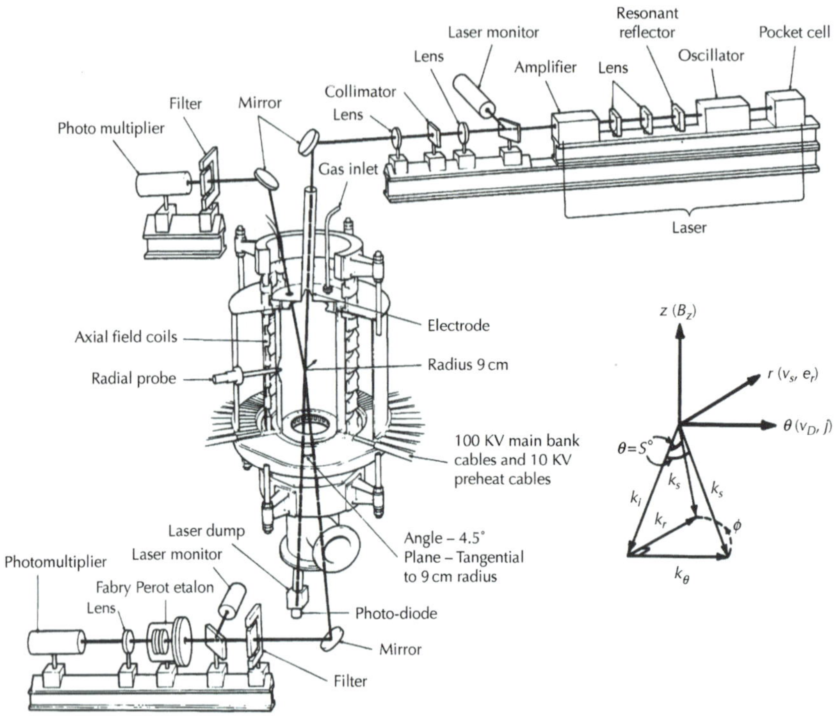

FIGURE 12.8 Tarantula shock wave experiment and forward scattering equipment. The inset shows the vector directions. (Courtesy of Paul *et al.* (1971) and UKAEA Culham Lab., U.K.)

Through the shock, we can find $T_e/T_i \cong 1-10$ and $v_d/a \lesssim 1$, and where the effect of the magnetic field may be neglected, we would expect the system to drive the ion-acoustic instability. The magnetic field will play a role mainly for waves propagating very close to perpendicular to $\mathbf{B}_z(\pm 2^\circ)$; here, Bernstein modes will be driven to instability (see Section 10.8.1).

The unstable waves have wavelengths from $\sim\lambda_{De}$ to a few λ_{De}. Measurements have been made with ruby laser sources, and the experimental difficulties rest in the small scattering angle $\theta < 10^\circ$ required in order to scatter with $k\lambda_{De} \lesssim 1$, in the shock plasmas. In the Tarantula experiment (Paul *et al.*, 1969; Daughney *et al.*, 1970; Muraoka *et al.*, 1973), θ was in the range 3.3–6.9°, allowing measurements down to $k\lambda_{De} \cong 0.7$ (note that resolution within the shock was not possible and the Debye length is an average value). The laser pulse $\lambda_i = 6943$ Å had a power of 50 MW and 35 ns halfwidth, which is greater than the time ~6 ns for the shock to cross the field of view.

The first important result was the observation of a spectrum enhanced considerably over the thermal level during the passage of the shock front through the laser beam.

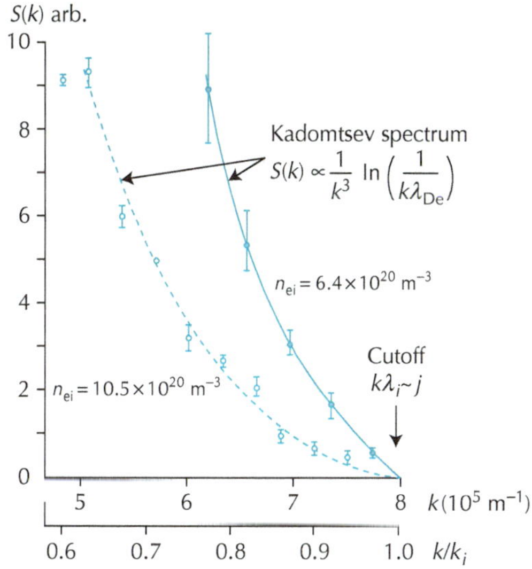

FIGURE 12.9 Wave number spectrum $S(k)$; experimental points are mean of five measurements, and error bars are standard deviation of the mean. The curve is a Kadomstev spectrum. (Courtesy of Paul *et al.* (1971).)

Subsequently, the spectrum was measured at different scattering angles, i.e., different k values, and the spectrum for $k\lambda_{De} < 1.0$ was observed to fit the theoretical form predicted by Kadomtsev (1965, p. 71)

$$S(k) \propto (1/k^3)\ln(1/k\lambda_{De})$$

(see Fig. 12.9). Similar spectra were observed by Keilhacker *et al.* (1971), Keilhacker & Steuer (1971a), and Machalek & Nielsen (1973). Unfortunately, the result is not a proof of the validity of the theory because this spectral shape was predicted only for smaller values of $k\lambda_{De}$.

The turbulent spectral density function was found to be two orders of magnitude above the thermal level. The enhanced fluctuations increase the strength of the velocity diffusion term, and this increases the rate at which energy is randomized. The effective collision frequency, Paul *et al.* (1969), has the form

$$\nu_{ei}^* = \frac{1}{6\pi^2}\left(\frac{e}{m_e}\right)^2 \frac{1}{a^3} \int_0^{1/\lambda_{De}} \left\langle E^2 \right\rangle k\,dk$$

The value of the integral is mainly dependent on the spectrum near λ_{De} is of course where the measurements were made. The measured spectrum $\nu_{ei}^* = 4.5 \times 10^9$ Hz is in good agreement with the value $\overline{\nu_{ei}} = 3.4 \times 10^9$ Hz, which is required to explain the anomalously large resistance.

The frequency spectrum was analyzed using a Fabry–Perot spectrometer. The waves were observed to propagate in the drift direction, and the peak frequency was $\omega \cong \omega_{\text{pi}}$, and while $\omega_{\text{pi}} \cong \omega_{\text{ce}}$, it was found to scale with ω_{pi} not ω_{ce}; Keilhacker & Steuer (1971a) found $\omega \cong 0.5\,\omega_{\text{pi}}$; Machalek & Nielsen (1973) found $\omega \cong 0.2\omega_{\text{pi}}$. This evidence suggests that at least in one case the Bernstein modes may not be an important factor.

b. Detailed measurements of the variation of $S(k)$ with angle in the plane perpendicular to $\mathbf{B}_{z0}$ (θ, z plane) have been reported by Machalek & Nielsen (1973). The shock waves are produced in a theta pinch apparatus with initial conditions similar to those in the Tarantula experiment. Similar measurements were made by Keilhacker & Steuer (1971a) and Muraoka *et al.* (1973).

c. In the experiments of Keilhacker *et al.* (1971) and Keilhacker & Steuer (1971a), shock waves were driven radially in a theta pinch discharge, but the initial plasma differed from that in the other two experiments. This lead to wider shock fronts, which made it possible to resolve the structure of the shock wave and to measure the spectra at a number of positions in the shock. The initial conditions were $n_e \cong 2-5 \times 10^{14}\,\text{cm}^{-3}$, $T_{\text{e}0} \cong 2-9\,\text{eV}$, $T_{\text{i}} \cong 10-40\,\text{eV}$, $T_{\text{i}}/T_{\text{e}} > 1$. The scattering angles were in the range $2-6°$, and a 500MW ruby laser source was used. Scattered spectra are shown in Fig. 12.10. Note that these unstable waves occur in a region $T_{\text{i}}/T_{\text{e}} \lesssim 1$, and $\text{v}_0/a < 1$, in which the ion-acoustic waves are stable; therefore, for this experiment, unstable Bernstein modes are a possible contender.

Summary

Turbulent spectra have been observed, with the turbulent fluctuation level 100 times the thermal level ($S(k)_{turb} = 10^2 S(k)_{th}$) thermal and, $\omega \lesssim \omega_{\text{pi}}$, using the scattering of ruby laser light. The short wavelength part of the spectrum, has the form $S(k) \propto (1/k^3)\ln(1/k\lambda_{\text{De}})$; in two of the experiments, the enhanced waves propagate in a region where ion-acoustic waves are linearly unstable. These results are not sufficient to differentiate between the various theories of this turbulent regime. Two key measurements are outstanding:

a. The variation of $S(k)$ with angle in the r, z plane is needed in order to see whether enhanced waves propagate far from the perpendicular to B_z, i.e., are not predominantly Bernstein modes.

b. Measurements of $S(k)$ to smaller values of $k\lambda_{\text{De}}$, i.e., < 0.5 are required.

The shape of the long wavelength part of the spectrum should help distinguish between the theories of Kadomstev (1965) in which the turbulent quasi-stationary state is a result of nonlinear Landau damping, of Tsytovich (1972) in which wave decay to regions of linear Landau damping is invoked, and of "strong" turbulence in which the nonlinear perturbations to the particle trajectories act to stabilize the turbulent state, as for example in the work of Wesson *et al.* (1973).

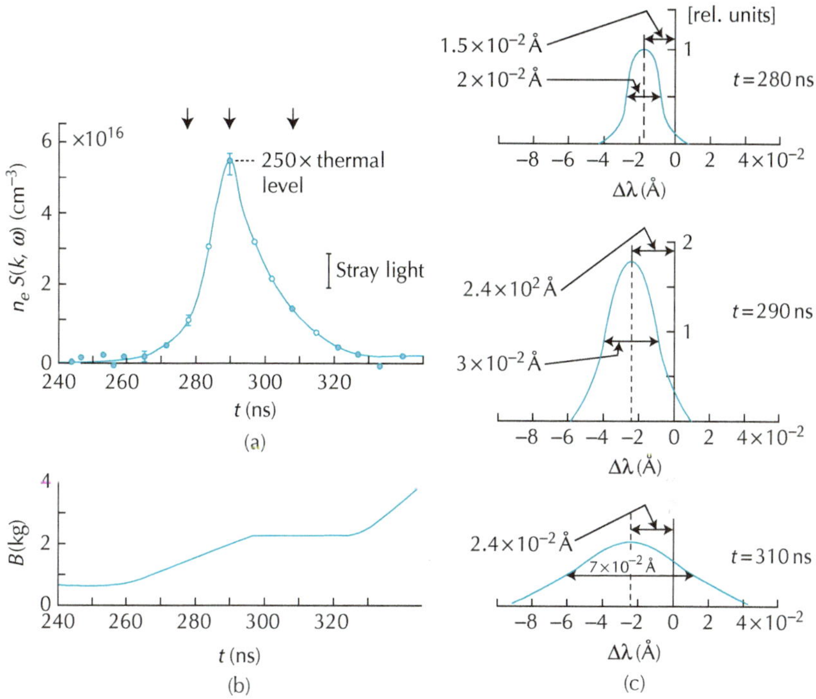

FIGURE 12.10 (a) Intensity of density fluctuations $n_e S(k, \omega)$ within a collisionless shock wave ($M_A = 2.5$, deuterium plasma) reaching 250 times the thermal level; (b) magnetic field B for comparison; (c) frequency spectra $S(k, \omega)$ of enhanced fluctuations for the three time points indicated in (a). (Courtesy of Keilhacker & Steuer (1971a).)

These measurements will certainly require the use of incident wavelengths longer than the ruby wavelength, to permit the use of larger scattering angles. The CO_2 laser source is a natural choice, and tests were made by Kornherr *et al.* (1972) and Bretz (1973).

12.6 INSTABILITIES IN MAGNETIC CONFINEMENT PLASMAS

Reviews of the theory of instabilities in magnetic plasmas have been written by Wesson (1978), Liewer (1985), Horton (1999), Freidberg (1982, 1987), and Weiland (2000).

In Fig. 12.11, we show a classification of modes given by Liewer and in Fig. 12.12 one by Doyle *et al.* (2002). In regard to scattering as a diagnostic, the shorter wavelength modes from drift instabilities are the most significant, because they are believed to be responsible for enhanced transport in many toroidal confinement devices (Hinton & Hazeltine, 1976). Ion and electron pressure gradients drive ion and electron diamagnetic currents. In the absence of

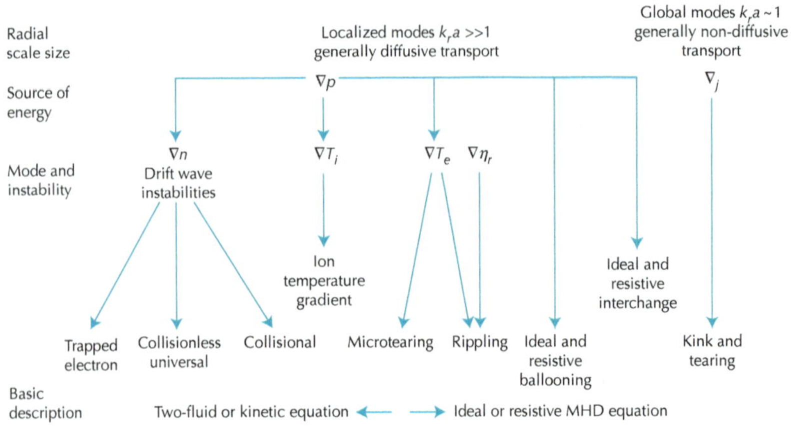

FIGURE 12.11 Classification of instabilities in magnetized plasmas. (Courtesy of Liewer (1985) and Nuclear Fusion.)

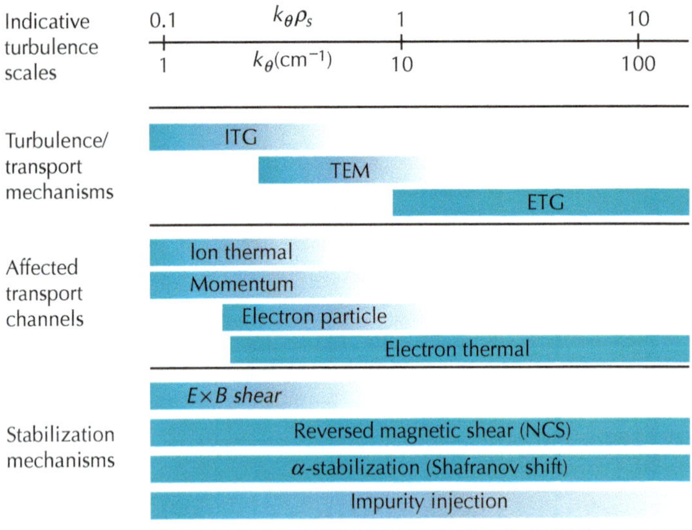

FIGURE 12.12 Indicative turbulence scales and mechanisms in toroidal systems. (Courtesy of Doyle *et al.* (2002) and Nuclear Fusion.)

dissipation of the electron motion parallel to the magnetic field, any density fluctuations will be in phase with the potential fluctuations resulting in no net flow $\langle nE \rangle = 0$.

In the presence of dissipation, owing to collisions or to trapped electrons, for example, a phase shift will occur and $\langle nE \rangle \neq 0$. This leads to a flux discussed by

Bretz (1997) and references therein,

$$\Gamma = -\langle n\nabla\phi \rangle \frac{\mathbf{B}}{B^2} + \langle n\mathrm{v}_{||} \rangle \mathbf{B}_0. \tag{12.6.1}$$

At high enough pressure, the fluctuations can perturb the magnetic field and lead to additional loss mechanisms.

An extensive review of diagnostic instrumentation for microturbulence may be found in Bretz (1997). More recently, Tynan *et al.* (2009) have reviewed the progress in experimental drift turbulence studies. Early studies in a tokamak by Mazzucato (1976, 1978) showed scattered spectra that were consistent with drift instabilities. But, at the time, the theory of instability growth and saturation was not developed sufficiently to clearly identify whether these modes were responsible for the observed anomalous transport. Since then, there has been growing evidence that in the case of toroidal plasmas, such as the tokamak and stellarator, the ion temperature gradient instability (ITG), trapped electron modes (TEM), and electron temperature gradient instability (ETG) cause anomalous transport in the plasma core.

The ITG instability occurs when $\eta_i = \frac{\partial_r \ln T_i}{\partial_r \ln n_i} > \eta_{\mathrm{crit}}$, typically 1. The modes have $k\rho_i = 0.1 - 0.5$ with a phase velocity around the ion diamagnetic drift velocity. TEM modes are driven by the electron pressure gradient and transition to the ETG mode as their wave number increases to $k\rho_i > 1$. The ETG instability occurs when $\eta_i = \frac{\partial_r \ln T_i}{\partial_r \ln n_i} > \eta_{\mathrm{crit}} \sim 1$.

From a theoretical standpoint, the main challenge in determining the evolution of instabilities is determining the stabilizing and destabilizing effects as the modes grow and modify their environment. Important stabilizing effects can include the rotational transform that connects regions of good and bad magnetic curvature, sheared flows, and magnetic shear and wells. An important effect, discovered in numerical simulations of collisional drift turbulence by Hasegawa *et al.* (1979), was that the turbulence itself can create stabilizing shear flows. These are known as "zonal flows" [see Diamond *et al.* (2005)].

The modes evolve, owing to the changes in the background plasma and through 3-wave coupling, to give a broad wave number spectrum. At some point, the turbulent energy cascades to smaller scales and modes where it may be damped by viscosity or Landau damping. Eventually, a new equilibrium balance is reached at a balance between the destabilizing and stabilizing forces.

Substantial progress, reviewed by Hammett (2009), has been made both theoretically and computationally using gyrokinetic theory coupled to the Particle-in-Cell (PIC) approach. The gyrokinetic approach, Rutherford & Frieman (1968) and Taylor & Hastie (1968), averages over the fast gyro-motion of the particles while retaining key nonlinearity and finite Larmor radius effects. PIC codes give insight into which effects are important and allow simplifications to be made. Hammett points out that a factor of 1023 speedup in the computations has occurred as a result of these efforts. Among modern codes making

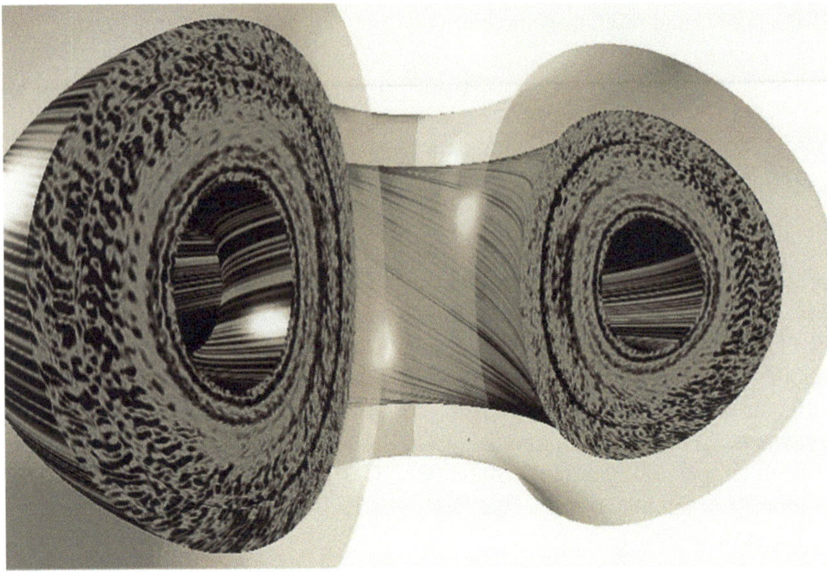

FIGURE 12.13 Global GYRO simulation of DIII-D shot 121717 with full physics: plasma shape, profile variation and $E \times B$ shear, collisions, and electromagnetic effects. Contours of electron density fluctuations. Shot has inverted q-profile for which q-min is slightly less than 2 (Courtesy of R. Waltz.)

important contributions to the plasma turbulence and transport area are GS2 (Dorland *et al.*, 2000), GENE (Jenko, 2000), GYRO (Candy & Waltz, 2003), and GEM (Chen *et al.*, 2003). A computed example of the turbulent density structure in a tokmak is shown in Fig. 12.13.

Experimental diagnostics measure conditions under the constraints of viewing sight-line and time and space. Because of these limitations, synthetic data, allowing for the experimental realities, are constructed from code calculations to compare with the experimental data. An example of this approach for phase contrast imaging (see Section 8.2) is shown for Alcator C-Mod in Fig. 12.14 (Lin *et al.*, 2009).

Experimental evidence for the existence of η_i-modes was reported by Brower *et al.* (1987, 1988a,b, 1989). Their data were obtained using a far-infrared laser ($P_i = 14$ mW, $\lambda_i = 1222\,\mu$m) based multichannel scattering system on the TEXT tokamak. This system could scan the scattering volume vertically and horizontally and permitted the mapping of the turbulence distribution ($0 < k < 15\,\text{cm}^{-1}$) over the entire poloidal cross section.

Among earlier work was the study of microturbulence on the KT-5 tokamak also using CO_2 light scattering (Chang *et al.*, 1990b).

An important contribution was made on Tore Supra using CO_2 light in the ALTAIR scattering system (Truc *et al.*, 1992). This system does scattering close

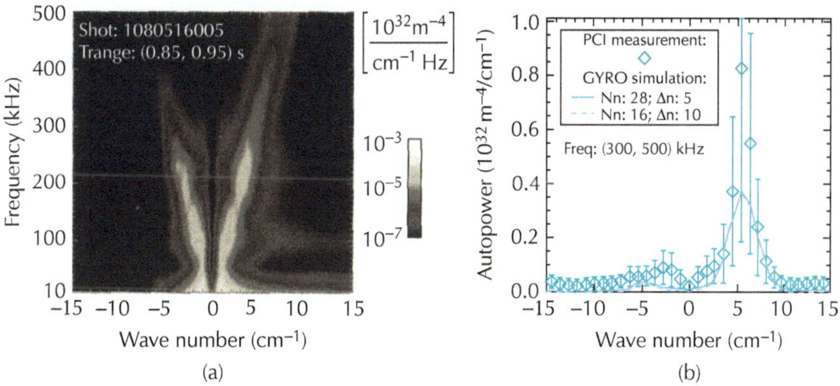

FIGURE 12.14 (a) Synthetic phase contrast image frequency/wave number spectra in the H-mode before the ITB formation. A system response function has been implemented to emulate the top view configuration of the masked phase plate. (b) Comparison of the synthetic and experimental PCI wave number spectra in the frequency range of 300–500 kHz. (Courtesy of L. Lin (2009) and The American Institude of Physics.)

to perpendicular to the magnetic field; the shear in the field helps localize the fluctuation measurements. Hennequin *et al.* (2004) reported measurements of wave number fluctuations on the range $3\,\mathrm{cm}^{-1} < k < 26\,\mathrm{cm}^{-1}$. The measurements showed how the spectrum and decorrelation times scaled with normalized poloidal gyroradius (see Figs. 12.15 and 12.16).

Using the techniques described in Section 7.7.4, Mazzucato (2003, 2006) made localized measurements at 280 GHz of fluctuations in tokamaks (Mazzucato *et al.*, 2008); see Fig. 12.17). Here, they performed scattering investigations in L-mode discharges in the National Spherical Torus Experiment (NSTX) device and found electron gyroscale fluctuations consistent with η_e-mode turbulence and observed enhanced fluctuations when the electron temperature gradient exceeded the ETG linear critical gradient. The results shown in Fig. 12.18 are consistent with the behavior of ETG modes. Smith *et al.* (2009) have reported further investigations of ETG turbulence in the NSTX device in H-mode plasmas. These plasmas had large toroidal rotation and accompanying $E \times B$ shear rates, and conditions such that the electron temperature gradient was marginally stable with respect to the ETG linear critical gradient. The data supported the view that the $E \times B$ flow shear can be an effective suppression mechanism for ETG turbulence.

Peebles *et al.* (2004) and Rhodes *et al.* (2006, 2007b) describe measurements made in the DIII-D tokamak during ohmic, neutral beam, and electron cyclotron heated discharges. The use of an FIR (288 GHz) forward scattering system plus an 84 GHz back scattering system allowed them to cover a wide range of wave numbers (k) from $0.2\,\mathrm{cm}^{-1}$ ($k_\perp \rho_i < 1$), and $7-11\,\mathrm{cm}^{-1}$ ($k_\perp \rho_i \sim 1-3$), to

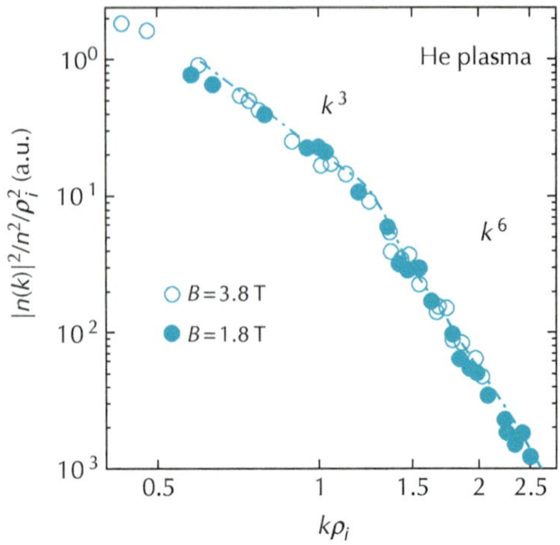

FIGURE 12.15 k spectrum of density fluctuations $|n(k)|^2$ (normalized to the squared density) divided by (ρ_i^2) as a function of $k\rho_i$ for the He series. (Courtesy of Hennequin *et al.* (2004) and the Institute of Physics Publications.)

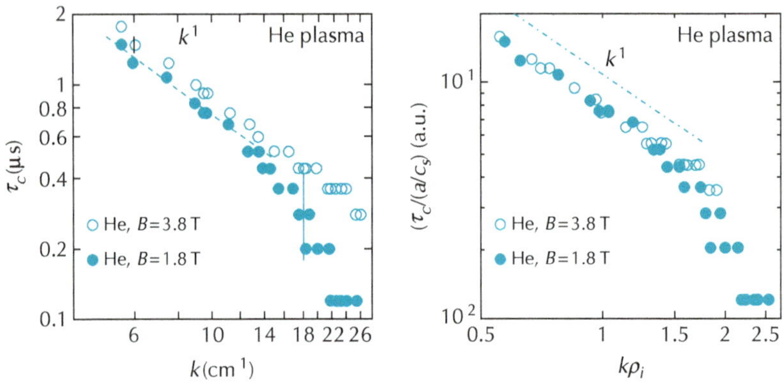

FIGURE 12.16 (a) Signal modulus correlation time as a function of the wave number spectrum and (b) normalized to (a/c_s) as a function of $k\rho_i$. (Courtesy of Hennequin *et al.* (2004) and the Institute of Physics Publications.)

$35-40\,\mathrm{cm}^{-1}$ ($k_\perp\rho_i \sim 4-10$). The diagnostic geometry is shown in Fig. 12.19, and the power spectra for the different k ranges is shown in Fig. 12.20. Code calculations of the instabilities appeared to be in agreement with the low- and intermediate-k measurements.

Measurements of ETG modes in the FT-2 tokamak have been reported by Gusakov *et al.* (2006). They used the enhancement of the probing beam at the upper hybrid resonance to obtain high localization and enhanced sensitivity of scattering from the internal region of the plasma. The scattered signal exhibited a huge frequency shift in the electron diamagnetic direction. van der Meiden

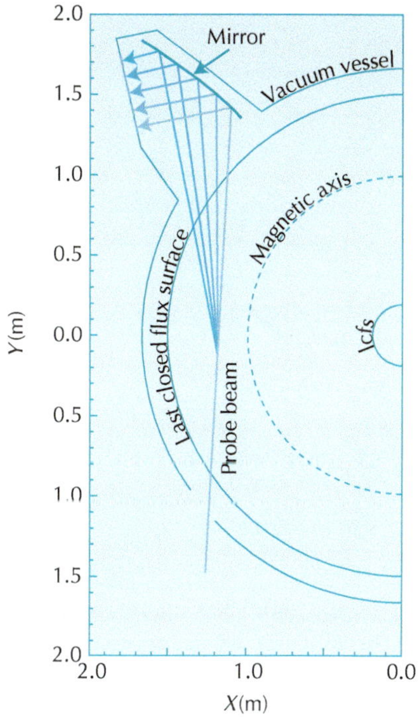

FIGURE 12.17 Probe beam (blue line) and scattered waves (red line for detection of core (left) and edge (right) fluctuations. (Courtesy of Mazzucato *et al.* (2008) and the American Physical Society.)

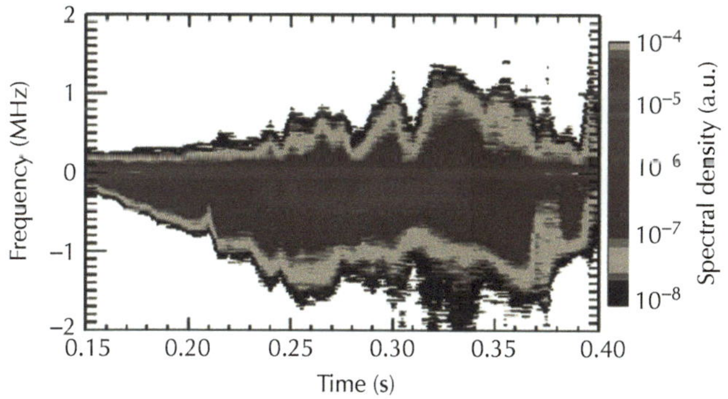

FIGURE 12.18 Logarithmic plot of spectral density of fluctuations with $k_\perp \rho_e = 0.2-0.4$ at $R = 1.4$m. Negative frequencies correspond to wave propagation in the electron diamagnetic direction. (Courtesy of Mazzucato *et al.* (2008) and the American Physical Society.)

et al. (2006) used a 10-Hz repetitive high-resolution TV Thomson scattering system on the TEXTOR tokamak to make detailed enough measurements that they could detect the $m = 2$ magnetic island (Fig. 12.21).

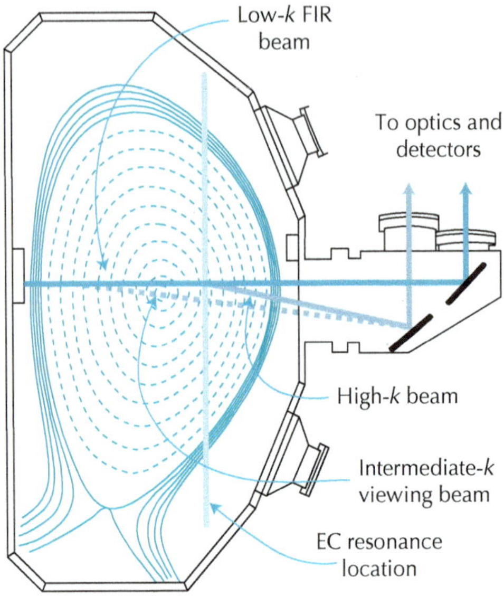

FIGURE 12.19 Diagnostic geometry for the low- and intermediate-k FIR forward scattering and the high-k, millimeter-wave backscattering systems. The flux surfaces for the discharges studied, as well as the EC resonance location (which acts as a beam dump for the high-k probe), are shown. The intersection of the intermediate-k viewing beam and the low-k beam defines the intermediate-k sample volume (approximately the minor radius – a – for the wave numbers presented here). (Courtesy of Rhodes *et al.* (2007a) and Nuclear Fusion.)

12.7 INSTABILITIES IN INERTIAL CONFINEMENT PLASMAS

An important concern for inertial fusion is the backscattering of the laser beams by waves in the plasma created or enhanced by the laser beams. Examples are stimulated Raman scattering (SRS) and stimulated Brillouin scattering (SBS) (see Kruer, 2003). Experimental studies in support of the National Ignition Facility program, Moses & Wuest (2004), Lindl *et al.* (2004), and Drake (2006), have quantified the effects of density, temperature, intensity, and laser beam smooth on laser-plasma instabilities (MacGowan *et al.*, 1996; Lefebvre *et al.*, 1998; Fernandez *et al.*, 2000; Seka *et al.*, 2002). Most recently, at the Omega Laser Facility (LLE/U. Rochester) (Froula *et al.*, 2007b, 2010), experiments were undertaken in gas-filled hohlraums, in which uniform multi-millimeter long plasmas were formed with electron temperatures between 1.5 and 3.5 keV and densities from 5 to 13% of the critical density n_{cr} for 0.35 micron light (see Fig. 12.22). Measurements were made of the backscatter and laser beam propagation (i.e., transmission and beam spray), and the impact on intensity thresholds of phase plates and polarization smoothing was evaluated. The results provide strong guidance for the design of ignition hohlraums and the choice of laser parameters. Specifically, to insure propagation of 0.35-micron light through millimeters of plasma, keep the plasma temperature above

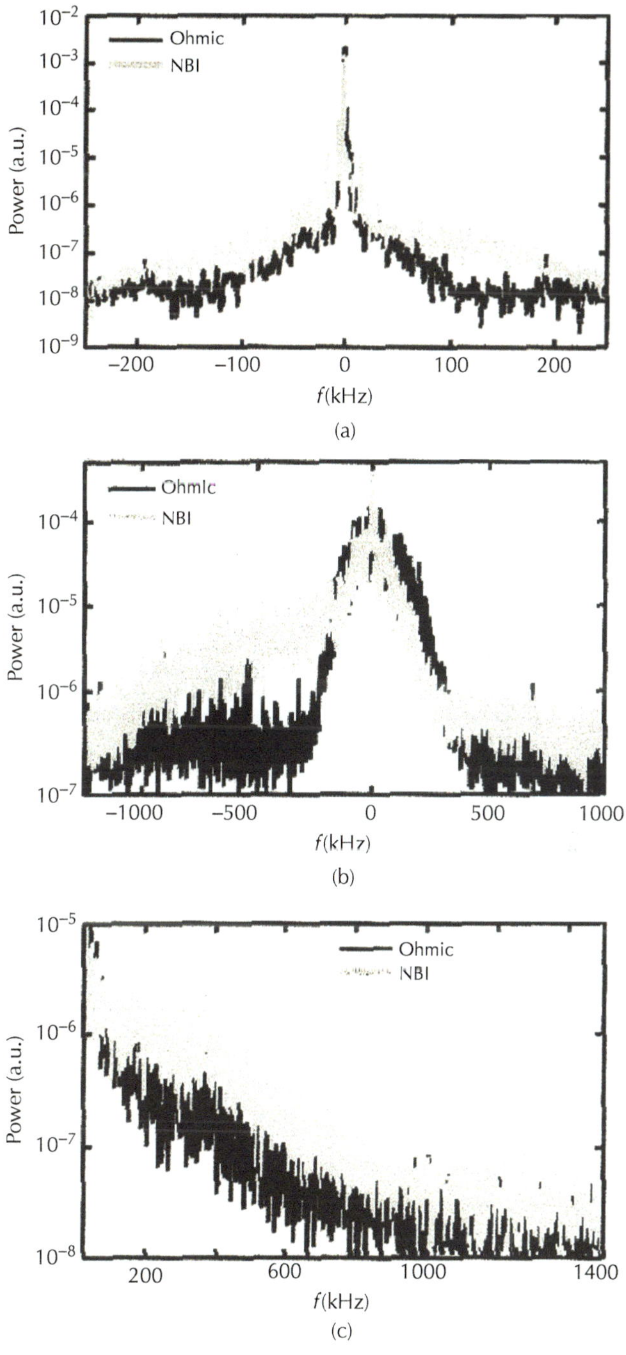

FIGURE 12.20 Power spectra for (a) low k, (b) intermediate k, and (c) high k for two times: Ohmic (black) and peak response during third neutral beam pulse (gray). (Courtesy of Rhodes *et al.* (2007a) and Nuclear Fusion.)

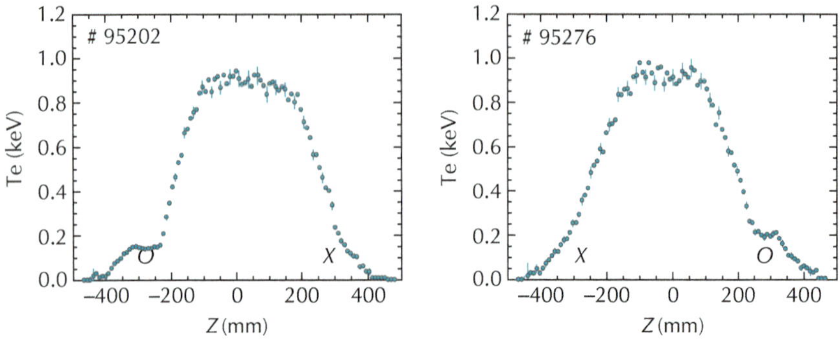

FIGURE 12.21 Temperature profiles through different phases of a stationary $m/n = 2/1$ magnetic island, measured in two different TEXTOR shots with static dynamic ergodic diverter. (Courtesy of Donné & Barth (2006) and the Institute of Physics Publishing.)

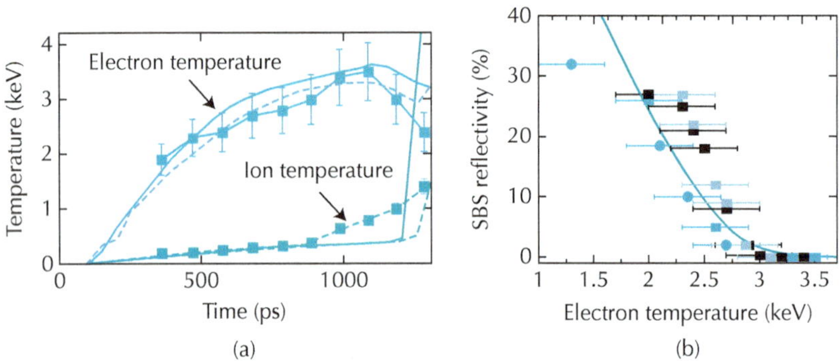

FIGURE 12.22 (a) The electron and ion temperatures are measured as a function of time using Thomson scattering when the hohlraum was heated by 16 kJ (see Fig. 5.11 for spectral data). (b) The measured SBS reflectivity is compared directly to the electron temperature by temporally correlating the measurements for an interaction beam intensity of $I = 1.7 \times 10^{15}$ W cm^{-2}. The total heater beam energy is varied from 8 kJ (circles) to 16 kJ (squares). As the electron temperature reaches 3 keV, the SBS backscatter drops below detection levels (< 1%). (Courtesy of D. H. Froula *et al.* (2006b, 2007))

2.5 keV and limit the laser intensity to less than 10^{15} W cm^{-2} in the 10–15% n_{cr} regimes.

The experimental setup is shown in Fig. 12.23 where a probe beam propagates along the hohlraum axis and therefore interacts with a long-scale underdense plasma. This geometry allows the measurements of both the backscattered and forward scattered light and provides a complete accounting of the energy in the interaction beam (Froula *et al.*, 2006a; Niemann *et al.*, 2008). The probe beam was converted to 0.26 microns for characterization experiments using Thomson scattering. Figure 12.22a shows the results of multi-ion species Thomson scattering (see Section 5.4.2) where the electron and ion temperatures are measured and compared with hydrodynamic simulations (see Froula *et al.*, 2006b). These measurements allowed the comparison of the backscattered light

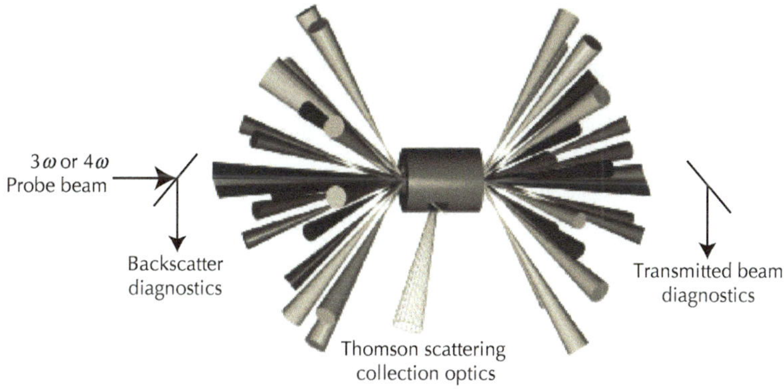

FIGURE 12.23 The 37 heater beams are distributed into three cones: 18 58° cone 3 beams, 9 42° cone 2 beams, and 10 21° cone 1 beams. The probe beam is aligned along the hohlraum axis. When characterizing the plasma conditions, the probe beam is converted to 0.26 micron light, and a 500 μm square window is cut into the side of the hohlraum wall to allow light scattered from a small scattering volume located at the center of the hohlraum to be collected by an optical Thomson scattering system. For laser-plasma studies, the probe beam is converted to 0.35 micron light, and the forward and back reflected light is collected by a transmitted beam diagnostic and a suite of backscatter diagnostics. (Courtesy of D. H. Froula *et al.* (2010).)

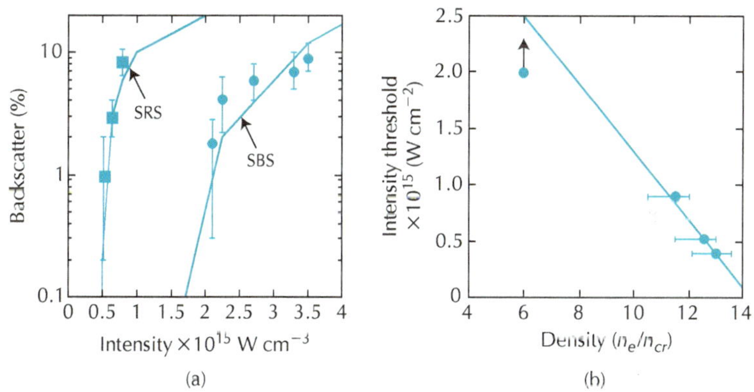

FIGURE 12.24 (a) The instantaneous SBS (circles) and SRS (squares) are plotted as a function of the interaction beam intensity. The simulated reflectivies calculated by pF3d are shown (solid curves). The SRS data are for a density of $n_e/n_{cr} = 11.5\%$, and the SBS data are for a density of $n_e/n_{cr} = 6\%$. (b) The intensity at which the SRS reaches 5% (intensity threshold) is plotted as a function of the plasma density. (Courtesy of D. H. Froula *et al.* (2010).)

with the measured plasma conditions; Fig. 12.22b shows that in a 6% n_{cr} plasma, the simulated Brillouin scattering is reduced below 1% for electron temperatures above 3 keV.

Experimental results presented in Fig. 12.24 show that both stimulated Brillouin and stimulated Raman scattering have an intensity threshold and experiments where backscatter will adversely affect the results must remain below these thresholds.

PROBLEMS

12.1. Many of the cases of interest to us involve unstable waves. As a beginning, we consider the simple case of an unmagnetized plasma consisting of a fixed neutralizing ion background and two cold electron components drifting relative to each other, i.e.,

$$f_{e0} = n_{01} f_{e01} + n_{02} f_{e02}$$

with

$$f_{e01} = \delta(\mathrm{v} - \mathrm{v}_{01}) \quad \text{and} \quad f_{e02} = \delta(\mathrm{v} - \mathrm{v}_{02})$$

(a) Derive the electrostatic dispersion relation for this system.
(b) For simplicity, let $\mathrm{v}_{01} = u$ and $\mathrm{v}_{02} = -u$ and $\omega_{p1}^2 = \omega_{p2}^2$. Find the condition for instability, i.e., an exponentially *growing* solution.
(c) What is the condition for maximum growth? What is the maximum growth rate?

12.2. To proceed with our investigation of plasma instability, we consider a slightly more realistic case of drifting Cauchy distributions to model the effects of temperature. Specifically, consider the following electron distribution function for an unmagnetized plasma:

$$\hat{f}_0 = \frac{a}{2\pi}\left[\frac{1}{(\mathrm{v} - \mathrm{v}_0)^2 + a^2} + \frac{1}{(\mathrm{v} + \mathrm{v}_0)^2 + a^2}\right]$$

Derive the high-frequency electrostatic dispersion for this system. For what ratio of v_0/a is the system stable?

12.3. Consider an unmagnetized plasma with a fixed, neutralizing ion background. We wish to consider the excitation of electron plasma oscillations in a warm, electron plasma by means of a cold electron beam. The one-dimensional velocity distribution function is given by

$$f_0(\mathrm{v}) = n_p g_0(\mathrm{v}) + n_b h_0(\mathrm{v})$$

In the above,

$$g_0 = \frac{1}{\mathrm{v}_{th}\pi^{1/2}} e^{-\mathrm{v}^2/\mathrm{v}_{th}^2}, \; h_0 = \delta(\mathrm{v} - \mathrm{v}_0) \quad \text{and} \quad n_b = \alpha n_p.$$

Derive the dispersion relation for high-frequency electrostatic perturbations in the limit that $\frac{\omega/k}{\mathrm{v}_{th}} \ll 1$. Show that a solution exists in which $\operatorname{Im}\omega > 0$ (i.e., growing oscillations). Obtain an explicit expression for the growth rate γ.

12.4. Temperature anisotropy can lead to instability. As an example, we consider the case of electron Bernstein waves propagating in the $\hat{x} - \hat{z}$ plane ($\mathbf{B}_0 = \hat{z} B_0$) in a Maxwellian plasma with $T_\perp \neq T_{||}$. Show that an instability occurs if $T_\perp / T_{||} > 1$.

12.5. Remember that we showed that ion-acoustic waves propagate and that they are only weakly damped in a plasma with $T_e/T_i \gg 1$. However, for $T_e = T_i$, they are heavily damped. In this problem, we wish to examine the effect of a current. We therefore assume a uniform, unmagnetized plasma comprised of ions and electrons with *drifting* electrons, as in the following figure. We see that for sufficiently large relative drift velocity (electron current) that ion waves see $f'_{eo} > 0$ and thus can be driven unstable.

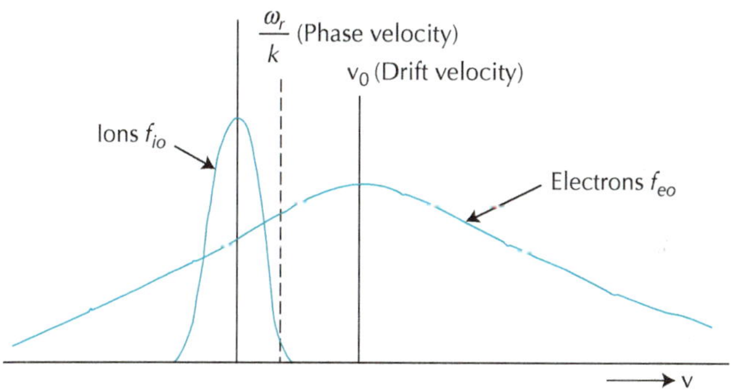

In this problem, you are going to derive the dispersion relation for unstable ion waves. To do this, you may make the following assumptions:

$$\left|\frac{\omega_r}{k}\right| \gg \sqrt{\frac{2\kappa T_i}{m_i}} = v_{thi}$$

and

$$\left|\frac{\omega_r}{k} - v_0\right| \ll \sqrt{\frac{2\kappa T_e}{m_e}} = v_{the},$$

where v_0 is the drift of the electrons governed by a Maxwellian distribution, and

$$\hat{f}_{eo} = \frac{1}{\sqrt{\pi}v_{te}} \exp[-(v - v_0)^2/v_{te}^2],$$

and

$$\hat{f}_{io} = \frac{1}{\sqrt{\pi}v_{ti}} \exp[-v^2/v_{ti}^2].$$

SOLUTIONS TO ODD PROBLEMS

12.1. To obtain insight into instabilities as well as continued experience with Vlasov dispersion relations, we will investigate the solution in excruciating detail. We are given a fixed ion background and two drifting electron components:

$$f_{e0} = n_{01}\hat{f}_{e01} + n_{02}\hat{f}_{e02} \quad \text{with} \quad \hat{f}_{e01,2} = \delta(\mathrm{v} - \mathrm{v}_{01,2}).$$

(a) Our dispersion relation for electrostatic waves is then

$$\begin{aligned} 0 &= 1 - \sum_\alpha \frac{\omega_{p\alpha}^2}{k^2} \int \frac{\partial \hat{f}_{0\alpha}/\partial \mathrm{v}}{\mathrm{v} - \frac{\omega}{k}} d\mathrm{v} = 1 - \sum_\alpha \frac{\omega_{p\alpha}^2}{k^2} \frac{\partial}{\partial\left(\frac{\omega}{k}\right)} \int \frac{\hat{f}_{0\alpha} d\mathrm{v}}{\mathrm{v} - \frac{\omega}{k}} \\ &= 1 - \frac{\omega_{p1}^2}{k^2} \frac{\partial}{\partial\left(\frac{\omega}{k}\right)} \int \frac{\delta\left(\mathrm{v} - \mathrm{v}_{01}\right)}{\mathrm{v} - \frac{\omega}{k}} d\mathrm{v} - \frac{\omega_{p2}^2}{k^2} \frac{\partial}{\partial\left(\frac{\omega}{k}\right)} \int \frac{\partial(\mathrm{v} - \mathrm{v}_{02})}{\mathrm{v} - \frac{\omega}{k}} d\mathrm{v} \\ &= 1 - \frac{\omega_{p1}^2}{k^2} \frac{\partial}{\partial\left(\frac{\omega}{k}\right)} \frac{1}{\mathrm{v}_{01} - \frac{\omega}{k}} - \frac{\omega_{p2}^2}{k^2} \frac{\partial}{\partial\left(\frac{\omega}{k}\right)} \frac{1}{\mathrm{v}_{02} - \frac{\omega}{k}} \\ &= 1 - \frac{\omega_{p1}^2}{k^2} \frac{1}{\left(\mathrm{v}_{01} - \frac{\omega}{k}\right)^2} - \frac{\omega_{p2}^2}{k^2} \frac{1}{\left(\mathrm{v}_{02} - \frac{\omega}{k}\right)^2}. \end{aligned}$$

This yields $\boxed{k^2 = \frac{\omega_{p1}^2}{(\mathrm{v}_{01} - \frac{\omega}{k})^2} + \frac{\omega_{p2}^2}{(\mathrm{v}_{02} - \frac{\omega}{k})^2}}$.

(b) Letting $\omega_{p1}^2 = \omega_{p2}^2 = \omega_i^2$ and $\mathrm{v}_{01} = u$, $\mathrm{v}_{02} = -u$, we have

$$\boxed{1 = \frac{\omega_i^2}{(\omega - ku)^2} + \frac{\omega_i^2}{(\omega + ku)^2}}.$$

Now, we can also solve this dispersion relation graphically; we have

$$1 = \frac{\omega_i^2}{(\omega - ku)^2} + \frac{\omega_i^2}{(\omega + ku)^2}.$$

Plotting this, we have

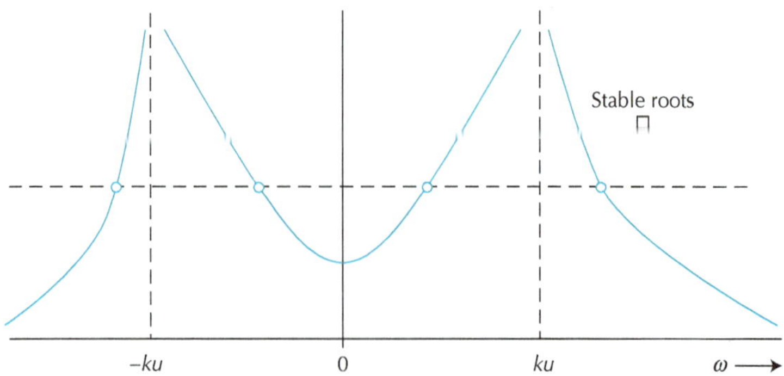

For a minimum greater than unity, we have two complex conjugate roots – one of which gives an instability.

For instability, we want the RHS to be >1. This we get for the minimum at $\omega = 0$ for $\frac{\partial\omega_i^2}{(ku)^2} > 1$
or

$$\boxed{ku_< \sqrt{2}\omega_i}.$$

We can see from the above that instability is due to long-wavelength oscillations, by a small drift speed – or an accumulation of charge in layers $\|$ to the stream $\mathbf{k}\cdot\mathbf{u} \ll |k||u|$.

However, as $|k||u|$ reduced below threshold, the growth rate decreases. Also note that for small k, thermal effects become important. Let us now proceed to solve the complete dispersion relation.

$$0 = 1 - \frac{\omega_i^2}{k^2}\left\{\frac{1}{\left(u-\frac{\omega}{k}\right)^2} + \frac{1}{\left(u+\frac{\omega}{k}\right)^2}\right\}$$

or

$$0 = 1 - \frac{\omega_i^2}{k^2}\frac{2\left(u^2+\frac{\omega^2}{k^2}\right)}{\left(u-\frac{\omega}{k}\right)^2\left(u+\frac{\omega}{k}\right)^2}.$$

We can write this as

$$\left(u-\frac{\omega}{k}\right)^2\left(u+\frac{\omega}{k}\right)^2 = \frac{2\omega_i^2}{k^2}\left(u^2+\frac{\omega^2}{k^2}\right)$$

or letting $\mathrm{V} = \omega/k$

$$\mathrm{V}^4 - 2u^2\mathrm{V}^2 + u^2 = \frac{2\omega_i^2}{k^2}\left(u^2 + \mathrm{V}^2\right)$$

and

$$\mathrm{V}^4 - 2\left(\frac{\omega_i^2}{k^2} + u^2\right)\mathrm{V}^2 + u^2\left[u^2 - \frac{2\omega_i^2}{k^2}\right] = 0$$

and

$$\mathrm{V}^2 = \frac{2\left(\frac{\omega_i^2}{k^2}+u^2\right) \pm \sqrt{4\left(\frac{\omega_i^2}{k^2}+u^2\right)^2 - 4\left(u^2 - \frac{2\omega_i^2}{k^2}\right)}^{\,2}}{2}$$

so

$$\mathrm{V}^2 = \frac{\omega_i^2}{k^2} + u^2 \pm \sqrt{\frac{\omega_i^2}{k^4} + 4\frac{u^2\omega_i^2}{k^2}}$$

and

$$\frac{\omega^2}{k^2} = \frac{\omega_i^2}{k^2} + u^2 \pm \frac{\omega_i^2}{k^2}\sqrt{1 + \frac{4u^2k^2}{\omega_i^2}}.$$

This yields

$$\omega^2 = \omega_i^2 + k^2u^2 \pm \omega_i^2\sqrt{1 + \frac{4u^2k^2}{\omega_i^2}}.$$

Then, finding the extremum, we have

$$\frac{\partial\omega^2}{\partial(k^2u^2)} = 0 = 1 \pm \omega_i^2 \frac{1}{2}\frac{4}{\omega_i^2}\left[1 + \frac{4k^2u^2}{\omega_i^2}\right]^{-1/2}$$

$$\Rightarrow k^2u^2 = \frac{3}{4}\omega_i^2 \quad \text{or} \quad ku = \frac{\sqrt{3}}{2}\omega_i.$$

Substituting this back into our dispersion relation gives us $(\omega_i)_{\max} = \dfrac{\omega_i}{2}$. On the below page, we see the entire dispersion relation plotted versus k.

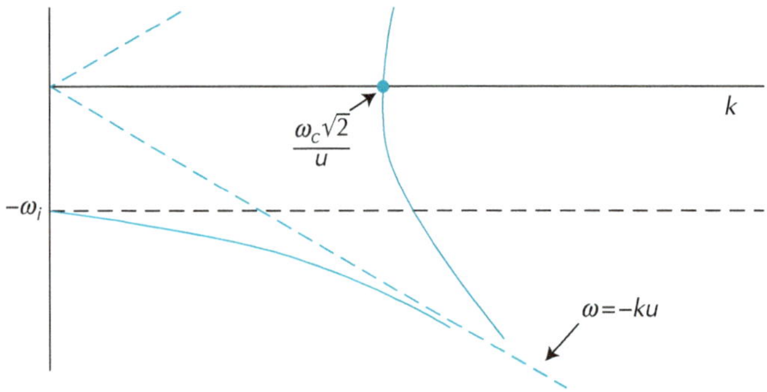

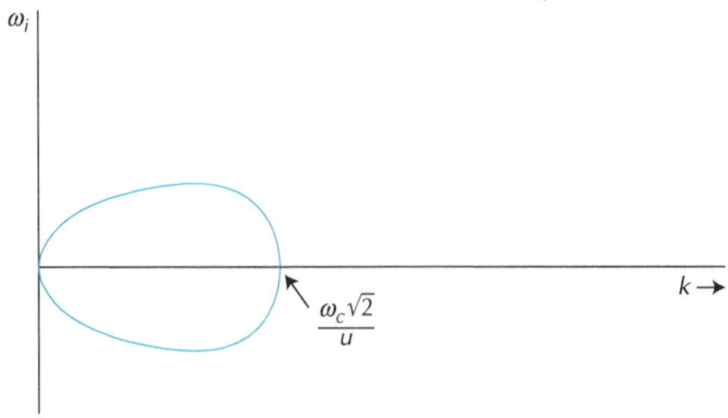

12.3. As with the following problems, we will investigate the solution in excruciating detail. We have the following distribution in equilibrium:

$$f_0(\mathrm{v}) = n_p g_0(\mathrm{v}) + n_b h_0(\mathrm{v}) \text{ with } g_0(\mathrm{v}) = \frac{1}{\mathrm{v}_{th}\sqrt{\pi}} e^{-\mathrm{v}^2/\mathrm{v}_{th}^2} \text{ and}$$

$$h_0 = \delta(\mathrm{v} - \mathrm{v}_0) \quad \text{where } n_b = \alpha n_p.$$

Now, for the warm electron (thermal) plasma, we have (one dimension)

$$\frac{\partial f_1}{\partial t} + \mathrm{v}\frac{\partial f_1}{\partial x} - \frac{e}{m} E f'_{0p} = 0 \quad \text{where} \quad f'_{0p} = \frac{\partial f_{0p}}{\partial \mathrm{v}}.$$

Then, looking for the time asymptotic solution, we have

$$-i\omega f_1 + ih\mathrm{v} f_1 - \frac{e}{m_\mathrm{e}} E f'_0 = 0$$

$$\text{or} \quad f_1 = \frac{ieE}{m_\mathrm{e}} \frac{f'_0}{\omega - k\mathrm{v}} \quad \text{and} \quad n_p = \frac{ieE}{m_\mathrm{e}} \int \frac{f'_{0p} d\mathrm{v}}{\omega - k\mathrm{v}} \text{ where } f_{0p} = n_p g_0.$$

For the cold beam, we have

$$\frac{\partial n_{1p}}{\partial t} + n_{b0} \nabla \cdot (\mathrm{v}) + \mathrm{v}_0 \cdot \nabla n_{1b} = 0 \text{ or after Fourier analyzing:}$$

$$\begin{aligned} i\omega n_{1b} &= n_{0b} \nabla \cdot \mathrm{v} + \mathrm{v}_0 \cdot \nabla n_{1b} \\ &= n_{0b} ik\mathrm{v} + \mathrm{v}_0 ik n_{1b}. \end{aligned}$$

From the momentum equation, we have

$$(-i\omega + ik\mathrm{v}_0)\mathrm{v} = \frac{-eE}{m_\mathrm{e}} \Rightarrow ik\mathrm{v} = \frac{ekE}{m_\mathrm{e}(\omega - k\mathrm{v}_0)}.$$

Therefore, we have

$$(\omega - k\mathrm{v}_0) n_{1b} = -in_{0b}(ik\mathrm{v}) = \frac{-in_{0b}ekE}{m_\mathrm{e}(\omega - k\mathrm{v}_0)}$$

or

$$n_{1b} = \frac{-in_{0b}ekE}{m_\mathrm{e}(\omega - k\mathrm{v}_0)^2}$$

Poisson's equation gives us

$$\begin{aligned} ikE &= -4\pi e[n_{1p} + n_{1b}] \\ &= -4\pi e \left[\frac{ieE}{m_\mathrm{e}} \int \frac{f'_{0p} d\mathrm{v}}{\omega - k\mathrm{v}} - \frac{in_{0b}ekE}{m_\mathrm{e}(\omega - k\mathrm{v}_0)^2} \right]. \end{aligned}$$

This can be written:

$$\begin{aligned} 1 &= \frac{\omega_{\mathrm{pe}}^2}{k^2} \int \frac{\hat{f}'_{0p}(\mathrm{v}) d\mathrm{v}}{\mathrm{v} - \frac{\omega}{k}} + \frac{\omega_b^2}{(\omega - k\mathrm{v}_0)^2} \\ &= \frac{\omega_{\mathrm{pe}}^2}{k^2} \int \frac{g'_0(\mathrm{v}) d\mathrm{v}}{\mathrm{v} - \frac{\omega}{k}} + \frac{\omega_b^2}{(\omega - k\mathrm{v}_0)^2} \end{aligned}$$

After integrating by parts, we have

$$\boxed{1=\frac{\omega_{pe}^2}{k^2}\int\frac{g_0(\mathrm{v})d\mathrm{v}}{\left(\mathrm{v}-\frac{\omega}{k}\right)^2}+\frac{\omega_b^2}{(\omega-k\mathrm{v}_0)^2}}.$$

Now, let us make use of the fact that $\frac{\omega}{k\mathrm{v}_{te}}\ll 1$ so that we can expand $\frac{1}{\mathrm{v}-\frac{\omega}{k}}$ for small $\frac{\omega}{k\mathrm{v}}$. We then obtain:

$$\frac{1}{\mathrm{v}-\frac{\omega}{k}}\simeq\frac{1}{\mathrm{v}}\left(1+\frac{\omega}{k\mathrm{v}}+\frac{\omega^2}{k^2\mathrm{v}^2}+\cdots\right)\simeq\frac{1}{\mathrm{v}}.$$

Then, using $g_0(\mathrm{v})=\dfrac{1}{\mathrm{v}_{th}\sqrt{\pi}}e^{-\mathrm{v}^2/\mathrm{v}_{th}^2}$, we have

$$g_0'=\frac{-2\mathrm{v}}{\mathrm{v}_{th}^3\sqrt{\pi}}e^{-\mathrm{v}^2}/\mathrm{v}_{th}^2=\frac{-2\mathrm{v}}{\mathrm{v}_{th}^2}g_0(\mathrm{v}).$$

This gives us

$\displaystyle\int\frac{g_0'd\mathrm{v}}{\mathrm{v}-\frac{\omega}{k}}\simeq\frac{-2}{\mathrm{v}_{th}^2}$ so that we have including the imaginary part given by

$$\frac{\omega_{pe}^2}{k^2}\pi i g_0'\left(\frac{\omega}{k}\right)=\frac{\omega_{pe}^2}{k^2}\pi i\left(\frac{-2v}{\mathrm{v}_{th}^2}\right)\left(\frac{1}{\mathrm{v}_{th}\sqrt{\pi}}\right)e^{-\mathrm{v}^2}/\mathrm{v}_{th}^2\Big|_{\mathrm{v}=\frac{\omega}{k}}$$
$$=-\sqrt{\pi}\frac{1}{k^2\lambda_{De}^2}\frac{\omega}{k\mathrm{v}_{th}},$$

where we have used $\lambda_{De}^2=\dfrac{1}{2}\dfrac{\mathrm{v}_{th}^2}{\omega_{pe}^2}$.

Putting all of the above together, we obtain

$$1=-\frac{1}{k^2\lambda_{De}^2}-\frac{1}{k^2\lambda_{De}^2}i\sqrt{\pi}\frac{\omega}{k\mathrm{v}_{th}}+\frac{\omega_b^2}{(\omega-k\mathrm{v}_0)^2}$$

or

$$\boxed{1+\frac{1}{k^2\lambda_{De}^2}\left(1+\frac{i\sqrt{\pi}\omega}{k\mathrm{v}_{th}}\right)-\frac{\omega_b^2}{(\omega-k\mathrm{v}_0)^2}=0}.$$

We now wish to solve for $\mathrm{Re}(\omega)$ and $\mathrm{Im}(\omega)=\gamma$.

We will use $\omega_b^2=\alpha\omega_{pe}^2$. Also, in $\mathrm{Re}\,\omega$, we will neglect the $i\sqrt{\pi}$ term. We then have

$$(\omega-k\mathrm{v}_0)^2\left(1+\frac{1}{k^2\lambda_{De}^2}\right)=\alpha\omega_{pe}^2.$$

Therefore, we have

$$\omega - k\mathrm{v}_0 = \frac{\pm\sqrt{\alpha}\omega_{\mathrm{pe}}}{\sqrt{1+\frac{1}{k^2\lambda_{\mathrm{De}}^2}}} \text{ and then,}$$

$$\boxed{\mathrm{Re}(\omega) = k\mathrm{v}_0 - \frac{\sqrt{\alpha}\omega_p}{\left(1+\frac{1}{k^2\lambda_{\mathrm{De}}^2}\right)}} \quad \text{Slow wave } (-\text{ sign})$$

Then, to solve for $\mathrm{Im}(\omega)$, we have

$$(\omega - k\mathrm{v}_0)^2 = \frac{\alpha\omega_{\mathrm{pe}}^2}{1+\frac{1}{k^2\lambda_{\mathrm{De}}}\left(1+\frac{i\sqrt{\pi}\omega}{k\sqrt{v_{te}}}\right)}$$

$$= \frac{\alpha\omega_{\mathrm{pe}}^2 k^2\lambda_{\mathrm{De}}^2}{1+\lambda^2\lambda_{\mathrm{De}}^2 + \frac{i\sqrt{\pi}\omega}{k\sqrt{\mathrm{v}_{th}}}}$$

$$= \frac{\frac{1}{2}\alpha k^2 \mathrm{v}_{th}^2}{1+k^2\lambda_{\mathrm{De}}^2 + \frac{i\sqrt{\pi}\omega}{k\sqrt{\mathrm{v}_{th}}}}$$

or

$$\omega - k\mathrm{v}_0 = -\left[\frac{\frac{1}{2}\alpha k^2 \mathrm{v}_{th}^2}{1+k^2\lambda_{\mathrm{De}}^2 + \frac{i\sqrt{\pi}\omega}{k\mathrm{v}_{th}}}\right]^{1/2}$$

Then, letting $G = 1+k^2\lambda_{\mathrm{De}}^2$, we have

$$\omega - k\mathrm{v}_0 = -|k|\mathrm{v}_{th}\left(\frac{\alpha}{2G}\right)^{1/2}\frac{1}{\left(1+\frac{i\sqrt{\pi}\omega}{|k|\mathrm{v}_{th}G}\right)^{1/2}}$$

$$\simeq -|k|\mathrm{v}_{th}\left(\frac{\alpha}{2G}\right)^{1/2}\left(1-\frac{i\sqrt{\pi}}{2|k|}\frac{\omega}{\mathrm{v}_{th}}\frac{1}{G}\right)$$

Then, we can write

$$\gamma = |k|\mathrm{v}_{th}\frac{\sqrt{\pi}}{2}\left(\frac{\alpha}{2G}\right)^{1/2}\frac{\omega_R}{|k|\mathrm{v}_{th}}\frac{1}{G}.$$

$\gamma \simeq \left(\frac{\pi\alpha}{8}\right)^{1/2}\frac{1}{G^{3/2}}\omega_R$, and finally,

$$\boxed{\gamma \simeq \left(\frac{\pi\alpha}{8}\right)^{1/2}\frac{|k|}{(1+k^2\lambda_{\mathrm{De}}^2)^{3/2}}\left\{\mathrm{v}_0 - \mathrm{v}_{th}\left[\frac{\alpha}{2(1+k^2\lambda_{\mathrm{De}}^2)}\right]^{1/2}\right\}}.$$

Then, solving for the critical drift velocity, we have

$$\mathrm{v}_{\text{crit}} = \mathrm{v}_{th}\left(\frac{\alpha}{2}\right)^{1/2}\frac{1}{G^{1/2}}$$
$$= \mathrm{v}_{th}\left(\frac{\alpha/2}{1+k^2\lambda_{\text{De}}^2}\right)^{1/2} \quad \text{and taking } k\lambda_{\text{De}} \simeq 1, \text{ we have}$$

$$\boxed{\mathrm{v}_{\text{crit}} \simeq \tfrac{1}{2}\mathrm{v}_{th}\sqrt{\alpha}}$$

Now, for $\mathrm{v} > \mathrm{v}_{\text{crit}}$, the $k_z\mathrm{v}_0$ term dominates so that we have

$$\gamma \simeq |k|\, \mathrm{v}_{0\sqrt{\alpha}},$$

where $|k| \sim \dfrac{1}{\lambda_{\text{De}}} = \dfrac{\omega_{\text{pe}}}{\mathrm{v}_{th}}$.

Then,

$$\boxed{\gamma \simeq \sqrt{\alpha}\mathrm{v}_0\frac{\omega_{\text{pe}}}{\mathrm{v}_{th}} = \omega_{\text{pe}}\sqrt{\alpha}\left(\frac{\mathrm{v}_0}{\mathrm{v}_{th}}\right)}.$$

Alternatively, we could have used the Z-function formalism:

$$1 = \sum_\alpha \frac{\omega_{p\alpha}^2}{k^2\mathrm{v}_{th\alpha}^2} Z'(\xi_\alpha) \quad \text{with } \xi_\alpha = \frac{\omega - k\mathrm{v}_{0\alpha}}{k\mathrm{v}_{th\alpha}}.$$

For the thermal plasma, $\mathrm{v}_0 = \mathrm{v}$ and $\xi \to 0$. For the cold beam, $\mathrm{v}_0 \neq 0$, $\mathrm{v}_{th} \to 0$, and $\xi \to \infty$. Then, using

$$\lim_{\xi\to\infty} Z'(\xi) \to \frac{1}{\xi^2}$$

and $\lim_{\xi\to 0} Z'(\xi) \to -2\left(1+i\sqrt{\pi}\right)$, we can immediately write:

$$1 = \frac{\omega_{\text{pe}}^2}{k^2\mathrm{v}_{th}^2}\left[-2\left(1+i\sqrt{\pi}\xi\right)\right] + \frac{\omega_b^2}{k^2\mathrm{v}_{th}^2}\frac{k^2\mathrm{v}_{th}^2}{(\omega - k\mathrm{v}_0)^2}$$

and then,

$$\boxed{1 + \frac{1}{k^2\lambda_{\text{De}}^2}\left(1 + i\sqrt{\pi}\frac{\omega}{k\mathrm{v}_{th}}\right) - \frac{\omega_b^2}{(\omega - k_z\mathrm{v}_0)^2} = 0}.$$

12.5 Our electrostatic dispersion relation is, of course, given by

$$\varepsilon(k,\omega) = 1 - \sum_\alpha \frac{\omega_{p\alpha}^2}{k^2}\int_{C_L} \frac{\partial \hat{f}_{0\alpha}/\partial \mathrm{v}}{\mathrm{v} - \omega/k} d\mathrm{v},$$

where

$$\int_{c_L} = \underline{\mathbf{P}}\int + \pi i\sigma \frac{\partial f_0}{\partial \mathrm{v}}.$$

We also assume $\omega_i < \omega_r$.
For the ions, we can approximate the integral, since $\mathrm{v}_{ph} \gg \mathrm{v}_{ti}$. So we have $\mathrm{v} \ll \omega/k$
wherever $\frac{\partial f_{0\alpha}}{\partial \mathrm{v}} \neq 0$.
Therefore

$$\underline{\mathbf{P}}\int_{-\infty}^{\infty} \frac{\partial \hat{f}_{0i}/\partial \mathrm{v}}{\mathrm{v}-\omega/k} d\mathrm{v} \approx \int_{-\infty}^{\infty} \frac{\partial \hat{f}_{0i}}{\partial \mathrm{v}} \left(\frac{k}{\omega} + \frac{k^2\mathrm{v}}{\omega^2} + \frac{k^3\mathrm{v}^2}{\omega^3} + \frac{k^4\mathrm{v}^4}{\omega^4} + \cdots\right) d\mathrm{v},$$

and for the electrons we have

$$\underline{\mathbf{P}}\int_{-\infty}^{\infty} \frac{\partial \hat{f}_{0e}/\partial \mathrm{v}}{\mathrm{v}-\omega/k} d\mathrm{v}.$$

Letting $\mathrm{v} - \mathrm{v}_0 = z$, we have

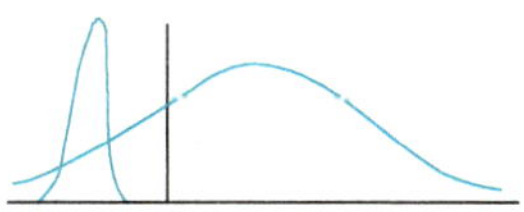

$$= \underline{\mathbf{P}}\int_{-\infty}^{\infty} \frac{\partial \hat{f}_{0e}/\partial z\, dz}{z + \mathrm{v}_0 - \omega/k} \approx \underline{\mathbf{P}}\int_{-\infty}^{\infty} \frac{\partial \hat{f}_{0e}/\partial z}{z}$$ ← a fair approximation, since we have $\mathrm{v}_0 \ll \mathrm{v}_t$

$$= \underline{\mathbf{P}}\int_{-\infty}^{\infty} 2\frac{\partial \hat{f}_{0e}}{\partial z^2} dz = \underline{\mathbf{P}}\frac{2}{\sqrt{\pi}\,\mathrm{v}_{te}} \int_{-\infty}^{\infty} \frac{\partial}{\partial z^2} \exp\left(-\frac{z^2}{\mathrm{v}_{te}^2}\right) dz$$

$$= \frac{2}{\sqrt{\pi}\,\mathrm{v}_{te}} \left(-\frac{1}{\mathrm{v}_{te}^2}\right) \underline{\mathbf{P}}\int_{-\infty}^{\infty} \exp\left(-z^2/\mathrm{v}_{te}^2\right) dz$$

$$= -\frac{2}{\mathrm{v}_{te}^2}$$

Therefore, we have

$$\varepsilon_r = 1 - \sum_\alpha \frac{\omega_{p\alpha}^2}{k^2} \underline{\mathbf{P}} \int_{-\infty}^{\infty} \frac{\partial \hat{f}_{0\alpha}}{\mathrm{v} - \omega/k} d\mathrm{v}$$

$$= 1 - \frac{\omega_{\mathrm{pe}}^2}{k^2}\left(\frac{-2}{\mathrm{v}_{te}^2}\right) - \frac{\Omega_p^2}{k^2}(-1)\underline{\mathbf{P}} \int_{-\infty}^{\infty} \frac{\partial \hat{f}_{0i}}{\partial \mathrm{v}}\left(\frac{k}{\omega} + \frac{k^2\mathrm{v}}{\omega^2} + \frac{k^4\mathrm{v}^3}{\omega^3} + \cdots\right) d\mathrm{v}$$

$$= 1 + \frac{2\omega_{\mathrm{pe}}^2}{k^2\mathrm{v}_{te}^2} + \frac{\Omega_p^2}{k^2}\underline{\mathbf{P}} \int_{-\infty}^{\infty} \frac{1}{\sqrt{\pi}\mathrm{v}_{ti}}\left(\frac{-2\mathrm{v}}{\mathrm{v}_{ti}^2}\right)\exp\left(\frac{\mathrm{v}^2}{\mathrm{v}_{ti}^2}\right)$$

$$\cdot\left(\frac{k}{\omega} + \frac{k^2\mathrm{v}}{\omega^2} + \frac{k^3\mathrm{v}^2}{\omega^2} + \frac{k^4\mathrm{v}^3}{\omega^4} + \cdots\right) d\mathrm{v}.$$

Discarding all odd functions of the full integral, we then have

$$\varepsilon_r = 1 + \frac{2\omega_{\mathrm{pe}}^2}{k^2\mathrm{v}_{te}^2} + \frac{\Omega_p^2}{k^2}\underline{\mathbf{P}} \int_{-\infty}^{\infty} \frac{(-2)}{\sqrt{\pi}}\left(\frac{k^2}{\omega^2} + \frac{2k^4\mathrm{v}_{ti}^2}{\omega^4}\left(\frac{\mathrm{v}^4}{\mathrm{v}_{ti}^4}\right) + \cdots\right) dx$$

$$= 1 + \frac{2\omega_{\mathrm{pe}}^2}{k^2\mathrm{v}_{te}^2} + \frac{\Omega_p^2}{k^2}\underline{\mathbf{P}} \int_{-\infty}^{\infty} \frac{-2}{\sqrt{\pi}} e^{-x^2}\left(\frac{k^2}{\omega^2} + \frac{2k^4\mathrm{v}_{ti}^2}{\omega^4}x^4 + \cdots\right) dx.$$

In the above, $x^2 = \mathrm{v}^2/\mathrm{v}_{ti}$. However, we have

$$\int_{-\infty}^{\infty} e^{-x^2} x^2 dx = \frac{\sqrt{\pi}}{2},$$

So

$$\varepsilon_r = 1 + \frac{2\omega_{\mathrm{pe}}^2}{k^2\mathrm{v}_{te}^2} - \frac{\Omega_p^2}{k^2}\frac{k^2}{\omega^2} \leftarrow$$ we have explicitly written the fact that we are considering small Im (ω) and evaluating $\underline{\mathbf{P}}$, i.e., along the real axis.

$$\boxed{\varepsilon_r = 1 + \frac{1}{k^2\lambda_{\mathrm{De}}^2} - \frac{\Omega_p^2}{\omega^2}}$$

Now we are ready to find ε_i.

$$\varepsilon_i = -\pi \sum_\alpha \frac{\omega_{p\alpha}^2}{k^2} \frac{\partial f_{\alpha 0}}{\partial \mathrm{v}}\bigg|_{\mathrm{v}=\omega/k},$$

and this gives us

$$\varepsilon_i = -\pi \left\{ \frac{\Omega_p^2}{k^2} \frac{(-2\mathrm{v})}{\sqrt{\pi}\mathrm{v}_{ti}\mathrm{v}_{te}^2} \exp\left(-\frac{\mathrm{v}^2}{\mathrm{v}_{ti}^2}\right) + \frac{\omega_{\mathrm{pe}}^2}{k^2} \frac{(-2)(\mathrm{v}-\mathrm{v}_0)}{\sqrt{\pi}\mathrm{v}_{te}\mathrm{v}_{te}^2} \exp\left(\frac{(\mathrm{v}-\mathrm{v}_0)^2}{\mathrm{v}_{te}^2}\right) \right\}_{\mathrm{v}=\omega/k}$$

$$= 2\pi \left\{ \frac{\Omega_p^2}{k^2} \frac{\omega_r/k}{\sqrt{\pi}\mathrm{v}_{ti}^3} \exp\left(-\frac{\omega_r^2}{k^2\mathrm{v}_{ti}^2}\right) + \frac{\omega_{\mathrm{pe}}^2}{k^2} \frac{\left(\frac{\omega_r}{k}-\mathrm{v}_0\right)}{\sqrt{\pi}\mathrm{v}_{te}^3} \exp\left(\frac{\left(\frac{\omega_r}{k}-\mathrm{v}_0\right)^2}{\mathrm{v}_{te}^2}\right) \right\}$$

Now we can easily solve for ω_r^2. As before, we get $\varepsilon_r = 0$, which gives us:

$$\omega_r^2 = \frac{\Omega_p^2}{1+\frac{1}{k^2\lambda_{\mathrm{De}}^2}} = \frac{\Omega_p^2}{1+k^2\lambda_{\mathrm{De}}^2} k^2\lambda_{\mathrm{De}}^2,$$

or

$$\Omega_p^2 k^2 \lambda_{\mathrm{De}}^2 = k^2 \frac{4\pi n_{\mathrm{e}} e^2}{m_{\mathrm{i}}} \frac{\kappa T_{\mathrm{e}}}{4\pi n_{\mathrm{e}} e^2} = k^2 \frac{KT_{\mathrm{e}}}{m_{\mathrm{i}}} = k^2 \mathrm{v}_s^2.$$

$$\boxed{\omega_r^2 = \frac{k^2\mathrm{v}^2}{1+k^2\lambda_{\mathrm{De}}^2}}$$

To solve for ω_i (which we have assumed is $\ll \omega_r$), we can use $\varepsilon_{r_i}\varepsilon_i$ and expand ε about $\omega = \omega_r$. We then have

$$\varepsilon(k,\omega) \simeq \varepsilon(k,\omega_r) + i\omega_i \frac{\partial \varepsilon(\mathbf{k},\omega_r)}{\partial \omega_r}.$$

Equating the real and imaginary parts, and using $\varepsilon_r = 0$, we have

$$\omega_i = \frac{-\varepsilon_i(k,\omega_r)}{\partial \varepsilon_r(k,\omega_r)/\partial\omega_r}.$$

Using this we have, first,

$$\frac{\partial \varepsilon_r}{\partial \omega_r} = \frac{2\Omega_p^2}{\omega_r^3}.$$

Then:

$$\varepsilon_i(k,\omega_r) = 2\pi \left\{ \frac{\Omega_p^2}{k^2} \frac{\omega_r/k}{\sqrt{\pi}\mathrm{v}_{ti}^3} \exp\left(-\frac{\omega_r^2}{k^2\mathrm{v}_{ti}^2}\right) + \frac{\omega_{\mathrm{pe}}^2}{k^2} \frac{\left(\frac{\omega_r}{k}-\mathrm{v}_0\right)}{\sqrt{\pi}\mathrm{v}_{te}^3} \exp\left(\frac{-\left(\frac{\omega_r}{k}-\mathrm{v}_0\right)^2}{\mathrm{v}_{te}^3}\right) \right\}$$

$$= \frac{2\pi}{\sqrt{\pi}} \frac{\Omega_p^2}{k^2} \left(\frac{1}{2\kappa T_{\mathrm{e}}}\right)^{3/2} m_{\mathrm{i}}^{3/2} \left\{ \left(\frac{T_{\mathrm{e}}}{T_{\mathrm{i}}}\right)^{3/2} \left(\frac{\omega_r}{k}\right) \exp\left(-\frac{\omega_r^2}{k^2\mathrm{v}_{ti}^2}\right) \right.$$

$$\left. + \underbrace{\frac{m_{\mathrm{i}}}{m_{\mathrm{e}}} \frac{m_{\mathrm{e}}^{3/2}}{m_{\mathrm{i}}^{3/2}}}_{\sqrt{\frac{m_{\mathrm{e}}}{m_{\mathrm{i}}}}} \left(\frac{\omega_r}{k} \mathrm{v}_0\right) \exp\left(\frac{-(\omega_r/k - \mathrm{v}_0)^2}{\mathrm{v}_{te}^3}\right) \right\}$$

$$= \frac{2\pi}{\sqrt{\pi}} \frac{\Omega_p^2}{k^2} \left(\frac{1}{2\kappa T_{\mathrm{e}}}\right)^{3/2} \left(\frac{\omega_r}{k}\right) m_{\mathrm{i}}^{3/2} \left\{ \left(\frac{T_{\mathrm{e}}}{T_{\mathrm{i}}}\right)^{3/2} \exp\left(-\frac{\omega_r^2}{k^2\mathrm{v}_{ti}^2}\right) + \frac{\sqrt{\frac{m_{\mathrm{e}}}{m_{\mathrm{i}}}}\left(\frac{\omega_r}{k}-\mathrm{v}_0\right)}{\omega_r/k} \right\}$$

Now, using then have $\omega_r = \frac{k\mathrm{v}_s}{\sqrt{1+k^2\lambda_{\mathrm{De}}^2}}$ and $\frac{\partial \varepsilon_r}{\partial \omega_r} = \frac{2\Omega_p^2}{\omega_r^3} = \frac{2\Omega_p^2[1+k^2\lambda_{\mathrm{De}}^2]^{3/2}}{k^3\mathrm{v}_s^3}$, we

$$\omega_i = \frac{-2\pi}{\sqrt{\pi}} \frac{\Omega_p^2}{k^2} m_{\mathrm{i}}^{3/2} \left(\frac{1}{2\kappa T_{\mathrm{e}}}\right)^{3/2} \frac{\omega_r}{k^3} \frac{k^3 \mathrm{v}_s^3}{2\Omega_p^2 \left[1+k^2\lambda_{\mathrm{De}}^2\right]^{3/2}}$$
$$\left\{ \left(\frac{T_{\mathrm{e}}}{T_{\mathrm{i}}}\right)^{3/2} \exp\left(-\omega_r^2/k^2\mathrm{v}_{ti}^2\right) + \frac{\sqrt{\frac{m_{\mathrm{e}}}{m_{\mathrm{i}}}}\left(\frac{\omega_r}{k} - \mathrm{v}_0\right)}{\omega_r/k} \right\}.$$

But $\mathrm{v}_s^3 = \left(\frac{\kappa T_{\mathrm{e}}}{m_{\mathrm{i}}}\right)^{3/2}$. Therefore, taking k^2 inside, this becomes

$$\omega_i = -2\sqrt{\pi} \frac{1}{2^{3/2}} \frac{1}{2} \frac{\omega_r}{[1+k^2\lambda_{\mathrm{De}}^2]^{3/2}} \left\{ \left(\frac{T_{\mathrm{e}}}{T_{\mathrm{i}}}\right)^{3/2} \exp\left(-\frac{T_{\mathrm{e}}/T_{\mathrm{i}}}{2\left(1+k^2\lambda_{\mathrm{De}}^2\right)}\right) \right.$$
$$\left. + \sqrt{\frac{m_{\mathrm{e}}}{m_{\mathrm{i}}}} \left(1 - \frac{\mathrm{v}_0}{\mathrm{v}_s}\sqrt{1+k^2\lambda_{\mathrm{De}}^2}\right) \right\}.$$

Finally, we obtain

$$\boxed{\omega_i = -\sqrt{\frac{\pi}{8}} \frac{\omega_r}{[1+k^2\lambda_{\mathrm{De}}^2]^{3/2}} \left\{ \left(\frac{T_{\mathrm{e}}}{T_{\mathrm{i}}}\right)^{3/2} \exp\left(\frac{-T_{\mathrm{e}}/T_{\mathrm{i}}}{2(1+k^2\lambda_{\mathrm{De}}^2)}\right) + \sqrt{\frac{m_{\mathrm{e}}}{m_{\mathrm{i}}}} \left(1 - \frac{\mathrm{v}_0}{\mathrm{v}_s}\sqrt{1+k^2\lambda_{\mathrm{De}}^2}\right) \right\}}$$

Note that for $\mathrm{v}_0 \to 0$, we get the right dispersion relation for ion acoustic waves, which are heavily damped unless $T_{\mathrm{e}} \gg T_{\mathrm{i}}$. For large enough v_0, you can see that $\omega_i \to 0$ and that we get an instability. For $T_{\mathrm{e}} \gg T_{\mathrm{i}}$, we can ignore ion Landau damping, and the instability condition is easily obtained, i.e.,

$$\mathrm{v}_0^2 > \frac{\mathrm{v}_s^2}{1+k^2\lambda_{\mathrm{De}}^2} = \frac{\kappa T_{\mathrm{e}}/m_{\mathrm{i}}}{1+k^2\lambda_{\mathrm{De}}^2} \gg \kappa T_{\mathrm{i}}/m_{\mathrm{i}}$$

Also, for large $T_{\mathrm{e}}/T_{\mathrm{i}}$, note that the growth rate is *proportional* to the streaming velocity

$$\omega_i = \frac{k\left(\mathrm{v}_0 - \frac{\omega_r}{k}\right)}{\left(1+k^2\lambda_{\mathrm{De}}^2\right)^{3/2}} \sqrt{\frac{\pi}{8}\frac{m_{\mathrm{e}}}{m_{\mathrm{i}}}}.$$

The general stability boundary for the problem has been treated by several authors.

(a) Plot after Fried and Gould, *Phys. Flds.* **4**, 139 (1961)

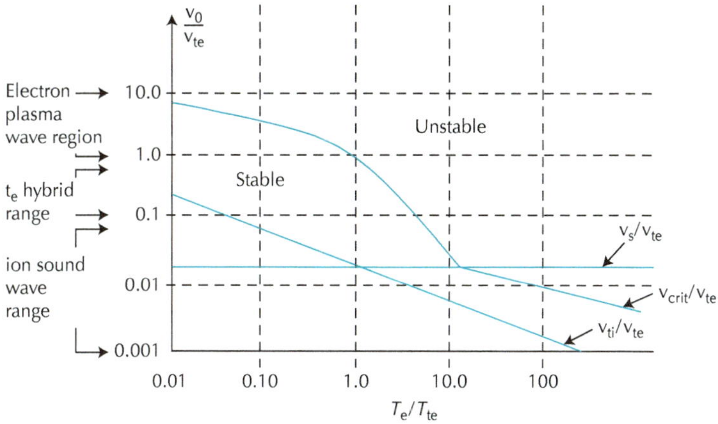

(b) Growth rate of the fastest growing wavelength in counter-streaming plasmas: Stringer has calculated the dispersion for two drifting Maxwellian plasmas.

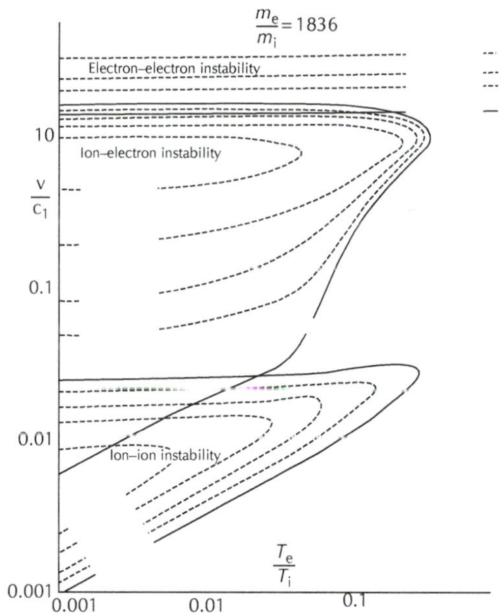

(c) In the plot below, Forme assumed that $T_e \gg T_i$ so that Landau damping is unimportant. He further assumed $k^2\lambda_{De}^2 \ll 1$ so that the growth rate $k^2\lambda_{De}^2 \ll 1$ assumes the simple form:

$$\gamma_{ia} = \left(\frac{\pi m_e}{8m_i}\right)^{1/2} \frac{\omega_{ia}}{\left(1+k^2\lambda_{De}^2\right)^{3/2}} \left(\frac{kv_d}{\omega_{ia}} - 1\right)$$

The figure below shows the schematic relation between the ion-acoustic dispersion relation and the growth rate of the instability. http://e7.eiscat.se/groups/Documentation/CourseMaterials/2003MenloPark/Forme.pdf

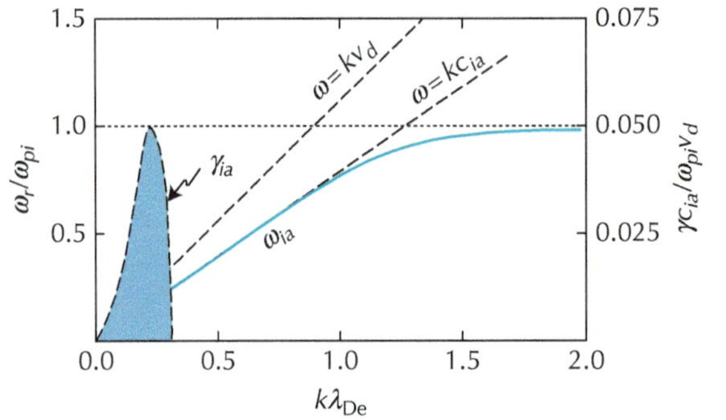

Appendix A

Mathematical Methods

Chapter Outline Head

A.1 COMPLEX VARIABLES AND INTEGRALS IN THE COMPLEX PLANE[1]

A.1.1 Complex Number

We define the complex number z by

$$z = x + iy, \tag{A.1.1}$$

where x and y are real and $i^2 = -1$. A convenient representation of z is in a rectangular coordinate system (Fig. A.1).

[1] See Copson (1960) and Butkov (1968)

DOI: 10.1016/B978-0-12-374877-5.00013-0

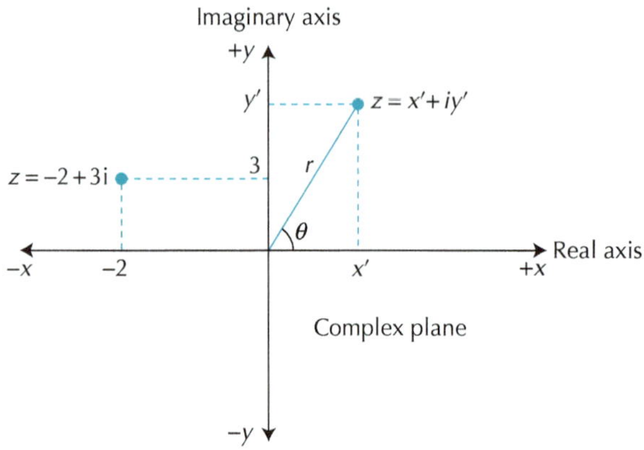

FIGURE A.1 Representation of complex number Z.

The complex conjugate is $z^* = x - iy$. The absolute value is

$$|z| = (z \cdot z^*)^{1/2} = (x^2 + y^2)^{1/2}. \tag{A.1.2}$$

Polar coordinates (r, θ) may also be used, and in this case,

$$z = r(\cos\theta + i\sin\theta) = re^{i\theta}, \quad |z| = r. \tag{A.1.3}$$

Our interest in the properties of complex variables is centered on their use in the evaluation of integrals of the form

$$\int_{-\infty}^{+\infty} \frac{f(z)}{z - z_0} dz.$$

These appear in the calculations of the fluctuation level in a plasma. The integral presents problems because the integrand goes to infinity when $z = z_0$. The theorems that apply in the evaluation of these integrals are reviewed below.

A.1.2 Cauchy–Goursat Theorem

If a function f is analytic at all points interior to and on a closed contour C, then

$$\int_C f(z) dz = 0. \tag{A.1.4}$$

For example, a function such as $f(z) = 1/(z^2 + 4)$ is not analytic at the singular points $z = \pm 2i$, therefore any contour which satisfies Eqn (A.1.4) must not enclose these points. The circle centered on the origin and of radius 1.8 would be satisfactory (see Fig. A.2).

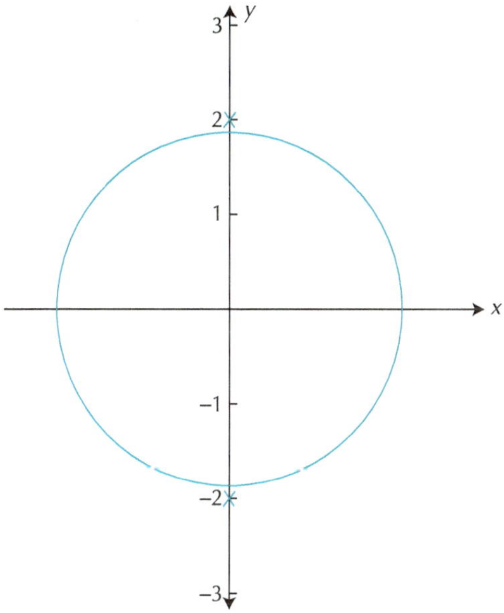

FIGURE A.2 Contour not including singular points.

Cauchy Integral Theorem

The function $f(z) = g(z)/(z - z_0)$, where $g(z)$ is analytic within and on a contour C, which encloses z_0. The contour is taken in the positive sense (counterclockwise); then

$$\int_C \frac{g(z)}{z - z_0} dz = 2\pi i g(z_0). \tag{A.1.5}$$

Clearly, we may draw a circle C_0 of radius $r_0 \equiv |z - z_0|$ within C and join it to C by adjacent parallel lines (see Fig. A.3). Since the integrals over the parallel lines cancel, we see by Eqn (A.1.4) that

$$\int_C f(z)\, dz = \int_{C_0} f(z)\, dz,$$

where C_0 is now in the clockwise direction.

Now, $g(z)$ is analytic within C; therefore, we may write it as

$$g(z) = g(z_0) + (z - z_0) g'(z_0) + (z - z_0)\eta, \tag{A.1.6}$$

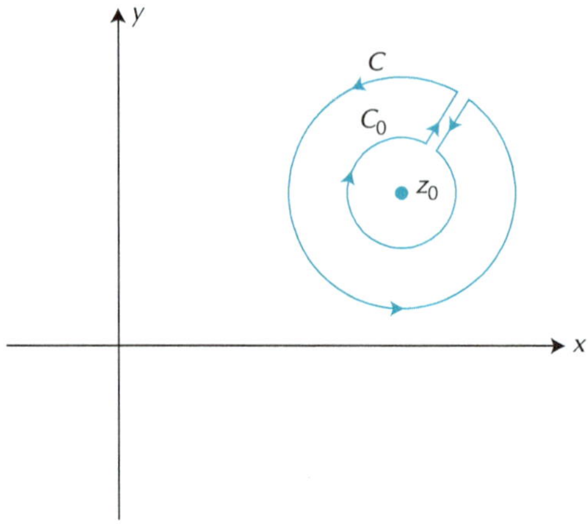

FIGURE A.3 Contour.

where $\eta \rightarrow 0$ as z_0 or $r_0 \rightarrow 0$. Further, we may use Eqn (A.1.3) and $z - z_0 = r_0 e^{i\theta}$, $dz = ir_0 e^{i\theta} d\theta$, substituting

$$\int_{C_0} f(z)dz = ig(z_0) \int_0^{2\pi} d\theta + ig'(z_0)r_0 \int_0^{2\pi} e^{i\theta} d\theta + ir_0 \int_0^{2\pi} \eta e^{i\theta} d\theta. \quad \text{(A.1.7)}$$

Now, the result is independent of r_0, and in the limit $r_0 \rightarrow 0$, the last term is zero, the second integral is zero, and the first integral is 2π, and thus, Eqn (A.1.5) is proved.

Integral Around a Semicircle

By a similar argument, we can see that the integral over a semicircle of radius $r_0 = |z - z_0|$ about the singular point z_0 of the function $f(z)$ (see Fig. A.4) is given by

$$\int_{C_1} \frac{g(z)}{z - z_0} dz = \pi i g(z_0). \quad \text{(A.1.8)}$$

Poles and Residues

The point $z = z_0$ of the function $f(z) = g(z)/(z - z_0)$ is termed a "simple pole." The quantity $g(z_0)$ is called the "residue."

More generally, the denominator may contain higher powers of $z - z_0$ such as $(z - z_0)^m$; such points are called "poles of order m." Consider $f(z) = g(z)/(z - z_0)^m$. The function $g(z)$ is analytic at and around $z = z_0$ and may be expanded in a Laurent series within some radius $r_0 = |z - z_0| < r_c$.

$$g(z) = g(z_0) + g'(z_0)(z - z_0) + \frac{g'(z_0)}{2!}(z - z_0)^2 + \cdots. \quad \text{(A.1.9)}$$

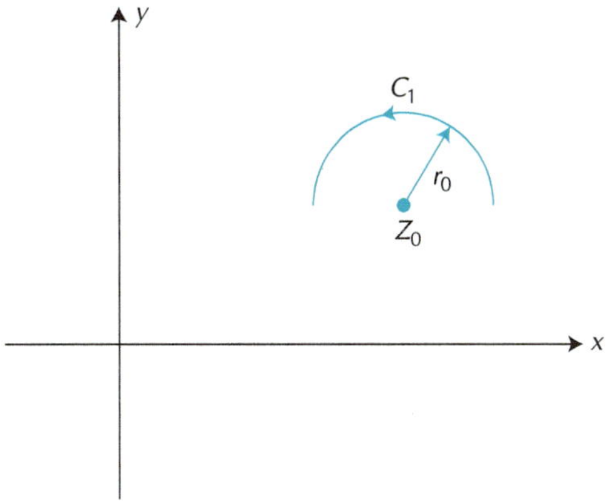

FIGURE A.4 Semicircle of radius $|z-z_0|$.

Then,

$$\int_C \frac{g(z)}{(z-z_0)^m} dz = \int_0^{2\pi} \frac{g(z) i\, d\theta}{r_0^{m-1} e^{(m-1)i\theta}},$$

where, from Eqn (A.1.3), $z-z_0 = r_0 e^{i\theta}$ and C is the circle of radius r_0 about z_0.

Now, $g(z)$ is substituted from Eqn (A.1.9), and clearly, the only nonzero term is the one containing $(z-z_0)^{m-1}$, i.e.,

$$\int_C \frac{g(z)}{(z-z_0)^m} dz = \frac{2\pi i}{(m-1)!} \frac{\partial^{m-1} g(z)}{\partial z^{m-1}}\bigg|_{z=z_0}. \tag{A.1.10}$$

Contour Containing a Number of Poles

If $f(z)$ is analytic inside C except for a finite number of singular points, $z_1, z_2, \ldots, z_N$ and $k_1, k_2, \ldots, k_N$ are the residues at these points; then,

$$\int_C f(z) dz = 2\pi i (k_1 + k_2 + \cdots + k_N). \tag{A.1.11}$$

A.1.3 Essential Singularity at Infinity

(a) The above results may now be applied in the evaluation of integrals of the form

$$\int_{-\infty}^{+\infty} \frac{\exp(-v^2)}{v - z_0} dv.$$

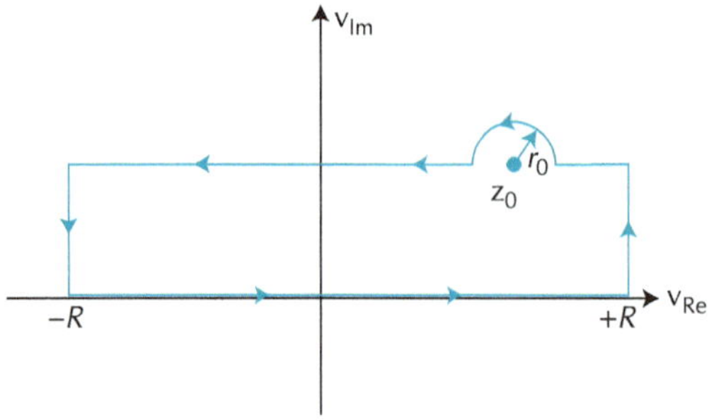

FIGURE A.5 Contour for z_0 in the upper half-plane.

Clearly, we may draw a contour in the complex plane that encloses the pole at $v = z_0$. The important point to note is that

$$\exp\left(-v^2\right) = \exp\left(-v_{Re}^2 - 2iv_{Re}v_{Im}\right) \cdot \exp\left(+v_{Im}^2\right),$$

and this function is infinite when $\mathrm{Im}(v) = v_{Im} \to \infty$. The function is said to have an essential singularity at infinity, and the contour must avoid this region. We use the contour shown in Fig. A.5 for the case where z_0 is in the upper half-plane. From (A.1.5), the integral around the complete contour is $2\pi i \exp\left[-(z_0)^2\right]$. In the limit $\mathbf{R} \to \infty$, the vertical contributions go to zero because of the factor $\exp\left[-(v_R)^2\right]$.

From (A.1.8), the semicircle of radius r_0 centered on z_0 contributes $-\pi i \exp\left[-(z_0)^2\right]$. In total, therefore,

$$\int_{-\infty}^{+\infty} \frac{\exp(-v^2)}{(v - z_0)} dv = \pi i \exp\left[-z_0^2\right] + \lim_{r_0 \to 0} \left\{ \left(\int_{-\infty + \mathrm{Im}(z_0)}^{z_0 - r} dv + \int_{z_0 + r_0}^{+\infty + \mathrm{Im}(z_0)} dv \right) \frac{\exp(-v^2)}{(v - z_0)} \right\}. \tag{A.1.12}$$

The latter two terms are called the "Principal value" and are usually written

$$\mathrm{P} \int_{-\infty}^{+\infty} \frac{\exp\left(-v^2\right)}{v - z_0} dv.$$

(b) In this example, the integral may be evaluated in terms of tabulated functions using the substitution $[\mathrm{Im}(z_0) > 0]$

$$\frac{1}{v - z_0} = i \int_0^{\infty} dt\, e^{-i(v - z_0)t}.$$

Note that this is chosen so that the integrand goes to zero as $t \to \infty$

$$\int_{-\infty}^{+\infty} \frac{\exp(-v^2)}{v - z_0} dv = i \int_0^{\infty} dt \exp\left(iz_0 t - \frac{t^2}{4}\right) \int_{-\infty}^{+\infty} dv \exp\left[-\left(v + \frac{it}{2}\right)^2\right] = I.$$

We are justified in reversing the order of integration because neither function in the integrand goes to infinity. We now substitute $p = v + \frac{1}{2}it$, $dp = dv$, and $v = \pm(-\infty), p = \pm\infty + \frac{1}{2}it$. We use the contour in Fig. A.6, for the second integral. There are no singularities between the axis through $\frac{1}{2}it$ and the real axis, so $\oint \exp(-p^2)dp = 0$. In the limit $\Re(p) \to \pm\infty$, then the vertical parts go to zero, and

$$\int_{-\infty}^{+\infty + it/2} \exp(p^2)dp = \int_{-\infty}^{+\infty} \exp(p^2)dp = \pi^{1/2}$$

and

$$\begin{aligned} I &= i\pi^{1/2} \exp(-z_0^2) \int_0^{\infty} dt \exp\left[-\left(\frac{1}{2}t - iz_0\right)^2\right] \\ &= 2i\pi^{1/2} \exp(-z_0^2) \left[\int_0^{\infty - iz_0} ds \exp(-s^2) - \int_0^{-iz_0} ds \exp(-s^2)\right] \end{aligned}$$

with the substitution $\frac{1}{2}t - iz_0 = s$.

Finally, we set $s = -ip$, and with the pole in the upper half-plane,

$$\int_{-\infty}^{+\infty} \frac{\exp(-v^2)}{z - z_0} dv = i\pi \exp(-z_0^2) - 2\pi^{1/2} \exp(-z_0^2) \int_0^{z_0} \exp(p^2)dp. \quad \text{(A.1.13)}$$

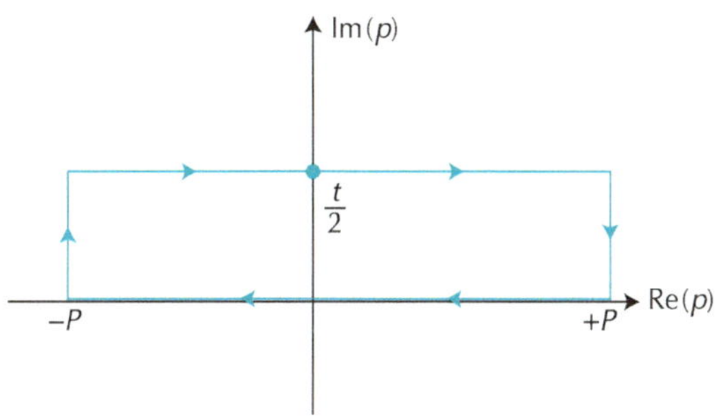

FIGURE A.6 Contour.

These or similar functions are tabulated by Fried & Conte (1961) and Fadeeva & Terent'ev (1954).

In the lower half-plane $\mathrm{Im}(z_0) < 0$, we use the substitution

$$\frac{1}{\mathrm{v}-z_0} = -i\int_0^{\infty} dt\, e^{i(\mathrm{v}-z_0)t}$$

and

$$\int_{-\infty}^{+\infty} \frac{\exp(-\mathrm{v}^2)}{\mathrm{v}-z_0} d\mathrm{v} = -i\pi \exp(-z_0^2) - 2\pi^{1/2}\exp(-z_0^2)\int_0^{z_0} \exp(p^2)dp.$$

(c) More generally, $z_{\mathrm{Im}} > 0$

$$\int_{-\infty}^{+\infty} \frac{f(\mathrm{v})d\mathrm{v}}{\mathrm{v}-(z_{\mathrm{Re}} \pm iz_{\mathrm{Im}})} = \pm i\pi f(z_{\mathrm{Re}}) - \mathrm{P}\int_{-\infty}^{+\infty} \frac{f(\mathrm{v})}{\mathrm{v}-z_{\mathrm{Re}}} d\mathrm{v}, \tag{A.1.14}$$

and as a consequence,

$$\int_{-\infty}^{+\infty} \frac{f(\mathrm{v})}{(\mathrm{v}-z_{\mathrm{Re}})^2 + z_{\mathrm{Im}}^2} d\mathrm{v} = \frac{\pi}{z_{\mathrm{Im}}} f(z_{\mathrm{Re}}). \tag{A.1.15}$$

A.2 FOURIER TRANSFORMS

A.2.1 Definition

Integral transforms are used in finding solutions to differential equations. Consider $f(x)$ to be the solution in rectangular coordinates of an equation, which describes the behavior of f in the region $a \leq x \leq b$. If $f(x)$ is well behaved, it is possible to construct this solution from a combination of sines and cosines terms, which are periodic in the interval $L = |b-a|$. In complex notation, we look for a solution of the form

$$f(x) = \sum_{n=-\infty}^{+\infty} a_n e^{-2\pi inx/L}. \tag{A.2.1}$$

The coefficients a_n are found by multiplying by $e^{2\pi imx/L}$ and integrating from $-L/2$ to $+L/2$.

$$\sum_{n=-\infty}^{+\infty} \int_{-L/2}^{+L/2} dx\, a_n e^{2\pi ix(m-n)/L} = \int_{-L/2}^{+L/2} dx f(x) e^{2\pi imx/L}. \tag{A.2.2}$$

The first integral is zero unless $m=n$, and for $m=n$, $I=La_n$

$$a_n=\frac{1}{L}\int_{-L/2}^{+L/2} dx f(x)e^{+2\pi inx/L}. \tag{A.2.3}$$

The series may be transformed to an integral when $L\to\infty$ by setting $f(k)=La_n$, and $k=2\pi n/L$

$$f(x)=\frac{1}{2\pi}\int_{-\infty}^{+\infty} dk f(k)e^{-ikx}, \quad f(k)=\int_{-\infty}^{+\infty} dx f(x)e^{+ikx}. \tag{A.2.4}$$

This may be extended to three dimensions, and we may include time transforms

$$\begin{aligned} f(\mathbf{r},t)&=\int_{-\infty}^{+\infty}\frac{d\mathbf{k}}{(2\pi)^3}e^{-i\mathbf{k}\cdot\mathbf{r}}\int_{-\infty}^{+\infty}\frac{d\omega}{2\pi}e^{i\omega t}f(\mathbf{k},\omega)\\ f(\mathbf{k},\omega)&=\int_{-\infty}^{+\infty}d\mathbf{r}\,e^{+i\mathbf{k}\cdot\mathbf{r}}\int_{-\infty}^{+\infty}dt\,e^{-i\omega t}f(\mathbf{r},t). \end{aligned} \tag{A.2.5}$$

A.2.2 Parseval's Theorem

In many situations, for example, when we are dealing with rapidly fluctuating quantities, we measure $|f(x)|^2$ rather than $f(x)$. Further, to this point, if we use a filter, it is $|f(k)|^2$ that is measured. The average values of the two quantities are related as

$$\begin{aligned} \int_{-\infty}^{+\infty} dx f(x)f^*(x)&=\int_{-\infty}^{+\infty}dx\int_{-\infty}^{+\infty}\frac{dk'}{2\pi}f(k')e^{-ik'x}\int_{-\infty}^{+\infty}\frac{dk}{2\pi}f^*(k)e^{ikx}\\ &=\int_{-\infty}^{+\infty}\frac{dk}{2\pi}f^*(k)\int_{-\infty}^{+\infty}dk' f(k')\frac{1}{2\pi}\int_{-\infty}^{+\infty}dx\,e^{i(k-k')x}\\ &=\frac{1}{2\pi}\int_{-\infty}^{+\infty}dk|f(k)|^2 \end{aligned} \tag{A.2.6}$$

[see Eqn (A.2.12)].

A.2.3 Spectral Density Function

The time average[2] of a quantity $|f(t)|^2$ is

$$\begin{aligned}\overline{|f(t)|^2} &= \lim_{T\to\infty}\frac{1}{T}\int_{-T/2}^{+T/2} dt f(t) f^*(t) \\ &= \lim_{T\to\infty}\int_{-\infty}^{+\infty} d\omega \frac{|f(\omega)|^2}{2\pi T} \quad \text{by (A.2.6).}\end{aligned} \tag{A.2.7}$$

The "spectral density function" is defined as

$$S(\omega) = \lim_{T\to\infty}\frac{|f(\omega)|^2}{2\pi T}.$$

This quantity is the transform of the "autocorrelation function"

$$\begin{aligned}C(\tau) &= \lim_{T\to\infty}\frac{1}{T}\int_{-T/2}^{+T/2} dt f(t) f(t+\tau) = \frac{1}{2\pi}\int_{-\infty}^{+\infty} d\omega\, e^{i\omega\tau} S(\omega) \\ S(\omega) &= \int_{-\infty}^{+\infty} d\tau\, e^{-i\omega\tau} C(\tau).\end{aligned} \tag{A.2.8}$$

For the comparison of theory and experiment, we usually act as if we have a large number of similar systems "an ensemble" on which we make measurements. The ensemble average value of $f(t)f(t')$ is given by

$$\langle f(t)f(t')\rangle = \int P_2(X,t;X',t')f(t)f(t')dX\,dX',$$

where P_2 is the probability that at t the system is in the state $X \to X+dX$ and at t' in $X' \to X'+dX'$. For a stationary random process, this depends only on $\tau = t'-t$, so applying this to Eqn (A.2.8), we see that

$$\langle C(\tau)\rangle = \langle f(t)f(t+\tau)\rangle.$$

---

[2]The averaging time T must be greater than the correlation time of fluctuations in order for this average to be independent of T (see Born & Wolf, 1965). Then, we may take $T \to \infty$, but to be consistent with Eqn (A.2.5), we must set

$$f(t) = \begin{cases} f(t), & -T/2 \le t \le T/2 \\ 0, & |T| > T/2 \end{cases}$$

(see Bekefi, 1966, section 1.4).

Similarly, the ensemble average value of the spectral density function of $S(\mathbf{k},\omega)$ density fluctuations in a plasma for a homogeneous, stationary random process may be defined for Fourier transforms as

$$S(\mathbf{k},\omega) \equiv \lim_{T,V\to\infty} \frac{1}{TV} \frac{\langle n(\mathbf{k},\omega) n^*(\mathbf{k},\omega)\rangle}{n_0}. \quad \text{(A.2.9)}$$

A.2.4 Delta Functions

The Kronecker delta $\delta_{\alpha\beta}$ has the properties

$$\delta_{\alpha\beta} = 1, \quad \alpha = \beta; \quad \delta_{\alpha\beta} = 0, \quad \alpha \neq \beta. \quad \text{(A.2.10)}$$

It enters in situations like that in Eqn (A.2.2).

The Dirac delta $\delta(x-x')$ has the properties

$$\delta(x-x') = 0, \quad x' \neq x; \quad \delta(x-x') = \infty, \quad x' = x,$$
$$\int_{-a}^{b} dx' \delta(x-x') = 1, \quad a < x < b. \quad \text{(A.2.11)}$$

This function may be represented in integral form as

$$\delta(x-x') = \frac{1}{2\pi} \int_{-\infty}^{+\infty} dt \exp\left[i(x-x')t\right], \quad \int_{-\infty}^{+\infty} dy g(y) \delta(x-y) = g(x). \quad \text{(A.2.12)}$$

Note that if c is a constant, then

$$\delta\,(c(x-y)) = \frac{1}{2\pi c} \int_{-\infty}^{+\infty} dt\, e^{ic(x-y)t} = \frac{1}{2\pi c} \int_{-\infty}^{+\infty} ds\, e^{i(x-y)s} = \frac{1}{c} \delta(x-y). \quad \text{(A.2.13)}$$

A.3 LAPLACE TRANSFORMS

A.3.1 Definition

This transform may be defined as

$$f(\omega) = \int_{0}^{\infty} dt f(t) e^{-(i\omega+\gamma)t}, \quad \text{(A.3.1)}$$

where $\gamma > 0$. The factor $e^{-\gamma t}$ is made large enough that the Fourier transform of $f(t)e^{-\gamma t}$ will converge, even though $f(t)$ might be such that the Fourier integral would not converge, e.g., $f(t) = e^{|a|t}$, than $\gamma > |a|$.

The inverse transform is

$$f(t)=\frac{1}{2\pi}\int_{-\infty}^{+\infty} d\omega' f(\omega')e^{+(i\omega'+\gamma)t}. \tag{A.3.2}$$

If we now set $\omega=\omega'-i\gamma$, this may be rewritten as

$$f(t)=\frac{1}{2\pi}\int_{-i\gamma-\infty}^{-i\gamma+\infty} d\omega f(\omega)e^{i\omega t}. \tag{A.3.3}$$

The contour is shown in Fig. A.7. This transform applies to the lower half of the ω plane below the line $\mathrm{Im}(\omega)=-\gamma$, since $f(\omega)$ is only defined for $\gamma>0$. For inverse transforms where it is necessary to close the contour in the upper half-plane, we must analytically continue $f(\omega)$ (see Appendix A.1.3).

Example

Consider the trivial equation

$$\frac{\partial f}{\partial t}=at,$$

with solution

$$f(t)=\frac{at^2}{2}+\text{const}=\frac{at^2}{2}+f(0).$$

Now,

$$\int_0^\infty dt\frac{df}{dt}e^{-(i\omega+\gamma)t}=-f(0)+(i\omega+\gamma)f(\omega)=\frac{a}{(i\omega+\gamma)^2}$$

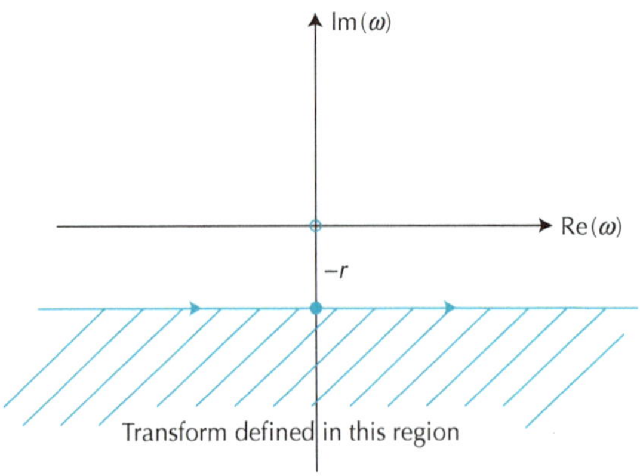

FIGURE A.7 Contour for Eqn (A.3.3).

or

$$f(\omega) = \frac{a}{(i\omega+\gamma)^3} + \frac{f(0)}{(i\omega+\gamma)}$$

$$f(t) = \frac{1}{2\pi} \int_{-i\gamma-\infty}^{-i\gamma+\infty} d\omega \left(\frac{ae^{i\omega t}}{(i\omega)^3} + \frac{f(0)e^{i\omega t}}{(i\omega)} \right).$$

The solution is obtained using Eqn (A.1.10) as

$$f(t) = \frac{at^2}{2} + f(0).$$

Similarly,

$$\int_0^\infty dt \frac{\partial^2 f}{\partial t^2} e^{-(i\omega+\gamma)t} = (i\omega+\gamma)f(0) - \left.\frac{\partial f}{\partial t}\right|_{t=0} + (i\omega+\gamma)^2 f(\omega). \qquad \text{(A.3.4)}$$

A.3.2 Spectral Density Function

The spectral density function $S(\mathbf{k},\omega)$ is defined in Eqn (A.2.9) in terms of Fourier time transforms. We want to use Laplace time transforms, and it is therefore important to find the corresponding definition. Now,

$$\langle n(\mathbf{k},\omega) \cdot n^*(\mathbf{k},\omega) \rangle_{\text{Lap}}$$
$$= \int_0^\infty dt\, e^{-(i\omega+\gamma)t} \int_0^\infty dt'\, e^{+(i\omega-\gamma)t'} \langle n(\mathbf{k},t) n^*(\mathbf{k},t) \rangle.$$

We put $t' = t + \tau$ and rearrange

$$= \int_0^\infty dt\, e^{-2\gamma t} \left[\int_0^\infty d\tau\, e^{(i\omega-\gamma)\tau} + \int_{-t}^0 d\tau\, e^{-(i\omega+\gamma)\tau} \right] \langle n(\mathbf{k},t) n^*(\mathbf{k},t) \rangle.$$

We may exchange the order of integration in the second term. Further, provided the system is stationary, $\langle n(\mathbf{k},t) n^*(\mathbf{k},t) \rangle$ is not a function of t, and we may complete the integration over t. We multiply by 2γ and take the limit $\gamma \to 0$ to obtain

$$\lim_{\gamma\to 0} 2\gamma \langle n(\mathbf{k},\omega) n^*(\mathbf{k},\omega) \rangle = \int_{-\infty}^{+\infty} d\tau\, e^{+i\omega\tau} \langle n(\mathbf{k},t) n^*(\mathbf{k},t+\tau) \rangle. \qquad \text{(A.3.5)}$$

With the aid of Eqns (A.2.8) and (A.2.9), we then see that

$$S(\mathbf{k},\omega)=\lim_{T,V\to\infty}\frac{1}{TV}\frac{\langle n(\mathbf{k},\omega)n^*(\mathbf{k},\omega)\rangle}{n_0},\quad \text{Fourier time transform}$$
$$=\lim_{\gamma\to 0,V\to\infty}\frac{2\gamma}{V}\frac{\langle n(\mathbf{k},\omega)n^*(\mathbf{k},\omega)\rangle}{n_0},\quad \text{Laplace time transform.} \tag{A.3.6}$$

Note that from Eqns (A.2.8) and (A.3.5), we obtain the result

$$\left\langle A(\mathbf{k},t)B^*(\mathbf{k},t+\tau)\right\rangle=\lim_{\gamma\to 0}2\gamma\int_{-i\gamma-\infty}^{-i\gamma+\infty}\frac{d\omega}{2\pi}e^{i\omega\tau}\left\langle A(\mathbf{k},\omega)B^*(\mathbf{k},\omega)\right\rangle. \tag{A.3.7}$$

A.4 STABILITY OF LONGITUDINAL PLASMA OSCILLATIONS

This topic, which has been discussed in detail by Jackson (1960), Penrose (1960), Bernstein *et al.* (1964) and others, provides a good example of the application of Fourier–Laplace transform techniques to plasmas. We investigate the simplest case, that in which the positive ion background is immobile, and there are no external fields and no collisions. In one dimension, the plasma is then described by

$$\frac{\partial F_1}{\partial t}+\mathrm{v}\frac{\partial F_1}{\partial x}-\frac{eE_1}{m_e}\frac{\partial F_0}{\partial \mathrm{v}}=0 \tag{A.4.1}$$

$$\frac{\partial E_1}{\partial x}=-4\pi e\int_{-\infty}^{+\infty}d\mathrm{v}\,F_1(x,\mathrm{v},t)=-4\pi e n_1(x,t). \tag{A.4.2}$$

We are given the initial distribution $F_0(\mathrm{v})=n_0 f_0(\mathrm{v})$, and we are looking for solutions of the form

$$F(x,\mathrm{v},t)=F_0(\mathrm{v})+F_1(\mathrm{v})e^{-ikx+i\omega t}. \tag{A.4.3}$$

From the solution of Eqns (A.4.1) and (A.4.2), we obtain a relationship between k and ω. Now, k must be real; otherwise, Eqn (A.4.3) would not be a satisfactory solution for all x in the range $-\infty - +\infty$. Then, if the imaginary part of ω is negative, we see that the system is unstable because there is a component of F proportional $e^{+\omega_i t}$. This calculation, using linearized equations, is only valid for times in which any unstable mode remains at low amplitude. For longer times, the nonlinear term $eE_1/m\cdot(\partial F_1/\partial \mathrm{v})$ must be included.

The Fourier–Laplace transforms are

$$\left.\begin{matrix}F_1(\mathrm{v},k,\omega)\\ E_1(k,\omega)\end{matrix}\right\}=\int_{-\infty}^{+\infty}dx\int_0^{\infty}dt\,e^{ikx-(i\omega+\gamma)t}\left\{\begin{matrix}F_1(x,\mathrm{v},t)\\ E_1(x,t).\end{matrix}\right. \tag{A.4.4}$$

The inverse transforms are

$$\left.\begin{matrix} E_1(x,\mathrm{v},t) \\ E_1(x,t) \end{matrix}\right\} = \frac{1}{(2\pi)^2} \int_{-\infty}^{+\infty} dk \int_{-i\gamma-\infty}^{-i\gamma+\infty} d\omega\, e^{ikx+i\omega t} \left\{\begin{matrix} E_1(\mathrm{v},k,\omega) \\ E_1(k,\omega). \end{matrix}\right. \tag{A.4.5}$$

Applying Eqn (A.4.4) to Eqns (A.4.1) and (A.4.2) leads to

$$F_1(\mathrm{v},k,\omega) = -\frac{\{i[F_1(\mathrm{v},k,o) + (e/m_e)E_1(k,\omega)\partial F_0/\partial \mathrm{v}]\}}{(\omega - k\mathrm{v} - i\gamma)} \tag{A.4.6}$$

$$E_1(k,\omega) = -\frac{i4\pi e n_1(k,\omega)}{k}. \tag{A.4.7}$$

Equation (A.4.6) is integrated over velocity, and the resulting $n_1(\mathbf{k},\omega)$ is substituted in Eqn (A.4.7), and we find

$$E_1(\mathbf{k},\omega) = \left[-\frac{4\pi e}{k^2}\int_{-\infty}^{+\infty} \frac{d\mathrm{v} F(\mathrm{v},k,o)}{(\omega/k) - \mathrm{v} - (i\gamma/k)}\right] \Big/ \varepsilon(k,\omega), \tag{A.4.8}$$

where the dielectric function is

$$\varepsilon(k,\omega) = 1 - \frac{\omega_{\mathrm{pe}}^2}{k^2}\int_{-\infty}^{+\infty} \frac{d\mathrm{v}\,\partial f_0/\partial \mathrm{v}}{\mathrm{v} - [(\omega - i\gamma)/k]}. \tag{A.4.9}$$

For simplicity, we now consider the temporal behavior of just one mode, and the inverse transform of Eqn (A.4.8) is

$$E_1(k,t) = -\frac{2e}{k^2}\int_{-i\gamma-\infty}^{-i\gamma+\infty} \left\{d\omega\, e^{i\omega t}\left[\int \frac{d\mathrm{v} F(\mathrm{v},k,o)}{(\omega/k) - \mathrm{v}}\right] \Big/ \varepsilon(k,\omega)\right\}. \tag{A.4.10}$$

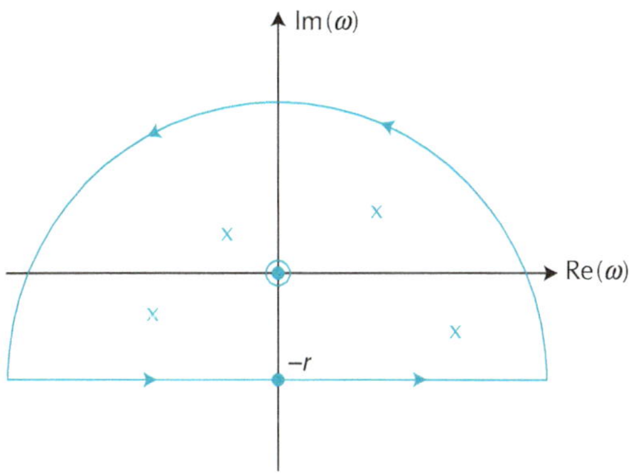

FIGURE A.8 Contour closed in the upper half-plane.

The line $\text{Im}(\omega) = -\gamma$ is chosen below any poles. The integral may be evaluated with a contour closed in the upper half-plane (see Fig. A.8). This overcomes problems with the factor $e^{i\omega t}$. The integral is then the sum of residues arising from zeros in $\varepsilon(k, \omega)$. We assume that $F(\text{v}, k, 0)$ is a sensible function so that the integral that includes it has no poles. We denote the roots of the "dispersion relation" $\varepsilon(k, \omega) = 0$ by ω_c and

$$E_1(k,t) = -\frac{4\pi ie}{k^2} \sum_c e^{i\omega_c t} \text{ Residue} \left[\int \frac{d\text{v} F(\text{v},k,0)}{(\omega/k) - \text{v}} \Big/ \varepsilon(k,\omega) \right]_{\omega=\omega_c}. \tag{A.4.11}$$

The tricky part of this is that $\varepsilon(k, \omega)$ is not defined in the upper half-plane. The source of the problem is readily seen if we consider the behavior of the pole in the v plane for the integral in Eqn (A.4.9) as $\text{Im}(\omega - i\gamma)$ changes from negative to positive and the pole in the v plane changes from Im(v) positive to negative [see Eqn (A.1.14)]

$$\text{Im}(\omega - i\gamma)\Big\} \begin{matrix} <0 \\ >0 \end{matrix} \; \varepsilon(k,\omega) = 1 + \frac{\omega_{\text{pe}}^2}{k} \left(\left[\pm i\pi \frac{\partial f_0(\text{v})}{\partial \text{v}} \right]_{\text{v}=\omega/k} + \int \frac{\partial \text{v} \partial f_0/\partial \text{v}}{\text{v} - (\omega/k)} \right). \tag{A.4.12}$$

Both results are obtained using the contour for the v integral shown in Fig. A.9. For $\text{Im}(\omega - i\gamma) < 0$, the contour is along the real v axis, while for $\text{Im}(\omega - i\gamma) > 0$, the contour is depressed to accommodate the pole, and this leads to the additional factor

$$+\frac{\omega_{\text{pe}}^2}{k} 2\pi i \frac{\partial f_0(\text{v})}{\partial \text{v}} \bigg|_{\text{v}=\omega/k}.$$

Finally, we can see that if ω_s has a negative imaginary part, then the appropriate mode $E_1(k_s, t)$, Eqn (A.4.11), will grow exponentially, i.e., the system is unstable. The system is stable only if $\text{Im}(\omega_s) > 0$ for *all* modes. Marginal stability occurs when $\text{Im}(\omega_s) = 0$ for the mode closest to instability.

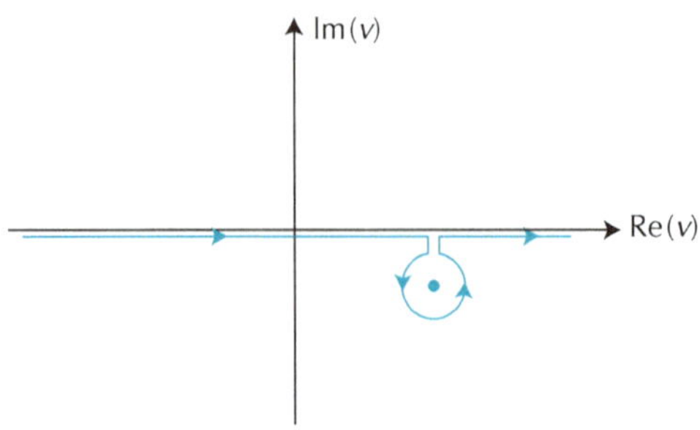

FIGURE A.9 Contour for Eqn (A.4.12).

A.5 TOTAL CROSS SECTION FOR A STABLE PLASMA

The total cross section is given by $S_T(\mathbf{k}) = \int_{-\infty}^{+\infty} d\omega S(\mathbf{k}, \omega)$, where the integral is along the real ω axis. In many cases,

$$S(\mathbf{k}, \omega) \propto \frac{\text{function}\,(\omega/k)}{(\omega/k)\varepsilon(\mathbf{k}, \omega/k)},$$

and it is convenient to transform the integral to a contour integral in the complex plane. For a stable plasma, the dielectric function for our definition of transforms Eqn (A.4.4) has no poles in the upper half-plane, and we therefore close the contour there (see Fig. A.10). The contour is indented to avoid the pole at ω/k. We treat the case discussed in Section 5.6.1.

$$-\mathrm{Im} \int_{-\infty}^{+\infty} \frac{dz}{\pi^{1/2} z \alpha^2 \varepsilon(\alpha, z)}$$

$$= -\mathrm{Im} \lim_{R\to\infty,\, r\to 0} \left\{ \left(\int_{-R}^{-r} dz + \int_{+r}^{+R} dz \right) \frac{1}{\pi^{1/2} z \alpha^2 \varepsilon(\alpha, z)} \right\}$$

$$= -\mathrm{Im} \left[\frac{2\pi i}{\pi^{1/2} \alpha^2 \varepsilon(\alpha, 0)} - \left\{ \left(\int_{C_r} dz + \int_{C_R} dz \right) \frac{1}{\pi^{1/2} z \alpha^2 \varepsilon(\alpha, z)} \right\} \right].$$

Now, from Eqn (A.1.8),

$$\int_{C_r} dz/\pi^{1/2} z \alpha^2 \varepsilon(\alpha, z) = \pi i/\pi^{1/2} \alpha^2 \varepsilon(\alpha, 0),$$

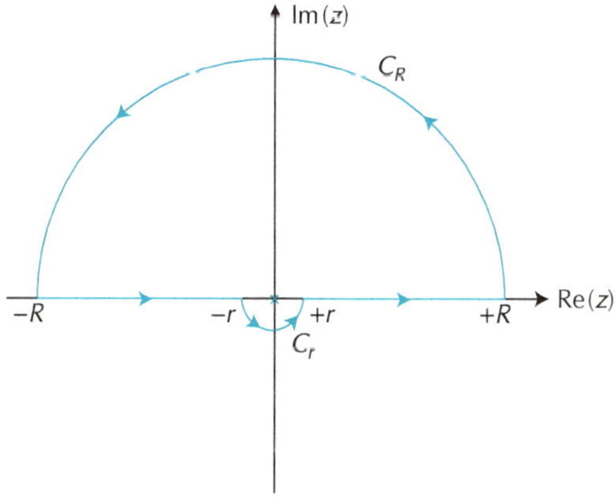

FIGURE A.10 Contour indented at $\omega/k = 0$.

and since $z = R\cos\theta - iR\sin\theta, dz = -i(R\sin\theta + R\cos\theta)d\theta$ and $\varepsilon(\alpha, z)|_{R\to\infty} = 1$; we find that

$$\int_{C_R} dz/\pi^{1/2} z\alpha^2\varepsilon = i\pi^{1/2}/\alpha^2.$$

In total,

$$-\mathrm{Im}\left[\int_{-\infty}^{+\infty} \frac{dz}{\pi^{1/2} z\alpha^2\varepsilon}\right] = \frac{\pi^{1/2}}{\alpha^2}\left[1 - \frac{1}{\varepsilon(\alpha, 0)}\right].$$

For the case discussed in Section 5.6, $\varepsilon(\alpha, 0) = 1 + \alpha^2$, and this leads to the result of Eqn (5.6.2).

Appendix B

Kinetic Theory of a Plasma

Chapter Outline Head

DOI: 10.1016/B978-0-12-374877-5.00014-2

B.1 INTRODUCTION

In this appendix, a brief review is given of the plasma theory, which underlies the calculations presented in this book. The intention is to give the reader a feel for the subject and enough background material that he or she can follow or initiate future developments.[1] We will see that the main problem is to make a realistic assessment of the effect of interparticle interactions (collisions). The calculations in this book are made with a Krook model collision term. The use of this form allows us to estimate the essential features of the effect of collisions or fluctuations in a plasma, with a minimum of mathematics. It is, however, important to appreciate the limitations of this approach, and a large section of this appendix is, therefore, devoted to a discussion of the derivation of more accurate collision terms.

In the derivation of the collision term, we are pushed, because of the long-range nature of the Coulomb force, into using different approximations in regions of small and large impact parameters b. The Boltzmann approach that considers binary encounters between particles leads to a collision term valid for $\sim 0 < b < \lambda_{\text{De}}$. The cutoff in b at the Debye length λ_{De} is somewhat arbitrary, and therefore, we turn to a different approach in order to deal with the large b range in a self-consistent fashion. The collision term derived from the Liouville equation (BBGKY and Klimontovich approaches) makes allowance for collective interactions (Debye shielding) and is valid for $e^2/\kappa T < b < \infty$. A complete form of the term is then obtained by joining the above two approximations at some intermediate distance $e^2/\kappa T < b_m < \lambda_{\text{De}}$.

B.2 CHARACTERISTIC LENGTHS AND TIMES IN A PLASMA

B.2.1 Characteristic Lengths

The major difference between the plasma state and other states of matter is clearly brought out by considering the net electric potential around a test charge q, which is placed in a vacuum, a solid, or gas, and finally a plasma.

---

[1] The reader is referred to the following list of excellent books and articles for a fuller and more rigorous treatment: Brittin *et al.* (1967); Clemmow & Dougherty (1969); Dawson & Nakayama (1966); Klimontovich (1967); Kunkel (1966); Montgomery & Tidman (1964); Montgomery (1971); Rogister & Oberman (1968); Rostoker & Rosenbluth (1960); Schmidt (1966); Spitzer (1962); Tanenbaum (1967); Thompson (1962); Wu (1966)

In a vacuum, the electric potential is

$$\varphi_{\text{vac}} = q/r. \tag{B.2.1}$$

In a simple, solid or gaseous, dielectric material, the net potential is modified because the bound charges that make up the neighboring atoms polarize under the influence of the electric field of the test charge. In general, the charges will act to reduce the potential, and this then becomes

$$\varphi_{\text{sol}} = q/\varepsilon r, \tag{B.2.2}$$

where ε the dielectric constant of the material is a measure of the extent of the polarizability.

In a plasma, the situation is more complicated because the neighboring charges are not bound. Nevertheless there is again an adjustment in the position of each charge. Like charges are repelled, and unlike ones are attracted. In equilibrium, the electron and ion densities have the Boltzmann distribution

$$n_{\text{e}}(r) = n_{\text{e}0}\, e^{+e\varphi/\kappa T_{\text{e}}}, \quad n_{\text{i}}(r) = n_{\text{i}0} e^{-e\varphi/\kappa T_{\text{i}}}. \tag{B.2.3}$$

For strict equilibrium, $T_{\text{e}} = T_{\text{i}} = T$, and if we have overall charge neutrality, $n_{\text{e}0} \cong n_{\text{i}0} = n_0$. The net potential around the charge q including the effect of the neighboring charges is given by Poisson's equation

$$\nabla^2\varphi = -4\pi q\,\delta(\mathbf{r}) + 4\pi e n_0 e^{+e\varphi/\kappa T} - 4\pi e n_0 e^{-e\varphi/\kappa T}. \tag{B.2.4}$$

For large enough $|\mathbf{r}|$ that $e\varphi/\kappa T \ll 1$, we have in spherical coordinates

$$\frac{1}{r}\frac{d}{dr}(r\varphi) = \frac{8\pi e^2 n_0 \varphi}{\kappa T}. \tag{B.2.5}$$

The solution is

$$\varphi_{\text{p}} \cong (q/r)\exp(-r/2^{1/2}\lambda_{\text{De}}). \tag{B.2.6}$$

The potential of the test charge q is screened by the other charges, and the range of the net potential is reduced to about λ_{De} the Debye screening length, where

$$\lambda_{\text{De}} = (\kappa T_{\text{e}}/4\pi e^2 n_{\text{e}})^{1/2}\text{cm} \quad \text{(Gaussian units)}. \tag{B.2.7}$$

Certain requirements must be met for this calculation to have significance. First, this solution is valid only for

$$r \gg eq/\kappa T = r_{\text{c}}. \tag{B.2.8}$$

In order, therefore, that the screening charges see the potential Eqn (B.2.6), we require $r_{\text{e}} \ll n_{\text{e}}^{-1/3}$, the mean interparticle separation. Second, this fluid type of calculation only makes sense if there are a large number of screening charges. Therefore, we require

$$n_{\text{e}}^{1/3} \ll \lambda_{\text{De}} \qquad \text{or} \qquad N_{\text{D}} = (4\pi/3)n_{\text{e}}\lambda_{\text{De}}^3 \gg 1. \tag{B.2.9}$$

Up to this point, we have looked at the collective interaction of the charges, but in addition, there can be simple collisions between individual pairs of charges. The collision length for 90° scattering in a simple binary collision is given by

$$\lambda_{\text{c}} = (1/4\pi r_{\text{c}}^2 n_{\text{e}}). \tag{B.2.10}$$

Rearranging gives

$$\lambda_{\mathrm{De}}^2 = \lambda_{\mathrm{c}} r_{\mathrm{c}}.$$

Combining all these relationships shows us that a plasma is characterized by the following ordering of scale lengths:

$$r_{\mathrm{c}} \ll n_{\mathrm{e}}^{-1/3} \ll \lambda_{\mathrm{De}} \ll \lambda_{\mathrm{c}}. \tag{B.2.11}$$

B.2.2 Characteristic Times

When some charge in a plasma is displaced, the net electrostatic force, which acts on the charge, is a restoring force. The subsequent motion of the charge is simple harmonic at the "plasma frequency" ω_{p}

$$\omega_{\mathrm{p}} = (4\pi n_0 q^2/m)^{1/2}\,\mathrm{rad/s} \quad \text{(Gaussian)}. \tag{B.2.12}$$

This is easily shown by the following fluid calculation for the case where electrons are displaced.

The ion density $n_{\mathrm{i}} = n_0$, the electron density $n_{\mathrm{e}} = n_0 + n_1(\mathbf{r}, t)$, where $|n_1| \ll n_0$ is a small perturbation.

A restoring electric field is set up:

$$\nabla \cdot \mathbf{E}_1 = -4\pi n_1 e. \tag{B.2.13}$$

The charges start to flow back to their original positions with velocity $\mathbf{u}_1$. We must satisfy conservation of mass, and therefore,

$$\left(\frac{\partial n_1}{\partial t}\right) + n_0 \nabla \cdot \mathbf{u}_1 = 0. \tag{B.2.14}$$

From conservation of momentum,

$$m_{\mathrm{e}} \frac{\partial \mathbf{u}_1}{\partial t} = -e\mathbf{E}_1. \tag{B.2.15}$$

We combine the three equations to obtain

$$\left(\frac{\partial^2 n_1}{\partial t^2}\right) + \omega_{\mathrm{pe}}^2 n_1 = 0. \tag{B.2.16}$$

That is, the displaced electrons oscillate at the electron plasma frequency. A similar result would clearly occur for ions.

We note in passing one important feature, namely, that $2\pi/\omega_{\mathrm{pe}}$ is a measure of the timescale on which electrons can respond to an electric perturbation. Electromagnetic waves of frequency $\omega < \omega_{\mathrm{pe}}$ cannot penetrate a plasma precisely because the electrons can respond rapidly enough to cancel out the electric field of the wave. As a result, scattering measurements within plasmas require incident radiation of frequency $\omega > \omega_{\mathrm{pe}}$.

There is a simple relationship between all the characteristic lengths and times in a plasma, for example,

$$\omega_{\mathrm{pe}} \lambda_{\mathrm{De}} = (\kappa T_{\mathrm{e}}/m_{\mathrm{e}})^{1/2} = \overline{v_{\mathrm{th}}} \quad \text{mean thermal speed.}$$

Similarly for collisions $\omega_{\mathrm{c}} \lambda_{\mathrm{c}} = \overline{v_{\mathrm{th}}}$.

B.2.3 Characteristic Parameters Associated with a Magnetic Field

Most plasmas of interest contain a magnetic field. The force on each charge is perpendicular to the direction of motion of the charge and to the magnetic field, and the charge follows an helical orbit, and the basic features of the orbit can be seen from a calculation for a steady homogeneous field:

$$m_q \frac{d\mathrm{v}}{dt} = q(\mathrm{v}/c) \times B. \tag{B.2.17}$$

Let $\mathrm{v} = \mathrm{v}_{\|} + \mathrm{v}_{\perp}$, where $\mathrm{v} \cdot \mathbf{B} = \mathrm{v}_{\|}$,

$$m_q \frac{d(\mathrm{v} \times \mathbf{B})}{dt} = -q\mathrm{v}_{\perp} \frac{B^2}{c} = \frac{cm^2}{q} \frac{d^2\mathrm{v}}{dt^2},$$

or

$$\mathrm{v}_{\|} = \mathrm{v}_{\|0};\ \mathrm{v}_{\perp} = \mathrm{v}_{\perp 0} \cos(\Omega t + \varphi_0), \tag{B.2.18}$$

where the cyclotron or gyrofrequency $\Omega_q = qB/m_q c$. It is simple to show that the radius of the orbit is

$$\rho_q = \mathrm{v}_{\perp}/\Omega_q. \tag{B.2.19}$$

B.3 THE BOLTZMANN EQUATION

B.3.1 The Boltzmann Equation

Consider a volume V that contains N_α particles of a species α. The distribution function $f_\alpha(\mathbf{r}, \mathrm{v}, t)$ gives the average number of particles dN_α which have the position $\mathbf{r} \to \mathbf{r} + d\mathbf{r}$ and velocity $\mathrm{v} \to \mathrm{v} + d\mathrm{v}$ at time t

$$f_\alpha(\mathbf{r}, \mathrm{v}, t) = \frac{dN_\alpha}{d\mathbf{r}} d\mathrm{v}. \tag{B.3.1}$$

The particle density is

$$n_a(\mathbf{r}, t) = \int f_\alpha d\mathrm{v}. \tag{B.3.2}$$

The rate of change of the number of particles in a given element ($\mathbf{r} \to \mathbf{r} + d\mathbf{r}$, $\mathrm{v} \to \mathrm{v} + d\mathrm{v}$) in phase space is simply the difference between the rate at which particles enter and leave the element. Particles move from one element to another for two reasons:

(i) because of their basic orbit, that is, the part determined by externally applied fields,
(ii) because of interactions with other particles (collisions).

Let the rate of change of $f_{(\alpha)}$ due to collisions be denoted by $(\partial f_\alpha/\partial t)_c$ and the acceleration due to external forces by $\mathbf{a}$. In the short interval $t \to t + \Delta t$, the charges at $(\mathbf{r}, \mathrm{v})$ will move to $(\mathbf{r}+\mathrm{v}, \Delta t$ and $\mathrm{v}+\mathbf{a}, \Delta t)$; all other changes in the orbit are bound up in the collision term and

$$f_\alpha(\mathbf{r}+\mathrm{v}\Delta t,\ \mathrm{v}+\mathbf{a}\Delta t;\, t+\Delta t)-f_\alpha(\mathbf{r},\ \mathrm{v},\ t)=(\partial f_\alpha/\partial t)_{\mathrm{c}}\Delta t.$$

The final distribution function may now be expanded about the initial value, and in the limit $\Delta t \to 0$, we obtain the Boltzmann equation

$$\frac{\partial f_\alpha}{\partial t}+\mathrm{v}\cdot\frac{\partial f_\alpha}{\partial \mathbf{r}}+\mathbf{a}\cdot\frac{\partial f_\alpha}{\partial \mathrm{v}}=\left(\frac{\partial f_\alpha}{\partial t}\right)_{\mathrm{c}}. \tag{B.3.3}$$

B.3.2 Neutral Gas: Short-Range Force

For a short-range force, two particle encounters are normally dominant, and for a given interparticle force law, we may calculate the trajectories of two interacting particles. This calculation allows us to identify the types of encounter (collision) in which particles are lost or gained by a given element. Figure B.1 provides an illustration of a collision with a repulsive force in the rest frame of particle β whose laboratory velocity is initially $\mathrm{v}_{1\beta}$ and finally $\mathrm{v}'_{1\beta}$. We see that the particle α is lost from the range $\mathrm{v}_\alpha \to \mathrm{v}_\alpha + d\mathrm{v}_\alpha$ and gained by the range $\mathrm{v}'_\alpha \to \mathrm{v}'_\alpha + d^{\mathrm{v}'}_{\alpha}$. In the inverse collision, the opposite would be true. (Note that the particle β would then go from $\mathrm{v}'_{1\beta} \to \mathrm{v}_{1\beta}$.) The rate of the reaction forward or backward is obviously proportional to the probability that the two particles have their respective trajectories.

For a neutral gas, the set of charges, which constitutes a molecule, are so grouped that the interparticle force is short range, and in the Boltzmann treatment, it is therefore assumed that

(a) since the duration of a collision is short compared to the time between collisions, we may treat the distribution functions as constant on the microscopic timescale of an encounter,
(b) since the particles do not "know" about each other until a collision is imminent, we may write the probability that a particle α is at $(\mathbf{r}, \mathrm{v})$ when a

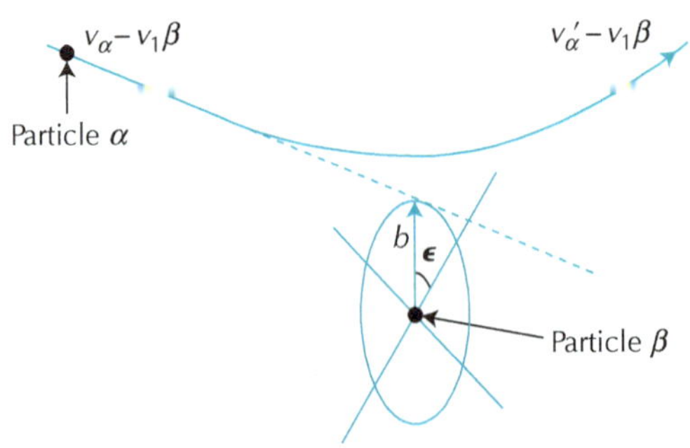

FIGURE B.1 Collision with repulsive force.

particle β is at $(\mathbf{r}_1, \text{v}_1)$ as the probabilities that each is there independently, i.e., in the notation we will discuss later

$$f_{\alpha\beta}^{(2)}(r, \text{v}, r_1, \text{v}_1; t) = f_\alpha^{(1)}(r, \text{v}, t) f_\beta^{(1)}(r_1, \text{v}_1; t).$$

B.3.3 Plasma–Long-Range Force

At first sight, it would appear that the Boltzmann treatment is not appropriate for a plasma because the Coulomb force is long range and many charges can interact simultaneously. In fact due to the Debye shielding effect, each charge is aware of another charge only if it is within about a Debye length λ_{De} of it. The duration of a "collision" is therefore typically $\tau \cong \lambda_{\text{De}}/\bar{\text{v}} \cong 1/\omega_{\text{pe}}$, where ω_{pe} is the plasma frequency. Therefore, in applying the Boltzmann treatment to a plasma, we are assuming that f_α and f_β do not change significantly on this timescale. Further, the integral Eqn (B.3.4) diverges for large impact parameter for the bare Coulomb force; therefore, we limit the impact parameter b to distances $b \lesssim \lambda_{\text{De}}$, and the collision term then becomes (see, for example, Tanenbaum, 1967, chapter 6)

$$(\partial f_\alpha/\partial t)_c = \sum_b \int_0^{\lambda_{\text{De}}} b\,db \int d\varepsilon\, d\text{v}_1 \{ f'_\alpha(\mathbf{r}, \text{v}'; t) f'_\beta(\mathbf{r}, \text{v}'; t) - f_\alpha(\mathbf{r}, \text{v}, t) f_\beta(\mathbf{r}, \text{v}_1; t)\} |\text{v} - \text{v}_1|. \quad \text{(B.3.4)}$$

The impact parameter b is the distance of closest approach if there were no collision. The angle ε gives the orientation of the scattering plane relative to the plane perpendicular to the initial relative direction of motion.

Alternatively, we may use the screened Coulomb potential in calculating the particle trajectories. This has the effect of cutting off b.

$$F_{\text{scr}} = (q_\alpha q_\beta / r_{\alpha\beta}^2) \exp(-r_{\alpha\beta}/2^{1/2}\lambda_{\text{De}}).$$

The result of these calculations is a collision term, which is a fair representation of charge interactions for short distances $\sim 0 < b < \lambda_{\text{De}}$. The unsatisfactory part of the treatment is the somewhat arbitrary choice of the cutoff at (λ_{De}). A collision term that copes with large impact parameters in a self-consistent fashion can be obtained from the Liouville equation. This is discussed in Sections B.5 and B.6.

B.4 COMMENTS ON THE COLLISION TERM

B.4.1 Inelastic Collisions

We must remember that other kinds of collision are possible. For unlike charges, there can be recombination, for neutrals, excitation, ionization, dissociation, and attachment. Obviously, these processes must be taken into account if they significantly change the energy, momentum, or species balance in the system. It is beyond the scope of this book to deal with such questions, and therefore, in the calculations below, we will only consider charge–charge or charge–neutral elastic scattering collisions.

B.4.2 Collision Frequency

The collision time $\tau = 1/\nu$ is the time for a charge to experience a significant deflection (change in momentum). For a plasma, this deflection is normally achieved as the result of many small angle collisions rather than in a single collision as in the case for neutrals.

The equation of conservation of momentum for the species α is obtained by multiplying Eqn (B.3.3) by $m\mathrm{v}_\alpha$ and then integrating over v_α. The collision term for momentum transfer can be evaluated for drifting Maxwellian distribution functions, and we find (Tanenbaum, 1967, p. 251) that

$$\int_{-\infty}^{+\infty} d\mathrm{v}_\alpha m\mathrm{v}_\alpha (\partial f_\alpha/\partial t)_c = \sum_\beta m_\alpha n_\alpha \nu_{\alpha\beta}(\mathbf{u}_\alpha - \mathbf{u}_\beta), \tag{B.4.1}$$

where $\mathbf{u}_\alpha$ and $\mathbf{u}_\beta$ are the drift velocities of species α and β. The *charge–charge* collision frequency is given by

$$\nu_{\alpha\beta} = \frac{(m_\alpha + m_\beta)}{3\pi^{3/2} m_\alpha^2 m_\beta} \cdot \frac{q_\alpha^2 q_\beta^2}{\varepsilon_0^2} n_\beta \left(\frac{2\kappa T_\alpha}{m_\alpha} + \frac{2\kappa T_\beta}{m_\beta}\right)^{-3/2} \ln\Lambda (\text{mks})$$

$$\Lambda \cong \frac{12\pi\varepsilon_0 \kappa T}{q_\alpha q_\beta}\left(\frac{\varepsilon_0 \kappa T}{e^2 n}\right)^{1/2} \text{mks}, \qquad T = T_\alpha \cong T_\beta. \tag{B.4.2}$$

For *charge–neutral* collisions,

$$\nu_{q\beta} = \frac{8\pi^{1/2}}{3} \frac{m_\beta}{(m_q + m_\beta)} n_\beta \cdot \sigma^2 \cdot \left(\frac{2\kappa T_\alpha}{m_\alpha} + \frac{2\kappa T_\beta}{m_\beta}\right)^{1/2}, \tag{B.4.3}$$

where σ is the sum of the effective radii of the interacting particles.

B.4.3 Fokker–Planck Collision Term

For the case where small angle deflections are important, we may rewrite the Boltzmann collision term by expanding the final velocity of each charge about the initial velocity $\Delta\mathrm{v} = \mathrm{v}' - \mathrm{v}$. This leads to the Fokker–Planck term

$$\left(\frac{\partial f_\alpha}{\partial t}\right)_c = -\frac{\partial}{\partial \mathrm{v}_i} \underbrace{\left(f_\alpha \langle \Delta \mathrm{v}_i \rangle_{\mathrm{av}}\right)}_{\text{friction term}} + \underbrace{\frac{1}{2}\frac{\partial^2 \left(f_\alpha \langle \Delta \mathrm{v}_i \Delta \mathrm{v}_j \rangle_{\mathrm{av}}\right)}{\partial \mathrm{v}_i \partial \mathrm{v}_j}}_{\text{velocity diffusion term}}. \tag{B.4.4}$$

Estimates of the coefficients $<\Delta\mathrm{v}_i>_{\mathrm{av}}$ and $<\Delta\mathrm{v}_i\Delta\mathrm{v}_j>_{\mathrm{av}}$ can be made for some cases (see Rosenbluth *et al.* (1957); Dougherty (1964); Tanenbaum (1967)).

B.4.4 Krook Collision Term

A simpler form for the collision term was proposed by Bhatnager *et al.* (1954); this is normally called the *Krook term*

$$(\partial f_\alpha/\partial t)_c = -\nu_{\alpha\beta}(f_\alpha - f_{\alpha m}), \tag{B.4.5}$$

where $f_{\alpha m}$ is the distribution function toward, which f_α is relaxing and $\nu_{\alpha\beta}$ is the collision frequency for momentum transfer [Eqns (B.4.2) and (B.4.3)]. Our interest centers on the effect of collisions on the microscopic fluctuations in the plasmas. We therefore set $f_\alpha = F_{0\alpha} + F_{1\alpha}$, where the ensemble average function $F_{0\alpha} = n_{0\alpha} f_{0\alpha}$ [see Eqn (A.2.11)]. Now, the number density $n_\alpha(\mathbf{r}, t) = n_{0\alpha} + \int F_{1\alpha} d\mathrm{v} = n_{0\alpha} + N_{1\alpha}$. To satisfy conservation of particles, we require that $\int (\partial f_\alpha/\partial t)_c d\mathrm{v} = 0$, and for $\nu_{\alpha\beta}$ independent of velocity, this means that $n_\alpha f_\alpha = n_0 f_{\alpha m}$, so

$$(\partial f_\alpha/\partial t)_c = -\nu_{\alpha\beta}(F_{1\alpha} - N_{1\alpha} f_{0\alpha}). \tag{B.4.6}$$

This is discussed by Dougherty (1963) and Tanenbaum (1967). While this term is strictly more appropriate to large angle collisions, it may be used in a rough assessment of the main collisional effects even in situations in which small angle deflections are dominant. The main weakness of the Krook term is that it cannot in the above form accurately and simultaneously represent both the transfer of momentum and of energy. For our purposes, however, it enables us to obtain the basic form of the collisional spectra in a simple fashion.

Some authors have simplified the term further to

$$(\partial f_\alpha/\partial t) = -\nu_{\alpha\beta} f_\alpha.$$

As we have seen above, this would not satisfy particle conservation, and for example, in application to the effect of collisions on electron plasma frequency waves amounts to the omission of the collision term which regenerates the waves (see Section 5.5).

B.5 KINETIC DESCRIPTION OF SCATTERING FROM A PLASMA

B.5.1 Introduction

The spatial and spectral distribution of the radiation scattered by a group of N electrons in a plasma of volume V can be determined, provided that the orbit of each electron is known. The time-averaged scattered power in unit solid angle is[2] [see Eqn (1.8.4)]

$$\frac{\overline{dP}}{d\Omega} = \frac{cR^2}{4\pi}\left|\sum_{j=1}^{N} E_j(r_j(t), t) \sum_{g=1}^{N} E_g(\mathbf{r}_g(t), t)\right|. \tag{B.5.1}$$

[2]Note time average in this case can in principle be over one period. In practice, for the meaningful comparison of theory and experiment, it must in general be many periods.

For each electron j, we may in principle write down an equation of motion, and when all the forces $\mathbf{F}_j$ are known, the system is completely described by the N equations.

$$\sum_{j=1}^{N}\left[\frac{d(m\mathrm{v}_j)}{dt}-\mathbf{F}_j\right]=0. \tag{B.5.2}$$

For a plasma, the relevant forces are generally electric and magnetic and

$$\mathbf{F}_j=\frac{q_j}{m_j}\left(\mathbf{E}+\frac{\mathrm{v}_j}{c}\times\mathbf{B}\right). \tag{B.5.3}$$

These fields may be externally applied or internally generated. For a nonrelativistic plasma, the case treated below, the only significant internal force, is the electric field $\mathbf{E}=-\nabla\varphi$, and for example, the potential seen by the electron m is

$$\varphi(\mathbf{r}_m)=\sum_{j=1,j\neq m}^{N}-\frac{e}{\left|\mathbf{r}_m-\mathbf{r}_j(t)\right|}+\sum_{l=1}^{N/Z}\frac{Ze}{\left|\mathbf{r}_m-\mathbf{r}_l(t)\right|}. \tag{B.5.4}$$

Overall the system is neutral, and the electron charge is balanced by N/Z positive ions each of charge Ze.

The interparticle Coulomb force drops away slowly with distance, and therefore, the electron m will interact significantly with some or all the N electrons and N/Z ions. In practice, we may be dealing with $N\cong 10^{10}-10^{20}$ charges, and we cannot hope to follow their detailed motions as they interact with each other and with externally applied forces.

Fortunately, we do not need an exact description of the system. In experiments, the diagnostic devices study finite volumes, integrate the results over finite times, and therefore measure average values of parameters such as density, temperature, and scattered power. From the theoretical standpoint, the equivalent results are obtained by studying the probable behavior of a large number (an ensemble) of similar systems, which represent the possible states of the given system.

B.5.2 Ensemble Average and N Particle Probability Distribution Function

The time- and space-averaged measurement is equated with the calculated ensemble average. The ensemble average of a quantity $0(x)$, which is a parameter of the system, is defined as

$$\langle 0(x)\rangle \equiv \frac{\int_{\mathrm{all}\,x} 0(x)P(x)dx}{\int P(x)dx}, \tag{B.5.5}$$

where $P(x)$ is the probability that the system will be in the state x. For our situation, the state of the system is defined if we specify the position and velocity of each of the N electrons and N/Z ions. For simplicity, consider just the electrons for the moment. We denote the probability that the system is in the state at time t in which electron (1) is at $\mathbf{r}_1 \to \mathbf{r}_1 + d\mathbf{r}_1$ with velocity $\mathrm{v}_1 \to \mathrm{v}_1 + d\mathrm{v}_1$ when (2) is at $\mathbf{r}_2$, v_2, etc. by

$$f^{(\mathrm{N})}(\mathbf{r}_1, \mathrm{v}_1, \mathbf{r}_2, \mathrm{v}_2, \ldots, \mathbf{r}_\mathrm{N}, \mathbf{r}_\mathrm{N}; t)d\mathbf{r}_1, d\mathrm{v}_1, d\mathbf{r}_2, d\mathrm{v}_2, \ldots, d\mathrm{v}_{\mathrm{N}-1}, d\mathbf{r}_\mathrm{N}, d\mathrm{v}_\mathrm{N}. \tag{B.5.6}$$

We restrict to situations where the N particles remain within the system, therefore

$$\int f^{(\mathrm{N})} d\mathbf{r}_1, d\mathrm{v}_1, \ldots, d\mathrm{v}_\mathrm{N} = 1.$$

B.5.3 The Liouville Equation

Consider how the system evolves in the time interval $t \to t + \Delta t$. Electron (1) moves from $(\mathbf{r}_1, \mathrm{v}_1)$ to $(\mathbf{r}_1 + \mathrm{v}_1 \Delta t, \mathrm{v}_1 + \mathbf{a}_1 \Delta t)$, while (2) moves from $(\mathbf{r}_2, \mathrm{v}_2)$ to $(\mathbf{r}_2 + \mathrm{v}_2 \Delta t, \mathrm{v}_2 + \mathbf{a}_2 \Delta t)$, etc. ($\mathbf{a}_j$ represents the acceleration of electron j owing to all the forces acting upon it.) Now, we have followed the exact particle orbits; therefore, the probability of finding the new state at $t + \Delta t$ must be the same as the probability of finding the old state at t or

$$\begin{aligned} &F^{(\mathrm{N})}(r_1 + \mathrm{v}_1 \Delta t; \mathrm{v}_1 + a_1 \Delta t; \ldots; \mathrm{v}_1 + a_1 \Delta t; t + \Delta t) d\mathbf{r}'_1, d\mathrm{v}'_1, \ldots, d\mathrm{v}'_\mathrm{N} \\ &\quad = f^{(\mathrm{N})}(\mathbf{r}_1, \mathrm{v}_1, \ldots, \mathrm{v}_\mathrm{N}; t) d\mathbf{r}_1, d\mathrm{v}_1, \ldots, d\mathrm{v}_\mathrm{N}. \end{aligned}$$

We now expand $f^{(\mathrm{N})}\,(t + \Delta t)$ about the value at the time t, and in the limit $\Delta t \to 0$, we obtain the Liouville equation[3]:

$$\frac{\partial f^{(\mathrm{N})}}{\partial t} + \sum_{j=1}^{\mathrm{N}} \left(\mathbf{v}_j \cdot \frac{\partial f^{(\mathrm{N})}}{\partial \mathbf{r}_j} + \mathbf{a}_j \cdot \frac{\partial f^{(\mathrm{N})}}{\partial \mathbf{v}_j} \right) = 0. \tag{B.5.7}$$

A similar equation can be written for the N/Z ions. The equations for each species involve through the acceleration $\mathbf{a}_j$, all the multiparticle interactions, and therefore, they cannot be solved.

In the plasma case, we have the bare long range Coulomb force applying, and at first sight, it might appear that each charge would interact significantly with each other charge. In fact, as we have seen in Section B.2, the particles group locally to restrict the effective range of each other's influence, and only the charges within about a Debye length λ_{De} of a given charge are affected by it. Conversely, a given charge forms part of the Debye shield for each of the charges which are within about a Debye length of it. The critical point in the treatment of these collective interactions is the assumption that the level of the individual two-particle interactions (correlations) is small. Therefore, we do not have to allow for the interaction of charges j and g while computing the interaction

[3]The abbreviation $\mathbf{r}_j$, v_j for $\mathbf{r}_j \to \mathbf{r}_j + d\mathbf{r}_j$ and $\mathrm{v}_j \to \mathrm{v}_j + d\mathrm{v}_j$ is used here.

of charges j and m. That is, three-particle or higher correlations play a negligible role.

In addition, we argue that the timescale for an interaction is $\sim 1/\omega_{pe}$, and this is a much shorter time than the time for the average distribution function to change significantly. In comparing the plasma with a neutral gas, the analogy to the group of charges which form a molecule is a charge, and its shield and the assumption is therefore that these shielded (or dressed) charges are uncorrelated. The tricky part is that any given charge is a part of many of these quasi-molecules.

B.5.4 The S Particle Distribution Function

The dual nature of the plasma interaction makes it a difficult case to treat. In some respects, it behaves somewhat like a neutral gas with the charges colliding like billiard balls. In other respects, it behaves like a jelly; when one charge moves, all the neighbors respond.

For both types of interaction, however, we can make arguments for neglecting the three or more particle correlations. We therefore look for equations that describe the evolution of some restricted probability function. The most drastic approximation is to neglect all interparticles correlations; in this case, our interest centers on the one-particle distribution function. This function gives the probability that a particle (1) is at $\mathbf{r}_1 \rightarrow \mathbf{r}_1 + d\mathbf{r}_1$ with velocity $\mathrm{v}_1 \rightarrow \mathrm{v}_1 + d\mathrm{v}_1$ at time t irrespective of the whereabouts of the other N-1 particles. It is obtained by simply summing up all the arrangements of the N particles in which this situation is met. It is denoted by

$$\frac{f^{(1)}(\mathbf{r}_1, \mathrm{v}_1, t)}{V} \equiv \int f^{(N)} d\mathbf{r}_2 \, d\mathrm{v}_2 \, d\mathbf{r}_2 \cdots d\mathbf{r}_N \, d\mathrm{v}_2. \tag{B.5.8}$$

More generally, the probability that particles (1)–(s) have the respective positions $\mathbf{r}_1$, v_1; $\mathbf{r}_2$, v_2; ...; $\mathbf{r}_s$, v_s at t irrespective of the whereabouts of the other $s+1 \rightarrow N$ particles is written as

$$\frac{f^{(s)}(\mathbf{r}_1, \mathrm{v}_1, \ldots, \mathbf{r}_s, \mathrm{v}_s, t)}{V^{(s)}} = \int f_N^{(N)} d\mathbf{r}_{s+1}, d\mathrm{v}_{s+1}, \ldots, d\mathrm{v}_N. \tag{B.5.9}$$

The form of the definition with volume dependence explicitly written is a convention.

B.5.5 The Scattered Power Spectrum and the Distribution Function

At this point, it is important to make the connection between these probability distribution functions and the scattered power Eqn (B.5.1). The ensemble average value of Eqn (B.5.1) may be divided into the terms for which $j = g$ and $j \neq g$. Now, the N electrons are indistinguishable one from another, $N \gg 1$, and for a homogeneous stationary system, we obtain[4]

$$\left\langle \frac{\overline{dP}}{d\Omega} \right\rangle = \frac{cR^2}{4\pi} \left|\overline{E}_j\right|^2_{j=g} + \frac{cR^2}{4\pi} N^2 \left|\overline{E_j(\mathbf{r}_j,\, t),\, E_g(\mathbf{r}_g,\, t)}\right|_{j\neq g}. \tag{B.5.10}$$

(a) The first term gives the scattered power we would obtain from N randomly distributed electrons. As we know from the discussions in Chapters 1–3, this power is spread over a band of frequencies set about the incident frequency. The frequency spread is the result of the Doppler shift induced by the motion of the electrons. The spectrum reflects the velocity distribution of the electron, i.e., the one-particle distribution function $f^{(1)}(\mathbf{r},\, \mathrm{v},\, t)$. For a homogeneous plasma, there is of course no $\mathbf{r}$ dependence; for a stationary plasma in equilibrium, the distribution is Maxwellian

$$f^{(1)}(\mathrm{v}) = \exp(-m\mathrm{v}^2/\kappa T)/(2\pi\kappa T/m)^{1/2}. \tag{B.5.11}$$

If the system is not in equilibrium, then as a result of interparticle interactions $f^{(1)}$ will change with time.

(b) The second term includes all the contributions to the scattered spectrum, which occur because the charges are not randomly distributed. The form of the spectrum from this term depends on the joint probability that one electron is at $\mathbf{r}_j$, v_j when another is at $\mathbf{r}_g$, v_g, i.e., it depends on the two-particle distribution function. We drop the labels j and g and denote this by $f^{(2)}(\mathbf{r},\, \mathrm{v},\, \mathbf{r}',\, \mathrm{v}',\, t)$. The situation is somewhat analogous to the Bragg scattering from the regular array of electrons in a crystal lattice. The individual scattered components are in phase in certain directions and cancel in others. If we crush the crystal, laying aside the probable loss of transmission, we would eliminate the correlations between electron positions and consequently lose this coherence in the scattering and obtain an noncollective spectrum.

There are a number of ways of establishing and solving equations for $f^{(1)}$ and $f^{(2)}$. Two of these methods have particular significance in the development of the theory of scattering from nonequilibrium plasmas. The BBGKY hierarchy of equations is discussed briefly first, and then, a more detailed discussion is given of the Klimontovich hierarchy of equations. The latter are the basis of the derivations presented in this book.

B.6 THE BBGKY HIERARCHY

An equation that describes the evolution of $f^{(1)}$ is obtained by integrating the Liouville equation (A.4.7) over the coordinates of the other $N-1$ charges.

[4]It must again be emphasized that by "stationary" we mean that the system is evolving on a timescale slow compared to both the characteristic time of microscopic interactions in the plasma $1/\omega_{\mathrm{pe}}$ and the period of the incident radiation $1/\omega_{\mathrm{i}}$. It is not a restriction to no time changes at all. Equally, inhomogeneities are permitted provided that they do not lead to excessively rapid changes in the system.

Let the acceleration be divided into the part due to external forces $\mathbf{a}_{\text{ext}}$ and the part due to internal forces $\mathbf{a}_{\text{int}}$. For N identical charges, we obtain ($N \gg 1$)

$$\frac{\partial f^{(1)}}{\partial t} + \text{v} \cdot \frac{\partial f(1)}{\partial \mathbf{r}} + \mathbf{a}_{\text{ext}} \cdot \frac{\partial f^{(1)}}{\partial \text{v}} + N \frac{\partial}{\partial \text{v}} \cdot \int \mathbf{a}_{\text{int}} f^{(2)}(\mathbf{r}, \text{v}, \mathbf{r}', \text{v}', t) d\mathbf{r}' d\text{v}' = 0. \tag{B.6.1}$$

$\mathbf{a}_{\text{int}}$ is the acceleration of a charge at $\mathbf{r}, \mathbf{v}$ owing to the presence of another charge at $\mathbf{r}', \mathbf{v}'$. The extension to more than one species is reasonably clear. The important point to note is that $f^{(1)}$ depends on $f^{(2)}$. Similarly, $f^{(2)}$ depends on $f^{(3)}$.

$$\frac{\partial f^{(2)}}{\partial t} + \sum_{\alpha=1}^{2} \left(\text{v}_\alpha \cdot \frac{\partial f^{(2)}}{\partial \mathbf{r}_\alpha} + \mathbf{a}_{\alpha\text{ext}} \cdot \frac{\partial f^{(2)}}{\partial_{\text{v}\alpha}} + \sum_{\gamma=1,\gamma\neq\alpha}^{2} \mathbf{a}_{\alpha\gamma} \cdot \frac{\partial f^{(2)}}{\partial_{\text{v}\alpha}} \right)$$
$$+ \sum_{\alpha=1}^{2} \sum_{\gamma=3,\gamma\neq\alpha}^{N} \frac{\partial}{\partial_{\text{v}\alpha}} \cdot \int \mathbf{a}_{\alpha\gamma} f^{(3)} d\mathbf{r}_\gamma d\text{v}_\gamma = 0. \tag{B.6.2}$$

To truncate the hierarchy of equations $\left(f^{(\text{s})}\right.$ depends on $\left.f^{(\text{s}+1)}\right)$, we must stop at some level. This requires that we justify writing some $f^{(\text{s}+1)}$ in terms of $f^{(\text{s})}$, solve that equation, and then work back down the hierarchy to $f^{(1)}$.

B.6.1 The Vlasov Equation

The most drastic approximation we can make is to assume that the probabilities of one charge species α being at $\mathbf{r}, \mathbf{v}$ when another of species β is at $\mathbf{r}', \mathbf{v}'$ are independent, i.e.,

$$f_{\alpha\beta}^{(2)}(\mathbf{r}, \text{v}, \mathbf{r}', \text{v}'; t) = f_\alpha^{(1)}(\mathbf{r}, \text{v}; t) f_\beta^{(1)}(\mathbf{r}', \mathbf{v}'; t). \tag{B.6.3}$$

This is substituted in Eqn (B.6.1), and for the case of N electrons and N/Z ions of charge Ze, we obtain for the electrons, for example,

$$\frac{\partial f_\text{e}^{(1)}}{\partial t} + \text{v} \cdot \frac{\partial f_\text{e}^{(1)}}{\partial \mathbf{r}} + \mathbf{a}_{\text{ext}} \cdot \frac{\partial f_\text{e}^{(1)}}{\partial \text{v}} - \frac{e}{m} E \cdot \frac{\partial f_\text{e}^{(1)}}{\partial \text{v}} = 0, \tag{B.6.4}$$

where the microscopic electric field is given by Poisson's equation as

$$\nabla \cdot \mathbf{E} = 4\pi \int d\text{v} \left[Zef_\text{i}^{(1)} - ef_\text{e}^{(1)} \right]. \tag{B.6.5}$$

There is a similar equation to Eqn (B.6.4) for the ions.

The limitations of these equations become clear when we consider their application to a homogeneous system so that $\nabla \cdot \mathbf{E} = 0$, but $T_\text{e} \neq T_\text{i}$; therefore, it is not in equilibrium. Intuitively, it is clear that by interactions (collisions of electrons and ions) energy will be transferred to bring T_e and T_i together, but equally it is clear that these equations contain no term that can accomplish this. The equations are therefore only valid for times much less than the equilibration time, and to first determine this time, we must include correlations.

B.6.2 Two-Particle Correlations

The effect of two-particle correlations may be included in a perfectly general manner by setting

$$f_{\alpha\beta}^{(2)}(\mathbf{r},\mathrm{v},\mathbf{r}',\mathrm{v}',t)=f_{\alpha}^{(1)}(\mathbf{r},\mathrm{v},t)f_{\beta}^{(1)}(\mathbf{r}',\mathbf{v}',t)+g_{\alpha\beta}(\mathbf{r},\mathrm{v},\mathbf{r}',\mathrm{v}',t). \tag{B.6.6}$$

In order to calculate $g_{\alpha\beta}$, we must solve Eqn (B.6.3), but this involves $f^{(3)}$. We set

$$\begin{aligned} f_{\alpha\beta\gamma}^{(3)}(\mathbf{r},\mathrm{v},\mathbf{r}',\mathrm{v}',\mathbf{r}'',\mathrm{v}'';t) &= f_{\alpha}^{(1)}(\mathbf{r},\mathrm{v},t)f_{\beta}^{(1)}(\mathbf{r}',\mathbf{v}',t)f_{\gamma}^{(1)}(\mathbf{r}'',\mathrm{v}'',t) \\ &\quad +f_{\alpha}^{(1)}g_{\beta\gamma}+f_{\beta}^{(1)}g_{\alpha\gamma}+f_{\gamma}^{(1)}g_{\alpha\beta}+g_{\alpha\beta\gamma}. \end{aligned} \tag{B.6.7}$$

A closed equation for $f^{(2)}$ or $g_{\alpha\beta}$ is obtained by neglecting the triple correlation function $g_{\alpha\beta\gamma}$. This procedure is justified on the basis of the form of these correlation functions in the equilibrium case with the assumption that $g_{\alpha\beta}$ relaxes to the steady-state value in $\sim 1\omega_{\mathrm{pe}}$, which is fast compared to changes in $f^{(1)}$.

It can be shown (Section B.8.1) that for an equilibrium plasma, this approximation yields

$$g_{\mathrm{ee}}(r,\mathrm{v},r',\mathrm{v}',t)=-\frac{e^2}{\kappa T_{\mathrm{e}}}\frac{\exp\left(-\left|\mathbf{r}-\mathbf{r}'\right|/\lambda_{\mathrm{De}}\right)}{|\mathbf{r}-\mathbf{r}'|}f_{\mathrm{e}}^{(1)}(\mathrm{v},t)f_{\mathrm{e}}^{(1)}(\mathrm{v}',t) \tag{B.6.8}$$

and

$$\frac{e^2}{\kappa T_{\mathrm{e}}\,|\mathbf{r}-\mathbf{r}'|}=\frac{\lambda_{\mathrm{De}}}{|\mathbf{r}-\mathbf{r}'|}\frac{1}{4\pi n_0\lambda_{\mathrm{De}}^3};$$

therefore, $g_{\mathrm{ee}} \propto 1/n_0\lambda^3$, and similarly $g_{\mathrm{eee}} \propto 1/(n_0\lambda_{\mathrm{De}}^3)^2$. Now, a plasma is customarily defined as a state for which $n_0\lambda^3 \gg 1$; therefore, this expansion procedure makes sense.[5]

It must be remembered that for short distances,

$$\left|\mathbf{r}-\mathbf{r}'\right| \lesssim e^2/\kappa T_{\mathrm{e}} = \lambda_{\mathrm{De}}/4\pi n_0\lambda_{\mathrm{De}}^3;$$

the calculation will break down, and then, the Boltzmann approach is appropriate.

In this book, an alternative formulation, proposed by Klimontovich, is used. In this approach, the ensemble averaging of the microscopic fluctuations is postponed to the end of the calculation, and this helps a little to make the physics clearer.

---

[5]In fact, the neglect of $g_{\alpha\beta\gamma}$ appears to be satisfactory even for $n_0\lambda_{\mathrm{De}}^3 \cong 1$, and for scattered spectra in this situation are observed to agree with the theory which was based on this procedure (Boyd *et al.*, 1966; Theimer, 1966; Röhr, 1968; Kato, 1972).

B.7 THE KLIMONTOVICH HIERARCHY[6]

The starting point is the microscopic distribution function, which is defined for each species q as

$$F_q(\mathbf{r}, \mathrm{v}, t) = \sum_{j=1}^{N_q} \delta(\mathbf{r} - \mathbf{r}_j(t))\delta(\mathrm{v} - \mathrm{v}_j(t)). \tag{B.7.1}$$

This is not a probability distribution function because $\mathbf{r}_j(t)$ and $\mathrm{v}_j(t)$ are the exact position and velocity of the particle j; $F_q(\mathbf{r}, \mathrm{v}, t)d\mathbf{r}, d\mathrm{v}$ is the number of particles of type q, which are at $\mathbf{r} \to \mathbf{r} + d\mathbf{r}$ and have the velocity $\mathrm{v} \to \mathrm{v} + d\mathrm{v}$ at time t; $\int d\mathbf{r}d\mathrm{v}F_q(\mathbf{r}, \mathrm{v}, t) = N_q$ the total number of type q particles in the system; $\int d\mathrm{v}F_q(\mathbf{r}, \mathrm{v}, \mathbf{t}) = n(\mathbf{r}, t)$ the exact number density. Now,[7]

$$\begin{aligned}\frac{d}{dt}F_q(\mathbf{r}, \mathrm{v}, t) = \frac{\partial F_q}{\partial t} &= -\sum_{j=1}^{N_q} \mathrm{v}_j \cdot \frac{\partial}{\partial \mathbf{r}}\delta(\mathbf{r} - \mathbf{r}_j)\delta(\mathrm{v} - \mathrm{v}_j) \\ &\quad - \sum_j \mathbf{a}_j \cdot \frac{\partial}{\partial \mathrm{v}}\delta(\mathbf{r} - \mathbf{r}_j)\delta(\mathrm{v} - \mathrm{v}_j) = \mathrm{v} \cdot \frac{\partial F_q}{\partial \mathbf{r}} - \mathbf{a} \cdot \frac{\partial F_q}{\partial \mathrm{v}}\end{aligned}$$

or

$$\frac{\partial F_q}{\partial t} + \mathrm{v} \cdot \frac{\partial F_q}{\partial \mathbf{r}} + \mathbf{a} \cdot \frac{\partial F_q}{\partial \mathrm{v}} = 0. \tag{B.7.2}$$

The acceleration $\mathbf{a}$ may be due to external or internal forces. For a plasma, it is reasonable to set

$$\mathbf{a} = \frac{q}{m}\left(\mathbf{E} + \frac{\mathrm{v}}{c} \times \mathbf{B}\right). \tag{B.7.3}$$

Let $\mathbf{E}$ and $\mathbf{B}$ be the fields generated by the plasma; then,

$$\begin{aligned}&\nabla \cdot \mathbf{E}_\mathrm{p} = 4\pi \sum_q q \int d\mathrm{v}F_q, \quad \nabla \times \mathbf{E}_\mathrm{p} = -\frac{1}{c}\frac{\partial \mathbf{B}_\mathrm{p}}{\partial t} \\ &\nabla \cdot \mathbf{B}_\mathrm{p} = 0, \quad \nabla \times \mathbf{B}_\mathrm{p} = \frac{1}{c}\frac{\partial \mathbf{E}_\mathrm{p}}{\partial t} + \frac{4\pi}{c}\sum_q q \int d\mathrm{v}\mathrm{v}F_q,\end{aligned} \tag{B.7.4}$$

where $m = m_\mathrm{e}, q = -e$ for electrons; $m = m_\mathrm{i}, q = Ze$ for positive ions.

From now on the discussion is restricted to a plasma which is

(1) nonrelativistic, therefore the effect of the internal magnetic field is negligible, $B_\mathrm{p} = 0$;
(2) homogeneous.

---

[6] Klimontovich (1967).

[7] Note:

$$\frac{\partial}{\partial t}[\mathbf{r} - \mathbf{r}_j(t)] = \frac{\partial \mathbf{r}_j}{\partial t} \cdot \frac{\partial}{\partial \mathbf{r}_j}\delta(\mathbf{r} - \mathbf{r}_j) = -\mathrm{v}_j \cdot \frac{\partial}{\partial \mathbf{r}}\delta(\mathbf{r} - \mathbf{r}_j).$$

Even under these circumstances, we cannot solve the Klimontovich Eqn (B.7.2) because the distribution function F_q follows the detailed motion of all the charges, and all the multiparticle interactions (correlations) are contained in **a**. To deal with this problem, we appeal to two arguments.

(1) The system as a whole will change on a much slower time than the time of microscopic fluctuations. We therefore set

$$F_q = F_{0q} + F_{1q}, \tag{B.7.5}$$

where $F_{0q} = \langle F_q \rangle$, the ensemble average distribution function. It is this function that we expect to vary on a slower timescale than F_1, which represents the local microscopic fluctuations about this average state.

(2) From this expanded distribution function, we form a hierarchy of equations, where each successive equation brings in a higher order correlation. We terminate this hierarchy by using the argument, based on the equilibrium result, that the two-particle correlations will be dominant.

The discussion is restricted to quasi-neutral systems, i.e., $N_e e = N_i Ze$. In addition, there is no applied electric field, and any applied magnetic field $\mathbf{B}_0$ is constant and uniform. The ensemble-averaged internal electric field is given by

$$\nabla \cdot \mathbf{E}_0 = 4\pi \sum_q q \int d\mathrm{v}(F_{0q}). \tag{B.7.6}$$

The microscopic fluctuating electric field is given by

$$\nabla \cdot \mathbf{E}_1 = 4\pi \sum_q q \int d\mathrm{v}(F_{1q}). \tag{B.7.7}$$

The quantities defined by Eqns (B.7.3), (B.7.5)–(B.7.7) are substituted into Eqn (B.7.2), and we obtain

$$\begin{aligned} &\frac{\partial F_{0q}}{\partial t} + \mathrm{v} \cdot \frac{\partial F_{0q}}{\partial \mathbf{r}} + \frac{q}{m}\left(\frac{\mathrm{v}}{c} \times \mathbf{B}_0\right) \cdot \frac{\partial F_{0q}}{\partial \mathrm{v}} + \frac{q}{m}\mathbf{E}_0 \cdot \frac{\partial F_{0q}}{\partial \mathrm{v}} + \frac{q}{m}\mathbf{E}_1 \cdot \frac{\partial F_{0q}}{\partial \mathrm{v}} \\ &+ \frac{\partial F_{1q}}{\partial t} + \mathrm{v} \cdot \frac{\partial F_{1q}}{\partial \mathbf{r}} + \frac{q}{m}\left(\frac{\mathrm{v}}{c} \times \mathbf{B}_0\right) \cdot \frac{\partial F_{1q}}{\partial \mathrm{v}} + \frac{q}{m}\mathbf{E}_0 \cdot \frac{\partial F_{1q}}{\partial \mathrm{v}} + \frac{q}{m}\mathbf{E}_1 \cdot \frac{\partial F_{1q}}{\partial \mathrm{v}} = 0. \end{aligned} \tag{B.7.8}$$

The beginnings of a hierarchy are obtained by taking an ensemble average of Eqn (B.7.8). (Note that by definition, $\langle F_{1q} \rangle = \langle \mathbf{E}_{1q} \rangle = 0$.) A second equation is obtained by men substituting this ensemble average from Eqn (B.7.8). The resultant equations are

$$\frac{\partial F_{0q}}{\partial t} + \mathrm{v} \cdot \frac{\partial F_{0q}}{\partial \mathbf{r}} + \frac{q}{m}\left(\frac{\mathrm{v}}{c} \times \mathbf{B}_0\right) \cdot \frac{\partial F_{0q}}{\partial \mathrm{v}} + \frac{q}{m}\mathbf{E}_0 \cdot \frac{\partial F_{0q}}{\partial \mathrm{v}} + \frac{q}{m}\left\langle \mathbf{E}_1 \cdot \frac{\partial F_1}{\partial \mathrm{v}} \right\rangle = 0 \tag{B.7.9}$$

and

$$\frac{\partial F_{1q}}{\partial t}+\mathrm{v}\cdot\frac{\partial F_{1q}}{\partial \mathbf{r}}+\frac{q}{m}\cdot\left(\frac{\mathrm{v}}{c}\times\mathbf{B}_0\right)\cdot\frac{\partial F_{1q}}{\partial \mathrm{v}}+\frac{q}{m}\mathbf{E}_0\cdot\frac{\partial F_{1q}}{\partial \mathrm{v}}+\frac{q}{m}\mathbf{E}_1\cdot\frac{\partial F_{0q}}{\partial \mathrm{v}}$$
$$+\frac{q}{m}\mathbf{E}_1\cdot\frac{\partial F_{1q}}{\partial \mathrm{v}}-\frac{q}{m}\left\langle\mathbf{E}_1\cdot\frac{\partial F_{1q}}{\partial \mathrm{v}}\right\rangle=0. \quad \text{(B.7.10)}$$

B.7.1 The Ensemble: Averaged Distribution Function

From the definition of ensemble average Eqn (B.5.5), with the aid of Eqns (B.5.6) and (B.5.8), we see that for N identical particles,

$$\begin{aligned}\langle F\rangle \equiv F_0 &= \int\sum_{j=1}^{N}\delta(\mathbf{r}-\mathbf{r}_j)\delta(\mathrm{v}-\mathrm{v}_j)f^{(N)}d\mathbf{r}_1 d\mathrm{v}_1\cdots d\mathrm{v}_N\\ &= N\int\delta(\mathbf{r}-\mathbf{r}_1)\delta(\mathrm{v}-\mathrm{v}_1)d\mathbf{r}_1 d\mathrm{v}_1 f^{(N)}d\mathbf{r}_2 d\mathrm{v}_2\cdots d\mathrm{v}_N\\ &= N\int\delta(\mathbf{r}-\mathbf{r}_1)\delta(\mathrm{v}-\mathrm{v}_1)[f^{(1)}/(\mathbf{r}_1,\mathrm{v}_1,t)/V]\end{aligned}$$

or

$$F_0 = n_0 f_0(\mathbf{r},\mathrm{v},t), \quad \text{(B.7.11)}$$

where $n_0 = N/V$, and to avoid confusion with F_1, and because for identical particles $n_0 f^{(1)}(\mathbf{r},\mathrm{v},t)$ is the probable number of particles at $(\mathbf{r},\mathrm{v})$, we rewrite $f^{(1)} \equiv f_0$.

B.7.2 The Ensemble Average $\langle F_1(x)F_1(x')\rangle$

$$\begin{aligned}&\langle F(\mathbf{r},\mathrm{v},t)F(\mathbf{r}',\mathrm{v}',t)\rangle\\ &\quad = F_0(\mathbf{r},\mathrm{v},t)F_0(\mathbf{r}',\mathrm{v}',t)+\langle F_1(\mathbf{r},\mathrm{v},t)F_1(\mathbf{r}',\mathrm{v}',t)\rangle\\ &\quad = \int\sum_{j=1}^{N}\delta(\mathbf{r}-\mathbf{r}_j)\delta(\mathrm{v}-\mathrm{v}_j)\sum_{g=1}^{N}\delta(\mathbf{r}'-\mathbf{r}_g)\delta(\mathrm{v}'-\mathrm{v}_g)f^N d\mathbf{r}_1\cdots d\mathrm{v}_N\\ &\quad = n_0 f_0(\mathbf{r},\mathrm{v},t)\delta(\mathbf{r}-\mathbf{r}')\delta(\mathrm{v}-\mathrm{v}')+N(N-1)\int_{j\neq g}\delta(\mathbf{r}-\mathbf{r}_j)\delta(\mathrm{v}-\mathrm{v}_j)\\ &\qquad\times\delta(\mathbf{r}'-\mathbf{r}_g)\delta(\mathrm{v}'-\mathrm{v}_g)d\mathbf{r}_j d\mathrm{v}_j d\mathbf{r}_g d\mathrm{v}_g f^{(2)}(\mathbf{r}_j,\mathrm{v}_j,\mathbf{r}_g,\mathrm{v}_g,t)\\ &\quad = n_0 f_0(\mathbf{r},\mathrm{v},t)\delta(\mathbf{r}-\mathbf{r}')\delta(\mathrm{v}-\mathrm{v}')+n_0^2 f^{(2)}(\mathbf{r},\mathrm{v},\mathbf{r}',\mathrm{v}',t).\end{aligned} \quad \text{(B.7.12)}$$

With substitutions from Eqn (B.6.6) and (B.7.11), we see that

$$\langle F_1(\mathbf{r},\mathrm{v},t)F_1(\mathbf{r}',\mathrm{v}',t)\rangle = n_0^2 g(\mathbf{r},\mathrm{v},\mathbf{r}',\mathrm{v}',t)+n_0 f_0\delta(\mathbf{r}-\mathbf{r}')\delta(\mathrm{v}-\mathrm{v}'). \quad \text{(B.7.13)}$$

The first term is proportional to the two-particle correlation function, and the second term is simply a self-correlation term. The latter term gives no contribution to $\langle F_1(\mathbf{r}, \mathrm{v}, t)E_1(\mathbf{r}', \mathrm{v}', t)\rangle$ because a given charge does not interact with its own field.

In a similar manner, we find that $\langle F_1, F_1, F_1\rangle$ is proportional to the three-particle correlation function.

In order to evaluate F_0, we must evaluate $\langle \mathbf{E}_1 \cdot \partial F_{1q}/\partial \mathrm{v}\rangle$, that is, a term of the form $\langle F_{1q}(x)F_{1q}(x')\rangle$. Such terms involve two-particle correlations. To evaluate them, we could multiply (B.7.10) by F_{1q} and take an ensemble average. This brings in terms of the form $\langle F_{1q}(x)F_{1q}(x')F_{1q}(x'')\rangle$, and these involve three-particle correlations. The analogy to the BBGKY hierarchy is obvious. Again, we terminate the hierarchy at this level by dropping the three-particle correlation terms. We are then able to solve Eqn (B.7.10) for F_{1q} in terms of F_{0q}, which is substituted into Eqns (B.7.7) and (B.7.9). The result is a closed equation for F_{0q}, the "Lenard–Balescu" equation. In addition, we are able to evaluate the form of $\langle F_{1q}F_{1q}\rangle$ and can therefore determine the conditions under which the approximations are valid.

B.8 STABLE, HOMOGENEOUS, QUASI-STATIONARY PLASMAS

B.8.1 Two-Particle Correlation Function

For a homogeneous plasma, a quantity such as $\langle A(\mathbf{r})B(\mathbf{r})\rangle$ is a function only of $\mathbf{r}-\mathbf{r}'$; therefore,

$$\langle F_{1q}(\mathbf{r}, \mathrm{v}, t)F_{1q}(\mathbf{r}', \mathbf{v}', t)\rangle = \int \frac{d\mathbf{k}}{(2\pi)^3}\langle F_{1q}(\mathbf{k}, \mathrm{v}, t)F_{1q}(-\mathbf{k}, \mathbf{v}', t)\rangle e^{i\mathbf{k}\cdot(\mathbf{r}-\mathbf{r}')}, \tag{B.8.1}$$

and using Eqn (A.3.7), we may rewrite this for a quasi-stationary plasma as

$$\langle F_{1q}, F_{1q}\rangle = \mathrm{Re}\int \frac{d\mathbf{k}}{(2\pi)^3} \lim_{\gamma\to 0}\frac{\gamma}{\pi}\int_{-i\gamma-\infty}^{-i\gamma+\infty} d\omega \left\langle F_{1q}(\mathbf{k}, \mathrm{v}, \omega)F^*_{1q}(\mathbf{k}, \mathbf{v}', \omega)\right\rangle. \tag{B.8.2}$$

The Fourier–Laplace transforms of F_1 and other pertinent quantities are determined in Chapter 5. Therefore, we can use these results.

$$F_{1q}(\mathbf{k}, \mathrm{v}, \omega) = -\left[\frac{iF_{1q}(\mathbf{k}, \mathrm{v}, 0) + (4\pi q/mk^2)\rho_1(\mathbf{k}, \omega)\mathbf{k}\cdot \partial F_{0q}/\partial \mathrm{v})}{\omega - \mathbf{k}\cdot \mathrm{v} - i\gamma}\right] \tag{B.8.3}$$

$$\rho_1(\mathbf{k}, \omega) = \frac{iZ_e}{\varepsilon(\mathbf{k}, \omega)}\sum_{l=1}^{N/Z}\frac{e^{-i\mathbf{k}\cdot\mathbf{r}_l(0)}}{\omega - \mathbf{k}\cdot \mathrm{v}_l(0) - i\gamma} - \frac{ie}{\varepsilon(\mathbf{k}, \omega)}\sum_{j=1}^{N}\frac{e^{-i\mathbf{k}\cdot\mathbf{r}_j(0)}}{\omega - \mathbf{k}\cdot \mathrm{v}_j(0) - i\gamma} \tag{B.8.4}$$

$$F_{1q}(\mathbf{k}, \mathrm{v}, 0) = \sum_{j=1}^{N_q} \delta\left(\mathrm{v} - \mathrm{v}_j(0)\right) e^{-i\mathbf{k}\cdot\mathbf{r}_j(0)} \tag{B.8.5}$$

$$E_1(\mathbf{k}, \omega) = -(4\pi i\mathbf{k}/k^2)\rho_1(\mathbf{k}, \omega), \tag{B.8.6}$$

where we consider a plasma of volume V containing N electrons and N/Z ions of charge Ze, $n_0 = N/V$; $F_{0e} = n_0 f_{0e}$.

For the calculation of the two-particle correlation function, we will only consider the electron–electron term and therefore drop the ion contribution in Eqns (B.8.3)–(B.8.5). We consider stable plasma, so $\varepsilon \neq 0$ in the upper half of the complex ω plane.

We substitute in Eqn (B.8.2) and find if initially the charges are uncorrelated (see Section 3.6.1) that

$$\begin{aligned}
&\left\langle F_{1e}(\mathbf{k}, \mathrm{v}, \omega) F_{1e}^*(\mathbf{k}, \mathbf{v}', \omega)\right\rangle \\
&= \frac{\pi}{\gamma} n_0 \delta(\omega - \mathbf{k}\cdot\mathrm{v})\delta(\mathrm{v} - \mathbf{v}') f_{0e}(\mathrm{v}) \\
&+ \left\langle \sum_j \sum_s \frac{e^{i\mathbf{k}\cdot\left(\mathbf{r}_j(0) - \mathbf{r}_s(0)\right)}}{(\omega - \mathbf{k}\cdot\mathrm{v} - i\gamma)(\omega - \mathbf{k}\cdot\mathbf{v}' + i\gamma)} \frac{4\pi e^2}{mk^2} \left\{ \frac{\delta(\mathrm{v} - \mathrm{v}_j(0))\mathbf{k}\cdot\partial F_{0e}/\partial\mathbf{v}'}{[\omega - \mathbf{k}\cdot\mathbf{v}'_s(0) + i\gamma]\varepsilon^*} \right.\right. \\
&\left.\left. + \frac{\delta(\mathbf{v}' - \mathrm{v}_s(0))\mathbf{k}\cdot\partial F_{0e}/\partial\mathrm{v}}{\left[\omega - \mathbf{k}\cdot\mathrm{v}_j(0) - i\gamma\right]\varepsilon} \right\} \right\rangle + \text{product of shielding terms.}
\end{aligned} \tag{B.8.7}$$

Now, the expansion in a hierarchy requires that the correlation function g_{ee} contributes only a small correction to $\left(f^{(1)} f^{(1)}\right)$; the last term that represents higher order correlations is therefore dropped.

In the integration over position in the ensemble average, only the N terms with $j = s$ survive. Reference to Eqn (B.5.4) shows that a factor V will appear underneath, and in total, this leads to a factor n_0. Further, we then have $\mathrm{v}_s(0) = \mathrm{v}_j(0) \equiv \mathrm{v}(0)$, and the integrals weighted by the initial velocity distributions $f_{0e}(\mathrm{v})$ and $f_{0e}(\mathbf{v}')$ are readily obtained because of the delta function so that

$$\begin{aligned}
\left\langle F_{1e}(\mathbf{k}, \mathrm{v}, \omega) F_{1e}^*(\mathbf{k}, \mathbf{v}', \omega)\right\rangle &= \frac{\pi}{\gamma} n_0 \delta(\omega - \mathbf{k}\cdot\mathrm{v})\delta(\mathrm{v} - \mathbf{v}') f_{0e} \\
&+ \frac{4\pi e^2}{mk^2} n_0^2 \left\{ \frac{f_{0e}(\mathrm{v})\mathbf{k}\cdot\partial f_{0e}/\partial\mathbf{v}'}{(\omega - \mathbf{k}\cdot\mathbf{v}' + i\gamma)\varepsilon^*\left[(\omega - \mathbf{k}\cdot\mathrm{v})^2 + \gamma^2\right]} \right. \\
&\left. + \frac{f_{0e}(\mathbf{v}')\mathbf{k}\cdot\partial f_{0e}/\partial\mathrm{v}}{(\omega - \mathbf{k}\cdot\mathrm{v} - i\gamma)\varepsilon\left[(\omega - \mathbf{k}\cdot\mathrm{v})^2 + \gamma^2\right]} \right\}.
\end{aligned} \tag{B.8.8}$$

In the integral over ω, in the limit $\gamma \to 0$, only the contributions from the poles at $\omega = \mathbf{k}\cdot \mathrm{v}$ and $\omega = \mathbf{k}\cdot \mathbf{v}'$ survive. This is seen if we rewrite

$$\frac{1}{(\omega - \mathbf{k}\cdot \mathrm{v})^2 + \gamma^2} = \frac{1}{2i\gamma}\left\{\frac{1}{\omega - \mathbf{k}\cdot \mathrm{v} - i\gamma} - \frac{1}{\omega - \mathbf{k}\cdot \mathrm{v} + i\gamma}\right\}.$$

The principle parts of the two integrals cancel, and we are left with π/γ

$$\begin{aligned}&\langle F_{1\mathrm{e}}(\mathbf{k}, \mathrm{v}, t) F^*_{1\mathrm{e}}(\mathbf{k}, \mathbf{v}', t)\rangle \\ &\quad = n_0 \delta(\mathrm{v} - \mathbf{v}') f_{0\mathrm{e}} + \frac{4\pi e^2 n_0^2}{mk^2}\left\{\frac{f_{0\mathrm{e}}(\mathrm{v})\mathbf{k}\cdot \partial f_{0\mathrm{e}}/\partial \mathbf{v}'}{\varepsilon^*(\mathbf{k}, \mathbf{k}\cdot \mathrm{v})[\mathbf{k}\cdot(\mathrm{v} - \mathbf{v}') + i\gamma])}\right. \\ &\qquad \left. + \frac{f_{0\mathrm{e}}(\mathbf{v}')\mathbf{k}\cdot \partial f_{0\mathrm{e}}/\partial \mathrm{v}}{\varepsilon^*(\mathbf{k}, \mathbf{k}\cdot \mathrm{v})[\mathbf{k}\cdot(\mathbf{v}' - \mathrm{v}) - i\gamma])}\right\}. \qquad \text{(B.8.9)}\end{aligned}$$

For a plasma in equilibrium, we may use a Maxwellian distribution

$$f_{0\mathrm{e}} = \exp(-\mathrm{v}^2/a^2)/(\pi a^2)^{3/2}, \quad a^2 = 2\kappa T_\mathrm{e}/m_\mathrm{e},$$

and then if we also use the static approximation for the dielectric function, $\varepsilon = 1 + (1/k^2 \lambda_{\mathrm{De}}^2)$ (see Section 5.2),

$$\begin{aligned}\langle F_{1\mathrm{e}}(\mathbf{r}, \mathrm{v}, t) F_{1\mathrm{e}}(\mathbf{r}', \mathbf{v}', t)\rangle &\cong \int \frac{d\mathbf{k}}{(2\pi)^3} e^{i\mathbf{k}\cdot(\mathbf{r}-\mathbf{r}')} n_0 \delta(\mathrm{v} - \mathbf{v}') f_{0\mathrm{e}} \\ &\quad - \frac{8\pi e^2}{m_\mathrm{e}} n_0^2 f_{0\mathrm{e}}(\mathrm{v}) f_{0\mathrm{e}}(\mathbf{v}') \int \frac{d\mathbf{k}}{(2\pi)^3} \cdot \frac{e^{i\mathbf{k}\cdot(\mathbf{r}-\mathbf{r}')}}{a^2(k^2 + 1/\lambda_{\mathrm{De}}^2)}. \qquad \text{(B.8.10)}\end{aligned}$$

With the aid of Eqn (A.2.12), we see that the first term reduces to $n_0 \delta(\mathbf{r} - \mathbf{r}')\delta(\mathrm{v} - \mathbf{v}') f_{0\mathrm{e}}$. The second integral is evaluated by setting $\mathbf{k}\cdot(\mathbf{r} - \mathbf{r}') = k_z(r - r')$ and denoting the magnitude of the perpendicular component by $k_\perp$ at the angle φ so that $d\mathbf{k} = d\mathbf{k}_z k_\perp dk_\perp d\varphi$, $k^2 = k_z^2 + k_\perp^2$. Comparison with Eqn (B.7.13) then shows that

$$\begin{aligned} f_{0\mathrm{e}}^2(\mathbf{r}, \mathrm{v}, \mathbf{r}', \mathbf{v}', t) &= f_\mathrm{e}^{(1)}(\mathbf{r}, \mathrm{v}, t) f_\mathrm{e}^{(1)}(\mathbf{r}', \mathbf{v}', t) + g_{\mathrm{cc}} \\ &= f_\mathrm{e}^{(1)}(\mathbf{r}, \mathrm{v}, t) f_\mathrm{e}^{(1)}(\mathbf{r}', \mathbf{v}', t)\left(1 - \frac{e^2}{\kappa T_\mathrm{e}} \frac{e^{-|\mathbf{r}-\mathbf{r}'|/\lambda_{\mathrm{De}}}}{|\mathbf{r} - \mathbf{r}'|}\right). \qquad \text{(B.8.11)}\end{aligned}$$

We see that the condition, that g_{ee} represents a small correction, is met only for short distances such that $\mathbf{r} - \mathbf{r}' > e^2/\kappa T_\mathrm{e}$ or alternatively for wave numbers $k < e^2/\kappa T_\mathrm{e}$.

B.8.2 The Lenard–Balescu Equation

We will evaluate the collision term for the electrons

$$-(e/m_\mathrm{e})\mathbf{E}_1 \cdot \partial F_{1\mathrm{e}}/\partial \mathrm{v}$$

for the simplest case, that of a homogeneous, quasi-stationary (stable) non-relativistic plasma with $B_0 = 0$. In this case, (Eqns B.7.9) and the linearized Eqn (B.7.10) reduce to

$$\frac{\partial F_{0q}}{\partial t} = \frac{e}{m}\left\langle \mathbf{E}_1 \cdot \frac{\partial F_{1e}}{\partial \mathrm{v}} \right\rangle \tag{B.8.12}$$

$$\frac{\partial F_{1e}}{\partial t} + \mathrm{v} \cdot \frac{\partial F_{1e}}{\partial \mathbf{r}} + \frac{q}{m}\mathbf{E}_1 \cdot \frac{\partial F_{0e}}{\partial \mathrm{v}} = 0 \tag{B.8.13}$$

$$\nabla \cdot \mathbf{E}_1 = 4\pi \sum_q q \int d\mathrm{v} F_{1q}. \tag{B.8.14}$$

We may write

$$\frac{e}{m}\left\langle \mathbf{E}_1(\mathbf{r}, t) \cdot \frac{\partial F_{ie}}{\partial \mathrm{v}}(\mathbf{r}, \mathrm{v}, t) \right\rangle$$
$$= \frac{e}{m}\mathrm{Re} \int \frac{d\mathbf{k}}{(2\pi)^3} \mathbf{k} \cdot \frac{\partial}{\partial \mathrm{v}} \left(\lim_{\gamma \to 0} \frac{\gamma}{\pi} \int d\omega \langle E^*(\mathbf{k}, \omega) F_{1e}(\mathbf{k}, \mathrm{v}, \omega) \rangle \right). \tag{B.8.15}$$

We use the Fourier–Laplace transforms of F_{1e} and $\mathbf{E}_1$, which are given in Eqns (B.8.3)–(B.8.6).

(a) We will assume that there are no correlations[8] at $t = 0$. In this situation, when we take an ensemble average over position, we only pick up contributions from the double sums when we are dealing with the same charges; thus,

$$\left\langle \sum_l^{N/Z} e^{+i\mathbf{k}\cdot\mathbf{r}_j(0)} \sum_g^{N/Z} e^{-i\mathbf{k}\cdot\mathbf{r}_g(0)} \right\rangle = N/ZV = n_0/z; \qquad l = g, \text{ etc.}$$
$$= 0; \qquad l \neq g.$$

Further, we may then set

$$\mathrm{v}_j(0) = \mathrm{v}_g(0) = \mathbf{v}'(0), \text{ etc.}$$

(b) In the integration over ω, we pick up the poles at $\omega = \mathbf{k} \cdot \mathbf{v}'(0)$ and $\omega = \mathbf{k} \cdot \mathrm{v}(0)$. In addition, there is the possibility of poles associated with zeros in ε. These occur because of natural resonances of the plasma such as plasma waves. Provided the plasma is stable, these are Landau damped, and there are resonances but no zeros.

Finally, we must remember that we will eventually have an integral over $\mathbf{v}'$ to evaluate, because of the ensemble average over the initial velocity distribution, and we must note how the pole in the $\mathbf{v}'$ plane moves in the limit $\gamma \to 0$. The ω integral involves

$$1/(\omega - \mathbf{k} \cdot \mathbf{v}' - i\gamma).$$

---

[8] The effect of initial correlations on the long-term solution is discussed in Section 3.6 where it is shown that average quantities are unaffected for say simple shielding correlations for averaging time $T \gg 1/\omega_{pe}$.

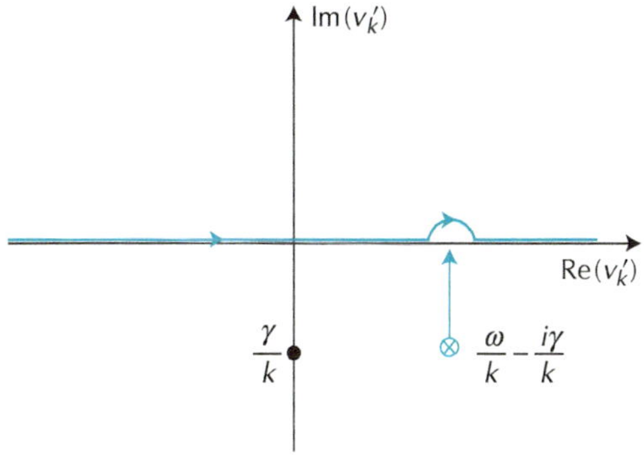

FIGURE B.2 Contour deformed around the pole.

This is illustrated in Fig. B.2. For this example for the component of $\mathbf{v}'$ in the direction of $\mathbf{k}$, we must deform the contour over the pole and set [see Eqns (A.1.8) and (A.1.14)].

$$\int_C f(\mathrm{v}_k')d\mathrm{v}_k' = -i\pi f(\omega/k) + \mathbf{P}\int f(\mathrm{v}_k')d\mathrm{v}_k'. \tag{B.8.16}$$

(c) Finally, we take the ensemble average over the initial velocity distribution $f_{0q}\left(\mathrm{v}'(0)\right)$. There are three terms arising from

$$\left\langle E^*(\mathbf{k},\omega)F_{1\mathrm{e}}(\mathbf{k},\mathrm{v},\omega)\right\rangle$$

in Eqn (B.8.15); the first is a product of the ion components of Eqns (B.8.3) and (B.8.6), the second comes from the product of the electron components, and the third from the product of the $F_{1\mathrm{e}}(\mathbf{k},\mathrm{v},0)$ term with the $n_{1\mathrm{e}}(\mathbf{k},\omega)$ component of Eqn (B.8.6). For example, the ion term after averaging over position is

$$\lim_{\gamma\to 0} i\frac{\gamma}{\pi}\int d\omega\int d\mathbf{v}' f_{0\mathrm{i}}\left(\mathrm{v}'(0)\right)\frac{N}{ZV}\frac{4\pi}{k^3}\frac{Ze^2 4\pi q}{m|\varepsilon(\mathbf{k},\omega)|^2}\cdot n_0\mathbf{k}\cdot\frac{\partial f_{0\mathrm{e}}}{\partial\mathrm{v}}$$
$$\times\left\{\frac{1}{(\omega-\mathbf{k}\cdot\mathrm{v}-i\gamma)\left[(\omega-\mathbf{k}\cdot\mathrm{v}(0))^2+\gamma^2\right]}\right\}. \tag{B.8.17}$$

In the integral over ω, the only nonzero term is from the pole at $\omega=\mathbf{k}\cdot\mathbf{v}'(0)$, and the ion term reduces to

$$i\int\frac{d\mathbf{v}' f_{0\mathrm{i}}(\mathbf{v}')(4\pi)^2 Ze^3 n_0^2}{\{\mathbf{k}\cdot(\mathbf{v}'-\mathrm{v})\}k^3 m_\mathrm{e}|\varepsilon(\mathbf{k},\mathbf{k}\cdot\mathbf{v}')|^2}\mathbf{k}\cdot\frac{\partial f_{0\mathrm{e}}}{\partial\mathrm{v}}. \tag{B.8.18}$$

Similarly, the electron term is

$$i\int\frac{d\mathbf{v}' f_{0\mathrm{e}}(\mathbf{v}')(4\pi)^2 Ze^3 n_0^2}{\{\mathbf{k}\cdot(\mathbf{v}'-\mathrm{v})\}k^3 m_\mathrm{e}|\varepsilon(\mathbf{k},\mathbf{k}\cdot\mathbf{v}')|^2}\mathbf{k}\cdot\frac{\partial f_{0\mathrm{e}}}{\partial\mathrm{v}}. \tag{B.8.19}$$

The final term becomes

$$-i\frac{4\pi e n_0 f_{0\mathrm{e}}(\mathrm{v})}{k\,|\varepsilon(\mathbf{k},\mathbf{k}\cdot\mathrm{v})|^2}\varepsilon(\mathbf{k},\mathbf{k}\cdot\mathrm{v}). \tag{B.8.20}$$

Now, from Eqns (3.3.9) and (3.3.10), the dielectric function is given by

$$\varepsilon(\mathbf{k},\mathbf{k}\cdot\mathrm{v}) = 1 + \int\frac{d\mathbf{v}' 4\pi_{\mathrm{e}}^2 n_0 \mathbf{k}\cdot\partial f_{0\mathrm{e}}/\partial\mathbf{v}'}{m_{\mathrm{e}}k^2[\mathbf{k}\cdot(\mathrm{v}-\mathbf{v}')-i\gamma]} + \int\frac{d\mathbf{v}' 4\pi Z e^2 n_0 \mathbf{k}\cdot\partial f_{0\mathrm{i}}/\partial\mathbf{v}'}{m_{\mathrm{i}}k^2[\mathbf{k}\cdot(\mathrm{v}-\mathbf{v}')-i\gamma]}. \tag{B.8.21}$$

We now combine Eqns (B.8.18)–(B.8.21) in Eqn (B.8.15) and use Eqn (B.8.7) to obtain the Lenard–Balescu equation

$$\begin{aligned}\frac{\partial f_{0\mathrm{e}}}{\partial t} = &-\frac{e}{m_{\mathrm{e}}}\int\frac{d\mathbf{k}}{k^2}\mathbf{k}\cdot\frac{\partial}{\partial\mathrm{v}}\left\{\frac{4\pi^2 e}{|\varepsilon(\mathbf{k},\mathbf{k}\cdot\mathrm{v})|^2}\right.\\ &\times\left(\frac{4\pi e^2}{k^2}\int d\mathbf{v}'\left[\frac{f_{0\mathrm{e}}(\mathrm{v})}{m_{\mathrm{e}}}\mathbf{k}\cdot\frac{\partial f_{0\mathrm{e}}}{\partial\mathrm{v}} - \frac{f_{0\mathrm{e}}(\mathbf{v}')}{m_{\mathrm{e}}}\mathbf{k}\cdot\frac{\partial f_{0\mathrm{e}}}{\partial\mathrm{v}}\right]\delta\left(\mathbf{k}\cdot(\mathbf{v}'-\mathrm{v})\right)\right.\\ &\left.\left.\times\frac{4\pi Z e^2}{k^2}\int d\mathbf{v}'\left[\frac{f_{0\mathrm{e}}(\mathrm{v})}{m_{\mathrm{i}}}\mathbf{k}\cdot\frac{\partial f_{0\mathrm{i}}}{\partial\mathbf{v}'} - \frac{f_{0\mathrm{i}}(\mathbf{v}')}{m_{\mathrm{e}}}\mathbf{k}\cdot\frac{\partial f_{0\mathrm{e}}}{\partial\mathrm{v}}\right]\delta\left(\mathbf{k}\cdot(\mathbf{v}'-\mathrm{v})\right)\right)\right\}.\end{aligned} \tag{B.8.22}$$

(i) The first term in brackets gives the change in the electron distribution function because of electron–electron interaction. It is zero for a Maxwellian electron velocity distribution, as we would expect. The second term in square brackets involves the interactions with ions. In equilibrium, when $f_{0\mathrm{e}}$ and $f_{0\mathrm{i}}$ are Maxwellian and $T_{\mathrm{e}} = T_{\mathrm{i}}$, it is also zero, as expected.

(ii) We have seen above that the expansion of the Liouville equation in a hierarchy is valid only for $k < e^2/\kappa T_{\mathrm{e}}$; therefore, the integral over k must be limited by this value. Fortunately, the divergence is logarithmic; therefore, the result is not sensitive to our choice of a limit.

(iii) The collision term in Eqn (B.8.22) may be reduced to the Fokker–Planck form Eqn (B.4.4), by using the static approximation to ε (see, for example, Montgomery & Tidman, 1964; Klimontovich, 1967; Clemmow & Dougherty, 1969).

(iv) It was assumed in the calculations above both that g was small and that correlations in the plasma decayed rapidly compared to the timescale for a change in the basic distribution functions. This is not correct for weakly damped modes in the stable case and certainly not correct for growing nodes in the unstable case. Kinetic equations that include the contribution from the dominant part of the wave spectrum have been derived by Rogister & Oberman (1967).

Appendix C

General Hot Plasma Dispersion Relation[1]

In most plasma physics courses, one obtains an expression for the perturbed distribution function *assuming electrostatic perturbations*. This permits one to obtain dispersion relations for ion-acoustic waves and electron plasma waves in unmagnetized plasmas, as well as the dispersion relation for Bernstein waves (for exact perpendicular propagation). Here, we wish to sketch the derivation of the general dispersion relation for a hot, magnetized uniform plasma.

The linearized Vlasov equation is given by

$$\frac{\partial f_{1\alpha}}{\partial t} + \mathrm{v}\cdot\nabla f_{1\alpha} - (q_\alpha/m_\alpha c)(\mathrm{v}\times\mathbf{B}_0)\cdot\frac{\partial f_{1\alpha}}{\partial \mathrm{v}} = (q_\alpha/m_\alpha c)\left(\mathbf{E}_1 + \frac{\mathrm{v}\times\mathbf{B}_1}{c}\right)\cdot\frac{\partial f_{0\alpha}}{\partial \mathrm{v}}$$

with

$$\mathbf{B}_0 = \hat{z}B_0.$$

We notice that even with Laplace and Fourier transforms, we have a differential equation in v to be solved. We will use *the method of characteristics*, which makes use of integration over the unperturbed orbits.

We define $\dfrac{dx'}{dt'} = \mathrm{v}'$

and $\dfrac{d\mathrm{v}'}{dt'} = \dfrac{-e}{m}\left[\dfrac{\mathrm{v}'\times\mathbf{B}_0\left(\mathbf{x}',t'\right)}{c}\right],$

with the boundary conditions $\mathbf{x}'(t'=t)=\mathbf{x}$ and $\mathrm{v}'(t'=t)=\mathrm{v}$, where $\mathbf{x}$ and v are points in phase space.

[1] Krall and Trivelpiece, Stix, and Montgomery & Tidman (1964).

DOI: 10.1016/B978-0-12-374877-5.00015-4

We then have

$$f_{\alpha 1}(\mathbf{r},\mathrm{v},t) = -\frac{q_\alpha}{m_\alpha}\int_{-\infty}^{t} dt'\left[\mathbf{E}_1(\mathbf{r}',t') + \frac{\mathrm{v}'\times\mathbf{B}_1(\mathbf{r}',t')}{c}\right].\nabla \mathrm{v}' f_{\alpha 0}(\mathbf{r},'\mathrm{v}')$$
$$+ f_{\alpha 1}\left[\mathbf{r}'(-\infty),\mathrm{v}'(-\infty),t'=-\infty\right],$$

where $\mathbf{r}', \mathrm{v}'$ correspond to the unperturbed orbits and coincide with $\mathbf{r}'$ and v' at $t = t'$. In the above, we have solved the Vlasov equation for $f_1(\mathbf{x},\mathrm{v},t)$ by integrating the equation from $-\infty$ to t along a path in $(\mathbf{x},\mathrm{v})$ phase space, which coincides with the orbit of a charged particle about $\mathbf{B}_0$. In cylindrical coordinates, the equilibrium orbits for $\mathbf{B}_0 = \hat{z}B_0$ are

$$\mathrm{v}_x = \mathrm{v}_\perp \cos\phi,$$
$$\mathrm{v}_y = \mathrm{v}_\perp \sin\phi,$$
$$\mathrm{v}_z = \mathrm{v}_\parallel$$

In terms of the primed variables, we have

$$\mathrm{v}_x' = \mathrm{v}_\perp \cos(\phi - \omega_c\tau) \qquad x' = x - \frac{\mathrm{v}_\perp}{\omega_c}\sin(\phi - \omega_c\tau) + \frac{\mathrm{v}_\perp}{\omega_c}\sin\phi$$
$$\mathrm{v}_y' = \mathrm{v}_\perp \sin(\phi - \omega_c\tau) \qquad y' = y - \frac{\mathrm{v}_\perp}{\omega_c}\cos(\phi - \omega_c\tau) + \frac{\mathrm{v}_\perp}{\omega_c}\cos\phi$$
$$\mathrm{v}_z' = \mathrm{v}_\parallel \qquad z' = \mathrm{v}_\parallel\tau + z$$

and for $\tau \to 0$, $\mathrm{v}' \to \mathrm{v}$, and $\mathbf{x}' \to x$.

We then obtain $n_{\alpha 1}(\mathbf{r},t)$ and $\mathbf{j}_1(\mathbf{r},t)$ from f_α and utilize them in the following Maxwell equations:

$$\nabla\times\mathbf{E}_1 = -\frac{1}{c}\frac{\partial\mathbf{B}_1}{\partial t}, \qquad \nabla\cdot\mathbf{E}_1 = 4\pi\sum_\alpha q_\alpha n_{\alpha 1} = 4\pi\sum_\alpha q_d\int d\mathrm{v} f_{\alpha 1}$$
$$\nabla\times\mathbf{B}_1 = \frac{1}{c}\frac{\partial\mathbf{E}_1}{\partial t} + \frac{4\pi}{c}\mathbf{j}_1,$$

where $\mathbf{j}_1 = \sum_\alpha q_\alpha(n_\alpha \mathrm{v}_\alpha)_1 = \sum_\alpha q_\alpha \int \mathrm{v} f_{\alpha 1} d_\mathrm{v}$.

From our introductory plasma course, we know that, rather than solve the initial value problem, all we really need to do is to solve for the time asymptotic solution in order to obtain the desired dispersion relation.

Therefore, we may assume (as before) a plane wave solution of the form

$$\mathbf{E}_1(\mathbf{r},t) = \mathbf{E}_1\exp(i\mathbf{k}\cdot\mathbf{r} - i\omega t).$$

We also assume that

$$f_{\alpha 1}(\mathbf{r}',\mathrm{v}',t' \to -\infty) = 0.$$

Then, we have with

$$f_{\alpha 1} = f_{\alpha 1}\exp(i\mathbf{k}\cdot\mathbf{r} - i\omega t)$$

$$f_{\alpha 1} = -\frac{q_\alpha}{m_\alpha}\int_{-\infty}^{0}\left(\mathbf{E}_1 + \frac{\mathrm{v}'\times\mathbf{B}_1}{c}\right)\cdot\nabla_{\mathrm{v}'}f_{\alpha 0}(\mathrm{v}')\exp\left[i(\mathbf{k}\cdot\mathbf{X} - \omega r)\right]dr,$$

where $\operatorname{Im}\omega > 0,\ \tau = t' - t$, and $\mathbf{X} = \mathbf{r}'\cdot\mathbf{r}$.

Now, we also have that the equilibrium distribution satisfies

$$\left(\mathrm{v}\cdot\nabla + \frac{q_\alpha}{m_\alpha}\frac{\mathrm{v}'\times\mathbf{B}_1}{c}\nabla_{\mathrm{v}}\right)f_{\alpha 0} = 0,$$

with

$$\sum_\alpha n_{\alpha 0}q_\alpha\int\hat{f}_{\alpha 0}d\mathrm{v} = 0 \qquad \text{(no net charge)}$$

and

$$\sum_\alpha n_{\alpha 0}q_\alpha\int\mathrm{v}\hat{f}_{\alpha 0}d\mathrm{v} = 0 \qquad \text{(no net current).}$$

We again assume a spatially uniform equilibrium distribution, i.e.,

$$\hat{f}_{\alpha 0} = \hat{f}_{\alpha 0}\left(\mathrm{v}_\perp^2, \mathrm{v}_z\right).$$

The zero-order distribution function $f_{\alpha 0}(\mathrm{v})$ is a function of the constants of motion $\mathrm{v}_\perp$ and v_z; hence, using polar coordinates in v-space,

$$\nabla_{\mathrm{v}}f_{\alpha 0}(\mathrm{v}_\perp, \mathrm{v}_z) = \frac{\partial f_{\alpha 0}}{\partial \mathrm{v}_\perp}\overset{\frac{\mathrm{v}_\perp}{\mathrm{v}_\perp}\Downarrow}{\mathbf{e}_\perp} + \frac{1}{\mathrm{v}_\perp}\overbrace{\frac{\partial f_{\alpha 0}}{\partial\phi}}^{0}\mathbf{e}_\varphi + \frac{\partial f_{\alpha 0}}{\partial \mathrm{v}_z}\mathbf{e}_z.$$

$$= 2\frac{\partial f_{\alpha 0}}{\partial \mathrm{v}_\perp^2}\mathrm{v}_\perp + 2\mathrm{v}_z\frac{\partial f_{\alpha 0}}{\partial \mathrm{v}_z^2}\mathbf{e}_z.$$

Then, since $\mathbf{B}_0 = \hat{z}B_0$ is a constant, we have $\mathrm{v}_\perp^2$ and v_z as constants of the motion. Therefore, we can remove the terms $\frac{\partial f_0}{\partial \mathrm{v}_\perp^2}$ and $\frac{\partial f_0}{\partial \mathrm{v}_z^2}$ from the integral. We then must perform integrals that look like

$$\int_{-\infty}^{0}\left(\mathrm{v}'_x, \mathrm{v}'_y, 1\right)\exp\left[1(\mathbf{k}\cdot\mathbf{X} - \omega\tau)\right]dr.$$

We will assume (without loss of generality) that $k_\perp$ coincides with the x-axis, i.e., that $\mathbf{k}$ lies in the x-z plane,

$$\mathbf{k} = k_\perp\mathbf{e}_x + k_z\mathbf{e}_z.$$

Then,

$$\mathbf{k}\cdot\left[\mathbf{x}'(\tau)-\mathbf{x}\right]=k_\tau[x'(\tau)-x]+k_z z'(\tau)-k_z z$$
$$=-\frac{k_\perp \mathrm{v}_\perp}{\Omega_\alpha}[\sin(\phi-\Omega_\alpha\tau)-\sin\phi]+k_z \mathrm{v}_z\tau$$

and

$$e^{-i\omega\tau+i\mathbf{k}\cdot[\mathbf{x}'(\tau)-\mathbf{x}]}=e^{-i(\omega-k_z\mathrm{v}_z)\tau-i\frac{k_\perp \mathrm{v}_\perp}{\Omega_\alpha}[\sin(\phi-\Omega_\alpha\tau)-\sin\phi]}.$$

To evaluate these integrals, we use the Bessel function identities

$$\exp(iz\sin\phi)=\sum_{n=-\infty}^{\infty}e^{in\phi}J_n(z)$$

and

$$\exp[-iz\sin(\phi-\alpha)]=\sum_{m=-\infty}^{\infty}\exp[-im(\phi-\alpha)]J_m(z).$$

Then,

$$\exp\left[\frac{ik_\perp \mathrm{v}_\perp}{\omega_c}\sin(\phi-\omega_c\tau)\right]=\sum_{n=-\infty}^{\infty}J_n\left(\frac{k_\perp \mathrm{v}_\perp}{\omega_c}\right)\exp[in(\phi-\omega_c\tau)].$$

Hence,

$$e^{-i\omega\tau+i\mathbf{k}\cdot[\mathbf{x}'(\tau)-\mathbf{x}]}=e^{-i(\omega-k_z\mathrm{v}_z)\tau-i\frac{k_\perp \mathrm{v}_\perp}{\Omega_\alpha}[\sin(\phi-\Omega_\alpha\tau)-\sin\phi]}$$
$$=\sum_s\sum_{s'}J_s\left(\frac{k_\perp \mathrm{v}_\perp}{\Omega_\alpha}\right)J_{s'}\left(\frac{k_\perp \mathrm{v}_\perp}{\Omega_\alpha}\right)e^{-i(\omega-k_z\mathrm{v}_z-s\Omega_\alpha)\tau+i(s'-s)\phi}.$$

Now, consider the other factors in the integrand.

We have

$$\mathbf{B}_k=\frac{c}{\omega}\mathbf{k}\times E_k.$$

Hence,

$$\mathrm{v}\times\mathbf{B}_k=\frac{c}{\omega}\mathrm{v}\times(\mathbf{k}\times E_k)=\frac{c}{\omega}[(\mathrm{v}\cdot\mathbf{E}_k)\mathbf{k}-(\mathbf{k}\cdot\mathrm{v})E_k].$$

Thus, we have

$$\mathbf{E}_k\cdot\nabla_{\mathrm{v}}f_{\alpha0}=2\left(E_{kx}\mathbf{e}_x+E_{ky}\mathbf{e}_y+E_{kz}\mathbf{e}_z\right)\cdot\left(\frac{\partial f_{\alpha0}}{\partial \mathrm{v}_\perp^2}\mathrm{v}_\perp+\mathrm{v}_z\frac{\partial f_{\alpha0}}{\partial \mathrm{v}_z^2}\mathbf{e}_z\right)$$
$$=2\left(E_{kx}\mathrm{v}_x+E_{ky}\mathrm{v}_y\right)\frac{\partial f_{\alpha0}}{\partial \mathrm{v}_\perp^2}+2E_{kz}\mathrm{v}_z\frac{\partial f_{\alpha0}}{\partial \mathrm{v}_z^2}$$

and

$$\frac{1}{c}\,(\mathrm{v}\times\mathbf{B}_k)\cdot\nabla_{\mathrm{v}}f_{\alpha 0}$$

$$=\frac{2}{\omega}\left[(\mathrm{v}\cdot\mathbf{E}_k)\,\mathbf{k}-(\mathbf{k}\cdot\mathrm{v})\,\mathbf{E}_k\right]\cdot\left(\frac{\partial f_{\alpha 0}}{\partial \mathrm{v}_\perp^2}\mathrm{v}_\perp+\mathrm{v}_z\frac{\partial f_{\alpha 0}}{\partial \mathrm{v}_z^2}\mathbf{e}_z\right)$$

$$=\frac{2}{\omega}\left\{\left[(\mathrm{v}\cdot\mathbf{E_k})\,(\mathbf{k}\cdot\mathrm{v}_\perp)-(\mathbf{k}\cdot\mathrm{v})\,(\mathrm{v}_\perp-\mathbf{E}_k)\right]\frac{\partial f_{\alpha 0}}{\partial \mathrm{v}_\perp^2}\right.$$

$$\left.+\left[(\mathrm{v}\cdot\mathbf{E}_k)\,k_z\mathrm{v}_z-(\mathbf{k}\cdot\mathrm{v})\,\mathrm{v}_z\mathbf{E}_k\right]\frac{\partial f_{\alpha 0}}{\partial \mathrm{v}_z^2}\right\}$$

$$-\frac{2}{\omega}\left\{\left[\left(\mathrm{v}_xE_{kx}+\mathrm{v}_yE_{ky}+\mathrm{v}_zE_{kz}\right)(k_\perp\mathrm{v}_x)-(k_\perp\mathrm{v}_x+k_z\mathrm{v}_z)\left(\mathrm{v}_xE_{kx}+\mathrm{v}_yE_{ky}\right)\right]\frac{\partial f_{\alpha 0}}{\partial \mathrm{v}_\perp^2}\right.$$

$$\left.+\left[\left(\mathrm{v}_xE_{kx}+\mathrm{v}_yE_{ky}+\mathrm{v}_zE_{kz}\right)k_z\mathrm{v}_z-(k_\perp\mathrm{v}_x+k_z\mathrm{v}_z)\,\mathrm{v}_zE_{kz}\right]\frac{\partial f_{\alpha 0}}{\partial \mathrm{v}_z^2}\right\}$$

$$=\frac{2}{\omega}\left[\left(-k_z\mathrm{v}_xE_{kx}-k_z\mathrm{v}_yE_{ky}+k_\perp\mathrm{v}_xE_{kz}\right)\mathrm{v}_y\frac{\partial f_{\alpha 0}}{\partial \mathrm{v}_\perp^2}\right.$$

$$\left.+\left(k_z\mathrm{v}_xE_{kx}+k_z\mathrm{v}_yE_{ky}-k_\perp\mathrm{v}_xE_{kz}\right)\mathrm{v}_z\frac{\partial f_{\alpha 0}}{\partial \mathrm{v}_z^2}\right].$$

Combining the above equations, we obtain

$$\left(\mathbf{E}_k+\tfrac{1}{c}\mathrm{v}'\times\mathbf{B}_k\right)\cdot\nabla_{\mathrm{v}'}f_{\alpha 0}=2\mathrm{v}_x'\underline{\overline{X}}+2\mathrm{v}_y'Y+2\overset{\substack{\mathrm{v}_z\\ \Downarrow}}{\mathrm{v}_z'}Z$$

$$=\mathrm{v}_\perp\left[e^{i(\phi-\Omega_\alpha\tau)}+e^{-i(\phi-\Omega_\alpha\tau)}\right]\underline{\overline{X}}$$

$$-i\mathrm{v}_\perp\left[e^{i(\phi-\Omega_\alpha\tau)}-e^{-i(\phi-\Omega_\alpha\tau)}\right]Y+2\mathrm{v}_zZ,$$

where

$$\underline{\overline{X}}=E_{kx}\frac{\partial f_{\alpha 0}}{\partial \mathrm{v}_\perp^2}+\frac{\mathrm{v}_z}{\omega}\,(k_zE_{kx}-k_\perp E_{kz})\left(\frac{\partial f_{\alpha 0}}{\partial \mathrm{v}_z^2}-\frac{\partial f_{\alpha 0}}{\partial \mathrm{v}_\perp^2}\right)$$

$$Y=E_{ky}\frac{\partial f_{\alpha 0}}{\partial \mathrm{v}_\perp^2}+\frac{\mathrm{v}_z}{\omega}k_zE_{ky}\left(\frac{\partial f_{\alpha 0}}{\partial \mathrm{v}_z^2}-\frac{\partial f_{\alpha 0}}{\partial \mathrm{v}_\perp^2}\right)$$

$$Z=E_{kz}\frac{\partial f_{\alpha 0}}{\partial \mathrm{v}_z^2}.$$

Note: $\mathrm{v}_\perp$, v_z, and v are constants of motion, but $\mathrm{v}_\perp$, v_x, and v_y are not. $\underline{\overline{X}}$, Y, and Z are functions of constants of motion; we have

$$\left(\mathbf{E}_k + \frac{1}{c}\mathrm{v}' \times \mathbf{B}_k\right) \cdot \nabla_{\mathrm{v}'} f_{\alpha 0} e^{-i\omega\tau + i\mathbf{k}\cdot[x'(\tau)-\mathbf{x}]}$$

$$= \mathrm{v}_\perp \underline{\overline{X}} \left[e^{i(\phi-\Omega_\alpha\tau)} + e^{-i(\phi-\Omega_\alpha\tau)}\right] \sum_{s,s'} J_s J_{s'} e^{-i(\omega-k_z\mathrm{v}_z-s\Omega_\alpha)\tau+i(s'-s)\phi}$$

$$- i\mathrm{v}_\perp Y \left[e^{i(\phi-\Omega_\alpha\tau)} - e^{-i(\phi-\Omega_\alpha\tau)}\right] \sum_{s,s'} J_s J_{s'} e^{-i(\omega-k_z\mathrm{v}_z-s\Omega_\alpha)\tau+i(s'-s)\phi}$$

$$+ 2\mathrm{v}_z Z \sum_{s,s'} J_s J_{s'} e^{-i(\omega-k_z\mathrm{v}_z-s\Omega_\alpha)\tau+i(s-s)\phi}.$$

We rewrite the terms in the above as follows:

$$\sum_{s,s'} J_s J_{s'} e^{i(\phi-\Omega_\alpha\tau)} e^{-i(\omega-k_z\mathrm{v}_z-s\Omega_\alpha)\tau+i(s'-s)\phi}$$

$$= \sum_{s,s'} J_s J_{s'} e^{-i[\omega-k_z\mathrm{v}_z-(s-1)\Omega_\alpha]\tau+i(s'-s+1)\phi}$$

$$(\text{let } n = s-1)$$

$$= \sum_{n,s'} J_{n+1} J_{s'} e^{-i(\omega-k_z\mathrm{v}_z-n\Omega_\alpha)\tau+i(s'-n)\phi}$$

$$\sum_{s,s'} J_s J_{s'} e^{-i(\phi-\Omega_\alpha\tau)} e^{-i(\omega-k_z\mathrm{v}_z-s\Omega_\alpha)\tau+i(s'-s)\phi}$$

$$= \sum_{s,s'} J_s J_{s'} e^{-i[\omega-k_z\mathrm{v}_z-(s+1)\Omega_\alpha]\tau+i(s'-s-1)\phi}$$

$$(\text{let } n = s+1)$$

$$= \sum_{n,s'} J_{n+1} J_{s'} e^{-i(\omega-k_z\mathrm{v}_z-n\Omega_\alpha)\tau+i(s'-n)\phi}.$$

Substituting, we obtain

$$(\mathbf{E}_k + \tfrac{1}{c}\mathrm{v}' \times \mathbf{B}_k) \cdot \nabla_{\mathrm{v}'} f_{\alpha 0} e^{-i\omega\tau + i\mathbf{k}\cdot[x'(\tau)-\mathbf{x}]} = \textstyle\sum_{n,s'} \left[\mathrm{v}_\perp \underline{\overline{X}} (J_{n+1} + J_{n-1})\right.$$

$$\left. - i\mathrm{v}_\perp Y (J_{n+1} - J_{n-1}) + 2\mathrm{v}_z Z J_n\right] J_{s'} \cdot \underbrace{e^{-i(\omega-k_z\mathrm{v}_z-n\Omega_\alpha)\tau} e^{i(s'-n)\phi}}_{\text{the only factor that depends on } \tau}$$

Again substituting and carrying out the τ-integration, we obtain

$$f_{\alpha k}(\mathrm{v}) = \frac{q_\alpha}{m_\alpha}\sum \frac{\mathrm{v}_\perp \underline{X}(J_{n+1}+J_{n-1}) - i\mathrm{v}_\perp Y(J_{n+1}-J_{n-1}) + 2\mathrm{v}_z Z I_n}{i(\omega - k_z \mathrm{v}_z - n\Omega_\alpha)} J_{s'} e^{i(s'-n)\phi}.$$

The perturbed current is given by

$$\begin{aligned}\mathbf{J}_k &= \sum_\alpha q_\alpha \int f_{\alpha k}(\mathrm{v})\mathrm{v} d^3\mathrm{v} \\ &= \sum_\alpha q_\alpha \int_0^\infty \mathrm{v}_\perp d\mathrm{v}_\perp \int_0^{2\pi} d\phi \int_{-\infty}^{\infty} d\mathrm{v}_z f_{\alpha k}(\mathrm{v}) \overbrace{\underset{=\mathrm{v}_\perp \cos\phi \mathbf{e}_x + \mathrm{v}_\perp \sin\phi \mathbf{e}_y + \mathrm{v}_z e_z}{=\mathrm{v}'(\tau=0)}}^{\mathrm{v}}.\end{aligned}$$

The x-component of $\mathbf{J}_k$ is

$$\begin{aligned}J_{kx} &= \sum_a q_\alpha \int_0^\infty \mathrm{v}_\perp d\mathrm{v}_\perp \int_0^{2\pi} d\phi \int_{-\infty}^{\infty} d\mathrm{v}_z f_{\alpha k}(\mathrm{v}) \overbrace{=\tfrac{1}{2}\mathrm{v}_\perp(e^{i\phi}+e^{-i\phi})}^{\mathrm{v}} \\ &= \sigma_{xx}E_{kx} + \sigma_{xy}E_{ky} + \sigma_{xz}E_{kz}.\end{aligned}$$

The first element, σ_{xx}, can be obtained by grouping the terms in $f_{\alpha k}$, which are proportional to E_{kx}. Thus,

$$\sigma_{xx} = \sum_\alpha \frac{q_\alpha^2}{m_\alpha} \int_0^\infty \mathrm{v}_\perp d\mathrm{v}_\perp \int_0^{2\pi} d\phi \int_{-\infty}^{\infty} d\mathrm{v}_z \frac{1}{2}\mathrm{v}_\perp \left(e^{i\phi} + e^{-i\phi}\right)$$

$$\sum_{n,s'} \frac{\mathrm{v}_\perp \left[\frac{\partial f_{\alpha 0}}{\partial \mathrm{v}_\perp^2}\left(1 - \frac{k_z \mathrm{v}_z}{\omega}\right) + \frac{k_z \mathrm{v}_z}{\omega}\frac{\partial f_{\alpha 0}}{\partial \mathrm{v}_z^2}\right]}{i(\omega - k_z \mathrm{v}_z - n\Omega_\alpha)} (J_{n+1} + J_{n-1}) J_{s'} e^{i(s'-n)\phi}.$$

Using the Bessel function identity,

$$\begin{cases} J_{n-1}(x) + J_{n+1}(x) = \frac{2n}{x} J_n(x) \\ J_{n-1}(x) - J_{n+1}(x) = 2J_n'(x). \end{cases}$$

We can write

$$J_{n+1}\left(\frac{k_\perp \mathrm{v}_\perp}{\Omega_\alpha}\right) + J_{n-1}\left(\frac{k_\perp \mathrm{v}_\perp}{\Omega_\alpha}\right) = \frac{2n\Omega_\alpha}{k_\perp \mathrm{v}_\perp} J_n\left(\frac{k_\perp \mathrm{v}_\perp}{\Omega_\alpha}\right).$$

Substituting, we obtain

$$\sigma_{xx} = \sum_\alpha \frac{q_\alpha^2}{m_\alpha} \int_0^\infty \mathrm{v}_\perp d\mathrm{v}_\perp \int_0^{2\pi} d\phi \int_{-\infty}^{\infty} d\mathrm{v}_z \frac{n\mathrm{v}_\perp \Omega_\alpha}{k_\perp} \sum_{n,s'} J_n J_{s'} \frac{\frac{\partial f_{\alpha 0}}{\partial \mathrm{v}_\perp^2}\left(1 - \frac{k_z \mathrm{v}_z}{\omega}\right) + \frac{k_z \mathrm{v}_z}{\omega}\frac{\partial f_{\alpha 0}}{\partial \mathrm{v}_z^2}}{i(\omega - k_z \mathrm{v}_z - n\Omega_\alpha)} \cdot \left[e^{i(s'-n+1)\phi} + e^{i(s'-n-1)\phi}\right].$$

We see immediately that only one term in the s'-sum will survive the ϕ-integration. Thus, upon ϕ-integration,

$$\sigma_{xx} = \sum_\alpha \frac{2\pi q_\alpha^2}{m_\alpha} \sum_n \int_0^\infty \mathrm{v}_\perp d\mathrm{v}_\perp \int_{-\infty}^{\infty} d\mathrm{v}_z \frac{n\mathrm{v}_\perp \Omega_\alpha}{k_\perp} \frac{\frac{\partial f_{\alpha 0}}{\partial \mathrm{v}_\perp^2}\left(1 - \frac{k_z \mathrm{v}_z}{\omega}\right) + \frac{k_z \mathrm{v}_z}{\omega}\frac{\partial f_{\alpha 0}}{\partial \mathrm{v}_z^2}}{i(\omega - k_z \mathrm{v}_z - n\Omega_\alpha)} \cdot J_n \underbrace{[J_{n-1} + J_{n+1}]}_{= \frac{2n\Omega\alpha}{k_\perp \mathrm{v}_\perp} J_n}$$

$$= \sum_\alpha \frac{2\pi q_\alpha^2}{m_\alpha} \sum_n \int_0^\infty 2\mathrm{v}_\perp d\mathrm{v}_\perp \int_{-\infty}^{\infty} d\mathrm{v}_z \frac{n^2 \Omega_\alpha^2}{k_\perp^2} J_n^2 \frac{\frac{\partial f_{\alpha 0}}{\partial \mathrm{v}_\perp^2}\left(1 - \frac{k_z \mathrm{v}_z}{\omega}\right) + \frac{k_z \mathrm{v}_z}{\omega}\frac{\partial f_{\alpha 0}}{\partial \mathrm{v}_z^2}}{i(\omega - k_z \mathrm{v}_z - n\Omega_\alpha)}.$$

We can employ σ_{xx} to obtain the D_{xx} element of the dispersion tensor:

$$D_{xx} = 1 - \frac{k_z^2 c^2}{\omega^2} + \frac{4\pi i}{\omega}\sigma_{xx}$$

$$= 1 - \frac{k_z^2 c^2}{\omega^2} - \frac{2\pi}{\omega} \sum_\alpha \frac{\omega_{p\alpha}^2}{\Omega_\alpha} \sum_n \int_0^\infty 2\mathrm{v}_\perp d\mathrm{v}_\perp \int_{-\infty}^{\infty} d\mathrm{v}_z \frac{n^2 \Omega_\alpha^2}{k_\perp^2} J_n^2 \left(\frac{k_\perp \mathrm{v}_\perp}{\Omega_\alpha}\right) \times \frac{\frac{\partial \bar{f}_{\alpha 0}}{\partial \mathrm{v}_\perp^2}\left(1 - \frac{k_z \mathrm{v}_z}{\omega}\right) + \frac{k_z \mathrm{v}_z}{\omega}\frac{\partial \bar{f}_{\alpha 0}}{\partial \mathrm{v}_z^2}}{k_z \mathrm{v}_z + n\Omega_2 - \omega}.$$

In the above, we have defined a normalized distribution function

$$\bar{f}_{\alpha 0} = \frac{1}{n_{0\alpha}} f_{\alpha 0} \qquad \left[\text{hence} \int \bar{f}_{\alpha 0} d^3\mathrm{v} = 1\right].$$

In a similar manner, we obtain the remaining elements of the dispersion tensor, which satisfies the relation

$$\begin{vmatrix} D_{xx} & D_{xy} & D_{xz} \\ D_{yx} & D_{yx} & D_{yz} \\ D_{zx} & D_{zy} & D_{zz} \end{vmatrix} = 0.$$

The results are

$$D_{xx}=1-\frac{k_z^2c^2}{\omega^2}-\frac{2\pi}{\omega}\sum_\alpha\left(\frac{\omega_{p\alpha}^2}{\Omega_\alpha}\right)\sum_\alpha\left\langle\frac{n^2\Omega_\alpha^3}{k_\perp^2}J_n^2\chi_\alpha\right\rangle$$

$$D_{xy}=-\frac{2\pi i}{\omega}\sum_\alpha\sum_n\left(\frac{\omega_{p\alpha}^2}{\Omega_\alpha}\right)_\alpha\left\langle\frac{n\Omega_\alpha^2}{k_\perp}J_n\frac{dJ_n}{d\left(k_\perp \mathrm{v}_\perp/\Omega_\alpha\right)}\chi_\alpha\right\rangle$$

$$D_{xz}=\frac{k_zk_\perp c^2}{\omega^2}-\frac{2\pi}{\omega}\sum_\alpha\sum_n\left(\frac{\omega_{p\alpha}^2}{\Omega_\alpha}\right)_\alpha\left\langle\frac{n\Omega_\alpha^2\mathrm{v}_zJ_n^2}{k_\perp}\Lambda_\alpha\right\rangle$$

$$D_{yx}=-D_{xy}$$

$$D_{yy}=1-\frac{\left(k_\perp^2+k_z^2\right)c^2}{\omega^2}-\frac{2\pi}{\omega}\sum_\alpha\sum_n\left(\frac{\omega_{p\alpha}^2}{\Omega_\alpha}\right)\left\langle\Omega_\alpha\left(\frac{dJ_n}{d(k_\perp \mathrm{v}_\perp/\Omega_\alpha)}\right)^2\mathrm{v}_\perp^2\chi_\alpha\right\rangle$$

$$D_{yz}=\frac{2\pi i}{\omega}\sum_\alpha\sum_n\left(\frac{\omega_{p\alpha}^2}{\Omega_\alpha}\right)\left\langle\Omega_\alpha \mathrm{v}_\perp \mathrm{v}_zJ_n\frac{dJ_n}{d\left(k_\perp \mathrm{v}_\perp/\Omega_\alpha\right)}\Lambda_\alpha\right\rangle$$

$$D_{zx}=\frac{k_zk_\perp c^2}{\omega^2}-\frac{2\pi}{\omega}\sum_\alpha\sum_n\left(\frac{\omega_{p\alpha}^2}{\Omega_\alpha}\right)\left\langle \mathrm{v}_z\frac{n\Omega_\alpha^2}{k_\perp}J_n^2\chi_\alpha\right\rangle$$

$$D_{zy}=-\frac{2\pi i}{\omega}\sum_\alpha\sum_n\left(\frac{\omega_{p\alpha}^2}{\Omega_\alpha}\right)_\alpha\left\langle \mathrm{v}_z\mathrm{v}_\perp\Omega_\alpha J_n\frac{dJ_n}{d\left(k_\perp \mathrm{v}_\perp/\Omega_\alpha\right)}\chi_\alpha\right\rangle$$

$$D_{zz}=1-\frac{k_\perp^2c^2}{\omega^2}-\frac{2\pi}{\omega}\sum_\alpha\sum_n\left(\frac{\omega_{p\alpha}^2}{\Omega_\alpha}\right)_\alpha\left\langle \mathrm{v}_z^2\Omega_\alpha J_n^2\Lambda_\alpha\right\rangle,$$

where the brackets, [ ], denote the integral operator

$$[F(\mathrm{v})]=\int_{-\infty}^{\infty}d\mathrm{v}_\parallel\int_0^{\infty}\frac{2\mathrm{v}_\perp F\left(\mathrm{v}_\perp,\mathrm{v}_\parallel\right)}{k_\parallel \mathrm{v}_\parallel+n\omega_{c\alpha}-\omega}d\mathrm{v}_\perp,$$

and where

$$\chi_\alpha=\frac{\partial f_{\alpha0}}{\partial \mathrm{v}_\perp^2}\left(1-\frac{k_\parallel \mathrm{v}_\parallel}{\omega}\right)+\frac{k_\parallel \mathrm{v}_\parallel}{\omega}\frac{\partial f_{\alpha0}}{\partial \mathrm{v}_\parallel^2}$$

and

$$\Lambda_\alpha=\frac{\partial f_{\alpha0}}{\partial \mathrm{v}_\parallel^2}-\frac{n\omega_{c\alpha}}{\omega}\left(\frac{\partial}{\partial \mathrm{v}_\parallel^2}-\frac{\partial}{\partial \mathrm{v}_\perp^2}\right)f_{\alpha0}.$$

Using our dispersion relation, we may look at specific cases (as customarily done for the case of cold-fluid waves); for instance, we may look at perpendicular propagation ($k_z=0$) or at parallel propagation ($k_\perp=0$).

Also, we note that we may recover the cold-fluid plasma results by letting the distribution function be given by

$$\bar{f}_{\alpha 0} = \frac{1}{2\pi \mathrm{v}_\perp}\delta(\mathrm{v}_\perp)\delta(\mathrm{v}_z)$$

$$\left[\Rightarrow \quad \int \bar{f}_{\alpha 0} d^3\mathrm{v} = \int_0^\infty \mathrm{v}_\perp d\mathrm{v}_\perp \int_0^{2\pi} d\phi \int_{-\infty}^{\infty} d\mathrm{v}_z f_{\alpha 0} = 1\right].$$

Another common equilibrium distribution function is the two-dimensional Maxwellian distribution with respect to the magnetic field axis:

$$f_{0\alpha} = \frac{1}{\pi^{3/2} a_{\perp\alpha}^2 a_{\|\alpha}} \exp\left[-\frac{\mathrm{v}_\perp^2}{a_{\perp\alpha}^2} - \frac{\mathrm{v}_\|^2}{a_{\|\alpha}^2}\right].$$

As an example of the resultant dispersion, we assume propagation along the magnetic field $\mathbf{B}_0$. After integration over $\mathrm{v}_\perp$, we find

$$\omega^2 = k_\|^2 c^2 + \sum_\alpha \omega_{p\alpha}^2 \left\{ \int_{-\infty}^{\infty} \frac{(\omega - k_\| \mathrm{v}_\|) \exp\left(-\frac{\mathrm{v}_\|^2}{a_{\|\alpha}^2}\right) d\mathrm{v}_\|}{\pi^{1/2} a_{\|\alpha} (\omega \pm \omega_{c\alpha} - k_\| \mathrm{v}_\|)} + \left(\frac{a_{\perp\alpha}}{a_{\|\alpha}}\right)^2 \int_{-\infty}^{\infty} \frac{k_\| \mathrm{v}_\| \exp\left(\frac{\mathrm{v}_\|^2}{a_{\|\alpha}^2}\right) d\mathrm{v}_\|}{\pi^{1/2} a_{\|\alpha} (\omega \pm \omega_{c\alpha} - k_\| \mathrm{v}_\|)} \right\}.$$

The resonant denominator gives rise to a collisionless damping (as we have seen in the case of Landau damping in a field-free plasma). A particle "sees" the wave at a Doppler-shifted frequency, $\omega' = \omega - k_\| \mathrm{v}_\| = n\omega_c\alpha$.

Letting $t = \frac{\mathrm{v}_\|}{a_{\|\alpha}}$ and $\phi_{\pm\alpha} = \frac{\omega \pm \omega_{c\alpha}}{k_\| a_{\|\alpha}}$, we have

$$\omega^2 = k_\|^2 c^2 - \sum_\alpha \omega_{p\alpha}^2 \left\{ \int_{-\infty}^{\infty} \frac{\left(\frac{\omega}{k_\| a_{\|\alpha}} - t\right) e^{-t^2}}{\pi^{1/2} (t - \phi_{\pm\alpha})} dt + \int_{-\infty}^{\infty} \frac{t a_{\perp\alpha}^2 e^{-t^2}}{\pi^{1/2} a_{\|\alpha}^2 (t - \phi_{\pm\alpha})} dt \right\}.$$

We can write this in terms of the Fried and Conte plasma dispersion function,

$$\omega^2 = k_\|^2 c^2 - \sum_\alpha \omega_{p\alpha}^2 \left\{ \frac{\omega}{k_\| a_{\|\alpha}} Z(\phi_{\pm\alpha}) + \left[\left(\frac{a_b}{a_{\|\alpha}}\right)^2 - 1\right] [1 + \phi_{\pm\alpha} Z(\phi_{\pm\alpha})] \right\},$$

where $Z(\phi_{\pm\alpha}) = \frac{1}{\pi^{1/2}} \int_{-\infty}^{\infty} \frac{e^{-t^2} dt}{(t - \phi_{\pm\alpha})}$, and the minus sign corresponds to the familiar R-wave.

It is instructive to demonstrate that this reduces to our familiar cold plasma results. We take $M_i \to \infty, a_\perp e = a_\parallel e$ and then use the fact that

$$\phi_{-e} = \frac{\omega - \omega_{ce}}{k_\parallel a_{\parallel e}} \to \infty$$

to write

$$\lim_{\phi_{-e} \to \infty} Z(\phi_{-e}) \to -\frac{1}{\phi_{-e}}.$$

Then, $\omega^2 = k_\parallel^2 c^2 + \frac{\omega_{pe}^2 \omega}{\omega - \omega_{ce}}$. This gives us our old cold plasma result.

Appendix D

Computation of the Form Factor

PART 1

In this appendix, a method for numerically evaluating the nonrelativistic ($v/c \ll 1$) Thomson scattering form factor and the phase-matching requirement is outlined. For a complete relativistic treatment see Palastro et al., 2010. Recall that the form factor can be expressed as follows (see Section 3.4):

$$S(k,\omega) = \left|\frac{1+\chi_i}{\epsilon}\right|^2 f_e(\omega/k) + Z\left|\frac{\chi_e}{\epsilon}\right|^2 f_i(\omega/k), \tag{D.1}$$

where $\varepsilon = 1 + \chi_e + \chi_i$,

$$\chi_\alpha(\omega,k) = -\frac{1}{2}\left(k\lambda_{d,\alpha}\right)^{-2}\frac{\partial}{\partial\xi}W(\xi), \tag{D.2}$$

W is the plasma dispersion function, $\xi = \omega/\sqrt{2}k\mathrm{v}_{T\alpha}$, and for a nonrelativistic plasma,

$$f_\alpha(\omega/k) = \left(\frac{1}{2\pi}\right)^{1/2}\left(\frac{c}{\mathrm{v}_{T\alpha}}\right)\mathrm{e}^{-\omega^2/2k^2\mathrm{v}_{T\alpha}^2}. \tag{D.3}$$

The fluctuation wave number k is the magnitude of the fluctuation wave vectors $\boldsymbol{k}$, which is determined by the phase-matching conditions

$$\omega = \omega_s - \omega_i \tag{D.4}$$

$$\boldsymbol{k}(\omega) = \boldsymbol{k}_s(\omega_s) - \boldsymbol{k}_i(\omega_i). \tag{D.5}$$

The frequencies and wave vectors are related through the zeroes of the electromagnetic dispersions relations

$$1 - \left(\frac{k_{i,s}}{c\omega_{i,s}}\right)^2 + \chi_e(\omega_{i,s}, k_{i,s}) + \chi_i(\omega_{i,s}, k_{i,s}) = 0. \tag{D.6}$$

DOI: 10.1016/B978-0-12-374877-5.00016-6

For most situations, the approximation $\chi_i \simeq 0$ can be used in Eqn (D.6), and for nonrelativistic plasmas, $\chi_e \simeq -\omega_{pe}^2/\omega_{i,s}^2$ is a valid approximation. These approximations provide the familiar quadratic dispersion relation: $\omega_{i,s}^2 = \omega_{pe}^2 + c^2 k_{i,s}^2$. For relativistic plasmas, $\chi_e(\omega_{i,s}, k_{i,s})$ should be calculated exactly, which can be done using an iterative method.

The primary difficulty in numerical evaluation of the form factor arises from the nonanalyticity of the plasma dispersion function's integrand, which has a simple pole at $z = \xi$,

$$W(\xi) = \left(\frac{1}{\pi}\right)^{1/2} \int_{-\infty}^{\infty} dz \frac{e^{-z^2}}{z-\xi}. \tag{D.7}$$

The integral can be rewritten symbolically using Plemlj's formula

$$W(\xi) = \left(\frac{1}{\pi}\right)^{1/2} \left[\mp i\pi e^{-\xi^2} + PV \int_{-\infty}^{\infty} dz \frac{e^{-z^2}}{z-\xi} \right]; \tag{D.8}$$

but no further symbolic simplification is possible. The plasma dispersion function can, however, be evaluated numerically using a numerical equivalent of Plemlj's formula.

To start, the integral in Eqn (D.7) is split into three parts

$$\int_{-\infty}^{\infty} dz \frac{e^{-z^2}}{z-\xi} = \left[\int_{-\infty}^{\xi-\varphi} + \int_{\xi+\varphi}^{\infty} + \int_{\xi-\varphi}^{\xi+\varphi} \right] \frac{e^{-z^2}}{z-\xi} dz, \tag{D.9}$$

where $\varphi << |\xi|$ is the standoff distance from the simple pole. The first two integrals in Eqn (D.9) represent the principle value, while the third integral provides the contribution from the pole. To minimize cancellation error (the integrand switches sign as z passes across the pole), two symmetric arrays should be defined for evaluating the first two integrals as follows: $z_{-,p} = \xi - \varphi - (p-1)\Delta$ and $z_{+,m} = \xi + \varphi + (m-1)\Delta$, where p and m are integers and Δ is the integration step size. Defining the maximum number of steps as P and M, the principle value integrals can be written as follows:

$$\left[\int_{-\infty}^{\xi-\varphi} + \int_{\xi+\varphi}^{\infty} \right] \frac{e^{-z^2}}{z-\xi} dz \simeq \Delta \left[\sum_{m=1}^{M} \frac{e^{-z_{+,m}^2}}{\varphi + (m-1)\Delta} - \sum_{p=1}^{P} \frac{e^{-z_{-,p}^2}}{\varphi + (p-1)\Delta} \right], \tag{D.10}$$

where in general P and M should be chosen such that $z_{-,P} << -\xi - \varphi$ and $z_{+,M} >> \xi + \varphi$. To evaluate the third integral in Eqn (D.9), we Taylor expand the exponential about $z=\xi$ and integrate, giving

$$\int_{\xi-\varphi}^{\xi+\varphi} \frac{e^{-z^2}}{z-\xi} dz \simeq -i\pi e^{-\xi^2} - 4\xi e^{-\xi^2}\varphi. \tag{D.11}$$

The first term is the traditional value obtained from Plemlj's theorem, while the second is a correction due to the fact that φ cannot be set to zero. The error in the integration will be determined by both Δ and φ, and in general, both should be chosen as small as possible. The lower limit on the size of Δ is determined by the desired computation time or precision with the error in Eqn (D.10) scaling as Δ^3. The lower limit on φ is based on $1/\varphi$ being a machine computable number. Altogether the plasma dispersion function can be written

$$W(\xi) \simeq \left(\frac{1}{\pi}\right)^{1/2} \left[\Delta \sum_{m=1}^{M} \frac{e^{-z_{+,m}^2}}{\varphi+(m-1)\Delta} - \Delta \sum_{p=1}^{P} \frac{e^{-z_{-,p}^2}}{\varphi+(p-1)\Delta} - e^{-\xi^2}(i\pi + 4\xi\varphi) \right]. \tag{D.12}$$

The derivate of Eqn (D.12) can then be taken to find χ_α,

$$\chi_\alpha \simeq \left(k\lambda_{d,\alpha}\right)^{-2} \left(\frac{1}{\pi}\right)^{1/2} \left[\sum_{m=1}^{M} \frac{\Delta z_{+,m} e^{-z_{+,m}^2}}{\varphi+(m-1)\Delta} - \sum_{p=1}^{P} \frac{\Delta z_{-,p} e^{-z_{-,p}^2}}{\varphi+(p-1)\Delta} + e^{-\xi^2}\left(2\varphi - i\pi\xi - 4\varphi\xi^2\right) \right]. \tag{D.13}$$

Because $(k\lambda_{d,\alpha})^2\chi_\alpha$ depends on only one parameter, ξ, a table only needs to be generated once, which can be used in a program or function that calculates Eqn (D.1).

After a table for $(k\lambda_{d,\alpha})^2\chi_\alpha$ vs. ξ has been generated, the general procedure for calculation of the form factor is as follows:

1. Choose a fixed $\boldsymbol{k}_i$ and calculate ω_i by finding the positive root of $\omega_{pe}^2 = \omega_{pe}^2 + c^2k_i^2$.
2. Define an array of $\boldsymbol{k}_s$ values and calculate the corresponding ω_s values by finding the roots of $\omega_s^2 = \omega_{pe}^2 + c^2k_s^2$.
3. Define the scattering angle, θ, and calculate the array of k and ω values using the law of cosines: $k^2 = k_s^2 + k_i^2 - 2k_sk_i\cos\theta$ and Eqn (D.4).
4. Load the table, and interpolated to find the values of $(k\lambda_{d,\alpha})^2\chi_\alpha$ for the electrons and ions.
5. Evaluate (3) for all values of k and ω and plug everything into the Eqn (D.1).

PART 2

In this section we apply the methods discussed above to calculate a scattering spectrum using MATLAB routines. The main script sets the plasma parameters, experimental geometry, and probe laser specifications. An ascii table of the derivative of the plasma dispersion function $dW(xi)/dxi$ as a function of $xi = \omega/(kv_T)/\sqrt{2}$ must be generated by running the script wtable.m below. This table is used by the Zprime.m function which is called from the main script.

```
%%% Thomson.m
%This file calculates the Thomson scattering form factor and requiresthe
%tables generated by wtable.m and function Zprime.mCompute the Thomson
%scattering, per unit scattering volume, per unit
%solid angle, per freq/2pi i.e. n r_e^2(1-sin^2(sa)*sin^2(dphi))S(k,omeg)

% Plasma Parameters
Te=1.35;                        % electron temperature KeV
Ti=.4;                          % ion temperatuer KeV
Z=[1, 6];                       % ionization level of plasma
A=[1, 12];                      % atomic mass
fract=[0.5, 0.5];               % fraction of each ion species
ne=1.e21;                       % plasma density in cm^(-3) 0.9e21
Vpar=0;                         % fluid velocity parallel to the Thomson
                                  laser (ki) in cm/s
Vperp=0;                        % fluid velocity perp to the Thomson laser
                                  (ki) in cm/s, in the scattering plane, in the
                                  direction of the observer
ud=0;                           % relative drift velocity between the
                                  electrons and ions in cm/s along k
gamma=0;                        % the angle between k and the drift velocity
                                  ud
sa=63;                          % scattering angle in degrees
dphi=90;                        % dphi is the angle between the plane of
                                  polarization and the scattering plane in
                                  degrees
%basic quantities
C=2.99792458e10;                % velocity of light cm/sec
Me=510.9896/C^2;                % electron mass KeV/C^2
Mp=Me*1836.1;                   % proton Mass KeV/C^2
Mi=A*Mp;                        % ion mass
re=2.8179e-13;                  % classical electron radius cm
e=1.6e-19;                      % electron charge
Esq = Me*C*C*re;                % sq of the electron charge KeV-cm
constants = sqrt(4*pi*Esq/Me);  % sqrt(4*pi*e^2/Me)
sarad=sa*2*pi/360;              % scattering angle in radians
dphirad = dphi*2*pi/360;        % dphi is the angle between the plane of
                                     polarization and the scattering plane
gammarad = gamma*2*pi/360;      % the angle between k and the drift velocity
                                     ud

%calculating k and omega vectors
omgpe=constants*sqrt(ne);       % plasma frequency Rad/s
omg = omgs - omgL;
ks=sqrt(omgs.^2-omgpe^2)/C;
kL=sqrt(omgL^2-omgpe^2)/C;      % laser wavenumber in Rad/cm
```

```
k=sqrt(ks.^2+kL^2-2*ks*kL*cos(sarad));
kdotv = (kL - ks*cos(sarad))*Vpar - ks*sin(sarad)*Vperp;
omgdop=omg - kdotv;

% plasma parameters

% electrons
vTe=sqrt(Te/Me);                  % electron thermal velocity
klde=(vTe/omgpe)*k;

% ions
Zbar = sum(Z.*fract);
ni = fract*ne./Zbar;
omgpi = constants.*Z.*sqrt(ni.*Me./Mi);
vTi=sqrt(Ti./Mi);                 % ion thermal velocity
kldi=transpose(vTi./omgpi)*k;

% electron susceptibility

%calculating normilized phase velcoity (xi's) for electrons
xd = ud/(sqrt(2.)*vTe)*cos(gammarad);
xie=omgdop./(k*sqrt(2.)*vTe) - xd;
Zpe = Zprime(xie);
chiE= -0.5./(klde.^2).* (Zpe(1,:) + sqrt(-1)*Zpe(2,:));

% ion susceptibilities

% finding derivative of plasma dispersion function along xii array
xii=1./transpose((sqrt(2.)*vTi))*(omgdop./k);
[num_species, num_pts] = size(xii);
chiI=zeros(num_species, num_pts);

for m = 1:num_species
Zpi = Zprime(xii(m,:));
chiI(m,:) =-0.5./(kldi(m,:).^2).* (Zpi(1,:)+sqrt(-1)*Zpi(2,:));
end
chiItot = sum(chiI,1);

% Formfactor

econtr = sqrt(2*pi)./(vTe*k).*exp(-
xie.^2).*abs((1+chiItot)./(1+chiItot+chiE)).^2;
icontr = 2*Ti/Te*klde.^2./omgdop .* abs(chiE).^2 ./abs(1+chiItot+chiE).^2
.*imag(chiItot);

FF = econtr + icontr;
r = ne*(1-sin(sarad)^2*cos(dphirad)^2)*FF*re^2;

formfactor = r;

%*****************************************************************%
%End of Thomson.m
%*****************************************************************%

% Zprime.m
% This function calculates the derivative of the Z-function given an array of
% normalized phase velocities (xi) as defined in Chapter 5. For values of xi
% between -10 and 10 a table is used that can be generated by running the
```

```
% script below (see Fig. 5.1). Outside of this range the asymptotic
% approximation (see Eqn. 5.2.10) isused.

functionZp = Zprime(xi)

load('rdWT');
load('idWT');

ai=find(xi<-10,1,'last');
bi=find(xi>10,1,'first');

[tmpnpts] = size(xi);
if ((isempty(ai)) && (isempty(bi)))
rZp=interp1(rdWT(1,:),rdWT(2,:),xi);
iZp=interp1(idWT(1,:),idWT(2,:),xi);

elseif ((isempty(ai)) && (isempty(bi)==0))
rZp(1:bi-1)=interp1(rdWT(1,:),rdWT(2,:),xi(1:bi-1));
iZp(1:bi-1)=interp1(idWT(1,:),idWT(2,:),xi(1:bi-1));
rZp(bi:npts)= xi(bi:npts).^(-2);
iZp(bi:npts)=0.0;

elseif ((isempty(ai)==0) && (isempty(bi)))
rZp(ai+1:npts)=interp1(rdWT(1,:),rdWT(2,:),xi(ai+1:npts));
iZp(ai+1:npts)=interp1(idWT(1,:),idWT(2,:),xi(ai+1:npts));
rZp(1:ai) = xi(1:ai).^(-2);
iZp(1:ai)=0.0;

else
rZp(ai+1:bi-1)=interp1(rdWT(1,:),rdWT(2,:),xi(ai+1:bi-1));
iZp(ai+1:bi-1)=interp1(idWT(1,:),idWT(2,:),xi(ai+1:bi-1));
rZp(1:ai)= xi(1:ai).^(-2);
iZp(1:ai)=0.0;
rZp(bi:npts)= xi(bi:npts).^(-2);
iZp(bi:npts)=0.0;
end

Zp(1,:) = rZp;
Zp(2,:) = iZp;

%******************************************************************%
%End of Z-prime function
%******************************************************************%

% wtable.m
% This file creates an ascii table for the derivative of the
% plasma dispersion function d/dxi W(xi)
% as a function of xi=omega/(k*vt)/sqrt(2).
% This only needs to be run once to create the files.

clear all;

ximin=-10;
ximax=10;
xi=ximin:.01:ximax;
L=length(xi);
```

```
N=4;              % determines the precision of the numerical calculation
                  (reduce if your system has a memory issue)

IPV=zeros(1,L);
RP=zeros(1,L);

for i=1:L

% defining how close to get to singularity
phi=.01*abs(xi(i))+1e-6;

% integration step size is N times smaller than phi
dz=phi/N;

% defining arrays symmetrically around xi
zm=xi(i)-phi:-dz:ximin-1;
zp=xi(i)+phi:dz:ximax+1;

nm=length(zm);
np=length(zp);

% performing integrals
    Ip=0;
Im=0;

Ip=dz*sum(zp.*exp(-zp.^2)./(zp-xi(i)));
Im=dz*sum(zm.*exp(-zm.^2)./(zm-xi(i)));

IPV(i)=Ip+Im;

%evaluating real pole contribution
RP(i)=2*phi*(1-2*xi(i)^2);

end

% putting together dW/dxi
dW=1/sqrt(pi)*(IPV+exp(-xi.^2).*(RP-sqrt(-1)*pi*xi));

% forming table
rdWT=[xi; -2*real(dW)];
idWT=[xi; 2*imag(dW)];

% saving as ascii file
save-asciirdWT'rdWT'
save-asciiidWT'idWT'

clear;


%*******************************************************************%
%End of table generation
%*******************************************************************%
```

Appendix E

Review of Work on the Scattering of Radiation from Plasmas

Chapter Outline Head

DOI: 10.1016/B978-0-12-374877-5.00017-8

E.1 INTRODUCTION

This review is based on the previous book Sheffield (1975) and more recent reviews by Bindslev (1992), DeSilva (2000), and Woskov (2005). The main emphasis in this book is on the scattering of electromagnetic waves from plasmas under conditions when $\omega_i \gg \omega_{pe}, \Omega_e, \nu$, when $\lambda_i \ll L$, the dimension of the plasma, and when multiple scattering and absorption are not important. The review in this appendix[1] is therefore slanted toward bringing out the development of this aspect of the subject.

The first interest in scattering arose from the observation by Bailey *et al.* (1952), Forsyth *et al.* (1953), and others of an anomalous reflection of vhf radio waves from the ionosphere. At low frequencies $\omega_i < 30$ mc, the radio waves are reflected because normally $\omega_i < \omega_{pe}$ (the plasma frequency in the ionosphere). However, it was observed that with vhf (30 MHz $< \omega_i <$ 300 MHz), there was a significant reflected signal even when $\omega_i > \omega_{pe}$. This phenomenon was interpreted by Villars & Weisskopf (1955) as evidence for scattering from periodic inhomogeneities (turbulence) in the ionosphere electron density. Their analysis was based on the earlier work of Ratcliffe (1948) and Booker & Gordon (1950) on scattering from atmospheric turbulence. It was this work that stimulated the development of scattering as a more general diagnostic tool for plasmas.

The subject has been reviewed in articles and sections of books by Bowles (1964), Kunze (1968), Evans & Katzenstein (1969a), Evans (1969), Farley (1970), DeSilva & Goldenbaum (1970), Bekefi (1966), Sitenko (1967), Granatstein (1968), Sheffield (1975), Hutchinson (1987), Bretz (1997), Donné *et al.* (2008), Luhmann *et al.* (2008), Glenzer & Redmer (2009).

E.2 SCATTERING FROM THE IONOSPHERE

In 1958, Gordon suggested the use of the backscattering of radio waves from the randomly distributed electrons in the ionosphere (noncollective scattering) as a method of measuring the electron density and temperature from the ground. At this point, we must remind the reader that in this book, "noncollective scattering" means scattering from randomly distributed charges. "Collective scattering" refers to scattering from charges which have positions that are correlated with other charges. The technique was tried by Bowles (1958), who found that the observed frequency spectrum reflected the ion temperature rather than

[1] The authors hope that they have included all the significant work in this field and offer their apologies to authors whose work they have overlooked. For completeness, some references to papers on absorption and scattering with $\omega_i \cong \omega_{pe}$, $\lambda_i \cong L$ are given.

the electron temperature, and further that the total scattered power corresponded to an electron density of half the expected value. The results were confirmed by Pineo *et al.* (1960). Concurrently, Kahn (1959) had calculated that the cross section was affected by the Coulomb interaction between the electrons and ions, and Bowles (1959) suggested that his results were due to this interaction.

The general form of the scattered spectrum for a plasma, including the Coulomb interaction, was determined independently by Fejer (1960), Renau (1960), Dougherty & Farley (1960), and Salpeter (1960). The latter paper covered small departures from equilibrium, such as unequal electron and ion temperatures, and also included multiply charged ions. These theoretical calculations showed that under the conditions of Bowle's experiment for which $\alpha = 1/k\lambda_{De} \gg 1$, the spectrum should in fact reflect the ion velocity distribution. This occurs because the incident wave interacts with the electrons which shield the ions, and further, this can lead to a reduction in the cross section. A simple outline of this phenomenon is given in Chapter 5. The further developments of the theory to include the effects of a magnetic field and collisions are discussed below.

Numerous other experimental investigations of the ionosphere were made. For example, Pineo & Hynek (1962), Evans (1962), and Misyura *et al.* (1969) determined the variation with altitude of the electron density and the electron and ion temperatures by fitting theoretical spectra to the experimental profiles. Pineo *et al.* (1963) observed the narrowing of the ion feature for scattering at directions close to the perpendicular to the magnetic field. Wand (1969) measured the time and space autocorrelation functions of the scattered signal using two incident radio beams and obtained from the results information about the ion–neutral collisions in the ionosphere. The effect of high-energy electrons in the generation and damping of plasma waves was studied by Perkins & Salpeter (1965a), and scattered spectra consistent with their predictions were observed by Perkins *et al.* (1965). Excellent reviews of the subject were given by Bowles (1964), Evans (1969), and Farley (1970).

E.3 SCATTERING FROM LABORATORY PLASMAS WITH $\lambda_i \cong L$ AND $\omega_i \cong \omega_{pe}$

The earliest laboratory experiments were of the scattering of microwaves from cylindrical arc plasmas under circumstances where the incident wavelength λ_i was comparable with the plasma dimensions L and/or the incident frequency ω_i was close to the electron plasma frequency ω_{pe}. The scattered spectrum when the incident wave and incident electric vector are normal to the cylindrical axis contains numerous resonant peaks. These peaks were investigated by, for example, Denno *et al.* (1950), Dattner (1957), Bowles (1958), Akao & Ida (1963), Vandenplas & Gould (1964), and Bryant & Franklin (1963, 1964). The main resonant peak at $\omega \approx \omega_{pe}/\sqrt{2}$ is the natural dipole resonance expected of a cold plasma cylinder in a vacuum. This result and the corrections due to inhomogeneities and finite temperature are discussed by Vandenplas & Gould (1964). The effect of collisions on the resonance was investigated by Platzman & Ozaki (1960) and Bryant & Franklin (1963). In the former paper, the effect of a magnetic field was included.

The peaks that occur at frequencies less than the mean resonance were shown by Trivelpiece & Gould (1959) to be surface waves at the plasma boundary.

The resonance with frequency greater than the main resonance, Tonks–Dattner resonances, were shown by Crawford (1963), to come from longitudinal standing waves, which fit between the wall and the high-density region near the axis where $\omega_i \cong \omega_{pe}$.

E.4 SCATTERING FROM A PLASMA CLOSE TO EQUILIBRIUM, $B=0, \mathrm{v}=0, \lambda_i \ll L, \omega_i \gg \omega_{pe}$

The general form of the equilibrium scattered spectrum, including the Coulomb interaction and therefore correlation effects, was determined independently by Dougherty & Farley (1960), using a generalized version of the Nyquist fluctuation dissipation theory and by Fejer (1960) as an initial value problem using the Boltzmann–Vlasov equation to establish the form of the equilibrium microscopic density fluctuations. At the same time, Salpeter (1960), using what is in essence the Klimontovich hierarchy, gave a treatment which is valid for small departures from equilibrium, as for example in the situation with unequal ion and electron temperatures. Subsequently, the same results were obtained by Rosenbluth & Rostoker (1962) using the BBGKY hierarchy of equations. In this treatment, the effect of a small relative drift of the electrons and ions was considered. Computed spectral profiles for a range of values of the parameters $\alpha, T_e/T_i$, ion charge Z, and normalized drift velocity are to be found in these papers. Spectra for hydrogen plasmas that contain a small percentage of an impurity have been computed by Evans & Carolan (1970).

A quantum mechanical treatment of scattering has been given by Dubois & Gilinsky (1964). The scattering of radiation from electrons in the conduction bands of semiconductors has been discussed by Platzman (1965).

E.4.1 The Non-Collective Spectrum $\alpha \ll 1$

Among the first reports of noncollective scattering from laboratory plasmas were papers by Funfer *et al.* (1963), Thompson & Fiocco (1963), and Schwarz (1963). In all cases, ruby lasers were used. A detailed spectrum, giving a clear-cut electron velocity distribution, was first reported by Davies & Ramsden (1964). Gerry & Rose (1966) compared scattering measurements in a hollow cathode argon arc of the electron temperature and density with those from a Langmuir probe. The technique is now widely applied to the measurement of electron temperature and density. Among the interesting applications was that of Peacock *et al.* (1969) to the T-3 Tokamak plasma (see Chapter 4).

An argon–ion laser was used with synchronous detection by Koons & Fiocco (1968), and a measurement using a CO_2 laser source was reported by Yokoyama *et al.* (1971).

The technique has contributed significantly to the understanding of shock waves, for which good time and space resolution is important. Measurements were made by Patrick (1965), Paul *et al.* (1967), Keilhacker *et al.* (1969), Martone & Segre (1969), and Sheffield *et al.* (1970).

Methods for unfolding the noncollective spectrum in order to obtain the velocity distribution function were described by Brown & Rose (1966) and Williamson & Clarke (1971).

E.4.2 Ion Feature of Collective Spectrum $\alpha \gtrsim 1, \omega_s - \omega_i \ll \omega_{pe}$

The first collective scattering results showing the ion feature of the scattered spectrum were obtained by Bowles (1958) and Pineo *et al.* (1960) by the backscattering of radio waves from the ionosphere. More detailed measurements were later reported by Pineo & Hynek (1962) and Misyura *et al.* (1969). They fitted theoretical spectra to the experimental profiles and determined the variation with altitude of n_e, T_e, and T_i. They observed the ion-acoustic wave satellites, which occur when $T_e \gg T_i$.

The first measurements of the ion feature in the laboratory were made by DeSilva *et al.* (1964) showing the expected narrowing of the spectrum for $\alpha > 1$. Measurements were also reported by Kunze *et al.* (1964), Ascoli-Bartoli *et al.* (1964), and Kronast *et al.* (1966). Scattering of microwaves from ion-acoustic waves in arc plasma was observed by Little & Hamberger (1966). Anderson (1966) made the first fully resolved measurements of the ion feature, using a theoretical fit to find equal electron and ion temperatures and using a comparison with Rayleigh scattering to obtain the electron density. Measurements for α near unity were made by Izawa *et al.* (1967). The frequency, wave number, and amplitude evolution of ion acoustic fluctuations was measured in a laser produced plasma by Bauer *et al.* (1994).

The ion temperature in the high-density plasma ($n > 10^{18}$ cm^{-3}) of a plasma focus was measured by collective scattering by Baconnet *et al.* (1969a,b), Bernard *et al.* (1971), and Peacock & Forrest (1973).

Bulk ion measurements were made first on low-temperature laboratory experiments, for example, Holzhauer (1977), Kasparek & Holzhauer (1983a), and Lachambre & Decoste (1985). The first successful measurement in a tokamak was by Behn *et al.* (1989).

E.4.3 Electron Plasma Frequency Satellite $\alpha \gtrsim 1, \omega_s - \omega_i \cong \omega_{pe}$

The first theoretical treatment of the scattering of radiation from electron plasma waves was presented by Akhiezer *et al.* (1958). The satellites were resolved in scattering from the ionosphere by Perkins & Salpeter (1965a). In the laboratory, Ramsden *et al.* (1966a) and Ramsden & Davies (1966) obtained very clear spectra showing the central ion peak, the ion-acoustic wave effects, and the plasma frequency satellites. The satellites were also observed by Chan & Nodwell (1966). Measurements of the scattering of microwaves from the satellites when $\omega_i \cong \omega_{pe}$ were made by Perepelkin (1966).

Broadening of the satellite peaks due to plasma density inhomogeneities was observed by Röhr (1967), Nodwell & Vanderka (1968), and Chan (1971). The suggestion that such spectra might be unfolded to give information about the density profile in the scattering volume was made by Kronast & Benesch (1969) and is now standard in the analysis of Thomson scattering spectrum, see e.g., Rozmus *et al.* (2000). The effect of finite collection optics on the resolution of the peaks was considered by Pyatnitski & Korobkin (1971).

Resonant peaks at multiples of the plasma frequency and enhancement of the zero frequency in the ion feature were observed by Ringler & Nodwell (1969) in scattering from a hydrogen arc. Similar peaks were reported by Besshaposhnikov *et al.* (1967) and later by Neufeld (1970) and Ludwig & Mahn (1971). In the latter paper, it was shown that these peaks were not a function of the orientation of the scattering vector with respect to the axial magnetic field. Numerous explanations of this phenomenon have been given, among them Kegel (1970) computed similar spectra using a plasma with a two component velocity distribution. Infeld & Zakowicz (1971) suggested that multiples of the plasma frequency will appear for noncollective scattering from the regular electron fluctuations at the plasma frequency; this is by analogy with the appearance of peaks at multiples of the cyclotron frequency in the magnetized case.

E.4.4 Ion Feature in the Presence of a Relative Drift of Ions and Electrons

In the presence of a relative drift, the ion-acoustic satellite on one side of the ion feature is enhanced. Asymmetric spectra that were consistent with this effect have been observed by Evans *et al.* (1966), Baconnet *et al.* (1969b), Evans & Forrest (1969), and Kronast & Pietrzyk (1971). This technique was further used by Hawreliak *et al.* (2004) to study the heat flux across the critical surface of a laser produced plasma. For drift velocities less than the ion acoustic speed, the system is stable. The unstable case is discussed in Section 12.3.

E.4.5 Small $n_e\lambda_{De}^3$

The majority of theoretical calculations of plasma effects are based on an expansion of the general kinetic equations in powers of $\left(n_e\lambda_{De}^3\right)^{-1}$. The corrections to the cooperative electron fluctuations when $n_e\lambda_{De}^3 \cong 1$ have been considered by Theimer (1966). He concludes that deviations from the normal spectra will occur when $(4\pi/3)n_e\lambda_{De}^3 < 0.5$. Comments on this regime have also been made by Boyd *et al.* (1966).

Measurements of collective spectra including both the ion feature and the plasma features were made in the range $n\lambda_{De}^3 = 2 - 10$ by Röhr (1968). He found no significant deviations from the standard (large $n_e\lambda_{De}^3$) results. Similar agreement with standard theory was obtained by Kato (1972), who made measurements at $n_e\lambda_{De}^3 = 2$ and 12. Further work was done by Landen & Winfield (1985) showing good agreement with $n_e\lambda_{De}^3$ as small as 1.5. An anomalous result for $n_e\lambda_{De}^3 \cong 5, \alpha \cong 0.45$ showing an unexpected resonant peak near the plasma frequency has been reported by John *et al.* (1971).

The effect of local field corrections has been predicted to be significant in the collective x-ray Thomson scattering regime by Gregori *et al.* (2003) and the theory of local field corrections has been expanded to include collisions by Fortmann *et al.* (2010).

E.4.6 Nonstationary Inhomogeneous and Nonlinear Effects

The extension of the theory to a nonstationary, inhomogeneous, but nevertheless stable, plasma was discussed by Weinstock (1967). Nonlinear resonant scattering from plasma waves was discussed by Matsuura (1966). Nonthermal plasmas occur, for example, when the electrons have a mean drift, as reported by Fiocco & Thompson (1963) and Wrubel *et al.* (1996). Unequal electron and ion temperatures modify the ion feature as seen by Kronast *et al.* (1966). DeSilva & Stamper (1967) observed a two-electron temperature plasma in a theta pinch shock wave. Non-Maxwellian electron distributions were also reported by Ringler & Nodwell (1969) and Lins (1981). A calculation of the effect of non-Maxwellian velocity distributions, with $\alpha > 1$, on laser-produced plasmas by Zheng *et al.* (1997), showed that significant changes would occur in both the electron and ion features.

E.4.7 Impurities

Fejer (1961) considered the effect of an impurity on scattering from a hydrogen-oxygen plasma. Measurements were made by Evans & Yeoman (1974), Bretz (1973), Sharp *et al.* (1981), Kasparek & Holzhauer (1983a), and Orsitto (1990). DeSilva *et al.* (1992) pointed out that the additional narrow feature due to the impurity had an integrated intensity proportional to the square of the charge state. Glenzer *et al.* (1996) observed the two-ion waves features in Au/Be plasmas and Froula *et al.* (2002) measured the nonlinear growth of ion acoustic waves in two ion-species plasmas.

E.4.8 Energetic Particle Measurements

Energetic particle measurements were proposed for fusion alpha particle detection by Hutchinson *et al.* (1985), using a CO_2 laser, and by Woskoboinikow

(1986), using gyrotron radiation. A design study of a diagnostic based on a 152 μm free-electron laser was made by Shefer *et al.* (1990). A review of experiments in the millimeter to submillimeter region for measurement of the fast ion population in large tokamaks was written by Machuzak *et al.* (1990).

Hutchinson *et al.* (1985), Vahala *et al.* (1986, 1988), and Hughes & Smith (1989), using the electrostatic approximation, showed that collective scattering could be a useful tool for measuring fusion alpha distributions. Subsequently, it was appreciated that a full electromagnetic treatment was required to calculate the expected spectrum (Aamodt & Russell, 1990, 1992; Chiu, 1991; Bindslev, 1993, 1996).

In the millimeter scattering range, accessibility is limited to a narrow range in X-mode or O-mode propagation around the electron cyclotron emission harmonics. Bretz (1997) derived the geometrical form factor for X-mode to X-mode scattering perpendicular to the magnetic field. Hughes & Smith (1989) generalized Bretz's calculations to include X and O mode couplings and directions away from the perpendicular to the magnetic field. Bindslev (1999) has done additional work to model the effect of uncertainties in plasma parameters.

E.5 SCATTERING FROM A MAGNETIZED PLASMA CLOSE TO EQUILIBRIUM

The first calculation of the effect of a steady magnetic field on the noncollective scattering of radiation from a plasma was made by Laaspere (1960). The general theory of scattering from a magnetized plasma, with neglect of the coupling between longitudinal and transverse modes, was presented separately by Hagfors (1961), Farley *et al.* (1961), Fejer (1961), Salpeter (1961a,b), and Rosenbluth & Rostoker (1962). The results of the latter two papers are valid for quasi-equilibrium states, e.g., $T_e \neq T_i$. A detailed discussion of scattering from hybrid resonance has been given by Platzman *et al.* (1968). Contributions from transverse modes were included by Akhiezer *et al.* (1962), Weinstock (1965a,b), and Gorbunov & Silin (1966).

The main influence of this magnetic field is upon the scattered spectrum when the scattering vector **k** is close to normal to the field. Particularly, striking is the modulation of the spectrum at the electron cyclotron frequency. This is the result of scattering from cyclotron oscillations in the noncollective spectrum and from Bernstein modes in the collective spectrum. The first observations of the effect of the magnetic field on the ion feature were made by Pineo *et al.* (1963). The conditions for the detection of the modulation so that the magnitude of the magnetic field could be measured were discussed by Lehner & Pohl (1970) and Theimer & Theimer (1970). The first laboratory measurements of the modulation were made by Kellerer (1970) and Evans & Carolan (1970). The cyclotron peaks have also been observed by Ludwig & Mahn (1971). The effect on the modulation in the scattered spectrum of a finite scattering volume has been investigated by Carolan & Evans (1971).

Methods for measuring the direction of the magnetic field have been suggested by Murakami *et al.* (1970), Perkins (1970), and Sheffield (1972b). Bretz (1974) proposed scanning the change in modulation with the direction of observation. Forrest *et al.* (1978) measured the direction of the magnetic field in a tokamak using the approach suggested by Sheffield (1972b). Siegrist & Salomaa (1982) suggested studying modification to the ion feature when k was near perpendicular to field. Kasparek & Holzhauer (1983b) gave experimental verification. Woskov & Rhee (1992) suggested using the spectral location of the lower hybrid resonance as a way of determining the magnetic fields direction.

The possibility of measuring magnetic field fluctuations was considered by Lehner *et al.* (1989), Haas & Evans (1990), and Vahala *et al.* (1990, 1992).

E.6 SCATTERING FROM LASER-PRODUCED PLASMAS

Thomson scattering measurements of temperatures, densities, ionization state and plasma flow have been applied in numerous laser-produced plasma experiments to study plasma production and heating. Tracy *et al.* (1992) and LaFontaine *et al.* (1994) performed ultraviolet Thomson scattering experiments from laser heated solid disks to test radiation hydrodynamic modeling. Discrepancies between modeling and experiments were resolved by Glenzer *et al.* (1999) when simultaneous measurements of the ion and electron feature showed rapid cooling and recombination consistent with including two electron processes in the calculation of the corona plasma conditions. Hawreliak *et al.* (2004) performed spatially resolved ultraviolet Thomson scattering measurements for studying heat transport into the corona. Young (1994) used Thomson scattering to characterize blow-off plasmas for studies of stimulated Brillouin scattering and Montgomery *et al.* (1999) used single-shot spatially resolved Thomson scattering to characterize blow off plasmas for single hot spot laser-plasma interaction studies.

Laser-produced exploding foils plasmas were characterized for studies of parametric laser backscatter instabilities by Labaune *et al.* (1995) and for multiple beam interactions by Wharton *et al.* (1998). Gregori *et al.* (2003) performed Thomson scattering of nonlocal heat transport processes in laser-produced gas jet plasmas which have been shown to localize by application of large magnetic fields by Froula *et al.* (2007). Malka *et al.* (2001) characterized gas jet plasmas with Thomson scattering to investigate laser-based electron acceleration. Glenzer *et al.* (2000) performed experiments in large-scale length gas bags with an ultraviolet probe laser while Gregori *et al.* (2006) demonstrated the scattering of 9 keV x-ray from 280 eV large scale gas-bag plasmas.

Glenzer *et al.* (1997a) performed the first Thomson scattering measurements of the temperature in the interior of an inertial confinement fusion hohlraums and Glenzer *et al.* (2001b) used Thomson scattering to characterize the conditions of the high-Z hohlraum wall plasma.

E.7 COLLISIONAL EFFECTS

In the theories mentioned above, straight interparticle collisions were ignored, and the Landau damping of the plasma resonances was the only damping mechanism included. In real life, there are Coulomb collisions, and in weakly ionized plasmas, such as the ionosphere, electron–neutral and ion–neutral collisions occur. In addition, both the excitation and damping of density fluctuations are modified in nonequilibrium situations. For weakly ionized plasmas, the BGK collisions term has been used to represent charge–neutral collisions. Fejer (1960), Hagfors (1961), and Gorog (1969) have used the simple nonparticle conserving form, while more precise calculations have been made with the particle conserving form by Dougherty (1963), Dougherty & Farley (1963b), Taylor & Comisar (1963), and Waldteufel (1965). Dougherty and Farley include a treatment for a magnetized plasma. Williamson (1968) and Hagfors & Brockelm (1971) have used a Brownian motion model for charge–neutral collisions and compared it with the BGK model.

For a fully ionized plasma, a number of authors have employed a Fokker–Planck type collision term, for example, Taylor & Comisar (1963), Dougherty (1964), and Grewal (1964). The quasi-equilibrium regime ($T_e \neq T_i$) has been discussed by Grewal (1964), and Chappell & Williams (1971). A quantum-mechanical treatment has been given by Dubois & Gilinsky (1964).

Calculations of the effect of these collisions on the plasma resonance have been made by Ron & Tzoar (1963), Boyd (1965), and Boyd *et al.* (1966) and the dynamic form factor has been derived for ion acoustic fluctuations valid in the entire regime of collisionality by Myatt *et al.* (1998). The effect of ion–ion collisions on the ion–cyclotron resonances is discussed by Farley (1964). The effect of an increased level of high-velocity electrons on the shape of the plasma frequency satellites was investigated by Perkins & Salpeter (1965a). Such velocity distribution functions occur in the ionosphere, and the predicted spectra were observed by Perkins *et al.* (1965). The time and space autocorrelation function of the backscattered signal from the ionosphere was obtained by Wand (1969) using two incident radio beams. He used calculations of Dougherty & Farley (1963b) to determine the ion–neutral collision frequency.

The scattered spectra have been derived using continuum equations by Seasholtz & Tanenbaum (1969), Tanenbaum (1968), and Seasholtz (1971). This approach is valid when the ion neutral mean free path is less than the incident wavelength. One interesting feature is the appearance of resonances, which correspond to ordinary sound waves.

The early measurements of collisional effects include work by Masters & Rye (1976), and Holzhauer (1977). Kasparek & Holzhauer (1984) observed that the two broad ion-acoustic peaks merged into a single peak when the collision rate was sufficiently large. Subsequently, Zhang & DeSilva (1991) showed that in a highly collisional plasma, the ion-acoustic peaks became sharper and a zero frequency feature, called an entropy fluctuation, appeared. This was observed experimentally in optically ionized plasmas by Offenberger *et al.* (1993).

Redmer *et al.* (2005), Höll *et al.* (2004), Thiele *et al.* (2008) developed the dynamic form factor for high density plasma experiments utilizing short wavelength x-ray sources or free-electron lasers and the first experiments on plasmans were performed by Glenzer *et al.* (2007) and Fäustlin *et al.* (2010).

E.8 RELATIVISTIC EFFECTS

The first calculations of the noncollective spectrum for an unmagnetized plasma, to first order in $\bar{\beta}$, were presented by Papas & Lee (1962) and Pogutse (1963). The relativistic noncollective spectrum was derived by Zhuravlev & Petrov (1972). The general spectrum from an unmagnetized plasma to first order in $\bar{\beta}$ was derived by Pappert (1963) and subsequently rederived including drift by Theimer & Sollid (1968a). The ratio of the total cross sections, polarized perpendicular and parallel to the incident polarization, was calculated to second order in β by Theimer & Hicks (1968). Pechacek & Trivelpiece (1967) computed the noncollective spectrum in a complete relativistic treatment. The most significant effect of the high-temperature corrections is the shift of the noncollective spectrum toward shorter wavelengths. The first measurement of such a spectrum was made by Gondhalekar & Kronast (1971). Computations of the noncollective spectra for a relativistic magnetized plasma were made by Nee *et al.* (1969), and Stewart (1972).

Trivelpiece and Pechacek introduced a correction factor $(1-\beta_s)$ to account for the so-called finite transit time effect. Kukushkin (1981) points out that this additional factor only served to correct an error used in the formula for the square of a delta function.

The analysis by Sheffield (1972a, 1975) is corrected in Chapter 4. The effect is quite small. Zhuravlev & Petrov (1972, 1979) derived an analytic solution using the approximation that the depolarization term varied slowly with v. However, using this approach requires a calibration for each temperature and density, as pointed out by Beausang & Prunty (2008), who derived a solution for the case of backward scattering.

Matoba *et al.* (1979) computed the spectra and extended the work of Pogutse and Sheffield to second order in beta for the scattered spectrum. Subsequently, Naito *et al.* (1993) derived an analytic formula for the depolarization term requiring no calibration. Salzmann (1986) used the equations of special relativity to transform between the frame of reference of the electron at rest and the observer. Naito *et al.* (1997) derived a formula for reconstructing relativistic electron distributions from noncollective scattering data.

Palastro *et al.* (2010) derived a fully relativistic general expression for scattered power which is valid in both the collective and noncollecitve regime and for all scattering angles perpendicular to the incident polarization. These calculations were used to derive an experiment by Ross *et al.* (2010) that demonstrated relativistic effects in the collective Thomson scattering spectrum at low temperature ($T_e \ll 1\,\text{keV}$).

E.9 TOTAL SCATTERING CROSS SECTION

The total scattering cross section is $S(\mathbf{k}) = \int d\omega S(\mathbf{k}, \omega)$. For an equilibrium plasma, this quantity is of order 1, but as we recollect, from the introduction to this appendix, it can vary because of the Coulomb screening effects. The total scattered intensity at a given angle is proportional to $n_e S(\mathbf{k})$; therefore, it is important to know the relationship of $S(\mathbf{k})$ to the scattering parameters if we want to measure the electron density. Calculations of $S(\mathbf{k})$ have been made by Dougherty & Farley (1960) and Farley *et al.* (1961), who showed that the cross section is unaffected by the magnetic field; Rosenbluth & Rostoker (1962), Salpeter (1963), and Moorcroft (1963), who presented some detailed computations of $S(\mathbf{k})$; and Farley (1966) and Pilija (1967). Polarization effects in scattering from high-temperature plasmas are discussed by Theimer & Hicks (1968).

E.10 UNSTABLE AND TURBULENT PLASMA

It is possible to analyze the behavior of the plasma in some unstable situations. The basic requirement for such a calculation is usually that the plasma evolves on a timescale slow compared to microscopic timescales so that each phase of the evolution may be treated as a quasi-equilibrium state. A particularly interesting case occurs when the relative drift velocity of electrons and ions approaches and then exceeds a critical value. The critical value depends on T_e/T_i, but is usually in the region of the ion-acoustic speed. The marginal stability case was analyzed by Ichimaru (1962) and Ichimaru *et al.* (1962a). It was shown that the level of ion-acoustic fluctuations on one side of the ion feature increased rapidly as the drift velocity approached the critical value. This leads to a dramatic increase in the scattering cross section because of the N^2 dependence on the fluctuation level of the coherent component. Good agreement with the theoretical predictions was obtained by Arunasalam & Brown (1965), who scattered microwaves from an arc plasma which was in the marginally stable regime. Anomalous results were obtained by Kronast & Pietrzyk (1971) in scattering from ion-acoustic waves. This is discussed by Infeld *et al.* (1972).

Some general comments on the progress of an instability through the unstable region are to be found in articles by Rogister & Oberman (1968) and Joyce & Salat (1971). The case of a marginally unstable partially ionized gas has been investigated by Akhiezer & Angeleik (1969).

Iannuzzi *et al.* (1968) induced ion waves in a alkali metal plasma and then scattered microwaves from them. The scattered signal followed the theoretical predictions, and at large fluctuation amplitude, they detected harmonics of the main frequency. Kalinin *et al.* (1969) have observed microwave scattering from waves in a mirror trapped plasma.

As the plasma is driven into a more unstable regime, the amplitude of the critical wave increases. In some situations, the system reaches a new quasi-equilibrium with a more or less harmonic single large amplitude fluctuation.

An example of this situation occurs in the positive column of an arc, as the axial magnetic field is raised. Kubo & Inuishi (1968) have obtained scattered spectra from the helical instability in the positive column. In the most common case, as the system becomes more unstable, other waves grow and a broad wavespectrum (turbulence) develops.

Akhiezer *et al.* (1958) predicted collective scattering of electromagnetic radiation before Bowles' ionospheric radar scattering measurements.

The first measurements of low-frequency density fluctuations were made by Mazzucato (1976) on ATC tokamak, using microwaves. Also on ATC, collective scattering with CO_2 laser light was performed by Surko & Slusher (1976). Surko & Slusher (1980) described a setup with two intersecting CO_2 laser beams. Evans *et al.* (1982) discussed a Fourier optics approach for understanding far forward scattering. The consequence of scattering in a regime intermediate to the Bragg and Raman–Narath regimes was discussed by Doyle & Evans (1988).

E.10.1 Microscopic Turbulence

The situation when a broad spectrum of waves has developed but the wave amplitude and growth rate are small is called "weak turbulence." General comments on the development of this situation are to be found in a paper by Drummond (1962). The form of the quasi steady state reached in a driven ion acoustic instability has been studied, notably by Kadomstev (1965), Akhieser (1965), and Tsytovich (1970). Their treatments differ in the damping mechanism which they propose to balance the growth of the wave. Possible spectra for a range of unstable situations are discussed by Sakhokiyja & Tsytovich (1968).

Enhanced scattering from the broad spectrum of waves, which develops when a small group of electrons have supra thermal velocities (beam–plasma instability), has been observed by Bohmer & Raether (1966), Malmberg & Wharton (1969), Bollinger & Bohmer (1972), and Arunasalam *et al.* (1971). Enhanced scattering from a high-density theta pinch plasma has been reported by Daehler & Ribe (1967) and Daehler *et al.* (1969). Spectra of the weakly turbulent ion-acoustic type have been observed in collisionless shock waves by Paul *et al.* (1969), Daughney *et al.* (1970), Keilhacker *et al.* (1971), Machalek & Nielsen (1973), Muraoka *et al.* (1973), and Bretz (1973). Enhanced scattering of microwaves from a turbulently heated plasma has been reported by Sharp & Hamberger (1970).

E.10.2 Fluid Turbulence

Turbulence in plasmas has also been studied from the fluid turbulence point of view. In fact, we recollect from the introduction to this appendix that an interest in scattering from plasmas developed from the turbulence studies of Ratcliffe (1948), Booker & Gordon (1950), and Villars & Weisskopf (1955). In this fully turbulent situation, the scattering cross section for the critical waves can be very large, and it is necessary to include multiple scattering processes. The scattered photons have a sort of Brownian motion as they scatter from one turbulent eddy to another. This situation has been discussed by Salpeter & Treiman (1964), Ruffine & Dewolf (1965), Stott (1968), Granatstein (1968), Feinstein & Granatstein (1969), Watson (1970), and Pieroni & Bremmer (1970) and is reviewed by Granatstein and Feinstein. Measurements in this regime have been reported by Guthart *et al.* (1966), Wort (1966), Granatstein (1968), and Graf *et al.* (1971).

E.11 THOMSON SCATTERING APPLIED TO LASER-DRIVEN PLASMA WAVES

Thomson scattering was applied in early studies of the growth of plasma waves by parametric laser-plasma instabilities driven by long wavelength lasers. Villeneuve *et al.* (1991) fielded collective Thomson scattering resolved in time, space, frequency, or wave number. Walsh *et al.* (1984) and Baldis *et al.* (1991) applied k-vector resolved Thomson scattering to measure competition and interactions, respectively, between Langmuir waves excited by stimulated Raman scattering and the onset of large ion acoustic waves driven by the stimulated Brillouin scattering instability. Clayton *et al.* (1983) directly measured the saturation of stimulated Brillouin instability driven ion acoustic waves from a CO_2 laser with ruby Thomson scattering.

Extensive investigations of plasma waves driven by short wavelength glass lasers include Thomson scattering experiments by Renard *et al.* (1996) and Labaune *et al.* (1996) to measure the location of Langmuir waves driven by stimulated Raman scattering and stimulated Brillouin scattering, respectively. Seka *et al.* (2009) applied Thomson scattering to measure Langmuir waves driven by the two plasmon decay instability. Thomson scattering was further applied to measure the Langmuir wave spectra generated by stimulated Raman scattering and the Langmuir decay products, i.e., a second Langmuir wave (Baker *et al.*, 1996) and an ion acoustic wave (Depierreux *et al.*, 2000). Kline *et al.* (2005) measured the transition from fluid to kinetic Langmuir wave behavior.

Glenzer *et al.* (2001) measured saturation of stimulated Brillouin instability driven ion acoustic waves in large-scale length inertial confinement fusion plasmas and Froula *et al.* (2003) observed saturation by velocity gradients and ion trapping induced frequency shifts (Froula *et al.*, 2004). Niemann *et al.* (2004) and Bandulet *et al.* (2004) measured ion acoustic decay waves with k-vector resolved Thomson scattering.

E.12 ABSORPTION OF THE INCIDENT BEAM AND TWO-BEAM SCATTERING

Calculation of the absorption of the incident beam by numerous mechanisms has been made by Ron & Tzoar (1963), Rand (1964), Albini & Rand (1965),

Bornatici *et al.* (1969a,b), Kaw *et al.* (1970), Nicholson-Florence (1971), Martineau & Pepin (1972), Yamanaka *et al.* (1972), and Freidberg (1982).

In many cases, absorption leads to enhancement of waves at the plasma frequency and ion-acoustic frequency. Numerous suggestions have been made for employing a second beam to scatter from these induced fluctuations to obtain information about the plasma. Among the theoretical work in this field are papers by Kroll *et al.* (1964), Kegel (1965), Cheng & Lee (1965), Dubois (1965), Goldman & Dubois (1965), and Dubois & Goldman (1967). On the experimental side, Stern & Tzoar (1966) have used a microwave bean $\omega_0 \cong \omega_{pe}$ to excite oscillation and a second beam $\omega_i \gg \omega_{pe}$ to scatter from it. Stansfield *et al.* (1971) have used organic lasers to achieve this end.

E.13 INDUSTRIAL PLASMAS

Reviews of work on industrial plasmas have been given by Muraoka *et al.* (1998), Warner & Hieftje (2002), and van de Sande (2002). As discussed in Section 9.1, measurements have been made on plasmas with temperatures in the $\sim$0.1–30 eV range and electron densities in the range of 10^{10} to 10^{18} cm^{-3}, often with $n_n \gg n_e$.

Appendix F

Physical Constants and Formulas

PHYSICAL CONSTANTS

		Gaussian	MKS
Electronic charge	e	4.81×10^{-10} stat C	1.602×10^{-19} C
Electron mass	m_e	9.109×10^{-28} g	9.109×10^{-31} kg
Proton mass	m_i	1.673×10^{-24} g	1.673×10^{-27} kg
Speed of light *in vacuo*	c	2.998×10^{10} cm s^{-1}	2.998×10^{8} m s^{-1}
Planck's constant	h	6.626×10^{-27} erg s	6.626×10^{-34} J s
Boltzmann's constant	κ	1.380×10^{-16} erg°K^{-1}	1.380×10^{-23} J°K^{-1}
Avogadro's number			6.022×10^{23} particles mole^{-1}
Permeability of free space	μ_0		$4\pi \times 10^{-7}$ H m^{-1}
Permittivity of free space	ε_0		8.854×10^{-12} F m^{-1}

CONVERSION FACTORS

1 amp $= 3 \times 10^9$ stat A

1 V $= \frac{1}{300}$ stat V

1 Ω $= \frac{1}{9} \times 10^{11}$ s cm^{-1}

1 N $= 10^5$ dyn

1 J $= 10^7$ ergs

1 eV $= 1.602 \times 10^{-12}$ ergs is equivalent to κT for $T = 1.160 \times 10^4$ °K

1 Wb m^{-2} $= 10^4$ G

1 Torr $=$ 1mm Hg contains 3.54×10^{16} particles/cm^3 at 273°K

DOI: 10.1016/B978-0-12-374877-5.00018-X

FORMULAS

Classical electron radius r_0 $\quad r_0^2 = e^4/m_0^2c^4 = 7.95 \times 10^{-26}\,\text{cm}^2$

Electron mean thermal speed
$$a = (2\kappa T_e/m_e)^{1/2} = 6 \times 10^7\,[T_e(\text{eV})]^{1/2}\,\text{cm}\,\text{s}^{-1}$$

Electron Debye length
$$\lambda_{De} = \left(\frac{\kappa T_e}{4\pi e^2 n_e}\right)^{1/2}\text{cm} \quad \text{(Gaussian)}$$
$$= \left(\frac{\varepsilon_0 \kappa T_e}{e^2 n_e}\right)^{1/2}\text{m} \quad \text{(MKS)}$$
$$= 740\left[T(\text{eV})/n_e(\text{cm}^{-3})\right]^{1/2}\text{cm}$$

Electron plasma frequency
$$\omega_{pe} = (4\pi n_e e^2/m_e)^{1/2} \quad \text{(Gaussian)}$$
$$= (n_e e^2/m_e\varepsilon_0)^{1/2} \quad \text{(MKS)}$$
$$= 5.65 \times 10^4\left[n_e(\text{cm}^{-3})\right]^{1/2}\text{rad}\,\text{s}^{-1}.$$

Electron cyclotron frequency
$$\Omega_e = eB/m_e c \quad \text{(Gaussian)}$$
$$= eB/m_e \quad \text{(MKS)}$$
$$= 1.76 \times 10^7 B(\text{G})\,\text{rad}\,\text{s}^{-1}$$

Electron cyclotron radius
$$\rho_e = \frac{v_\perp}{\Omega_e} = \frac{0.57 \times 10^{-7} v_\perp(\text{cm}\,\text{s}^{-1})}{B(\text{G})}\text{cm}$$

Number of electrons in Debye sphere
$$N_D = \frac{4}{3}\pi\lambda_{De}^2 n_e = \frac{1.7 \times 10^9\,[T_e(\text{eV})]^{3/2}}{\left[n_e(\text{cm}^{-3})\right]^{1/2}}\text{electrons}$$

Coulomb collisions See Eqn (2.2.6)

Collision frequency, electrons and singly charge ions ($m_e \ll m_i$), $T_e \cong T_i$

$$\nu_{ei} = 2.92 \times 10^{-6} n_i(\text{cm}^{-3}) \cdot [T_e(\text{eV})]^{-3/2}\ln\Lambda\,\text{s}^{-1}$$

$$\Lambda = \frac{1.53 \times 10^{10}\,[T_e(\text{eV})]^{1/2}}{\left[n_e(\text{cm}^{-3})\right]^{1/2}}$$

Charge neutral collisions See Eqn (2.2.9)

Collision frequency, electron–neutral collisions (r_n is effective radius of neutral)

$$\nu_{en} = 2.8 \times 10^8\,[r_n(\text{cm})]^2 \cdot \left[n_n(\text{cm}^{-3})\right][T_e(\text{eV})]^{1/2}\,\text{s}^{-1}$$

SYMBOLS

A	Area
$a=(2kT_e/m_e)^{1/2}$	Mean electron thermal speed
$\mathbf{B}$	Magnetic field
$b=(2kT_i/m_i)^{1/2}$	Mean ion thermal speed
c	Speed of light in vacuum
$\mathbf{E}$	Electric field
$-e$	Charge for electrons
F, f	Distribution functions
h	Planck's constant
$\mathbf{k}$	Wave number, $k_{\parallel}=\mathbf{k}\cdot\hat{B}_0$
L	Length
m_q	Mass
n_q	Density
P_i	Incident power
$P_s(\mathbf{R}, \omega_s)d\Omega\, d\omega_s$	Scattered power in solid angle $d\Omega$, at $\mathbf{R}$, in $\omega_s \to \omega_s + d\omega_s$
q	Charge
$r_0=e^2/m_e c^2$	Classical electron radius
$\mathbf{r}$	Position
$S(\mathbf{k}, \omega)$	Spectral density function
T_e	Electron temperature
t	Time
V	Volume
$\mathrm{v}_{ph}\equiv\omega/k$	Phase velocity
$\mathbf{v}$	Velocity

$x_e\equiv\omega/ka,\quad x_i=\omega/kb,\quad x_{el}=(\omega-l\Omega_e)/k_{\parallel}a,\quad x_{im}=(\omega-m\Omega_i)/k_{\parallel}b$

$y_e\equiv(\omega-i\nu_e)/ka,\quad y_i=(\omega-i\nu_i)/kb,\quad y_{el}=(\omega-l\Omega_e-i\nu_e)/k_{\parallel}a,$

$y_{im}=(\omega-m\Omega_i-i\nu_i)/k_{\parallel}b$

Ze	Charge for ions
$\alpha\equiv1/k\lambda_{De}$	
$\varepsilon(\mathbf{k}, \omega)$	Dielectric function
θ	Angle
κ	Boltzmann's constant
$\lambda=2\pi/k$	Wavelength
λ_D	Debye length
$\nu_{q\beta}$	Collision frequency
ρ_q	Gyroradius, $\bar{\rho}_e=a/\sqrt{2}\Omega e,\quad \bar{\rho}_i=b/\sqrt{2}\Omega_i$
ϕ	Angle
φ	Angle

$d\Omega$	Element of a solid angle
Ω_q	Gyrofrequency
ω	Angular frequency
ω_{pq}	Plasma frequency

Subscripts

i	Incident wave ($\mathbf{k}_i$, ω_i, $\mathbf{E}_i$, $\mathbf{B}_i$)
q	Charge type; $q=e$ for electrons, $q=$ i for ions
s	Scattered wave ($\mathbf{k}_s$, ω_s, $\mathbf{E}_s$, $\mathbf{B}_s$)

SCATTERING FORMULAS

$$P_s(\mathbf{R}, \omega_s) = P_i \cdot r_0^2 \cdot L\, d\Omega/2\pi\, d\omega_s \left(1 + 2\frac{\omega}{\omega_i}\right) \left|\hat{s} \times (\hat{s} \times \hat{E}_{i0})\right|^2 \cdot n_e \cdot S(\mathbf{k}, \omega) \tag{2.3.17}$$

The scattered power at $\mathbf{R}$ in solid angle $d\Omega$ and frequency range $\omega_s \to \omega_s + d\omega_s$. ($\mathbf{k} = \mathbf{k}_s - \mathbf{k}_i$, $\omega = \omega_s - \omega_i$).

Low-temperature plasma, spectral density function

$$S(\mathbf{k}, \omega) = \lim_{T\to\infty, V\to\infty} \frac{1}{TV} \left\langle \frac{|n_e(\mathbf{k}, \omega)|^2}{n_{e0}} \right\rangle \quad \text{Fourier time transforms} \tag{2.3.14}$$

$$S(\mathbf{k}, \omega) = \lim_{\gamma\to 0} \frac{2\gamma}{V} \left\langle \frac{|n_e(\mathbf{k}, \omega - i\gamma)|^2}{n_{e0}} \right\rangle \quad \text{Laplace time transforms} \tag{2.3.19}$$

High-temperature plasmas [see Eqns (3.5.15), (3.5.23), (4.2.2), (4.2.1), (4.5.3), and (4.6.12)].

For example, Eqn (4.5.1) is the noncollective spectral density function for an equilibrium plasma

$$S(\mathbf{k}, \omega)\, d\omega_s = 2\pi^{1/2} \exp[-(\omega/ka)^2]\, d\omega_s/ka \qquad \text{(dimensionless)}$$

UNITS

P_s and P_i	W or ergs s^{-1}
r_0, L, n_e	m or cm
$d\Omega$	sr
$d\omega_s$	rad s^{-1}

Bibliography

Aamodt, R. E., & Drummond, W. E. (1965). Wave-wave scattering of beam and plasma oscillations. *Physics of Fluids, 8*(1), 171.

Aamodt, R. E., & Russell, D. A. (1990). Scattering of radiation by alpha-particle induced fluctuations. *Review of Scientific Instruments, 61*(10), 3211–3213.

Aamodt, R. E., & Russell, D. A. (1992). Alpha-particle detection by electromagnetic scattering off of plasma fluctuations. *Nuclear Fusion, 32*(5), 745–755.

Abramowitz, M., & Stegun, I. A. (1965). *Handbook of mathematical functions, with formulas, graphs, and mathematical tables*. New York: Dover Publications.

Ackermann, W., Asova, G., Ayvazyan, V., Azima, A., Baboi, N., Bahr, J., *et al.* (2007). Operation of a free-electron laser from the extreme ultraviolet to the water window. *Nature Photonics, 1*(6), 336–342.

Afeyan, B. B., Chou, A. E., Matte, J. P., Town, R. P. J., & Kruer, W. J. (1998). Kinetic theory of electron-plasma and ion-acoustic waves in nonuniformly heated laser plasmas. *Physical Review Letters, 80*(11), 2322–2325.

Akao, Y., & Ida, Y. (1963). Resonance of surface waves on a cylindrical plasma column. *Journal of Applied Physics, 34*(7), 2119.

Akhiezer, A. I., Prokhoda, I. G., & Sitenko, A. G. (1958). Scattering of electromagnetic waves in a plasma. *Soviet Physics JETP-USSR, 6*(3), 576–582.

Akhiezer, A. I., Akhiezer, I. A., & Sitenko, A. G. (1962). Contribution to the theory of plasma fluctuations. *Soviet Physics JETP-USSR, 14*(2), 462–468.

Akhieser, I. A. (1965). Dielectric effects on Thomson scattering in a relativistic magnetized plasma. *Soviet Physics JETP, 21*, 774.

Akhiezer, I. A., & Angeleik, V. V. (1969). Fluctuations and wave scattering in a plasma subject to strong electric and magnetic fields. *Soviet Physics JETP-USSR, 28*(6), 1216.

Akre, R. (2008). *Physical Review Special Topics – Accelerators and Beams, 11*, 030703.

Albini, F., & Rand, S. (1965). Nonlinear plasma absorption of high-frequency radiation. *Physics of Fluids, 8*(1), 134.

Aliev, Y. M., Silin, V. P., & Watson, C. (1966). *Soviet Physics JETP, 13*, 626.

Amemiya, Y., & Miyahara, J. (1988). Imaging plate illuminates many fields. *Nature, 336*(6194), 89–90.

Anderson, O. A. (1966). Measurement of electron correlation spectrum in a plasma. *Physical Review Letters, 16*(22), 978.

Artsimovich, L. A., Bobrovsky, G. A., Gorbunov, E. P., Ivanov, D. P., Kirillov, V. D., Kuznetsov, E. I., *et al.* (1969). *Plasma physics and controlled nuclear fusion research (International Atomic Energy Agency)*, 3rd, Novosibirsk, Soviet Union. IAEA, Vienna, *1*, 157. Paper CN24/B1.

Arunasalam, V., & Brown, S. C. (1965). Microwave scattering due to acoustic ion plasma wave instability. *Physical Review, 140*(2A), A471.

Arunasalam, V., Heald, M. A., & Sinnis, J. (1971). Microwave scattering from unstable electron plasma waves. *Physics of Fluids, 14*(6), 1194.

DOI: 10.1016/B978-0-12-374877-5.00019-1

Ascoli-Bartoli, U., Katzenstein, J., & Lovisetto, L. (1964). Forward scattering of light from a laboratory plasma. *Nature (London), 204*, 672–673.

Ascoli-Bartoli, U., Benedetti-Michelangele, G., & DeMarco, F. (1967). *An Improvement in Fabry-Perot Spectrometry Applied Optics, 6*, 467.

Baconnet, J. P., Cesari, G., Coudeville, A., & Watteau, J. P. (1969a). *Proceedings of the 9th international conference on phenomena in ionized gases*.

Baconnet, J. P., Cesari, G., Coudevil, A., & Watteau, J. P. (1969b). 90 degrees laser light scattering by a dense plasma focus. *Physics Letters A, 29*(1), 19.

Bailey, D. K., Bateman, R., Berkner, L., Booker, H. G., Montgomery, G. F., Purcell, E. M., *et al.* (1952). A new kind of radio propagation at very high frequencies observable over long distances. *Physical Review, 86*, 141.

Bakker, L. P., & Kroesen, G. M. W. (2001). Thomson scattering in a low-pressure neon mercury positive column. *Journal of Applied Physics, 90*(8), 3720–3725.

Baldis, H. A., Villeneuve, D. M., Labaune, C., *et al.* (1991). Coexistence of stimulated Raman and Brillouin-scattering in laser-produced plasmas. *Physics of Fluids B-Plasma Physics, 3*(8), 2341–2348.

Baldwin, D. E., & Rowlands, G. (1966). Plasma oscillations perpendicular to a weak magnetic field. *Physics of Fluids, 9*(12), 2444.

Bandulet, H. C., Labaune, C., Lewis, K., *et al.* (2004). Thomson-scattering study of the subharmonic decay of ion-acoustic waves driven by the Brillouin instability. *Physical Review Letters, 93*(3), 035002.

Barbrel, B., Koenig, M., Benuzzi-Mounaix, A., Brambrink, E., Brown, C. R. D., Gericke, D. O., *et al.* (2009). Measurement of short-range correlations in shock-compressed plastic by short-pulse x-ray scattering. *Physical Review Letters, 102*(16), 165004.

Barth, C. J., van der Meiden, H. J., Oyevaar, T., & Cardozo, N. J. L. (2001). High resolution multiposition Thomson scattering for the TEXTOR Tokamak. *Review of Scientific Instruments, 72*(1), 1138–1142.

Bassan, M., Giudicotti, L., & Pasqualotto, R. (1993). Nonlinear-optical effects in Raman calibrations of a Thomson scattering system. *Applied Optics, 32*(27), 5313–5323.

Bateman, J. E. (1977). Detection of hard x-rays (10-140 keV) by channel plate electron multipliers. *Nuclear Instruments & Methods, 144*(3), 537–545.

Bauer, B. S., Drake, R. P., Estabrook, K. G., *et al.* (1995). Detection of ion-plasma waves by collective Thomson scattering. *Physical Review Letters, 74*(18), 3604–3607.

Beausang, K. V., & Prunty, S. L. (2008). An analytic formula for the relativistic Thomson scattering spectrum for a Maxwellian velocity distribution. *Plasma Physics and Controlled Fusion, 50*(9), 095001. Figure reprinted courtesy of K.V. Beausang. Copyright 2008 by the Institue of Physics.

Behn, R., Dicken, D., Hackmann, J., Salito, S.A., Siegrist, M.R., Krug, P.A., Kjelberg, I., Duval, B., Joye, B., and Pochelon A. (1989). Ion temperature meaurement of collective thomson scattering of D_2O laser radiation, *Phys. Rev. Lett. 62*, 2833. Figures reprinted with permission from R. Behn. Copyright 1989 by the American Institute of Physics.

Beiersdorfer, P., Lopez-Urrutia, J. R. C., Springer, P., Utter, S. B., & Wong, K. L. (1999). Spectroscopy in the extreme ultraviolet on an electron beam ion trap. *Review of Scientific Instruments, 70*(1), 276–279.

Beiersdorfer, P., Brown, G. V., Goddard, R., & Wargelin, B. J. (2004). High-resolution crystal spectrometer for the 10-60 angstrom extreme ultraviolet region. *Review of Scientific Instruments, 75*(10), 3720–3722.

Bekefi, G. (1966). *Radiation processes in plasmas*. Wiley series in plasma physics. New York: Wiley.

Bernard, A., Coudevil, A., Garconne, J. P., Jolas, A., Demascur, J., & Watteau, J. P. (1971). Forward laser scattering by plasma focus during neutron emission. *Physics Letters A, 35*(1), 7.

Bernstein, I. B. (1958). Waves in a plasma in a magnetic field. *Physical Review, 109*(1), 10–21.

Bernstein, I. B., Trehane, S. K., & Weenink, M. (1964), *Nuclear Fusion, 4*, 61.

Besshaposhnikov, A. A., Voloshin, A. E., Kuchuberya, I. Kh., Sidorov, V. P., & Simonova, N. N. (1967). *Zhurnal Prikladnoi Spektrosk, 6*, 172.

Bhatnagar, A. L., Gross, E. P., & Krook, M. (1954). *Physical Review, 94*, 511.

Bindslev, H. (1991). Dielectric effects on Thomson scattering in a relativistic magnetized plasma. *Plasma Physics and Controlled Fusion, 33*, 1775.

Bindslev, H. (1992). On the theory of Thomson scattering and reflectometry in a relativistic magnetized plasma. Technical Report, Roskilde, Denmark: Risø National Laboratory.

Bindslev, H. (1993). 3-wave mixing and Thomson scattering in plasmas. *Plasma Physics and Controlled Fusion, 35*(11), 1615–1640.

Bindslev, H. (1996). A quantitative study of scattering from electromagnetic fluctuations in plasmas. *Journal of Atmospheric and Terrestrial Physics, 58*(8–9), 983–989.

Bindslev, H. (1999). Methods for optimizing and assessing diagnostic capability, demonstrated for collective Thomson scattering (invited). *Review of Scientific Instruments, 70*(1), 1093–1099.

Bindslev, H., Hoekzema, J. A., Egedal, J., Fessey, J. A., Hughes, T. P., & Machuzak, J. S. (1999a). Fast-ion velocity distributions in jet measured by collective Thomson scattering. *Physical Review Letters, 83*(16), 3206–3209.

Bindslev, H., Meo, F., & Korsholm, S. (2003). ITER fast ion collective scattering. Technical Report, Risø National Laboratory, Roskilde, Denmark. 20 November. http://www.risoe.dtu.dk/Research/sustainable_energy/Fusion_energy/projects/fusion_CTS/ITER.aspx.

Bindslev, H., Meo, F., Tsakadze, E. L., Korsholm, S. B., & Woskov, P. (2004). Feasibility study of fast ion diagnosis in iter by collective Thomson scattering, millimeter waves to CO_2 laser. *Review of Scientific Instruments, 75*(10), 3598–3600.

Bindslev, H., Nielsen, S. K., Porte, L., Hoekzema, J. A., Korsholm, S. B., Meo, F., *et al.* (2006). Fast-ion dynamics in the TEXTOR Tokamak measured by collective Thomson scattering. *Physical Review Letters, 97*(20), 205005.

Bindslev, H., Nielsen, S. K., Korsholm, S. B., Meo, F., Michelsen, P. K., Michelsen, S., *et al.* (2007). Fast ion dynamics in magnetically confined plasma measured by collective Thomson scattering. *Journal of Plasma and Fusion Research, 2*, S1023.

Bohmer, H., & Raether, M. (1966). Incoherent scattering of microwaves by unstable electron plasma oscillations. *Physical Review Letters, 16*(25), 1145.

Bollinger, L. D., & Bohmer, H. (1972). Microwave combination scattering from plasma-waves in an electron beam-plasma system. *Physics of Fluids, 15*(4), 693.

Booker, H. G., & Gordon, W. E. (1950). A theory of radio scattering in the troposphere. *Proceedings of the IRE, 38*, 22.

Born, M., & Wolf, E. (1965). *Principles of optics; electromagnetic theory of propagation, interference and diffraction of light* (3rd Rev. ed.). Oxford, NY: Pergamon Press.

Bornatici, M., Cavalier, A., & Engelmann, B. E. (1969a). *Lettere Al Nuovo Cimento, 1*, 713.

Bornatici, M., Cavalier, A., & Engelman, F. (1969b). Enhanced scattering and anomalous absorption of light due to decay into plasma waves. *Physics of Fluids, 12*(11), 2362.

Boutot, J. P., Lavoute, P., & Eschard, G. (1987). Multianode photomultiplier for detection and localization of low light level events. *IEEE Transactions on Nuclear Science, 34*(1), 449–452.

Bowden, M. D., Okamoto, T., Kimura, F., Muta, H., Uchino, K., Muraoka, K., *et al.* (1993). Thomson scattering measurements of electron-temperature and density in an electron-cyclotron resonance plasma. *Journal of Applied Physics, 73*(6), 2732–2738.

Bowden, M. D., Goto, Y., Yanaga, H., Howarth, P. J. A., Uchino, K., & Muraoka, K. (1999). A Thomson scattering diagnostic system for measurement of electron properties of processing plasmas. *Plasma Sources Science and Technology, 8*(2), 203–209.

Bowles, K. L. (1958). Observation of vertical-incidence scatter from the ionosphere at 41 mc-sec. *Physical Review Letters, 1*(12), 454–455.

Bowles, K. L. (1959). Technical Report, National Bureau of Standards. Washington, DC.

Bowles, K. L. (1964). Radio waves scattering in the ionosphere. *Advances in Electronics and Electron Physics, 19*(55), 55–176.

Boyd, T. J. M. (1965). Technical Report CLM-R 52. Culham Laboratory, UK.

Boyd, T. J. M., Evans, D. E., & Katzenst, J. (1966). Light scattering by plasmas. *Physics Letters, 22*(5), 589.

Brau, J. E. (2004). *Modern problems in classical electrodynamics*. Oxford University Press Inc., New York, New York 10016 ISBN 0-19-514665-4.

Bretz, N. L. (1973). Technical Report 73-073. Technical Report, University of Maryland.

Bretz, N. (1974). Proposed method of measuring current distribution in a tokamak plasma. *Applied Optics, 13*(5), 1134–1140.

Bretz, N. (1997). Diagnostic instrumentation for microturbulence in tokamaks. *Review of Scientific Instruments, 68*(8), 2927–2964.

Brittin, W. E., Barut, A.O., & Guenin, M. (1967). Lectures in Theoretical Physics, Vol. 9 (PTC).

Brower, D. L., Peebles, W. A., & Luhmann, N. C. (1981). Cavity operation of a pulsed C-13H3F laser. *International Journal of Infrared and Millimeter Waves, 2*(1), 35–47.

Brower, D. L., Peebles, W. A., Kim, S. K., Luhmann, N. C., Tang, W. M., & Phillips, P. E. (1987). Observation of a high-density ion mode in tokamak microturbulence. *Physical Review Letters, 59*(1), 48–51.

Brower, D. L., Peebles, W. A., Kim, S. K., & Luhmann, N. C. (1988a). Application of fir heterodyne-detection to collective scattering measurements of tokamak microturbulence. *Review of Scientific Instruments, 59*(8), 1559–1561.

Brower, D. L., Park, H. K., Peebles, W. A., & Luhmann, N. C. (1988b). *Multichannel far-infrared collective scattering system for plasma wave studies. Topics in Millimeter Wave Technology* (Vol. 2). New York: Academic Press.

Brower, D. L., Redi, M. H., Tang, W. M., Bravenec, R. V., Durst, R. D., Fan, S. P., *et al.* (1989). Experimental-evidence for ion pressure-gradient driven turbulence in text. *Nuclear Fusion, 29*(8), 1247–1254.

Brown, T. S., & Rose, D. J. (1966). Plasma diagnostics using lasers – relations between scattered spectrum and electron-velocity distribution. *Journal of Applied Physics, 37*(7), 2709.

Bryant, G. H., & Franklin, R. N. (1963). Scattering of a plane wave by a bounded plasma. *Proceedings of the Physical Society of London, 81*(521), 531.

Bryant, G. H., & Franklin, R. N. (1964). Electrostatic plasma resonances. *Proceedings of the Physical Society of London, 83*(5366), 971.

Burgess, D. W. (1971). London: Imperial College. Private communication.

Callen, H. B., & Welton, T. A. (1951). Irreversibility and generalized noise. *Physical Review, 83*(1), 34–40.

Callen, I. D., & Guest, G. E. (1973). Electromagnetic effects on electrostatic modes in a magnetized plasma. *Nuclear Fusion, 13*(1), 87–110.

Candy, J., & Waltz, R. E. (2003). An Eulerian Gyrokinetic-Maxwell solver. *Journal of Computational Physics, 186*(2), 545–581.

Cantarini, J., Knauer, J. P., & Pasch, E. (2009). Studies for the design of the Wendelstein 7-x Thomson scattering polychromators. *Fusion Engineering and Design, 84*(2–6), 540–545.

Carlstrom, T. N., Deboo, J. C., Evanko, R., Greenfield, C. M., Hsieh, C. L., Snider, R. T., *et al.* (1990). A compact, low-cost, 7-channel polychromator for Thomson scattering measurements on DIII-D (invited). *Review of Scientific Instruments, 61*(10), 2858–2860.

Carlstrom, T., Campbell, G. L., DeBoo, J. C., Evanko, R., Evans, J., Greenfield, C. M., Haskovec, J., Hsieh, C. L., McKee, E., Snider, R. T., Stockdale, R., Trost, P. K., and Thomas, M.P. (1992). Design and operation of the multipulse Thomson scattering diagnostic on DIII-D (invited). *Rev. Sci. Instrum.* 63, 4901. Figures reprinted with permission from T. Carlstrom. Copyright 1992 by the American Institute of Physics.

Carolan, P. G., & Evans, D. E. (1971). Influence of symmetry about magnetic vector on spectrum of light scattered by a magnetized plasma. *Plasma Physics, 13*(10), 947.

Chan, P. W. (1971). Evidence of density inhomogeneity broadening of plasma satellites from light scattering. *Physics of Fluids, 14*(12), 2787.

Chan, P. W., & Nodwell, R. A. (1966). Collective scattering of laser light by a plasma. *Physical Review Letters, 16*(4), 122.

Chang, L. Y., Zeng, L., Cao, J. X., Yu, C. X., & Zu, Q. X. (1990a). Dual laser acoustooptical diffraction technique for the calibration of a CO_2 2-laser scattering apparatus. *Review of Scientific Instruments, 61*(9), 2314–2317.

Chang, L. Y., Zeng, L., Cao, J. X., Yu, C. X., & Zu, Q. X. (1990b). Study of microturbulence on the KT-5 Tokamak by CO_2-laser scattering. *Chinese Physics Letters, 7*(1), 16–19.

Chappell, W. R., & Williams, R. H. (1971). Density fluctuations in a magnetized, weakly ionized plasma. *Physics of Fluids, 14*(9), 1938.

Chen, L. (1987). *Waves and instabilities in plasmas*. World Scientific lecture notes in physics Vol. 12. Singapore: World Scientific.

Chen, S. T., & Chatterjee, M. R. (1996). A numerical analysis and expository interpretation of the diffraction of light by ultrasonic waves in the Bragg and Raman-Nath regimes using multiple scattering theory. *IEEE Transactions on Education, 39*(1), 56–68.

Chen, Y., Parker, S. E., Cohen, B. I., Dimits, A. M., Nevins, W. M., Shumaker, D., *et al.* (2003). Simulations of turbulent transport with kinetic electrons and electromagnetic effects. *Nuclear Fusion, 43*(10), 1121–1127.

Cheng, H., & Lee, Y. C. (1965). Interaction of light with light in a plasma. *Physical Review Letters, 14*(12), 426.

Chihara, J. (1987). Difference in x-ray-scattering between metallic and nonmetallic liquids due to conduction electrons. *Journal of Physics F: Metal Physics, 17*(2), 295–304.

Chihara, J. (2000). Interaction of photons with plasmas and liquid metals-photoabsorption and scattering. *Journal of Physics-Condensed Matter, 12*(3), 231–247.

Chiu, S. C. (1991). Electromagnetic effects in fluctuations of nonequilibrium plasmas and applications to alpha diagnostics. *Physics of Fluids B, 3*(6), 1374–1380.

Chung, H. K., Morgan, W. L., & Lee, R. W. (2003). Flychk: An extension to the K-shell spectroscopy kinetics model fly. *Journal of Quantitative Spectroscopy and Radiative Transfer, 81*(1–4), 107–115.

Clayton, C. E., Joshi, C., & Chen, F. F. (1983). Ion-trapping saturation of the brillouin instability. *Physical Review Letters, 51*(18), 1656–1659.

Clayton, C. E., Darrow, C., & Joshi, C. (1985). Novel small-angle collective Thomson scattering system. *Applied Optics, 24*(17), 2823–2828.

Clemmow, P. C., & Dougherty, J. P. (1969). *Electrodynamics of particles and plasmas. Addison-Wesley series in advanced physics*. Reading, MA: Addison-Wesley Pub. Co.

Coda, S., Porkolab, M., & Carlstrom, T. N. (1992). A phase-contrast interferometer on DIII-D. *Review of Scientific Instruments, 63*(10), 4974–4976.

Coda, S., Porkolab, M., & Burrell, K. H. (2000). Decorrelation of edge plasma turbulence at the transition from low- to high-confinement mode in the DIII-D Tokamak. *Physics Letters A, 273*(1–2), 125–131.

Cordey, J. G. (1976). Effects of particle trapping on slowing-down of fast ions in a toroidal plasma. *Nuclear Fusion, 16*(3), 499–507.

Crawford, F. W. (1963). The mechanism of Tonks-Dattner plasma resonances. *Physics Letters, 5*(4), 244–247.

Daehler, M., & Ribe, F. L. (1967). Cooperative light scattering from theta-pinch plasmas. *Physical Review, 161*(1), 117.

Daehler, M., Sawyer, G. A., & Thomas, K. S. (1969). Coordinated measurements of plasma density and cooperative light scattering in a theta-pinch plasma. *Physics of Fluids, 12*(1), 225.

D'Ambrosio, C., & Leutz, H. (2003). Hybrid photon detectors. *Nuclear Instruments & Methods in Physics Research Section A-Accelerators Spectrometers Detectors and Associated Equipment, 501*(2–3), 463–498.

Dattner, A. (1957). Plasma resonance. *Ericsson Technology, 3*(2), 309.

Daughney, C. C., Holmes, L. S., & Paul, J. W. M. (1970). Measurement of spectrum of turbulence within a collisionless shock by collective scattering of light. *Physical Review Letters, 25*(8), 497.

Davies, W. E. R., & Ramsden, S. A. (1964). Scattering of light from the electrons in a plasma. *Physics Letters, 8*(3), 179–180.

Dawson, J. M., & Oberman, C. (1962). High-Frequency Conductivity and the Emission and Absorption Coefficients of a Fully Ionized Plasma. *The Physics of Fluids, 5*(5), 517.

Dawson, J. M., & Nakayama, T. (1966). Kinetic structure of a plasma. *Physics of Fluids, 9*(2), 252.

de la Luna, E., Krivenski, V., Giruzzi, G., Gowers, C., Prentice, R., Travere, J. M., *et al.* (2003). Impact of bulk non-Maxwellian electrons on electron temperature measurements (invited). *Review of Scientific Instruments, 74*(3), 1414–1420.

Denno, S. N., Prime, H. A., & Craggs, J. D. (1950). The scattering of 3-cm radiation by ionized gases. *Proceedings of the Physical Society of London Section B, 63*(369), 726–727.

Depierreux, S., Labaune, C., Fuchs, J., Pesme, D., Tikhonchuk, V. T., & Baldis, H. A. (2002). Langmuir decay instability cascade in laser-plasma experiments. *Physical Review Letters, 89*(4), 045001.

DeSilva, A. W., Evans, D. E., & Forrest, M. J. (1964). Observation of Thomson + co-operative scattering of ruby laser light by plasma. *Nature, 203*(495), 1321.

DeSilva, A. W., & Stamper, J. A. (1967). Observation of anomalous electron heating in plasma shock waves. *Physical Review Letters, 19*(18), 1027.

DeSilva, A. W., & Goldenbaum, G. (1970). *Methods of experimental physics* (Vol. 19, Pt. A, Chap. 3). New York: Academic Press.

DeSilva, A. W., Baig, T. J., Olivares, I., & Kunze, H. J. (1992). The effects of impurity ions on the scattered-light spectrum of a plasma. *Physics of Fluids B-Plasma Physics, 4*(2), 458–464.

DeSilva, A. W. (2000). The evolution of light scattering as a plasma diagnostic. *Contributions to Plasma Physics, 40*(1–2), 23–35.

Dhawan, S., & Majka, R. (1977). Development status of microchannel plate photomultipliers. *IEEE Transactions on Nuclear Science, 24*(1), 270–275.

Diamond, P. H., Itoh, S. I., Itoh, K., & Hahm, T. S. (2005). Zonal flows in plasma – a review. *Plasma Physics and Controlled Fusion, 47*(5), R35–R161.

Dietlein, C., Luukanen, A., Penttilä, J. S., Sipola, H., Groönberg, L., Seppä, H., *et al.* (2007). Performance comparison of Nb and Nbn antenna-coupled microbolometers. In O. J. James (Ed.), *Terahertz for military and security applications* (Vol. 6549, p. 65490M). Proceedings of SPIE. http://dx.doi.org/10.1117/12.719586.

Donné, A. J. H., & Barth, C. J. (2006). Laser-aided plasma diagnostics. *Fusion Science and Technology, 49*(2T), 375–386.

Donné, A. J. H., Barth, C. J., & Weisen, H. (2008). Laser-aided plasma diagnostics. *Fusion Science and Technology, 53*(2), 397–430.

Döppner, T., Neumayer, P., Girard, F., Kugland, N. L., Landen, O. L., Niemann, C., *et al.* (2008). High order reflectivity of highly oriented pyrolytic graphite crystals for x-ray energies up to 22 keV. *Review of Scientific Instruments, 79*(10), 10E311.

Döppner, T., Landen, O. L., Lee, H. J., Neumayer, P., Regan, S. P., & Glenzer, S. H. (2009). Temperature measurement through detailed balance in x-ray Thomson scattering. *High Energy Density Physics, 5*(3), 182–186.

Dorland, W., Jenko, F., Kotschenreuther, M., & Rogers, B. N. (2000). Electron temperature gradient turbulence. *Physical Review Letters, 85*(26), 5579–5582.

Dougherty, J. P., & Farley, D. T. (1960). A theory of incoherent scattering of radio waves by a plasma. *Proceedings of the Royal Society of London Series A-Mathematical and Physical Sciences, 259*(1296), 79–99.

Dougherty, J. P. (1963). The conductivity of a partially ionized gas in alternating electric fields. *Journal of Fluid Mechanics, 16*(1), 126–137.

Dougherty, J. P., & Farley, D. T. (1963b). A theory of incoherent scattering of radio waves by a plasma .3. scattering in a partly ionized gas. *Journal of Geophysical Research, 68*(19), 5473.

Dougherty, J. P. (1964). Model Fokker-Planck equation for a plasma and its solution. *Physics of Fluids, 7*(11), 1788–1799.

Doyle, E. J., & Evans, D. E. (1988). Asymmetric light-scattering effects in tokamak plasmas. *Review of Scientific Instruments, 59*(8), 1574–1576.

Doyle, E. J., Greenfield, C. M., Austin, M. E., Baylor, L. R., Burrell, K. H., Casper, T. A., *et al.* (2002). Progress towards increased understanding and control of internal transport barriers in DIII-D. *Nuclear Fusion, 42*(3), 333–339.

Drake, R. P. (2006). *High-energy-density physics*. New York: Springer Berlin Heidelberg.

Drummond, W. E. (1962). Microwave scattering from unstable plasma waves. *Physics of Fluids, 5*(9), 1133–1134.

Drummond, W. E., & Pines, D. (1962b). Non-linear stability of plasma oscillations. *Nuclear Fusion*, Suppl. Pt. 3, 1049–1057.

Dubois, D. F., & Gilinsky, V. (1964). Incoherent scattering of radiation by plasmas. I. quantum mechanical calculation of scattering cross sections. *Physical Review A-General Physics, 133*(5A), 1308.

Dubois, D. F. (1965). Nonlinear scattering of radiation from plasmas. *Physical Review Letters, 14*(20), 818.

Dubois, D. F., & Goldman, M. V. (1967). Parametrically excited plasma fluctuations. *Physical Review, 164*(1), 207.

Dupree, T. H. (1966). A perturbation theory for strong plasma turbulence. *Physics of Fluids, 9*(9), 1773.

Dürr, W., & Schmidt, W. (1985). Measurement of acoustooptic interaction in germanium in the far infrared. *International Journal of Infrared and Millimeter Waves, 6*(10), 1043–1049.

Eidmann, K., Sachsenmaier, P., Salzmann, H., & Sigel R. (1972). Optical isolators for highpower giant-pulse lasers. *Journal of Scientific Instruments, 5*, 56.

Einaudi, F., & Sudan, R. N. (1969). A review of nonlinear theory of plasma oscillations. *Plasma Physics, 11*(5), 359.

Eisele, H., & Kamoua, R. (2004). Submillimeter-wave InP Gunn devices. *IEEE Transactions on Microwave Theory and Techniques, 52*(10), 2371–2378.

Evans, D. E., Forrest, M. J., & Katzenst, J. (1966). Asymmetric co-operative scattered light spectrum in a thetatron plasma. *Nature, 212*(5057), 21.

Evans, D. E., & Forrest, M. J. (1969). *Proceedings of the 9th international conference on phenomena in ionized gases* (p. 646). Bucharest, Romania: Edituura Academia Republicii Socialiste Romania.

Evans, D. E., & Katzenstein, A. (1969a). Laser light scattering in laboratory plasmas. *Reports on Progress in Physics, 32*(2), 207.

Evans, D. E., & Carolan, P. G. (1970). Measurement of magnetic field in a laboratory plasma by Thomson scattering of laser light. *Physical Review Letters, 25*(23), 1605.

Evans, D. E., & Yeoman, M. L. (1974). *Physical Review Letters, 33*(2), 76–79.

Evans, D. E., Vonhellermann, M., & Holzhauer, E. (1982). Fourier optics approach to far forward scattering and related refractive-index phenomena in laboratory plasmas. *Plasma Physics and Controlled Fusion, 24*(7), 819–834.

Evans, D. E., Doyle, E. J., Frigione, D., Vonhellermann, M., & Murdoch, A. (1983). Measurement of long wavelength turbulence in a tokamak by extreme far forward scattering. *Plasma Physics and Controlled Fusion, 25*(6), 617–640.

Evans, J. V. (1962). Diurnal variation of temperature of F region. *Journal of Geophysical Research, 67*(12), 4914.

Evans, J. V. (1969). Theory and practice of ionosphere study by Thomson scatter radar. *Proceedings of the IEEE, 57*(4), 496.

Everett, M. J., Lal, A., Clayton, C. E., Mori, W. B., Johnston, T. W., & Joshi, C. (1995). Coupling between high-frequency plasma-waves in laser-plasma interactions. *Physical Review Letters, 74*(12), 2236–2239.

Farley, D. T., Dougherty, J. P., & Barron, D. W. (1961). A theory of incoherent scattering of radio waves by a plasma .2. scattering in a magnetic field. *Proceedings of the Royal Society of London Series A-Mathematical and Physical Sciences, 263*(131), 238.

Farley, D. T. (1964). Effect of coulomb collisions on incoherent scattering of radio waves by plasma. *Journal of Geophysical Research, 69*(1), 197.

Farley, D. T. (1966). A theory of incoherent scattering of radio waves by a plasma .4. effect of unequal ion and electron temperatures. *Journal of Geophysical Research, 71*(17), 4091.

Farley, D. T. (1970). Incoherent scattering at radio frequencies. *Journal of Atmospheric and Terrestrial Physics, 32*(4), 693.

Fäustlin, R. R., Bornath, Th., Döppner, T., Dusterer, S., Förster, E., Fortmann, C., *et al.* (2010a). Observation of ultrafast non-equilibrium collective dynamics in warm dense hydrogen. *Physical Review Letters, 104*, 125002.

Fäustlin, R. R., Zastrau, U., Toleikis, S., Uschmann, I., Förster, E., & Tschentscher, Th. (2010b). A compact soft x-ray spectrograph combining efficiency and resolution. *Journal of Instrumentation, 5*.

Feinstein, D. L., & Granatstein, V. L. (1969). Scalar radiative transport model for microwave scattering from a turbulent plasma. *Physics of Fluids, 12*(12P1), 2658.

Fejer, J. A. (1960). Scattering of radio waves by an ionized gas in thermal equilibrium. *Canadian Journal of Physics, 38*(8), 1114–1133.

Fejer, J. A. (1961). Scattering of radio waves by an ionized gas in thermal equilibrium in presence of a uniform magnetic field. *Canadian Journal of Physics, 39*(5), 716.

Fernandez, J. C., Cobble, J. A., Montgomery, D. S., Wilke, M. D., & Afeyan, B. B. (2000). Observed insensitivity of stimulated Raman scattering on electron density. *Physics of Plasmas, 7*(9), 3743.

Filip, C. V., Tochitsky, S. Y., Narang, R., Clayton, C. E., Marsh, K. A., & Joshi, C. J. (2003). Collinear Thomson scattering diagnostic system for the detection of relativistic waves in low-density plasmas. *Review of Scientific Instruments, 74*(7), 3576–3578.

Filip, C. V., Narang, R., Tochitsky, S. Y., Clayton, C. E., Musumeci, P., Yoder, R. B., *et al.* (2004). Nonresonant beat-wave excitation of relativistic plasma waves with constant phase velocity for charged-particle acceleration. *Physical Review E, 69*(2), 026404.

Finkelburg, W., & Peters, T. (1957). *Handbuch der physik* (Vol. 28). Berlin: Springer-Verlag.

Fiocco, G., & Thompson, E. (1963). Thomson scattering of optical radiation from an electron beam. *Physical Review Letters, 10*(3), 89.

Flora, F., & Giudicotti, L. (1987). Complete calibration of a Thomson scattering spectrometer system by rotational Raman-scattering in H-2. *Applied Optics, 26*(18), 4001–4008.

Forrest, M. J. (1967). A ten-channel fiber optic Fabry-Pérot fringe splitter. *Journal of Scientific Instruments, 44*(1), 26.

Forrest, M. J., Carolan, P. G., & Peacock, N. J. (1978). Measurement of magnetic-fields in a tokamak using laser-light scattering. *Nature, 271*(5647), 718–722.

Forsythe, P. A., Currie, B. W., & Vawter, F. E. (1953). Scattering of 56-Mc./s. radio waves from the lower ionosphere. *Nature London, 171*, 352.

Fortmann, C., Redmer, R., Reinholz, H., Röpke, G., Wierling, A., & Rozmus, W. (2006). Bremsstrahlung vs. Thomson scattering in VUV-FEL plasma experiments. *High Energy Density Physics, 2*, 57–69. http://dx.doi.org/10.1016/j.hedp.2006.04.001.

Fortmann, C., Wierling, A., & Röpke, G. (2010). Influence of local-field corrections on Thomson scattering in collision-dominated two-component plasmas. *Physical Review E, 81*, 026405.

Fortmann, C., Wierling, A., & Röpke, G. (2010a). Influence of local-field corrections on Thomson scattering in collision-dominated two-component plasmas. *Physical Review E, 81*(2), 026405.

Fournier, K. B., Constantin, C., Back, C. A., Suter, L., Chung, H. K., Miller, M. C., *et al.* (2006). Electron-density scaling of conversion efficiency of laser energy into L-shell x-rays. *Journal of Quantitative Spectroscopy and Radiative Transfer, 99*(1–3), 186–198.

Freidberg, J. P. (1982). Ideal magneto-hydrodynamic theory of magnetic fusion systems. *Reviews of Modern Physics, 54*(3), 801–902.

Freidberg, J. P. (1987). *Ideal magneto-hydrodynamics*. New York: Plenum Press.

Fried, B. D., & Conte, S. D. (1961). *The plasma dispersion function; the Hilbert transform of the Gaussian*. New York: Academic Press.

Fried, B. D., White, R. B., & Samec, T. K. (1971). Ion acoustic waves in a multi-ion plasma. *Physics of Fluids, 14*(11), 2388.

Froula, D. H., Divol, L., & Glenzer, S. H. (2002). Measurements of nonlinear growth of ion-acoustic waves in two-ion-species plasmas with Thomson scattering. *Physical Review Letters, 88*(10), 105003.

Froula, D. H. (2002b). *Experimental studies of the stimulated Brillouin scattering instability in the saturated regime*. Ph.D. thesis. University of California at Davis, Davis, CA.

Froula, D. H., Divol, L., MacKinnon, A., *et al.* (2003). Direct observation of stimulated-Brillouin-scattering detuning by a velocity gradient. *Physical Review Letters, 90*(15), 155003.

Froula, D. H., Divol, L., Mackinnon, A., Gregori, G., & Glenzer, S. H. (2003a). Direct observation of stimulated-Brillouin-scattering detuning by a velocity gradient. *Physical Review Letters, 90*(15), 155003.

Froula, D. H., Divol, L., Braun, D. G., Cohen, B. I., Gregori, G., Mackinnon, A., *et al.* (2003b). Stimulated Brillouin scattering in the saturated regime. *Physics of Plasmas, 10*, 1846.

Froula, D. H., Divol, L., Offenberger, A., *et al.* (2004). Direct observation of the saturation of stimulated Brillouin scattering by ion-trapping-induced frequency shifts. *Physical Review Letters, 93*(3), 035001.

Froula, D. H., Davis, P., Divol, L., Ross, J. S., Meezan, N., Price, D., *et al.* (2005). Measurement of the dispersion of thermal ion-acoustic fluctuations in high-temperature laser plasmas using multiple-wavelength Thomson scattering. *Physical Review Letters, 95*(19), 195005.

Froula, D. H., Rekow, V., Sorce, C., Piston, K., Knight, R., Alvarez, S., *et al.* (2006a). 3 omega transmitted beam diagnostic at the Omega laser facility. *Review of Scientific Instruments, 77*(10), 10E507.

Froula, D. H., Ross, J. S., Divol, L., Meezan, N., MacKinnon, A. J., Wallace, R., *et al.* (2006b). Thomson-scattering measurements of high electron temperature hohlraum plasmas for laser-plasma interaction studies. *Physics of Plasmas, 13*(5), 052704.

Froula, D. H., Divol, L., Meezan, N., Dixit, S., Moody, J., Neumayer, P., *et al.* (2007). Ideal laser beam propagation through high temperature ignition hohlraum plasmas. *Physical Review Letters, 85*, 085001.

Froula, D. H., Ross, J. S., Pollock, B. B. *et al.* (2007a). Quenching of the nonlocal electron heat transport by large external magnetic fields in a laser-produced plasma measured with imaging Thomson scattering. *Physical Review Letters, 98*(13), 135001.

Froula, D. H., Divol, L., Meezan, N. B., Dixit, S., Neumayer, P., Moody, J. D., *et al.* (2007b). Laser beam propagation through inertial confinement fusion Hohlraum plasmas. *Physics of Plasmas, 14*(5).

Froula, D. H., Divol, L., London, R. A., Berger, R. L., Döppner, T., Meezan, N. B., *et al.* (2010). Experimental basis for laser-plasma interactions in ignition hohlraums at the national ignition facility (Invited). *Physics of Plasmas, 17*, 056201.

Funfer, E., Kronast, B., & Kunze, H. J. (1963). Experimental results on light scattering by a theta-pinch plasma using a ruby laser. *Physics Letters, 5*(2), 125–127.

Gabriel, A. H., Niblett, G. B. F., & Peacock, N. J. (1962). Vacuum ultraviolet radiation from a magnetically compressed plasma. *Journal of Quantitative Spectroscopy and Radiative Transfer, 2*(4), 491.

Gamez, G., Huang, M., Lehn, S. A., & Hieftje, G. M. (2003). Laser-scattering instrument for fundamental studies on a glow discharge. *Journal of Analytical Atomic Spectrometry, 18*(6), 680–684.

Gerry, E. T., & Rose, D. J. (1966). Combined anode-cathode feed of a hollow-cathode arc. *Journal of Applied Physics, 37*(7), 2725.

Giles, R., & Offenberger, A. A. (1983). Time-resolved Thomson-scattering measurements of ion fluctuations driven by stimulated Brillouin scattering. *Physical Review Letters, 50*(6), 421–424.

Glass, A., & Guenther, A., eds. (1970). Damage on Laser materials. *Journal of Research of the National Bureau of Standards, (U.S.) Special Publication*, 341.

Glasstone, S., & Lovberg, R. H. (1960). *Controlled thermonuclear reactions, an introduction to theory and experiment*. Princeton, NJ: Van Nostrand.

Glenzer, S. H., Back, C. A., Estabrook, K. G., Wallace, R., Baker, K., MacGowan, B. J., *et al.* (1996). Observation of two ion-acoustic waves in a two-species laser-produced plamsa with Thomson scattering. *Physical Review Letters, 77*(8), 1496.

Glenzer, S. H., Back, C. A., Estabrook, K. G., Kirkwood, R. K., Wallace, R., MacGowan, B. J., *et al.* (1997a). Thomson scattering from two-species laser-produced plasmas (invited). *Review of Scientific Instruments, 68*(1), 641–646.

Glenzer, S. H., Back, C. A., Estabrook, K. G., & MacGowan, B. J. (1997b). Thomson scattering in the corona of laser-produced gold plasmas. *Review of Scientific Instruments, 68*(1), 668–671.

Glenzer, S. H., Rozmus, W., MacGowan, B. J. *et al.* (1999). Thomson scattering from high-Z laser-produced plasmas. *Physical Review Letters, 82*(1), 97–100.

Glenzer, S. H., Weiland, T. L., Bower, J., MacKinnon, A. J., & MacGowan, B. J. (1999a). High-energy 4 omega probe laser for laser-plasma experiments at nova. *Review of Scientific Instruments, 70*(1), 1089–1092.

Glenzer, S. H., Alley, W. E., Estabrook, K. G., De Groot, J. S., Haines, M. G., Hammer, J. H., *et al.* (1999b). Thomson scattering from laser plasmas. *Physics of Plasmas, 6*(5), 2117–2128.

Glenzer, S. H. (2000). Thomson scattering in inertial confinement fusion research. *Contributions to Plasma Physics, 40*(1–2), 36–45.

Glenzer, S. H., Fournier, K. B., Decker, C. *et al.* (2000a). Accuracy of K-shell spectra modeling in high-density plasmas. *Physical Review E, 62*(2), 2728–2738.

Glenzer, S. H., Estabrook, K. G., Lee, R. W., MacGowan, B. J., & Rozmus, W. (2000b). Detailed characterization of laser plasmas for benchmarking of radiation-hydrodynamic modeling. *Journal of Quantitative Spectroscopy and Radiative Transfer*, *65*(1–3), 253–271.

Glenzer, S. H., Divol, L. M., Berger, R. L. *et al.* (2001b). Thomson scattering measurements of saturated ion waves in laser fusion plasmas. *Physical Review Letters, 86*(12), 2565–2568.

Glenzer, S. H., Fournier, K. B., Wilson, B. G., Lee, R. W., & Suter, L. J. (2001a). Ionization balance in inertial confinement fusion Hohlraums. *Physical Review Letters, 87*(4), 045002.

Glenzer, S. H., Divol, L. M., Berger, R. L. *et al.* (2001). Thomson scattering measurements of saturated ion waves in laser fusion plasmas. *Physical Review Letters*, 86(12), 2565–2568.

Glenzer, S. H., Gregori, G., Rogers, F. J., Froula, D. H., Pollaine, S. W., Wallace, R. S., *et al.* (2003a). X-ray scattering from solid density plasmas. *Physics of Plasmas, 10*(6), 2433–2441.

Glenzer, S. H., Gregori, G., Lee, R. W., Rogers, F. J., Pollaine, S. W., & Landen, O. L. (2003b). Demonstration of spectrally resolved x-ray scattering in dense plasmas. *Physical Review Letters, 90*(17), 175002.

Glenzer, S. H., Landen, O. L., Neumayer, P., Lee, R. W., Widmann, K., Pollaine, S. W., *et al.* (2007). Observations of plasmons in warm dense matter. *Physical Review Letters, 98*(6), 065002.

Glenzer, S., & Redmer, R. (2009). X-ray Thomson scattering in high energy density plasmas. *Review of Modern Physics, 81*, 1625–1663.

Glenzer, S. H., MacGowan, B. J., Michel, P., Meezan, N. B., Suter, L. J., Dixit, S. N., *et al.* (2010). Symmetric inertial confinement fusion implosions at ultra-high laser energies. *Science*, science.1185634.

Goldman, M. V., & Dubois, D. F. (1965). Stimulated incoherent scattering of light from plasmas. *Physics of Fluids, 8*(7), 1404.

Goldston, R. J., & Rutherford, P. H. (1995). *Introduction to plasma physics*. Bristol, UK : Institute of Physics Pub.

Gondhalekar, A. M., & Kronast, B. (1971). National Research Council of Canada. Private communication.

Gondhalekar, A. M., & Kronast, B. (1973). Relativistic effect in light-scattering spectrum of a theta-pinch plasma. *Physical Review A, 8*(1), 441–445.

Gorbunov, L. M., & Silin, V. P. (1966). Scattering of waves in a plasma. *Soviet Physics JETP-USSR, 23*(4), 729.

Gorog, I. (1969). Effect of charged particel-neutral collisions on collective Thomson scattering. *Physics of Fluids, 12*(8), 1702.

Gowers, A., Gadd, A., Hirsch, K., Nielsen, P., & Salzmann, H. (1990). High power ruby and alexandrite lasers for LIDAR-Thomson scattering diangostics. *High-Power Solid State Lasers and Applications, SPIE 1277*, 162–173.

Graf, K. A., Guthart, H., & Douglas, D. G. (1971). Scattering from a turbulent laboratory plasma at 31 gigahertz. *Radio Science, 6*(7), 737.

Granatstein, V. L. (1968). Microwave scattering from anisotropic plasma turbulence. *Applied Physics Letters, 13*(1), 37.

Gregori, G., Glenzer, S. H., & Landen, O. L. (2003). Strong coupling corrections in the analysis of x-ray Thomson scattering measurements. *International conference on strongly coupled coulomb systems (SCCS)*, Santa FE, New Mexico, Sep 02–06, 2002. [*Journal of Physics A-Mathematical and general, 22*, 5971–5980].

Gregori, G., Glenzer, S. H., Rozmus, W., Lee, R. W., & Landen, O. L. (2003b). Theoretical model of x-ray scattering as a dense matter probe. *Physical Review E, 67*(2), 026412.

Gregori, G., Glenzer, S. H., Rogers, F. J., Pollaine, S. M., Landen, O. L., Blancard, C., *et al.* (2004). Electronic structure measurements of dense plasmas. *Physics of Plasmas, 11*(5), 2754–2762.

Gregori, G., Glenzer, S. H., Knight, J., *et al.* (2004a). Effect of nonlocal transport on heat-wave propagation. *Physical Review Letters, 92*(20), 205006.

Gregori, G., Tommasini, R., Landen, O. L., Lee, R. W., & Glenzer, S. H. (2006). Limits on collective X-ray scattering imposed by coherence. *Europhysics Letters, 74*(4), 637.

Gregori, G., Glenzer, S. H., & Landen, O. L. (2006a). Generalized x-ray scattering cross section from nonequilibrium plasmas. *Physical Review E, 74*(2), 026402.

Gregori, G., Glenzer, S. H., Chung, H. K., Froula, D. H., Lee, R. W., Meezan, N. B., *et al.* (2006b). Measurement of carbon ionization balance in high-temperature plasma mixtures by temporally resolved x-ray scattering. *Journal of Quantitative Spectroscopy and Radiative Transfer, 99*(1–3), 225–237.

Gregori, G., Glenzer, S. H., Fournier, K. B., Campbell, K. M., Dewald, E. L., Jones, O. S., *et al.* (2008). X-ray scattering measurements of radiative heating and cooling dynamics. *Physical Review Letters, 101*(4), 045003.

Gregori, G., & Gericke, D. O. (2009). Low frequency structural dynamics of warm dense matter. *Physics of Plasmas, 16*(5), 056306.

Grewal, M. S. (1964). Effects of collisions on electron density fluctuations in plasmas. *Physical Review A-General Physics, 134*(1A), A86.

Grossman, E. N., Luukanen, A., & Miller, A. (2004). Terahertz active direct detection imagers. In *Terahertz for military and security applications II* (Vol. 5411). Bellingham, WA: Proceedings of SPIE. http://dx.doi.org//10.1117/12.549135.

Gusakov, E. Z., Gurchenko, A. D., Altukhov, A. B., Stepanov, A. Y., Esipov, L. A., Kantor, M. Y., *et al.* (2006). Investigation of ETG mode-scale component of Tokamak plasma turbulence by correlative enhanced scattering diagnostics. *Plasma Physics and Controlled Fusion, 48*, A371–A376.

Guthart, H., Weissman, D. E., & Morita, T. (1966). Microwave scattering from an underdense turbulent plasma. *Radio Science, 1*(11), 1253.

Haas, F. A., & Evans, D. E. (1990). Proposal for measuring magnetic fluctuations in Tokamaks by Thomson scattering. *Review of Scientific Instruments, 61*(11), 3540–3543.

Hagfors, I. (1961). Density fluctuations in a plasma in a magnetic field, with applications to ionosphere. *Journal of Geophysical Research, 66*(6), 1699.

Hagfors, T., & Brockelm, R. A. (1971). Theory of collision dominated electron density fluctuations in a plasma with applications to incoherent scattering. *Physics of Fluids, 14*(6), 1143.

Haller, E. E., & Beeman, J. W. (2002). Recent advances and future prospects. *Proceedings of far-IR, sub-μm & mm detector technology.*

Hammett, G. W. (2009). *http://w3.pppl.gov/hammett/talks/2007/gyrokinetic_aps_review.pdf*. Private communication.

Han, S., Griffin, R. G., Hu, K., Joo, C., Joye, C. D., Mastovsky, I., *et al.* (2006). Continuous-wave submillimeter-wave gyrotrons. *Proceedings – Society of Photo-Optical Instrumentation Engineers, 6373*, 63730C.

Harada, T., Takahashi, K., Sakuma, H., & Osyczka, A. (1999). Optimum design of a grazing-incidence flat-field spectrograph with a spherical varied-line-space grating. *Applied Optics, 38*(13), 2743–2748.

Hasegawa, A., Maclennan, C. G., & Kodama, Y. (1979). Non-linear behavior and turbulence spectra of drift waves and Rossby waves. *Physics of Fluids, 22*(11), 2122–2129.

Hassaballa, S., Yakushiji, M., Kim, Y. K., Tomita, K., Uchino, K., & Muraoka, K. (2004). Two-dimensional structure of PDP micro-discharge plasmas obtained using laser Thomson scattering. *IEEE Transactions on Plasma Science, 32*(1), 127–134.

Hassaballa, S., Tomita, K., Kim, Y. K., Uchino, K., Hatanaka, H., Kim, Y. M., *et al.* (2005). Laser Thomson scattering measurements of electron density and temperature profiles of a striated plasma in a plasma display panel (PDP)-like discharge. *Japanese Journal of Applied Physics Part 2-Letters and Express Letters, 44*(12–15), L442–L444.

Hatae, T., Nagashima, A., Kondoh, T., Kitamura, S., Kashiwabara, T., Yoshida, H., *et al.* (1999). Yag laser Thomson scattering diagnostic on the JT-60U. *Review of Scientific Instruments, 70*(1), 772–775.

Hatae, T., Naito, O., Nakatsuka, M., & Yoshida, H. (2006). Applications of phase conjugate mirror to Thomson scattering diagnostics (invited). *Review of Scientific Instruments, 77*(10), 10E508. Figures reprinted with permission from T. Hatae. Copyright 2006 by the American Institute of Physics.

Hatae, T., Nakatsuka, M., Yoshida, H., Ebisawa, K., Kusama, Y., Sato, K., *et al.* (2007). Progress in development of edge Thomson scattering system for ITER. *Fusion Science and Technology, 51*(2T), 58–61.

Hawreliak, J., Chambers, D. M., Glenzer, S. H., *et al.* (2004). Thomson scattering measurements of heat flow in a laser-produced plasma. *Journal of Physics B-Atomic Molecular and Optical Physics, 37*(7), 1541–1551.

Hennequin, P., Sabot, R., Honore, C., Hoang, G. T., Garbet, X., Truc, A., *et al.* (2004). Scaling laws of density fluctuations at high-k on Tore Supra. *Plasma Physics and Controlled Fusion, 46*, B121–B133.

Heeter, R. F., Hansen, S. B., Fournier, K. B., Foord, M. E., Froula, D. H., *et al.* (2007). Benchmark measurements of the Ionization balance of non-local-thermodynamic-equilibrium gold plasmas. *Physical Review Letters, 99*, 195001.

Hinton, F. L., & Hazeltine, R. D. (1976). Theory of plasma transport in toroidal confinement systems. *Reviews of Modern Physics, 48*(2), 239–308.

Hirschberg, J. G. (1967). Colloque sur les méthodes nouvelles de spectroscopie instrumentale. *Journal de Physique (Paris), 28*, C2–226.

Hirschberg, J. G., & Platz, P. (1965). A multichannel Fabry-Perot interferometer. *Applied Optics, 4*(11), 1375.

Höll, A., Redmer, R., Ropke, G. (2004). X-ray Thomson scattering in warm dense matter. *European Physical Journal D, 29*(2), 159–162.

Höll, A., Bornath, Th., Cao, L., Döppner, T., Dšterer, S., Förster, E., *et al.* (2007). Thomson scattering from near-solid density plasmas using soft x-ray free electron lasers. *High Energy Density Physics, 3*, 120–130.

Holzhauer, E. (1977). Forward scattering at 10.6 μm using light mixing to measure ion temperature in a hydrogen arc plasma. *Physics Letters A, 62*(7), 495–497.

Hori, T., Bowden, M. D., Uchino, K., & Muraoka, K. (1996a). Measurement of non-Maxwellian electron energy distributions in an inductively coupled plasma. *Applied Physics Letters, 69*(24), 3683–3685.

Hori, T., Bowden, M. D., Uchino, K., Muraoka, K., & Maeda, M. (1996b). Measurements of electron temperature, electron density, and neutral density in a radio-frequency inductively coupled plasma. *Journal of Vacuum Science and Technology A-Vacuum Surfaces and Films, 14*(1), 144–151.

Hori, T., Kogano, M., Bowden, M. D., Uchino, K., & Muraoka, K. (1998). A study of electron energy distributions in an inductively coupled plasma by laser Thomson scattering. *Journal of Applied Physics, 83*(4), 1909–1916. Figures reprinted with permission from T. Hori. Copyright 1998 by the American Institute of Physics.

Horton, W. (1999). Drift waves and transport. *Reviews of Modern Physics, 71*(3), 735–778.

Howard, J. (2006). Application of polarization interferometers for Thomson scattering. *Plasma Physics and Controlled Fusion, 48*(6), 777–787.

Howard, J., James, B. W., & Smith, W. I. B. (1979). Rotational Raman calibration of Thomson scattering. *Journal of Physics D-Applied Physics, 12*(9), 1435–1440.

Howard, J., & Sharp, L. E. (1992). Diffraction analysis of forward-angle scattering in plasmas. *Plasma Physics and Controlled Fusion, 34*(6), 1133–1156.

Howard, J., & Hatae, T. (2008). Imaging interferometers for analysis of Thomson scattered spectra. *Review of Scientific Instruments, 79*(10), 10E704.

Huang, K. (1963). *Statistical mechanics*. New York: Wiley.

Hughes, J. W., Mossessian, D. A., Hubbard, A. E., Marmar, E. S., Johnson, D., & Simon, D. (2001). High-resolution edge Thomson scattering measurements on the Alcator C-Mod Tokamak. *Review of Scientific Instruments, 72*(1), 1107–1110. Figures reprinted with permission from J. W. Hughes. Copyright 2001 by the American Institute of Physics.

Hughes, T. P., & Smith, S. R. P. (1989). Effects of plasma dielectric-properties on Thomson scattering of millimeter waves in Tokamak plasmas. *Journal of Plasma Physics, 42*, 215–240.

Hungerford, G., & Birch, D. J. S. (1996). Single-photon timing detectors for fluorescence lifetime spectroscopy. *Measurement Science & Technology, 7*(2), 121–135.

Hutchinson, D. P., Vandersluis, K. L., Sheffield, J., & Sigmar, D. J. (1985). Feasibility of alpha-particle measurement by CO_2-laser Thomson scattering. *Review of Scientific Instruments, 56*(5), 1075–1077.

Hutchinson, I. H. (1987). *Principles of plasma diagnostics*. Cambridge [Cambridgeshire], New York: Cambridge University Press.

Iannuzzi, M., Magistre, F., & Piperno, F. (1968). Coherent forward scattering of microwaves from density fluctuations in alkali plasmas. *Physics of Fluids, 11*(8), 1822.

Ichimaru, S. (1962). Theory of fluctuations in a plasma. *Annals of Physics, 20*(1), 78–118.

Ichimaru, S., Rostoker, N., & Pines, D. (1962a). Observation of critical fluctuations associated with plasma-wave instabilities. *Physical Review Letters, 8*(6), 231.

Ichimaru, S., Tanaka, S., & Iyetomi, H. (1984). Screening potential and enhancement of the thermonuclear reaction-rate in dense-plasmas. *Physical Review A, 29*(4), 2033–2035.

Ichimaru, S. (1984a). Correction. *Physical Review A, 30*(3), 1548–1548.

Infeld, E., & Zakowicz, W. (1971). Explanation of anomalous scattering of laser light from an arc plasma. *Physics Letters A, 37*(2), 103.

Infeld, E., Skorupsk, A., & Zakowicz, W. (1972). Scattering of laser light from a marginally stable hydrogen plasma. *Plasma Physics, 14*(12), 1125.

Izawa, Y., Nakanish, Y., Yokoyama, M., & Yamanaka, C. (1967). Profile of shock wave plasma measured by laser light scattering. *Journal of the Physical Society of Japan, 23*(5), 1185.

Izumi, N., Snavely, R., Gregori, G., Koch, J. A., Park, H. S., & Remington, B. A. (2006). Application of imaging plates to x-ray imaging and spectroscopy in laser plasma experiments (invited). *Review of Scientific Instruments, 77*(10), 10E325.

Jackson, J. D. (1960). *Journal of Nuclear Energy Part C-Plasma Physics Accelerators Thermonuclear Research, 1*, 171.

Jackson, J. D. (1998). *Classical electrodynamics* (3rd ed.). New York: Wiely and Sons.

James, B. W., & Yu, C. X. (1985). Diffraction of laser-radiation by a plasma-wave – the near-field and far field limiting cases. *Plasma Physics and Controlled Fusion, 27*(5), 557–564.

Jasny, J., Teubner, U., Theobald, W., Wulker, C., Bergmann, J., & Schafer, F. P. (1994). A single-shot spectrograph for the soft-x-ray region. *Review of Scientific Instruments, 65*(5), 1631–1635.

Jenko, F. (2000). Particle pinch in collisionless drift-wave turbulence. *Physics of Plasmas, 7*(2), 514–518.

Jensen, R. V., Post, D. E., Grasberger, W. H., Tarter, C. B., & Lokke, W. A. (1977). Calculations of impurity radiation and its effects on Tokamak experiments. *Nuclear Fusion, 17*(6), 1187–1196.

John, P. K., Irisawa, J., & Ng, K. H. (1971). Observation of enhanced plasma waves by laser scattering. *Physics Letters A, 36*(4), 277.

Johnson, D., Dimock, D., Grek, B., Long, D., Mcneill, D., Palladino, R., *et al.* (1985). TFTR Thomson scattering system. *Review of Scientific Instruments, 56*(5), 1015–1017.

Johnson, L. C. (1992). Validation of spatial profile measurements of neutron emission in TFTR plasmas (invited). *Review of Scientific Instruments, 63*(10), 4517–4522.

Joyce, G., & Salat, A. (1971). Light scattering in a marginally stable plasma. *Plasma Physics, 13*(5), 359.

Jukes, T. H., & Shaffer, C. B. (1960). Antithyroid effects of aminotriazole. *Science, 132*(3422), 296–297.

Kadomstev, B. B. (1965). *Plasma turbulence*. London: Academic Press.

Kahn, F. D. (1959). Long-range interactions in ionized gases in thermal equilibrium. *Astrophysical Journal, 129*(1), 205–216.

Kajita, S., Hatae, T., & Naito, O. (2009). Optimization of optical filters for ITER edge Thomson scattering diagnostics. *Fusion Engineering and Design, 84*(12), 2214–2220.

Kalinin, Y. G., Lin, D. N., Ryutov, V. D., & Skoryupi, V. A. (1969). Scattering of electromagnetic waves by turbulent plasma fluctuations. *Soviet Physics JETP-USSR, 28*(1), 61.

Karasik, B. S., Delaet, B., McGrath, W. R., Wei, J., Gershenson, M. E., & Sergeev, A. V. (2003). Experimental study of superconducting hot-electron sensors for submm astronomy. *IEEE Transactions on Applied Superconductivity, 13*(2), 188–191.

Kasparek, W., & Holzhauer, E. (1983a). CO_2-laser scattering from thermal fluctuations in a plasma with 2 ion components. *Physical Review A, 27*(3), 1737–1740.

Kasparek, W., & Holzhauer, E. (1983b). Simultaneous measurement of magnetic-field direction and ion temperature in a plasma by collective scattering with a CO_2-laser. *Applied Physics Letters, 43*(7), 637–638.

Kasparek, W., & Holzhauer, E. (1984). Collective laser-light scattering from thermal fluctuations in a collision dominated plasma. *Plasma Physics and Controlled Fusion, 26*(9), 1133–1137.

Kato, M. (1972). Fluctuation spectrum of plasma with few electrons in debye volume observed by collective light-scattering. *Physics of Fluids, 15*(3), 460.

Katzenstein, J. (1971). *Private communication*. Culham Lab, UK: UKAEA Res. Group.

Kauffman, R. L. (1991). *Handbook of plasma physics* (Vol. 3). Amsterdam: Elsevier.

Kaw, P., Valeo, E., & Dawson, J. M. (1970). Interpretation of an experiment on anomalous absorption of an electromagnetic wave in a plasma. *Physical Review Letters, 25*(7), 430.

Kawahata, K., Tanaka, K., Akiyama, T., Okajima, S., & Nakayama, K. (2005). Development of key components related to the interferometry/polarimetry. In *9th Meeting of the ITPA topical group diagnostics, KBSI*.

Kazcor, M., & Soltwisch, H. (1999). Light scattering diagnostic system for investigation of reactive plasmas in a capacitatively coupled RF-discharge. In *Workshop on frontiers in low temperature plasma diagnostics III*.

Kazcor, M., & Soltwisch, H. (2001). Numerical simulation of a multi-channel light scattering diagnostic system for low temperature plasmas. In *Workshop on frontiers in low temperature plasma diagnostics III*.

Kegel, W. H. (1965). Light mixing and generation of 2nd harmonic in a plasma in an external magnetic field. *Zeitschrift fur Naturforschung Part A-Astrophysik Physik und Physikalische Chemie A, 20*(6), 793.

Kegel, W. H. (1970). Light scattering from plasmas with a non-Maxwellian velocity distribution. *Plasma Physics, 12*(5), 295.

Keilhacker, M., Kornherr, M., & Steuer, K-H. (1969). Observation of collisionless plasma heating by strong shock waves. *Zeitschrift fur Physik, 223*(4), 385.

Keilhacker, M., Kornherr, M., Niedermeyer, H., Steuer, K-H., & Chodura, R. (1971). *Proceedings of the 4th international conference on plasma physics controlled fusion* (Vol. 3, p. 265). Madison, WI: IAEA Vienna.

Keilhacker, M., and Steuer, K-H. (1971a). Time-resolved light-scattering mesurements of spectrum of turbulence within a high-beta collisionless shock wave. *Physical Review Letters, 26* (12), 694.

Kellerer, L. (1970). Measuring magnetic fields in plasmas by means of light scattering. *Zeitschrift fur Physik, 239*(2), 147.

Kenyon, M., Day, P. K., Bradford, C. M., Bock, J. J., & Leduc, H. G. (2006). Progress on background-limited membrane-isolated TES bolometers for far-IR/submillimeter spectroscopy. In J. Zmuidzinas, W. S. Holland, S. Withington, & W. D. Duncan (Eds.), *Detectors and instrumentation for astronomy III* (Vol. 6275, p. 627508). Proceedings of SPIE. http://dx.doi.org/10.1117/12.672036.

Kerr, A. R., Feldman, M. J., & Pan, S. K. (1997). Receiver noise temperature, the quantum noise limit, and the role of the zero-point fluctuations. *Proceedings of the eighth international symposium on space terahertz technology* (pp. 101–111), Harvard University.

Kerr, A. R. (1999). Suggestions for revised definitions of noise quantities, including quantum effects. *IEEE Transactions on Microwave Theory and Techniques, 47*(3), 325–329.

Kieft, E. R., van der Mullen, J. J. A. M., Kroesen, G. M. W., Banine, V., & Koshelev, K. N. (2004). Collective Thomson scattering experiments on a tin vapor discharge in the prepinch phase. *Phys. Rev. E, 70*(5), 056413. Figures reprinted courtesy of J. J. A. M van der Mullen. Copyright 2004 the American Physical Society.

Kieft, E. R., Groothuis, C. H. J. M., van der Mullen, J. J. A. M., & Banine, V. (2005). Subnanosecond Thomson scattering setup for space and time resolved measurements with reduced background signal. *Review of Scientific Instruments, 76*(9), 093503.

Klein, W. R., & Cook, B. D. (1967). Unified approach to ultrasonic light diffraction. *IEEE Transactions on Sonics and Ultrasonics, Su14*(3), 123.

Klimontovich, I. U. L. (1967). *The statistical theory of non-equilibrium processes in a plasma* (1st English ed.). International series of monographs in natural philosophy, Vol. 9. Oxford, NY: Pergamon Press.

Kline, J. L., Montgomery, D. S., Bezzerides, B., *et al.* (2005). Observation of a transition from fluid to kinetic nonlinearities for Langmuir waves driven by stimulated Raman backscatter. *Physical Review Letters, 94*(17), 175003.

Kondoh, T., Miura, Y., Lee, S., Richards, R. K., Hutchinson, D. P., & Bennett, C. A. (2003). Collective Thomson scattering based on CO_2 laser for ion energy spectrum measurements in JT-60U. *Review of Scientific Instruments, 74*(3), 1642–1645.

Kondoh, T., Hayashi, T., Kawano, Y., Kusama, Y., Sugie, T., Miura, Y., *et al.* (2006). High-repetition CO_2 laser for collective Thomson scattering diagnostic of alpha particles in burning plasmas. *Review of Scientific Instruments, 77*(10), 10E505.

Kondoh, T., Hayashi, T., Kawano, Y., Kusama, Y., Sugie, T., Hirata, M., *et al.* (2007). CO_2 laser collective Thomson scattering diagnostic of alpha-particles in burning plasmas. *Fusion Science and Technology, 51*(2T), 62–64.

Kono, A., & Nakatani, K. (2000). Efficient multichannel Thomson scattering measurement system for diagnostics of low-temperature plasmas. *Review of Scientific Instruments, 71*(7), 2716–2721.

Koons, H. C., & Fiocco, G. (1968). Measurements of density and temperature of electrons in a reflex discharge by scattering of cw Ar^+ laser light. *Journal of Applied Physics, 39*(7), 3389.

Kornherr, M., Keilhack, M., Rohr, H., Decker, G., & Lindenbe, F. (1972). CO_2-laser scattering measurements of turbulence in a high-beta collisionless shock-wave. *Physics Letters A, 39*(2), 95.

Korsholm, S. B., Bindslev, H., Meo, F., Leipold, F., Michelsen, P. K., Michelsen, S., *et al.* (2006). Current fast ion collective Thomson scattering diagnostics at TEXTOR and ASDEX upgrade, and ITER plans (invited). *Review of Scientific Instruments, 77*(10), 10E514.

Krall, N. A., & Trivelpiece, A. W. (1973). *Principles of plasma physics*. New York: McGraw-Hill.

Kritcher, A. L., Neumayer, P., Urry, M. K., Robey, H., Niemann, C., Landen, O. L., *et al.* (2007). K-alpha conversion efficiency measurements for X-ray scattering in inertial confinement fusion plasmas. *High Energy Density Physics, 3*, 156. http://dx.doi.org/10.1016/j.hedp.2007.02.012.

Kritcher, A. L., Neumayer, P., Castor, J., Döppner, T., Falcone, R. W., Landen, O. L., *et al.* (2008). Ultrafast x-ray Thomson scattering of shock-compresed matter. *Science, 322*(5898), 69–71.

Kritcher, A. L., Neumayer, P., Brown, C. R. D., Davis, P., Döppner, T., Falcone, R. W., *et al.* (2009). Measurements of ionic structure in shock compressed lithium hydride from ultrafast x-ray Thomson scattering. *Physical Review Letters, 103*(24), 245004.

Kroll, N. M., Rostoker, N., & Ron, A. (1964). Optical mixing as plasma density probe. *Physical Review Letters, 13*(3), 83.

Kronast, B., Rohr, H., Glock, E., Zwicker, H., & Funfer, E. (1966). Measurements of ion and electron temperature in theta-pinch plasma by forward scattering. *Physical Review Letters, 16*(24), 1082.

Kronast, B., & Benesch, R. (1969). A light scattering method to determine electron density distributions and accurate peak densities. *Proceedings of the 9th international conference on phenomena in ionized gases* (p. 651). Romania, Edituura Academia Republicii Socialiste (ed).

Kronast, B., & Pietrzyk, Z. A. (1971). Discrepancy between electron-drift velocities obtained from ion and electron features of light scattering in a z-pinch plasma. *Physical Review Letters, 26*(2), 67.

Kruer, W. L. (2003). *The physics of laser plasma interactions*. Boulder, Colorado: Addison-Wesley Publishing Company, Inc.

Kubo, U., & Inuishi, Y. (1968). Enhanced microwave scattering by plasma instabilty in a magnetic field. *Journal of the Physical Society of Japan, 25*(6), 1688.

Kugland, N. L., Gregori, G., Bandyopadhyay, S., Brenner, C. M., Brown, C. R. D., Constantin, C., *et al.* (2009). Evolution of elastic x-ray scattering in laser-shocked warm dense lithium. *Physical Review E, 80*(6).

Kuhlbrodt, S., Redmer, R., Reinholz, H., Ropke, G., Holst, B., Mintsev, V. B., *et al.* (2005). Electrical conductivity of noble gases at high pressures. *Contributions to Plasma Physics, 45*(1), 61–69.

Kukushkin, A. B. (1981). Incoherent scattering of light by a finite volume of a relativistic plasma. *Soviet Journal of Plasma Physics, 7*, 63.

Kunkel, W. B. (1966). *Plasma physics in theory and application*. New York: McGraw-Hill.

Kunze, H. J., Funfer, E., Kronast, B., & Kegel, W. H. (1964). Measurement of the spectral distribution of light scattered by a theta-pinch plasma. *Physics Letters, 11*(1), 42–43.

Kunze, D. (1968). *Plasma diagnstics*. Amsterdam: North Holland Publ.

Laaspere, T. (1960). On the effect of a magnetic field on the spectrum of incoherent scattering. *Journal of Geophysical Research, 65*(12), 3955–3959.

Labaune, C., Baldis, H. A., Renard, N., *et al.* (1995). Large-amplitude ion-acoustic-waves in a laser-produced plasma. *Physical Review Letters, 75*(2), 248–251.

Labaune, C., Baldis, H. A., Schifano, E., *et al.* (1996). Location of ion-acoustic waves from back and side stimulated Brillouin scattering. *Physical Review Letters, 76*(20), 3727–3730.

Lachambre, J. L., & Decoste, R. (1985). Ion temperature-measurement in a plasma-column using pulsed CO_2 collective scattering and coherent detection. *Review of Scientific Instruments, 56*(5), 1057–1059.

Lafontaine, B., Baldis, H. A., & Villeneuve, D. M., *et al.* (1994). Characterization of laser-produced plasmas by ultraviolet Thomson scattering. *Physics of Plasmas, 1*(7), 2329–2341.

Landen, O. L., & Windfield, R. J. (1985). Laser scattering from dense cesium plasmas. *Physical Review Letters, 54*(15), 1660–1663.

Landen, O. L., Lobban, A., Tutt, T., Bell, P. M., Costa, R., Hargrove, D. R., *et al.* (2001). Angular sensitivity of gated microchannel plate framing cameras. *Review of Scientific Instruments, 72*, 709.

Landen, O. L., Glenzer, S. H., Edwards, M. J., Lee, R. W., Collins, G. W., Cauble, R. C., *et al.* (2001b). Dense matter characterization by x-ray Thomson scattering. *Journal of Quantitative Spectroscopy and Radiative Transfer, 71*(2–6), 465–478.

Landau, B., & Lifshitz, E. M. (1958). *The classical theory of fields*. Oxford: Pergamon Press.

Landau, L. D., & Lifshitz, E. M. (1962). *The classical theory of fields* (2nd Rev. ed.). Their Course of theoretical physics. Oxford: Pergamon Press.

LeBlanc, B. P. (2008). Thomson scattering density calibration by Rayleigh and rotational Raman scattering on NSTX. *Review of Scientific Instruments, 79*(10), 10E737.

Lee, W., Park, H. K., Cho, M. H., Namkung, W., Smith, D. R., Domier, C. W., *et al.* (2008). Spatial resolution study and power calibration of the high-k scattering system on NSTX. *Review of Scientific Instruments, 79*(10), 10E723.

Lee, H. J., Neumayer, P., Castor, J., Döppner, T., Falcone, R. W., Fortmann, C., *et al.* (2009a). X-ray Thomson-scattering measurements of density and temperature in shock-compressed beryllium. *Physical Review Letters, 102*(11).

Lee, W., Smith, D. R., Park, H. K., Domier, C. W., & Luhmann, N. C. (2009b). Calibration of the collective scattering system on NSTX. *The 34th international conference on infrared, millimeter, and terahertz waves and 17th THz electronics conference. Proceedings of IRMMW-THz 2009.*

Lefebvre, E., Berger, R. L., Langdon, A. B., MacGowan, B., Rothenberg, J. E., & Williams, E. A. (1998). Reduction of laser self-focusing in plasma by polarization smoothing. *Physics of Plasmas, 5*(7), 2701–2705.

Lehecka, T., Luhmann, N. C., Peebles, W. A., Goldhar, J., & Obenschain, S. P. (1990). *The handbook of microwave and optical components* (Vol. 3). New York: John Wiley and Sons, Inc.

Lehner, G., & Pohl, F. (1970). On possibility of measuring magnetic fields by scattered light. *Zeitschrift fur Physik, 232*(5), 405.

Lehner, T., Rax, J. M., & Zou, X. L. (1989). Linear-mode conversion by magnetic fluctuations in inhomogeneous magnetized plasmas. *Europhysics Letters, 8*(8), 759–764.

LePape, S. (2010). X-ray radiography and scattering diagnosis of dense shock-compressed matter. *Physics of Plasmas, 17*, 056309.

Lerner, E. J. (2009). Infrared detectros offer high sensitivity. *Laser Focus World, 45*(6).

Liewer, P. C. (1985). Measurements of microturbulence in Tokamaks and comparisons with theories of turbulence and anomalous transport. *Nuclear Fusion, 25*(5), 543–621.

Lin, Y., Wukitch, S., Parisot, A., Wright, J. C., Basse, N., Bonoli, P., *et al.* (2005). Observation and modelling of ion cyclotron range of frequencies waves in the mode conversion region of Alcator C-Cod. *Plasma Physics and Controlled Fusion, 47*(8), 1207–1228.

Lin, L., Edlund, E. M., Porkolab, M., Lin, Y., & Wukitch, S. J. (2006). Vertical localization of phase contrast imaging diagnostic in Alcator C-Mod. *Review of Scientific Instruments, 77*(10), 10E918.

Lin, L., Porkolab, M., Edlund, E. M., Rost, J. C., Fiore, C. L., Greenwald, M., Lin, Y., Mikklesen, D. R., & Wukitch, S. J. (2009). Studies of turbulence and transport in Alcator C-Mod H-mode plasmas with phase contrast imaging and comparisons with GYRO. *Physics of Plasmas, 16*, 102502. Figures reprinted with permission from L. Lin. Copyright 2009 by the American Institute of Physics.

Lindl, J. D., Amendt, P. A., Berger, R. L., Glendinning, S. G., Glenzer, S. H., Haan, S. W., *et al.* (2004). The physics basis for ignition using indirect-drive targets on the national ignition facility. *Physics of Plasmas, 11*(2), 339.

Lins, G. A. W. (1981). Anomalous heat-conduction at the ends of a theta-pinch. *Physics of Fluids, 24*(11), 2090–2097.

Little, P. F., & Hamberger, S. M. (1966). Scattering of electromagnetic waves by electro-acoustic plasma waves. *Nature, 209*(5027), 972.

Liu, Z. J., Zheng, J., & Yu, C. X. (2002). Effects of super-Gaussian electron velocity distributions on the ion feature of Thomson scattering off two-ion plasmas. *Physics of Plasmas, 9*(4), 1073–1078.

Lowney, D. P., Heimann, P. A., Padmore, H. A., Gullikson, E. M., MacPhee, A. G., & Falcone, R. W. (2004). Characterization of CsI photocathodes at grazing incidence for use in a unit quantum efficiency x-ray streak camera. *Review of Scientific Instruments, 75*(10), 3131–3137.

Ludwig, D., & Mahn, C. (1971). Anomalous scattering of laser light from a magnetized arc plasma. *Physics Letters A, 35*(3), 191.

Luhmann, N. C., Bindslev, H., Park, H., Sanchez, J., Taylor, G., & Yu, C. X. (2008). Microwave diagnostics. *Fusion Science and Technology, 53*(2), 335–396.

MacGowan, B., Afeyan, B. B., Back, C. A., Berger, R. L., Bonanno, G., Casanova, M., *et al.* (1996). Laser-plasma interactions in ignition-scale Hohlraum plasmas. *Physics of Plasmas, 3*(5), 2029.

Machalek, M. D., & Nielsen, P. (1973). Light-scattering measurements of turbulence in a normal shock. *Physical Review Letters, 31*(7), 439–442.

Machuzak, J. S., Rhee, D. Y., Woskov, P. P., Cohn, D. R., Myer, R. C., Bretz, N., *et al.* (1990). Development of high-power millimeter and submillimeter wavelength collective Thomson scattering diagnostics for energetic ion measurements in Tokamaks. *Review of Scientific Instruments, 61*(11), 3544–3547.

Mack, J. E., McNutt, D. P., Roesler, F. L., & Chabbal, R. (1963). The PEPSIOS purely interferometric high-resolution scanning spectrometer. I. The pilot model. *Applied Optics, 2*, 873.

MacKinnon, A. J., Shiromizu, S., Antonini, G., Auerbach, J., Haney, H., Froula, D. H., *et al.* (2004). Implementation of a high energy 4ω probe beam on the omega laser facility. *Review of Scientific Instruments, 75*(10), 3906.

Malmberg, J. H., & Wharton, C. B. (1969). Spatial growth of waves in a beam-plasma system. *Physics of Fluids, 12*, 2600–2606.

Malka, V., Faure, J., & Amiranoff, F. (2001). Characterization of plasmas produced by laser-gas jet interaction. *Physics of Plasmas, 8*(7), 3467–3472.

Mannheimer, W. M. (1971). *Physics of Fluids, 14*, 579.

Marshall, F. J., & Oertel, J. A. (1997). Framed monochromatic x-ray microscope for ICF (invited). *Review of Scientific Instruments, 68*(1), 735–739.

Martineau, J., & Pepin, H. (1972). CO_2-laser heating of a theta-pinch plasma by inverse bremsstrahlung and induced compton processes. *Journal of Applied Physics, 43*(3), 917.

Martone, M., & Segre, S. E. (1969). Cooperative laser scattering at 90 degrees from a cold dense plasma. *Physics Letters A, 28*(9), 610.

Mase, A., & Tsukishima, T. (1975). Measurements of ion wave turbulence by microwave-scattering. *Physics of Fluids, 18*(4), 464–469. Figures reprinted with permission from A. Mase. Copyright 1975 by the American Institute of Physics.

Masters, D., & Rye, B. J. (1976). Heterodyne-detection of CO_2 laser radiation scattered from a plasma arc jet. *Physics Letters A, 58*(2), 108–110.

Matoba, T., Itagaki, T., Yamauchi, T., & Funahashi, A. (1979). Analytical approximations in the theory of relativistic Thomson scattering for high-temperature fusion plasma. *Japanese Journal of Applied Physics, 18*(6), 1127–1133.

Matsuura, K. (1966). Interaction between plasma oscillation and radiation field .3. nonlinear coupling processes. *Journal of the Physical Society of Japan, 21*(10), 2011.

Maurmann, S., Kadetov, V. A., Khalil, A. A. I., Kunze, H. J., & Czarnetzki, U. (2004). Thomson scattering in low temperature helium plasmas of a magnetic multipole plasma source. *Journal of Physics D-Applied Physics, 37*(19), 2677–2685.

Mazzucato, E. (1976). Small-scale density fluctuations in adiabatic toroidal compressor. *Physical Review Letters, 36*(14), 792–794.

Mazzucato, E. (1978). Low-frequency microinstabilities in PLT-Tokamak. *Physics of Fluids, 21*(6), 1063–1069.

Mazzucato, E. (2003). Localized measurement of turbulent fluctuations in Tokamaks with coherent scattering of electromagnetic waves. *Physics of Plasmas, 10*(3), 753–759.

Mazzucato, E. (2006). Detection of short-scale turbulence in the next generation of Tokamak burning plasma experiments. *Plasma Physics and Controlled Fusion, 48*(12), 1749–1763.

Mazzucato, E., Smith, D. R., Bell, R. E., Kaye, S. M., Hosea, J. C., LeBlanc, B. P., *et al.* (2008). Short-scale turbulent fluctuations driven by the electron-temperature gradient in the national spherical torus experiment. *Physical Review Letters, 101*(7), 075001, Copyright 2008 by the American Physical Society.

McCrory, R. L., Meyerhofer, D. D., Betti, R., Craxton, R. S., Delettrez, J. A., Edgell, D. H., *et al.* (2008). Progress in direct-drive inertial confinement fusion. *Physics of Plasmas, 15*(5), 055503.

Mehdi, I., Ward, J., Maestrini, A., Chattopadhyay, G., Schlecht, E., & Gill, J. (2008). Pushing the limits of multiplier-based local oscillator chains (invited). *Proceedings of the nineteenth international symposium on space terahertz technology* (pp. 196–200). Groningen, Netherlands: Space Research Organization of the Netherlands.

Meo, F., Bindslev, H., Korsholm, S. B., Tsakadze, E. L., Walker, C. I., Woskov, P., *et al.* (2004). Design of the collective Thomson scattering diagnostic for international thermonuclear experimental reactor at the 60 ghz frequency range. *Review of Scientific Instruments, 75*(10), 3585–3588.

Michael, C. A., Tanaka, K., Vyacheslavov, L., Sanin, A., Kawahata, K., & Okajima, S. (2006). Upgraded two-dimensional phase contrast imaging system for fluctuation profile measurement on LHD. *Review of Scientific Instruments, 77*(10), 10E923.

Michelsen, S., Korsholm, S. B., Bindslev, H., Meo, F., Michelsen, P. K., Tsakadze, E. L., *et al.* (2004). Fast ion millimeter wave collective Thomson scattering diagnostics on TEXTOR and ASDEX upgrades. *Review of Scientific Instruments, 75*(10), 3634–3636.

Middleton, C. F., & Boreman, G. D. (2006). Technique for thermal isolation of antenna-coupled infrared microbolometers. *Journal of Vacuum Science & Technology B, 24*(5), 2356–2359.

Mikhailovskii, A. B., & Laing, E. W. (1992). *Electromagnetic instabilities in an inhomogeneous plasma. Plasma physics series*. Bristol, PA: Institute of Physics Pub.

Mikhailovskii, A. B. (1998). *Instabilities in a confined plasma. Plasma physics series*. Bristol, PA: Institute of Physics Pub.

Miller, A. J., Luukanen, A., & Grossman, E. N. (2004). Micromachined antenna-coupled uncooled microbolometers for terahertz imaging arrays. In *Terahertz for military and security applications II* (Vol. 5411). Bellingham, WA: Proceedings of SPIE. http://dx.doi.org/10.1117/12.543236.

Mirnov, V. V., Ding, W. X., Brower, D. L., Van Zeeland, M. A., & Carlstrom, T. N. (2007). Finite electron temperature effects on interferometric and polarimetric measurements in fusion plasmas. *Physics of Plasmas, 14*(10), 102105.

Misyura, V. A., Tkachev, G. N., Yerokhin, Yu. G., Ivanov, V. I., Nisnevich, N. J., & Borodin, N. M. (1969). Regarding ionospheric measurements by the method of incoherent radio wave scattering. *Geomagnetism and Aeronomy, 9*, 60.

Montgomery, D. S., Johnson, R. P., Cobble, J. A., *et al.* (1999). Characterization of plasma and laser conditions for single hot spot experiments. *25th European Conference on Laser Interaction with Matter (ECLIM 1998)*, Formia, Italy, May 07, 1998 [*Laser and Particle Beams, 17*(3), 349–359].

Montgomery, D. S., Kline, J. L., & Tierney, T. E. (2004). Detailed characterization of plasma wave behavior using collective Thomson scattering (invited). *Review of Scientific Instruments, 75*(10), 3793–3799.

Montgomery, D. C., & Tidman, D. A. (1964). *Plasma kinetic theory. McGraw-Hill advanced physics monograph series*. New York: McGraw-Hill.

Montgomery, D. C. (1971). *Theory of the unmagnetized plasma*. New York: Gordon and Breach.

Moorcroft, D. R. (1963). On power scattered from density fluctuations in a plasma. *Journal of Geophysical Research, 68*(16), 4870.

More, R. M., Warren, K. H., Young, D. A., & Zimmerman, G. B. (1988). A new quotidian equation of state (QEOS) for hot dense matter. *Physics of Fluids, 31*(10), 3059–3078.

Moses, E. I., & Wuest, C. R. (2004). The national ignition facility: Laser performance and first experiments. *Fusion Science and Technology, 47*(3), 314–322.

Murakami, M., Clarke, J. F., Kelley, G. G., & Lubin, M. (1970). Method of measuring poloidal magnetic field in diffuse toroidal pinches. *Bulletin of the American Physical Society, 15*(11), 1411.

Muraoka, K., Murray, E. L., Paul, J. W. M., & Summers, D. D. R. (1973). Anisotropy of turbulence in a collisionless shock. *Journal of Plasma Physics, 10*, 135–140.

Muraoka, K. (2006). Problems of atomic science and technology. *Plasma Physics, 12*(No. 6 Series), 236.

Muraoka, K., Uchino, K., & Bowden, M. D. (1998). Diagnostics of low-density glow discharge plasmas using Thomson scattering. *Plasma Physics and Controlled Fusion, 40*(7), 1221–1239.

Naito, O., Yoshida, H., & Matoba, T. (1993). Analytic formula for fully relativistic Thomson scattering spectrum. *Physics of Fluids B-Plasma Physics, 5*(11), 4256–4258.

Naito, O., Yoshida, H., Hatae, T., Nagashima, A., & Matoba, T. (1996). Relativistic incoherent Thomson scattering spectrum for generalized Lorentzian distributions. *Physics of Plasmas, 3*(4), 1474–1476.

Naito, O., Yoshida, H., Hatae, T., & Nagashima, A. (1997). A formula for reconstructing fully relativistic electron distributions from incoherent Thomson scattering data. *Physics of Plasmas, 4*(4), 1171–1172.

Naito, O., Yoshida, H., Kitamura, S., Sakuma, T., & Onose, Y. (1999). How many wavelength channels do we need in Thomson scattering diagnostics? *Review of Scientific Instruments, 70*(9), 3780–3781.

Nakano, N., Kuroda, H., Kita, T., & Harada, T. (1984). Development of a flat-field grazing-incidence XUV spectrometer and its application in picosecond XUV spectroscopy. *Applied Optics, 23*(14), 2386–2392.

Nakayama, K., Tazawa, H., Okajima, S., Kawahata, K., Tanaka, K., Tokuzawa, T., *et al.* (2004). High-power 47.6 and 57.2 μm CH_3OD lasers pumped by continuous-wave 9r(8) CO_2 laser. *Review of Scientific Instruments, 75*(2), 329–332.

Nazikian, R., & Sharp, L. E. (1987). CO_2-laser scintillation interferometer for the measurement of density-fluctuations in plasma-confinement devices. *Review of Scientific Instruments, 58*(11), 2086–2091.

Nee, S. F., Pechacek, R. E., & Trivelpi, A. W. (1969). Electromagnetic wave scattering from a magnetically confined high-temperature plasma. *Physics of Fluids, 12*(12P1), 2651.

Neikirk, D. P., Lam, W. W., & Rutledge, D. B. (1984). Far-infrared microbolometer detectors. *International Journal of Infrared and Millimeter Waves, 5*(3), 245–278.

Neufeld, C. R. (1970). Enhanced scattering of laser light from a laboratory plasma. *Physics Letters A, 31*(1), 19.

Neumayer, P., *et al.* (2010). Plasmons in strongly coupled shock-compressed matter. *Physical Review Letters, 105*, 075003.

Nicholson-Florence, M. B. (1971). Intensity dependence of free-free absorption. *Journal of Physics Part A General, 4*(4), 574.

Nielsen, S. K., Bindslev, H., Porte, L., Hoekzema, J. A., Korsholm, S. B., Leipold, F., *et al.* (2008). Temporal evolution of confined fast-ion velocity distributions measured by collective Thomson scattering in TEXTOR. *Physical Review E, 77*(1), 016470.

Niemann, C., Glenzer, S. H., Knight, J., *et al.* (2004). Observation of the parametric two-ion decay instability with Thomson scattering. *Physical Review Letters, 93*(4), 045004.

Niemann, C., Berger, R. L., Divol, L., Froula, D. H., Jones, O., Kirkwood, R. K., *et al.* (2008). Green frequency-doubled laser-beam propagation in high-temperature hohlraum plasmas. *Physical Review Letters, 1*, 045002.

Nishiura, M., Tanaka, K., Kubo, S., Saito, T., Tatematsu, Y., Notake, T., *et al.* (2008). Design of collective Thomson scattering system using 77 ghz gyrotron for bulk and tail ion diagnostics in the large helical device. *Review of Scientific Instruments, 79*(10), 10E731.

Nodwell, R. A., & Vanderka, G. S. (1968). Electron-density profiles from laser scattering. *Canadian Journal of Physics, 46*(7), 833.

Noguchi, Y., Matsuoka, A., Bowden, M. D., Uchino, K., & Muraoka, K. (2001). Measurements of electron temperature and density of a micro-discharge plasma using laser Thomson scattering. *Japanese Journal of Applied Physics Part 1-Regular Papers Short Notes and Review Papers, 40*(1), 326–329.

Noguchi, Y., Matsuoka, A., Uchino, K., & Muraoka, K. (2002). Direct measurement of electron density and temperature distributions in a micro-discharge plasma for a plasma display panel. *Journal of Applied Physics, 91*(2), 613–616.

Nuckolls, J., Thiessen, A., Wood, L., & Zimmermann, G. (1972). Laser compression of matter to super-high densities-thermonuclear (CTR) applications. *Nature, 239*(5368), 139.

Nyquist, H. (1928). Thermal agitation of electric charge in conductors. *Physical Review, 31*, 110.

Offenberger, A. A., Blyth, W., Dangor, A. E., *et al.* (1993). Electron-temperature of optically ionized-gases produced by high-intensity 268 nm laser-radiation. *Physical Review Letters, 71*(24), 3983–3986.

Ogawa, I., Idehara, T., Saito, T., Park, H., & Mazzucato, E. (2006). Scattering measurement using a submillimeter wave gyrotron as a radiation source. In A. Zagorodny & O. Kocherga (Eds.), *Thirteenth international congress of plasma physics* (p. B118p).

Oliver, B. M. (1965). Thermal and quantum noise. *Proceedings/IEEE, 53*, 436. http://dx.doi.org/10.1109/PROC.1965.3814.

Orsitto, F. (1990). *JET Report, JET-R(90)05*. Culham, Torus, England: Joint European.

Pak, A., Gregori, G., Knight, J., Campbell, K., Price, D., Hammel, B., *et al.* (2004). X-ray line measurements with high efficiency Bragg crystals. *Review of Scientific Instruments, 75*(10), 3747–3749.

Palastro, J. P., Ross, J. S., Pollock, B. B., Divol, L., Froula, D. H., & Glenzer, S. H. (2010). Fully relativistic form factor for Thomson scattering. *Physical Review E, 81*, 036411.

Papas, C. H., & Lee, K. S. H. (1962). *Proceedings of the 5th conference on phenomena in ionized gases* (Vol. 2, p. 1204). Munich, Germany: North Holland, Amsterdam.

Pappert, R. A. (1963). Incoherent scatter from a hot plasma. *Physics of Fluids, 6*(10), 1452–1457.

Park, H., Yu, C. X., Peebles, W. A., Luhmann, N. C., & Savage, R. (1982). Multimixer far-infrared laser Thomson scattering apparatus. *Review of Scientific Instruments, 53*(10), 1535–1540.

Park, H. S., Chambers, D. M., Chung, H. K., Clarke, R. J., Eagleton, R., Giraldez, E., *et al.* (2006). High-energy K alpha radiography using high-intensity, short-pulse lasers. *Physics of Plasmas, 13*(5), 056309.

Patrick, R. M. (1965). Thomson scattering measurements of magnetic annular shock tube plasmas. *Physics of Fluids, 8*(11), 1985.

Paul, J. W. M., Goldenbaum, G. C., Iiyoshi, A., Holmes, L. S., & Hardcastle, R. A. (1967). Measurement of electron temperatures produced by collisionless shock waves in a magnetized plasma. *Nature, 216*(5113), 363.

Paul, J. W. M., Daughney, C. C., & Holmes, L. S. (1969). Measurement of light scattered from density fluctuations within a collisionless shock. *Nature, 223*(5208), 822.

Paul, J. W. M., Daughney, C. C., Holmes, L. S., Rumsby, P. T., Craig, A. D., Murray, E. L., Summers, D. D. R., Beaulieu, J. (1971). *Proceedings of IAEA conference*. Madison, WI: IAEA Vienna.

Peacock, N. J., Robinson, D. C., Forrest, M. J., Wilcock, P. D., & Sannikov, V. V. (1969). Measurement of electron temperature by Thomson scattering in Tokamak-T3. *Nature, 224*(5218), 488.

Peacock, N. J., & Forrest, M. J. (1973). UKAEA research group, Culham Laboratory, UK. Private communication.

Pechacek, R. E., & Trivelpiece, A. W. (1967). Electromagnetic wave scattering from a high-temperature plasma. *Physics of Fluids, 10*(8), 1688.

Peebles, W. A., Rhodes, T. L., Mikkelsen, D. R., Gilmore, M., Zeeland, M. Van, Dorland, W., *et al.* 2004. Investigation of broad spectrum turbulence on DIII-D via integrated microwave and far-infrared collective scattering. In Society, European Physical (Ed.), *Proceedings of the 31st European physical society conference on plasma physics* (Vol. 28G), 28 June–2 July, London, England, Vol 28G, P-2.174, 2004.

Penney, C. M., Stpeters, R. L., & Lapp, M. (1974). Absolute rotational Raman cross-sections for N_2, O_2, and CO_2. *Journal of the Optical Society of America, 64*(5), 712–716.

Perepelkin, N. F. (1966). Raman scattering of microwaves by plasma oscillations. *JETP Letters-USSR, 3*(6), 165.

Perkins, F. W., Salpeter, E. E., & Yngvesso, K. O. (1965). Incoherent scatter from plasma oscillations in ionosphere. *Physical Review Letters, 14*(15), 579.

Perkins, F., & Salpeter, E. E. (1965a). Enhancement of plasma density fluctuations by nonthermal electrons. *Physical Review, 1*(1A), A55.

Perkins, F. W. (1970). Measuring current distribution in a Tokamak plasma. *Bulletin of the American Physical Society, 15*(11), 1418.

Petrasso, R. D., Li, C. K., Seguin, F. H., Rygg, R. J., Frenje, J. A., Betti, R., *et al.* (2009). Lorentz mapping of magnetic fields in hot dense plasmas. *Physical Review Letters, 103*, 085001, http://link.aps.org/doi/10.1103/PhysRevLett.103.085001.

Pieroni, L., & Bremmer, H. (1970). Mutual coherence function of light scattered by a turbulent medium. *Journal of the Optical Society of America, 60*(7), 936.

Pilija, A. D. (1967). *Soviet Physics – Technical Physics, 11*, 1567.

Pineo, V. C., & Hynek, D. P. (1962). Spectral widths and shapes and other characteristics of incoherent backscatter from ionosphere observed at 440 megacycles per second during a 24-hour period in may 1961. *Journal of Geophysical Research, 67*(13), 5119.

Pineo, V. C., Kraft, L. G., & Briscoe, H. W. (1960). Ionospheric backscatter observation at 440 mc-s. *Journal of Geophysical Research, 65*(5), 1620–1621.

Pineo, V. C., Hynek, D. P., & Millman, G. H. (1963). Geomagnetic effects on frequency spectrum of incoherent backscatter observed at 425 megacycles per second at Trinidad. *Journal of Geophysical Research, 68*(9), 2695.

Platzman, P. M., & Ozaki, H. T. (1960). Scattering of electromagnetic waves from an infinitely long magnetized cylindrical plasma. *Journal of Applied Physics, 31*(9), 1597–1601.

Platzman, P. M. (1965). Incoherent scattering of light from anisotropic degenerate plasmas. *Physical Review, 139*(2A), A379.

Platzman, P. M., Wolff, P. A., & Tzoar, N. (1968). Light scattering from a plasma in a magnetic field. *Physical Review, 174*(2), 489.

Pogutse, O. (1963). Frequency displacement when light is scattered in a relativistic plasma. *Doklady Akademii Nauk SSSR, 153*(3), 578.

Pogutse, O. (1964). Frequency shift of light scattered in relativistic plasma. *Soviet Physics-Doklady, 8*, 1107.

Porkolab, M., Rost, J. C., Basse, N., Dorris, J., Edlund, E., Lin, L., *et al.* (2006). Phase contrast imaging of waves and instabilities in high temperature magnetized fusion plasmas. *IEEE Transactions on Plasma Science, 34*(2), 229–234.

Pyatnitski, L. N., & Korobkin, V. V. (1971). Limits in detection of radiation scattered by a plasma. *Soviet Physics Technical Physics-USSR, 15*(11), 1914.

Ramsden, S. A., & Davies, W. E. R. (1966). Observation of cooperative effects in scattering of a laser beam from a plasma. *Physical Review Letters, 16*(8), 303.

Ramsden, S. A., Benesch, R., Davies, W. E. R., & John, P. K. (1966a). Observation of cooperative effects and determination of electron and ion temperatures in a plasma from scattering of a ruby laser beam. *IEEE Journal of Quantum Electronics, QE 2*(8), 267.

Rand, S. (1964). *Physical Review B, 136*, 231.

Ratcliffe, J. A. (1948). Diffraction from the ionosphere and the fading of radio waves. *Nature, 162*(4105), 9–11.

Ravasio, A., Gregori, G., Benuzzi-Mounaix, A., Daligault, J., Delserieys, A., Faenov, A. Y., *et al.* (2007). Direct observation of strong ion coupling in laser-driven shock-compressed targets. *Physical Review Letters, 99*(13), 135006.

Rebeiz, G. M., Regehr, W. G., Rutledge, D. B., Savage, R. L., & Luhmann, N. C. (1987). Submillimeter-wave antennas on thin membranes. *International Journal of Infrared and Millimeter Waves, 8*(10), 1249–1255.

Redmer, R., Reinholz, H., Ropke, G., *et al.* (2005). Theory of x-ray Thomson scattering in dense plasmas. *31st International Conference on Plasma Science*, Baltimore, MD, Jun, 2003. [*IEEE Transactions on Plasma Science, 33*(1), 77–84].

Regan, S. P., Epstein, R., Goncharov, V. N., Igumenshchev, I. V., Li, D., Radha, P. B., *et al.* (2007). Laser absorption, mass ablation rate, and shock heating in direct-drive inertial confinement fusion. *Physics of Plasmas, 14*(5), 06305.

Renard, N., Labaune, C., Baldis, H. A., *et al.* (1996). Detailed characterization of electron plasma waves produced by stimulated Raman scattering. *Physical Review Letters, 77*(18), 3807–3810.

Renau, J. (1960). Scattering of electromagnetic waves from a nondegenerate ionized gas. *Journal of Geophysical Research, 65*(11), 3631–3640.

Rhee, D. Y., Cohn, D. R., Machuzak, J. S., Woskov, P., Bretz, N., Budny, R., *et al.* (1992). Performance evaluation of the TFTR gyrotron CTS diagnostic for alpha-particles. *Review of Scientific Instruments, 63*(10), 4644–4646.

Rhodes, T. L., Peebles, W. A., Nguyen, X., Van Zeeland, M. A., Degrassie, J. S., Doyle, E. J., *et al.* (2006). Millimeter-wave backscatter diagnostic for the study of short scale length plasma fluctuations (invited). *Review of Scientific Instruments, 77*(10), 10E922.

Rhodes, T. L., Peebles, W. A., Van Zeeland, M. A., deGrassie, J. S., Mckee, G. R., Staebler, G. M., *et al.* (2007a). Broad wavenumber turbulence measurements during neutral beam injection on the DIII-D Tokamak. *Nuclear Fusion, 47*(8), 936–942.

Rhodes, T. L., Peebles, W. A., Van Zeeland, M. A., Degrassie, J. S., Bravenec, R. V., Burrell, K. H., *et al.* (2007b). Response of multiscale turbulence to electron cyclotron heating in the DIII-D Tokamak. *Physics of Plasmas, 14*(5), 056117.

Rozmus, W., Glenzer, S. H., Estabrook, K. G., *et al.* (2000). Modeling of Thomson scattering spectra in high-Z, laser-produced plasmas. 2nd International workshop on laboratory astrophysics with intense lasers, Tucson, Arizona, MAR 19-21, 1998. [*Astrophysical Journal Supplement Series, 127(2)*, 459–463].

Richards, R. K., Hutchinson, D. P., Bennett, C. A., Hunter, H. T., & Ma, C. H. (1993). Measurement of CO_2-laser small-angle Thomson scattering on a magnetically confined plasma. *Applied Physics Letters, 62*(1), 28–30.

Richards, R. K., Hutchinson, D. P., Bennett, C. A., Kondoh, T., Miura, Y., & Lee, S. (2003). Applying the CO_2 laser collective Thomson scattering results from JT-60U to other machines. *Review of Scientific Instruments, 74*(3), 1646–1648.

Rieke, G. H. (1994). Detection of Light from the Ultraviolet to the Submillimeter by G. H. Rieke 1994 (Cambridge University Press, ISBN 0 521 41028 2).

Riley, D., Woolsey, N. C., McSherry, D., Weaver, I., Djaoui, A., & Nardi, E. (2000). X-ray diffraction from a dense plasma. *Physical Review Letters, 84*(8), 1704–1707.

Ringler, H., & Nodwell, R. A. (1969). Enhanced plasma oscillations observed with scattered laser light. *Physics Letters A, 29*(3), 151.

Rogalski, A. (2003). Quantum well photoconductors in infrared detector technology. *Journal of Applied Physics, 93*(8), 4355–4391.

Rogister, A. L., & Oberman, C. (1967). On the kinetic theory of stable and weakly unstable plasma. Part 1, *Plasma Physics, 2*, 33–49.

Rogister, A. L., & Oberman, C. (1968). On the kinetic theory of stable and weakly unstable plasma. Part 1. *Journal of Plasma Physics, 2*(1), 33–49.

Röhr, H. (1967). A 90 degree laser scattering experiment for measuring temperature and density of ions and electrons in a cold dense theta pinch plasma. *Physics Letters A, 25*(2), 167.

Röhr, H. (1968). A 90 degrees laser scattering experiment for measuring temperature and density of ions and electrons in a cold dense theta pinch plasma. *Zeitschrift fur Physik, 209*(3), 295.

Röhr, H. (1977). Raman-scattering – possibility of calibrating laser scattering devices. *Physics Letters A, 60*(3), 185–186.

Röhr, H. (1981). Rotational Raman-scattering of hydrogen and deuterium for calibrating Thomson scattering devices. *Physics Letters A, 81*(8), 451–453.

Röhr, H., Steuer, K. H., Schramm, G., Hirsch, K., & Salzmann, H. (1982). 1st high-repetition-rate Thomson scattering for fusion plasmas. *Nuclear Fusion, 22*(8), 1099–1102.

Röhr, H., Steuer, K. H., Murmann, H., & Meisel, D. (1987). *Periodic multi-channel Thomson scattering on ASDEX.* Technical Report IPP Report III/121B. Garching, Germany: Max-Planck-Institut fuer Plasmaphysik.

Ron, A., & Tzoar, N. (1963). Interaction of electromagnetic waves with quantum and classical plasmas. *Physical Review, 131*(1), 13.

Rosenbluth, M. N., Macdonald, W. M., & Judd, D. L. (1957). Fokker-Planck equation for an inverse-square force. *Physical Review, 107*(1), 1–6.

Rosenbluth, M. N., & Rostoker, N. (1962). Scattering of electromagnetic waves by a nonequilibrium plasma. *Physics of Fluids, 5*(7), 776–788.

Ross, D. W. (1989). Comments on *Plasma Physics and Controlled Fusion.*

Ross, J. S., Froula, D. H., Mackinnon, A. J., Sorce, C., Meezan, N., Glenzer, S. H., *et al.* (2006). Implementation of imaging Thomson scattering on the omega laser. *Review of Scientific Instruments, 77*(10), 10E520.

Ross, J. S., Glenzer, S. H., Palastro, J. P., Pollock, B. B., Price, D., Divol, L., *et al.* (2010). Observation of relativistic effects in collective Thomson scattering. *Physical Review Letters, 104*, 105001.

Rostoker, N., & Rosenbluth, M. N. (1960). Test particles in a completely ionized plasma. *Physics of Fluids, 3*(1), 1–14.

Rousseaux, C., Malka, G., Miquel, J. L., Amiranoff, F., Baton, S. D., & Mounaix, P. (2006). Experimental validation of the linear theory of stimulated Raman scattering driven by a 500-fs laser pulse in a preformed underdense plasma. *Physical Review Letters, 76*(24), 4649–4649.

Rousseaux, C., Baton, S. D., Benisti, D., Gremillet, L., Adam, J. C., Heron, A., *et al.* (2009). Experimental evidence of predominantly transverse electron plasma waves driven by stimulated Raman scattering of picosecond laser pulses. *Physical Review Letters, 102*(18), 185003.

Rozmus, W., Glenzer, S. H., Estabrook, K. G. *et al.* (2000). Modeling of Thomson scattering spectra in high-Z, laser-produced plasmas. *2nd International workshop on laboratory astrophysics with intense lasers*, Tucson, Arizona, MAR 19–21, 1998. [*Astrophysical Journal Supplement Series, 127*(2), 459–463].

Ruffine, R. S., & Dewolf, D. A. (1965). Cross-polarized electromagnetic backscatter from turbulent plasmas. *Journal of Geophysical Research, 70*(17), 4313.

Rutherford, P. H., & Frieman, E. A. (1968). Drift instabilities in general magnetic field configurations. *Physics of Fluids, 11*(3), 569.

Saito, T., Hamada, Y., Yamashita, T., Ikeda, M., & Tanaka, S. (1981). Plastic Bragg cell in far-infrared region. *Journal of Applied Physics, 52*(8), 5305–5307.

Saiz, E. G., Khattak, F. Y., Gregori, G., Bandyopadhyay, S., Clarke, R. J., Fell, B., *et al.* (2007). Wide angle crystal spectrometer for angularly and spectrally resolved x-ray scattering experiments. *Review of Scientific Instruments, 78*(9), 095101.

Saiz, E. G., Gregori, G., Khattak, F. Y., Kohanoff, J., Sahoo, S., Naz, G. S., *et al.* (2008a). Evidence of short-range screening in shock-compressed aluminum plasma. *Physical Review Letters, 101*(7), 075003.

Saiz, E. G., Gregori, G., Gericke, D. O., Vorberger, J., Barbrel, B., Clarke, R. J., *et al.* (2008b). Probing warm dense lithium by inelastic x-ray scattering. *Nature Physics, 4*(12), 940–944.

Sakhokiyja, D. M., & Tsytovich, V. N. (1968). Scattering of electromagnetic waves by turbulent pulsations in an inhomogeneous magnetoactive plasma. *Nuclear Fusion, 8*(3), 241.

Sakoda, T., Iwamiya, H., Uchino, K., Muraoka, K., Itoh, M., & Uchida, T. (1997). Electron temperature and density profiles in a neutral loop discharge plasma. *Japanese Journal of Applied Physics Part 2-Letters, 36*(1AB), L67–L69.

Salpeter, E. E. (1960). Electron density fluctuations in a plasma. *Physical Review, 1*(5), 1528–1535.

Salpeter, E. E. (1961a). Effect of magnetic field in ionospheric backscatter. *Journal of Geophysical Research, 66*(3), 982.

Salpeter, E. E. (1961b). Plasma density fluctuations in a magnetic field. *Physical Review, 122*(6), 1663.

Salpeter, E. E. (1963). Density fluctuations in a nonequilibrium plasma. *Journal of Geophysical Research, 68*(5), 1321.

Salpeter, E. E., & Treiman, S. B. (1964). Backscatter of electromagnetic radiation from turbulent plasma. *Journal of Geophysical Research, 69*(5), 869.

Salzmann, H. (1986). Basic and advanced diagnostic techniques for fusion plasmas. *Proceedings of course and workshop* (Vol. II) in Varenna, Italy, September 3–13. International School of Plasma Physics, Commission of the European Communities.

Salzmann, H., Hirsch, K., Nielsen, P., Gowers, C., Gadd, A., Gadeberg, M., *et al.* (1987). 1st results from the LIDAR Thomson scattering system on jet. *Nuclear Fusion, 27*(11), 1925–1928.

Sanin, A. L., Tanaka, K., Vyacheslavov, L. N., Kawahata, K., & Akiyama, T. (2004). Two-dimensional phase contrast interferometer for fluctuations study on LHD. *Review of Scientific Instruments, 75*(10), 3439–3441.

Schmidt, G. (1966). *Physics of high temperature plasmas; an introduction.* New York: Academic Press.

Schwarz, S. E. (1963). Scattering of optical pulses form a nonequilibrium plasma. *Proceedings of the IEEE, 51*(10), 1362.

Schwarz, V., Bornath, T., Kraeft, W. D., Glenzer, S. H., Holl, A., & Redmer, R. (2007). Hypernetted chain calculations for two-component plasmas. *Contributions to Plasma Physics, 47*(4–5), 324–330.

Scotti, F., & Kado, S. (2009). Comparative study of recombining He plasmas below 0.1 ev using laser Thomson scattering and spectroscopy in the divertor simulator MAP-II. *Journal of Nuclear Materials, 390–391*, 303–306.

Seasholtz, R. G., & Tanenbaum, B. S. (1969). Effect of collisions on Thomson scattering with unequal electron and ion temperatures. *Journal of Geophysical Research, 74*(9), 2271.

Seasholtz, R. G. (1971). Effect of collisions on Thomson scattering in a magnetic field with unequal electron and ion temperatures and electron drift. *Journal of Geophysical Research, 76*(7), 1793.

Seka, W., Baldis, H. A., Fuchs, J., Regan, S. P., Meyerhofer, D. D., Stoeckl, C., *et al.* (2002). Multi-beam stimulated Brillouin scattering from hot, solid-target plasmas. *Physical Review Letters, 89*(17), 175002.

Seka, W., Edgell, D. H., Myatt, J. F., *et al.* (2009). Two-plasmon-decay instability in direct-drive inertial confinement fusion experiments. *Physics of Plasmas, 16*(5), 052701.

Selden, A. C. (1980). Simple analytic form of the relativistic Thomson scattering spectrum. *Physics Letters A, 79*(5–6), 405–406.

Semet, A., Johnson, L. C., & Mansfield, D. K. (1983). A high-energy D_2O submillimeter laser for plasma diagnostics. *International Journal of Infrared and Millimeter Waves, 4*(2), 231–246.

Shapiro, V. D. (1963). Nonlinear theory of the interaction of a monoenergetic beam with a plasma. *Soviet Physics JETP-USSR, 17*(2), 416–423.

Sharp, L. E., Sanderson, A. D., & Evans, D. E. (1981). Signal-to-noise requirements for interpreting submillimeter laser scattering experiments in a Tokamak plasma. *Plasma Physics and Controlled Fusion, 23*(4), 357–370.

Sharp, L. E. (1983). The measurement of large-scale density-fluctuations in toroidal plasmas from the phase scintillation of a probing electromagnetic-wave. *Plasma Physics and Controlled Fusion, 25*(7), 781–792.

Sharp, L. E., & Hamberger, S. M. (1970). *4th European conference on controlled fusion plasma physics nuclear research* (p. 64) Rome, Italy: CNEN.

Shefer, R. E., Watterson, R. L., Goodman, D., & Klinkowstein, R. E. (1990). Design of a submillimeter-wave alpha-particle Thomson scattering diagnostic for fusion plasmas. *Review of Scientific Instruments, 61*(10), 3214–3216.

Sheffield, J., Decker, G., Macmahon, A. B., & Robson, A. E. (1970). Electron temperature measurements in oblique and normal collisionless shocks in magnetized plasmas. *4th European conference on controlled fusion plasma physics nuclear research.* European Physical Society, Rome, Italy, p. 58.

Sheffield, J. (1972a). Incoherent scattering of radiation from a high-temperature plasma. *Plasma Physics, 14*(8), 783.

Sheffield, J. (1972b). Method for measuring direction of magnetic-field in a Tokamak plasma by laser light-scattering. *Plasma Physics, 14*(4), 385.

Sheffield, J. (1975). *Plasma scattering of electromagnetic radiation.* New York: Academic Press.

Shur, M. (2008). Terahertz electronics. CS MANTECH Conference, Chicago, Illinois.

Siegel, P. H. (2002). Terahertz technology. *IEEE Transactions on Microwave Theory and Techniques, 50*(3), 910–928.

Siegmund, O. H. W. (1998). Methods of vacuum ultraviolet physics. In D. L. Ederer (Ed.), *Methods of vacuum ultraviolet physics* (2nd ed. (Chapter III)). New York: Academic Press.

Siegmund, O. H. W. (2004). High-performance microchannel plate detectors for UV/visible astronomy. *Nuclear Instruments & Methods in Physics Research Section A-Accelerators Spectrometers Detectors and Associated Equipment, 525*(1–2), 12–16.

Siegrist, M. R., & Salomaa, R. R. E. (1982). The measurement of plasma parameters by Thomson scattering of far-infrared laser-radiation. *Helvetica Physica Acta, 55*(2), 227.

Sitenko, A. G. (1967). *Electromagnetic fluctuations in plasma.* New York: Academic Press.

Smith, D. R., Mazzucato, E., Munsat, T., Park, H., Johnson, D., Lin, L., *et al.* (2004). Microwave scattering system design for rho(e) scale turbulence measurements on NSTX. *Review of Scientific Instruments, 75*(10), 3840–3842.

Smith, D. R., Kaye, S. M., Lee, W., Mazzucato, E., Park, H. K., Bell, R. E., *et al.* (2009). Observations of reduced electron gyroscale fluctuations in national spherical torus experiment H-mode plasmas with large $E \times B$ flow shear. *Physical Review Letters, 102*(22), 225005.

Smith, R. J. (2008). Nonperturbative measurement of the local magnetic field using pulsed polarimetry for fusion reactor conditions (invited). *Review of Scientific Instruments, 79*(10), 10E703 Figures reprinted with permission from R. J. Smith. Copyright 2008 by the American Physical Society.

Spitzer, L. (1962). *Physics of fully ionized gases* (2nd Rev. ed.). Interscience tracts on physics and astronomy 3. New York: Interscience Publishers.

Stansfield, B. I., Nodwell, R., & Meyer, J. (1971). Enhanced scattering of laser light by optical mixing in a plasma. *Physical Review Letters, 26*(20), 1219.

Stern, R. A., & Tzoar, N. (1966). Parametric coupling between electron-plasma and ion-acoustic oscillations. *Physical Review Letters, 17*(17), 903.

Stewart, I. N. (1972). PGP-15. Ph.D. thesis. Bangor, UK: University of North Wales.

Stix, T. H. (1992). *Waves in plasmas*. New York: American Institute of Physics, ISBN 0-88318-859-7.

Stott, P. E. (1968). A transport equation for multiple scattering of electromagnetic waves by a turbulent plasma. *Journal of Physics Part A General, 1*(6), 675.

Sumikawa, T., Yamashita, K., Onoda, M., Tokuzawa, T., Kawamori, E., & Ono, Y. (2007). Development of two-dimensional Thomson scattering diagnostic system involving use of multiple reflection and the time-of-flight of laser light. *Plasma and Fusion Research, 2*(S1108), 102105.

Surko, C. M., Slusher, R. E., Moler, D. R., & Porkolab, M. (1972). 10.6-mu laser scattering from cyclotron-harmonic waves in a plasma. *Physical Review Letters, 29*(2), 81.

Surko, C. M., & Slusher, R. E. (1976). Study of density fluctuations in adiabatic toroidal compressor scattering Tokamak using CO_2-laser. *Physical Review Letters, 37*(26), 1747–1750.

Surko, C. M., & Slusher, R. E. (1980). Study of plasma-density fluctuations by the correlation of crossed CO_2 laser-beams. *Physics of Fluids, 23*(12), 2425–2439.

Tanaka, K., Matsuo, K., Koda, S., Bowden, M., Muraoka, K., Kondo, K., *et al.* (1993). Characteristics of electron-density fluctuations in Heliotron-E measured using a wide beam laser phase-contrast method. *Journal of the Physical Society of Japan, 62*(9), 3092–3105.

Tanaka, K., Vyacheslavov, L. N., Akiyama, T., Sanin, A., Kawahata, K., Tokuzawa, T., *et al.* (2003). Phase contrast imaging interferometer for edge density fluctuation measurements on LHD. *Review of Scientific Instruments, 74*(3), 1633–1637.

Tanaka, K., Michael, C. A., Vyacheslavov, L. N., Sanin, A. L., Kawahata, K., Akiyama, T., *et al.* (2008). Two-dimensional phase contrast imaging for local turbulence measurements in large helical device (invited). *Review of Scientific Instruments, 79*(10), 10E702.

Tanenbaum, B. S. (1967). *Plasma physics. McGraw-Hill physical and quantum electronics series*. New York: McGraw-Hill.

Tanenbaum, B. S. (1968). Continuum theory of Thomson scattering. *Physical Review, 171*(1), 215.

Tartari, U., *et al.* (1999). Polarization in millimeter-wavelength collective Thomson scattering on FTU tokamak. *The Review of Scientific Instruments, 70*, 1162.

Taylor, E. C., & Comisar, G. G. (1963). Frequency spectrum of themal fluctuations in plasmas. *Physical Review, 132*(6), 2379.

Taylor, G., & Harvey, R. W. (2009). Assessment of an oblique ECE diagnostic for ITER. *Fusion Science and Technology, 55*(1), 64–75.

Taylor, J. B., & Hastie, R. J. (1968). Stability of general plasma equilibria .i. formal theory. *Plasma Physics, 10*(5), 479.

Teich, M. C. (1970). In R. K. Willardson, & A. C. Beer (Eds.), *Semiconductors and semimetals* (Vol 5, Infrared Detectors, Chapter 9), New York: Academic Press.

Theimer, O. (1966). High density corrections of scattering cross section of a plasma. *Physics Letters, 20*(6), 639.

Theimer, O., & Hicks, W. (1968). Depolarization of light scattered by a relativistic plasma. *Physics of Fluids, 11*(5), 1045.

Theimer, P., & Sollid, J. E. (1968a). Relativistic corrections to light-scattering spectrum of a plasma. *Physical Review, 176*(1), 198.

Theimer, O., & Theimer, R. (1970). *Laser Forsch, 3*, 53.

Thiele, R., Bornath, T., Fortmann, C., *et al.* (2008). Plasmon resonance in warm dense matter. *Physical Review E, 78*(2), 026411.

Thompson, A. R., Moran, J. M., & Swenson, G. W. (1986). *Interferometry and synthesis in radio astronomy*. New York: Wiley.

Thompson, E., & Fiocco, G. (1963). *Proceedings of the 6th international conference on ionization phenomena in gases* (Vol. 4, p. 111). Paris: Serma.

Thompson, W. B. (1962). *An introduction to plasma physics*. Oxford, NY: Pergamon Press.

Thumm, M. (2007). State-of-the-art of high power gyro-devices and free electron masers update 2008. *Wissenschaftliche Berichte FZKA*, Vol. 7467, p. 118. http://cat.inist.fr/?aModele=afficheN&cpsidt=21689172.

Tomita, K., Kagawa, T., Uchino, K., Katsuki, S., & Akiyama, H. (2009). Collective Thomson scattering diagnostics of EUV plasma. *Journal of Plasma and Fusion Research, 8*, 488. Figures reprinted with permission from K. Tomita and the JSPF.

Tracy, M. D., Degroot, J. S., Estabrook, K. G., *et al.* (1992). Detailed 266 nm Thomson scattering measurements of a laser-heated plasma. *Physics of Fluids B-Plasma Physics, 4*(6), 1576–1584.

Tremsin, A. S., & Siegmund, O. H. W. (2000). Polycrystalline diamond films as prospective UV photocathodes. *Proceedings of SPIE, 4139*, 16.

Trivelpiece, A. W., & Gould, R. W. (1959). Space charge waves in cylindrical plasma columns. *Journal of Applied Physics, 30*(11), 1784–1793.

Truc, A., Quemeneur, A., Hennequin, P., Gresillon, D., Gervais, F., Laviron, C., *et al.* (1992). ALTAIR – an infrared-laser scattering diagnostic on the Tore-Supra Tokamak. *Review of Scientific Instruments, 63*(7), 3716–3724.

Tsytovich, V. N. (1970). Technical Report. CLM-P 244. UKAEA, Culham Laboratory.

Tsytovich, V. N. (1972). *An introduction to the theory of plasma turbulence* (1st ed.). International series of monographs in natural philosophy. Oxford, NY: Pergamon Press.

Tynan, G. R., Fujisawa, A., & McKee, G. (2009). A review of experimental drift turbulence studies. *Plasma Physics and Controlled Fusion, 51*(11), 113001.

Uchino, K., Muraoka, T., Muraoka, K., & Akazaki, M. (1982). Studies of an impulse breakdown process in an atmospheric air using ruby-laser scattering diagnostics. *Japanese Journal of A Physics Part 2-Letters, 21*(11), L696–L698.

Urry, M. K., Gregori, G., Landen, O. L., Pak, A., & Glenzer, S. H. (2006). X-ray probe development for collective scattering measurements in dense plasmas. *Journal of Quantitative Spectroscopy and Radiative Transfer, 99*(1–3), 636–648.

Vahala, L., Vahala, G., & Sigmar, D. J. (1986). Effects of alpha-particles on the scattering function in CO_2-laser scattering. *Nuclear Fusion, 26*(1), 51–60.

Vahala, L., Vahala, G., & Sigmar, D. J. (1988). Effect of electrostatic scattering parameters on the direct detection of fusion alphas. *Nuclear Fusion, 28*(9), 1595–1602.

Vahala, L., Vahala, G., & Bretz, N. (1990). Measurement of magnetic fluctuations by O–>X mode conversion. *Review of Scientific Instruments, 61*(10), 3022–3024.

Vahala, L., Vahala, G., & Bretz, N. (1992). Electromagnetic-wave scattering from magnetic fluctuations in Tokamaks. *Physics of Fluids B-Plasma Physics, 4*(3), 619–629.

van de Sande, M. (2002). Laser scattering on low temperature plasmas High resolution and stray light rejection, Ph.D. thesis. Eindhoven: Technische Universiteit Eindhoven.

van der Meiden, H. J., Varshney, S. K., Barth, C. J., Oyevaar, T., Jaspers, R., Donne, A. J. H., *et al.* (2006). 10 khz repetitive high-resolution TV Thomson scattering on TEXTOR: Design and performance (invited). *Review of Scientific Instruments, 77*(10), 10E512, Figures reprinted with permission from A. J. Donn. Copyright 2006 by the American Institute of Physics.

van der Mullen, J., Boidin, J., & van de Sande, M. (2004). High-resolution electron density and temperature maps of a microwave plasma torch measured with a 2-D Thomson scattering system. *Spectrochimica Acta Part B, 59*, 929–940.

Vanandel, H. W. H., Boileau, A., & Vonhellerman, M. (1987). Study of microturbulence in the TEXTOR Tokamak using CO_2-laser scattering. *Plasma Physics and Controlled Fusion, 29*(1), 49–74.

Vandenplas, P. E., & Gould, R. W. (1964). Equations of hot inhomogeneous plasma model – i – resonance frequencies of cylindrical plasma column. *Journal of Nuclear Energy Part C-Plasma Physics Accelerators Thermonuclear Research, 6*(5pc), 449.

Vavriv, D. M., Volkov, V. A., & Chumak, V. G. (2007). Clinotron tubes: High-power THz sources. *Proceedings of the 37th European Microwave Conference* (pp. 826–829), Munich.

Vedenov, A. A., Velikhov, E. P., & Sagdeev, R. Z. (1961). Nonlinear oscillations of rarified plasma. *Nuclear Fusion, 1*(2), 82–100.

Vedenov, A. A. (1968). *Theory of turbulent plasma.* London: Iliffe.

Verhoeven, A. G. A., Bongers, W. A., Bratman, V. L., Caplan, M., Denisov, G. G., van der Geer, C. A. J., *et al.* (1998). First microwave generation in the FOM free-electron maser. *Plasma Physics and Controlled Fusion, 40*, A139–A156.

Villars, F., & Weisskopf, V. F. (1955). On the scattering of radio waves by turbulent fluctuations of the atmosphere. *Proceedings of the Institute of Radio Engineers, 43*(10), 1232–1239.

Villeneuve, D. M., Baldis, H. A., Bernard, J. E. *et al.* (1991). Collective Thomson scattering in a laser-produced plasma resolved in time, space, frequency, or wave number. *Journal of the Optical Society of America B-Optical Physics, 8*(4), 895–902.

Visco, A., Drake, R. P., Froula, D. H., Glenzer, S. H., & Pollock, B. B. (2008). Temporal dispersion of a spectrometer. *Review of Scientific Instruments, 79*(10), 10F545.

Vonhellermann, M., & Holzhauer, E. (1984). Far forward scattering from plasma fluctuations using a detector array and coherent signal-processing at 10-mu-m. *IEEE Transactions on Plasma Science, 12*(1), 5–11.

Vyacheslavov, L. N., Tanaka, K., Sanin, A. L., Kawahata, K., Michael, C., & Akiyama, T. (2005). 2-D phase contrast imaging of turbulence structure on LHD. *IEEE Transactions on Plasma Science, 33*(2), 464–465.

Waldteufel, P. (1965). Introduction de l'effet des collisions elastiques dans lequation de boltzmann dun gaz faiblement ionise application a la diffusion incoherente dune onde electromagnetique par lionosphere. *Annales de Qeophysique, 21*(1), 106.

Walsh, C. J., & Baldis, H. A. (1982). Non-linear ion-wave development and saturation of stimulated Brillouin scattering. *Physical Review Letters, 48*(21), 1483–1486.

Walsh, C. J., Villeneuve, D. M., & Baldis, H. A. (1984). Electron plasma-wave production by stimulated Raman-scattering – competition with stimulated Brillouin-scattering. *Physical Review Letters, 53*(15), 1445–1448.

Walsh, M. J., Beausang, K., Beurskens, M. N .A., Giudicotti, L., Hatae, T., Johnson, D., Kajita, S., Mukhin, E. E., Pasqualotto, R., Prunty, S. L., Scannell, R. D., & Vayakis, G. (2008). Performance evaluation of ITER Thomson scattering systems. *22nd IAEA fusion energy conference* (Vol. IT/P6-25, p. 171). ITER Diagnostic Division, Centre de Cadarache, 13108 St Paul Lez Durance, France.

Wand, R. H. (1969). Evidence for reversible heating in E region from radar Thomson scatter observations of ion temperature. *Journal of Geophysical Research, 74*(24), 5688.

Warner, K., & Hieftje, G. M. (2002). Thomson scattering from analytical plasmas. *Spectrochimica Acta Part B-Atomic Spectroscopy, 57*(2), 201–241.

Watanabe, K., Ueno, M., Wakaki, M., Abe, O., & Murakami, H. (2008). Gaas:se and gaas:te photoconductive detectors in 300 μm region for astronomical observations. *Japanese Journal of Applied Physics, 47*(11), 8261–8264.

Watson, G. N. (1958). *Theory of Bessel functions.* London: Cambridge University Press.

Watson, K. M. (1970). Electromagnetic wave scattering within a plasma in transport approximation. *Physics of Fluids, 13*(10), 2514.

Wei, J., Olaya, D., Karasik, B. S., Pereverzev, S. V., Sergeev, A. V., & Gershenson, M. E. (2008). Ultrasensitive hot-electron nanobolometers for terahertz astrophysics. *Nature Nanotechnology, 3*(8), 496–500.

Weiland, J. (2000). *Collective modes in inhomogeneous plasma: Kinetic and advanced fluid theory. Plasma physics series.* Bristol, PA: Institute of Physics.

Weinstock, J. (1965a). New approach to theory of fluctuations in a plasma. *Physical Review, 139*(2A), A388.

Weinstock, J. (1965b). Resonances in light scattered from a plasma. *Physics Letters, 18*(1), 21.

Weinstock, J. (1967). Correlation functions and scattering of electromagnetic waves by inhomogeneous and nonstationary plasmas. *Physics of Fluids, 10*(9p1), 2065.

Weisen, H. (1986). Imaging methods for the observation of plasma-density fluctuations. *Plasma Physics and Controlled Fusion, 28*(8), 1147–1159.

Weisen, H. (1988). The phase-contrast method as an imaging diagnostic for plasma-density fluctuations. *Review of Scientific Instruments, 59*(8), 1544–1549.

Weisen, H., Hollenstein, C., & Behn, R. (1988b). Turbulent density-fluctuations in the TCA Tokamak. *Plasma Physics and Controlled Fusion, 30*(3), 293–309.

Weisen, H., Borg, G., Joye, B., Knight, A. J., & Lister, J. B. (1989a). Measurements of the Tokamak-safety-factor profile by means of driven resonant Alfven waves. *Physical Review Letters, 62*(4), 434–437.

Weisen, H., Appert, K., Borg, G. G., Joye, B., Knight, A. J., Lister, J. B., *et al.* (1989b). Mode conversion to the kinetic Alfven-wave in low-frequency heating experiments in the TCA Tokamak. *Physical Review Letters, 63*(22), 2476–2479.

Weisen, H. (2006). Diagnostics des plasmas de fusion confinés magnétiquement diagnostics of magnetically confined fusion plasmas EPFL doctoral course PY-11 20.

Wentworth, S. M., & Neikirk, D. P. (1989). Far-infra-red microbolometers made with tellurium and bismuth. *Electronics Letters, 25*(23), 1558–1560. http://dx.doi.org/10.1117/12.21040.

Wentworth, S. M., & Neikirk, D. P. (1990). The transition-edge microbolometer (trembol). In *Superconductivity applications for infrared and micro wave devices* (Vol. 1292). Proceedings of SPIE.

Wesseling, H. J., & Kronast, B. (1996). Thomson light scattering measurements of electron temperature and density in the alpha-gamma transition of a capacitive RF discharge in helium. *Journal of Physics D-Applied Physics, 29*(4), 1035–1039.

Wesson, J. A., Sykes, A., & Lewis, H. R. (1973). Ion-sound instability. *Plasma Physics and Controlled Fusion, 15*(1), 49–55.

Wesson, J. A. (1978). Hydromagnetic-stability of Tokamaks. *Nuclear Fusion, 18*(1), 87–132.

Wharton, K. B., Kirkwood, R. K., Glenzer, S. H., *et al.* (1998). Observation of energy transfer between identical-frequency laser beams in a flowing plasma. *Physical Review Letters, 81*(11), 2248–2251.

Williams, E. A., Berger, R. L., Drake, R. P., Rubenchik, A. M., Bauer, B. S., Meyerhofer, D. D., *et al.* (1995). The frequency and damping of ion-acoustic-waves in hydrocarbon (CH) and 2-ion-species plasmas. *Physics of Plasmas, 2*(1), 129–138.

Williamson, J. H. (1968). Brownian motion of electrons. *Journal of Physics Part A General, 1*(6), 629.

Williamson, J. H., & Clarke, M. E. (1971). Construction of electron distribution functions from laser scattering spectra. *Journal of Plasma Physics, 6*, 211.

Wort, D. J. H. (1966). Microwave transmission through turbulent plasma. *Journal of Nuclear Energy Part C-Plasma Physics Accelerators Thermonuclear Research, 8*(1pc), 79.

Woskoboinikow, P., Erickson, R., & Mulligan, W. J. (1983). Submillimeter-wave dumps for fusion plasma diagnostics. *International Journal of Infrared and Millimeter Waves, 4*(6), 1045–1059.

Woskoboinikow, P. (1986). Development of gyrotrons for plasma diagnostics. *Review of Scientific Instruments, 57*(8), 2113–2118.

Woskov, P. (1987). *Suitability of millimeter-wave scattering for diagnostics of fusion alpha particles*. Technical Report PFC/RR-87-16. Plasma Science and Fusion Center, Massachusetts Institute of Technology, Cambridge, MA 02139.

Woskov, P., & Rhee, D. Y. (1992). Localized magnetci-field pitch angle measurements by collective Thomson scattering. *The Review of Scientific Instruments, 63*, 4641.

Woskov, P. P., *et al.* (1993). *ITER millimeter-wave CTS diagnostic option*. PFC/JA-93-25.

Woskov, P. P. (2005). *Overview of high power CTS experiments in magnetically confined plasmas.* Technical Report PSFC/JA-05-24. Plasma Science and Fusion Center, Massachusetts Institute of Technology, Cambridge, MA 02139.

Wrubel, Th., Glenzer, S., Büscher, S., & Kunze, H. J. (1996). Investigation of electron-proton drifts with Thomson scattering. *Journal of Atmospheric and Terrestrial Physics, 58,* 1077–1087.

Wu, Ta-you. (1966). *Kinetic equations of gases and plasmas. Addison-Wesley series in advanced physics.* Reading, MA: Addison-Wesley Pub. Co.

Wunsch, K., Hilse, P., Schlanges, M., & Gericke, D. O. (2008). Structure of strongly coupled multicomponent plasmas. *Physical Review E, 77*(5), 056404.

Wunsch, K., Vorberger, J., & Gericke, D. O. (2009). Ion structure in warm dense matter: Benchmarking solutions of hypernetted-chain equations by first-principle simulations. *Physical Review E, 79*(1), 010201.

Yaakobi, B., Kim, H., Soures, J. M., Deckman, H. W., & Dunsmuir, J. (1983). Sub-micron x-ray-lithography using laser-produced plasma as a source. *Applied Physics Letters, 43*(7), 686–688.

Yamanaka, C., Yamanaka, T., Shimamur, T., Waki, M., Yoshida, K., & Kang, H. (1972). Nonlinear-interaction of laser radiation and plasma. *Physics Letters A, 38*(7), 495.

Yatsu, Y., Kuramoto, Y., Kataoka, J., Kotoku, J., Saito, T., Ikagawa, T., *et al.* (2006). Study of avalanche photodiodes for soft x-ray detection below 20 keV. *Nuclear Instruments & Methods in Physics Research Section A-Accelerators Spectrometers Detectors and Associated Equipment, 564*(1), 134–143.

Yokoyama, M., Nakatsuk, M., Yamanaka, C., Izawa, Y., & Toyoda, K. (1971). Scattering diagnostics of plasma by CO_2 laser. *Physics Letters A, 36*(4), 317.

Yoshida, H., Naito, O., Matoba, T., Yamashita, O., Kitamura, S., Hatae, T., *et al.* (1997). Quantitative method for precise, quick, and reliable alignment of collection object fields in the JT-60U Thomson scattering diagnostic. *Review of Scientific Instruments, 68*(2), 1152–1161.

Yoshida, H., Naito, O., Sakuma, T., Kitamura, S., Hatae, T., & Nagashima, A. (1999a). A compact and high repetitive photodiode array detector for the JT-60U Thomson scattering diagnostic. *Review of Scientific Instruments, 70*(1), 747–750.

Yoshida, H., Naito, O., Yamashita, O., Kitamura, S., Sakuma, T., Onose, Y., *et al.* (1999b). Multilaser and high spatially resolved multipoint Thomson scattering system for the JT-60U Tokamak. *Review of Scientific Instruments, 70*(1), 751–754. Figures reprinted with permission from H.Yoshida. Copyright 1999 by the American Institute of Physics.

Young, E. T. (2000). Germanium detectors for the far-infrared. *Proceedings of the space astrophysics detectors and detector technologies conference.*

Young, P. E. (1994). Spatial profiles of stimulated brillouin-scattering ion waves in a laser-produced plasma. *Physical Review Letters, 73*(14), 1939–1942.

Yu, C. X., Cao, J. X., Shen, X. M., & Wang, Z. S. (1988). Diffraction of electromagnetic-wave by a damped plasma-wave. *Plasma Physics and Controlled Fusion, 30*(13), 1821–1831.

Zaidi, S. (2001). Filtered Thomson scattering in an argon plasma. *39th AIAA aerospace meeting and exhibit.* Reno, NV, January 8–11, 2001. American institute of Aeronautics and Astronautics (pp. AIAA–2001–0415).

Zhang, Y. Q., & DeSilva, A. W. (1991). Ion-acoustic and entropy fluctuations in collisional plasmas measured by laser light scattering. *Physical Review A, 44*(6), 3841–3855.

Zheng, J., Yu, C. X., & Zheng, Z. J. (1997). Effects of non-Maxwellian (super-Gaussian) electron velocity distribution on the spectrum of Thomson scattering. *Physics of Plasmas, 4*(7), 2736–2740.

Zheng, J., Yu, C. X., & Zheng, Z. J. (1999). The dynamic form factor for ion-collisional plasmas. *Physics of Plasmas, 6*(2), 435–443.

Zhuravlev, V. A., & Petrov, G. D. (1972). Light-scattering from electrons in a high-temperature plasma. *Optika i Spektroskopiya, 33*(1), 36.

Zhuravlev, V. A., & Petrov, G. D. (1979). Scattering of radiation by relativistic plasma streams. *Soviet Journal of Plasma Physics, 5*, 3.

Zhurovich, K., Mossessian, D. A., Hughes, J. W., Hubbard, A. E., Irby, J. H., & Marmar, E. S. (2005). Calibration of Thomson scattering systems using electron cyclotron emission cutoff data. *Review of Scientific Instruments, 76*(5), 053506. Figures reprinted with permission from K. Zhurovich. Copyright 2005 by the American Institute of Physics.

Zimmerman, G. B., & Kruer, W. L. (1975). Numerical simulations of laser-initiated fusion Comments. *Plasma Physics & Controlled Fusion, 2*(2), 51.

Index

A

B

C

D

Kolibri 360 GmbH
Ferdinand-Jühlke-Straße 7
99095 Erfurt, Deutschland
produktsicherheit@kolibri360.de